TETRAHEDRON ORGANIC CHEMISTRY SERIES
Series Editors: J E Baldwin, FRS & P D Magnus, FRS

VOLUME 10

Sulphones in Organic Synthesis

Related Pergamon Titles of Interest

BOOKS

Tetrahedron Organic Chemistry Series:

DAVIES: Organotransition Metal Chemistry: Applications to Organic Synthesis
DEROME: Modern NMR Techniques for Chemistry Research
DESLONGCHAMPS: Stereoelectronic Effects in Organic Chemistry
GAWLEY: Asymmetric Synthesis*
GIESE: Radicals in Organic Synthesis: Formation of Carbon - Carbon Bonds
HANESSIAN: Total Synthesis of Natural Products
HASSNER: Named Reactions and Unnamed Reactions*
PAULMIER: Selenium Reagents & Intermediates in Organic Synthesis
PERLMUTTER: Conjugate Addition Reactions in Organic Synthesis
WILLIAMS: Synthesis of Optically Active Alpha-Amino Acids
WONG: Enzymes in Synthetic Organic Chemistry*

JOURNALS

BIOORGANIC & MEDICINAL CHEMISTRY LETTERS
EUROPEAN POLYMER JOURNAL
JOURNAL OF PHARMACEUTICAL AND BIOMEDICAL ANALYSIS
TETRAHEDRON
TETRAHEDRON: ASYMMETRY
TETRAHEDRON LETTERS

Full details of all Pergamon publications/free specimen copy of any Pergamon journal available on request from your nearest Pergamon office

*In Preparation

Sulphones in Organic Synthesis

N.S. SIMPKINS
Department of Chemistry
University of Nottingham

PERGAMON PRESS

OXFORD · NEW YORK · SEOUL · TOKYO

U.K.	Pergamon Press Ltd, Headington Hill Hall, Oxford OX3 0BW
U.S.A.	Pergamon Press Inc. 660 White Plains Road, Tarrytown, New York 10591-5153, USA
KOREA	Pergamon Press Korea, KPO Box 315, Seoul 110-603, Korea
JAPAN	Pergamon Press Japan, Tsunashima Building Annex, 3-2-12 Yushima, Bunkyo-ku, Tokyo 113, Japan

First edition 1993

Library of Congress Cataloging in Publication Data

Simpkins, N.S.
Sulphones in organic synthesis / N.S. Simpkins. -- 1st ed.
p. cm.-- (Tetrahedron organic chemistry series; v. 10)
Includes index.
1. Sulphones. 2. Organic Compounds--Synthesis. I. Title. II. Series
QD305.S6S56 1993 547.2--dc20   92-34856

British Library Cataloguing in Publication Data

A catalogue record for this book is available from the British Library

ISBN 0 08 040283 6 Hardcover
ISBN 0 08 040284 4 Flexicover

Printed in Great Britain by BPCC Wheatons Ltd, Exeter

Preface

Over the last twenty years the use of sulphones in organic synthesis has increased dramatically, the synthetic repertoire of sulphones having been developed to such an extent as to rival the carbonyl functionality for versatility. Not only have sulphones been employed in a great many synthetic methodologies, enabling the preparation of a vast array of functionalised products, but the sulphone group has also proved of enormous value in many of the most demanding and sophisticated total syntheses carried out in recent years. The growing significance of the sulphone functional group in organic synthesis has been the impetus for the writing of this book.

This book is intended to highlight the synthetic aspects of the sulphone group, with a great emphasis on presenting the widest range of synthetic transformations in schematic form. Whilst it has also been impossible to be completely comprehensive, and to keep up with the flow of contributions whilst this book was in preparation, an effort has been made to deal with all of the synthetically important developments in sulphone chemistry appearing up until the end of 1990.

It is a pleasure to acknowledge those that have made the production of this book possible, in particular Mrs Melanie Dakin for her excellent typing, and Jeanette Eldridge for her help with detailed editing of the camera-ready copy. I would also like to thank all those friends and colleagues who kindly proof-read some or all of the manuscript.

Nigel S. Simpkins

The Department of Chemistry
The University of Nottingham
Nottingham NG7 2RD, UK

CONTENTS

CHAPTER 1

Introduction to Sulphone Chemistry

Over the last twenty years the use of sulphones in organic synthesis has increased dramatically, the synthetic repertoire of sulphones having been developed to such an extent as to rival the carbonyl functionality for versatility. Not only have sulphones been employed in a great many 'synthetic methodologies', enabling the preparation of a vast array of functionalised products, but the sulphone group has also proved of enormous value in many of the most demanding and sophisticated total syntheses carried out in recent years. The growing significance of the sulphone functional group in organic synthesis has been the impetus for the writing of this book.

The chemistry of sulphones has been described in various degrees of detail in the past. General texts on sulphur chemistry have included sections on the chemistry of sulphones,[1] as have more general works on organic chemistry.[2] More specialised reviews dealing with aspects of sulphone chemistry can be found scattered throughout the primary literature, and these have been included in the appropriate sections of this book. The most cited review of sulphone chemistry is the Tetrahedron Report by Magnus.[3] This is now sadly out of date, and more up-to-date and comprehensive coverage can be found in the later review by Schank,[4] as well as various chapters of a recent volume in the Patai series.[5]

In one important strategic aspect the role of the sulphone group in synthesis differs from that of the carbonyl function. Whereas the carbonyl group, or some oxygenated functionality derived from it, is frequently desired in the target molecule, the sulphone must almost always be disposed of.[6] In essence then, the sulphone is simply a synthetic tool. The importance of sulphones stems from their utility for C–C bond formation, and this in turn derives from several key aspects of their properties and reactivity.

Firstly, sulphones are easily prepared by a range of mild and high-yielding routes (Chapter 2). Also, the sulphone is a robust group and frequently confers useful properties such as crystallinity.

Secondly, and of paramount importance, is the ease of formation of carbanions α to the sulphone group. This enables efficient C–C bond formation via alkylation, acylation and aldol-like processes (Chapter 3), Scheme 1.

1

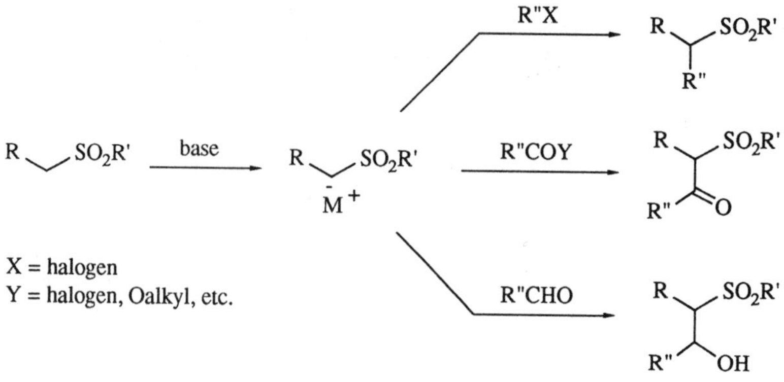

R"X

$\text{R} \diagup^{\text{SO}_2\text{R'}}_{\text{R"}}$

R SO₂R'   base   R SO₂R'   R"COY

M⁺

$\text{R} \diagup^{\text{SO}_2\text{R'}}_{\text{R"}} \diagdown_{\text{O}}$

X = halogen
Y = halogen, Oalkyl, etc.

R"CHO

$\text{R} \diagup^{\text{SO}_2\text{R'}}_{\text{R"}} \diagdown_{\text{OH}}$

Scheme 1

Most of this chemistry involves the use of sulphones which are designed to undergo deprotonation at only one site, otherwise ambiguity in the site of substitution would arise. The use of sulphones incorporating one blocking group, i.e. R' = phenyl, tolyl or *tert*-butyl is therefore commonplace (although it should be noted that aryl sulphones can also undergo deprotonation under certain circumstances).

Sulphones also allow C=C bond formation through elimination reactions, Scheme 2.

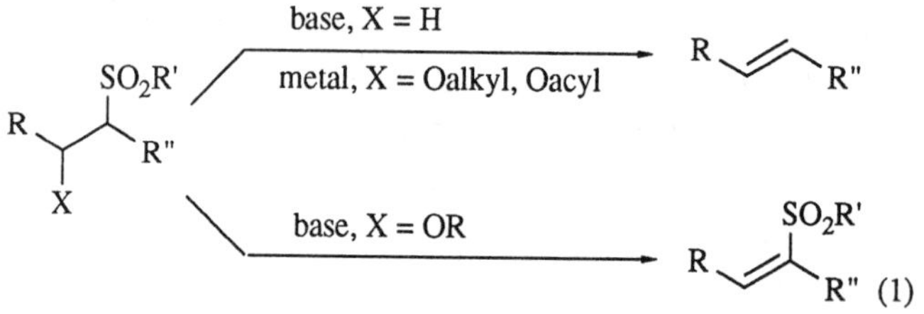

SO₂R'   base, X = H
R   metal, X = Oalkyl, Oacyl   R   R"
   R"
X   base, X = OR   SO₂R'
                                 R   R" (1)

Scheme 2

Thus, base-mediated elimination of the sulphone group (i.e. loss of the corresponding sulphinic acid), or metal-induced elimination from certain vicinally substituted sulphones (Julia reaction), leads to alkenes (Chapter 7). Eliminations are also a good route to vinyl sulphones (1), which engage in versatile chemistry including Michael additions and cycloadditions (Chapters 4 and 6).

Thirdly, the sulphone group is easily removed from the synthetic intermediate once its task has been achieved (Chapter 9). This most commonly involves simple replacement of the sulphone (usually $ArSO_2$) by hydrogen, although methods for oxidative and alkylative desulphonylation are becoming more plentiful, Scheme 3.

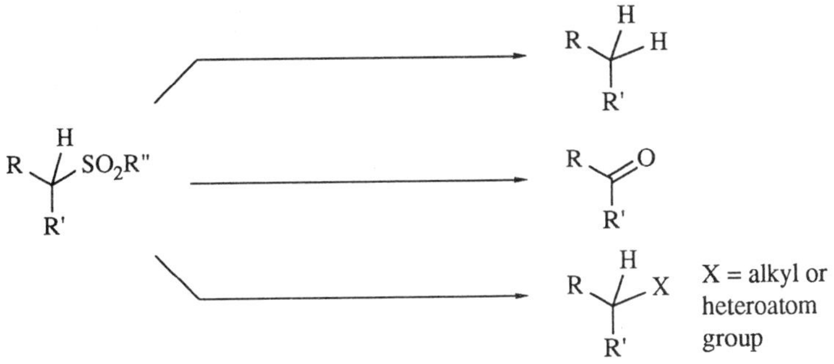

Scheme 3

This chemistry has been explored with all types of simple and functionalised sulphones including those mentioned in Schemes 1 and 2.

Other sections of this book describe rearrangement reactions of sulphones (Chapter 5), and the rather special chemistry exhibited by the sulphone group when it is incorporated into certain ring systems (Chapter 8).

Throughout the book an effort has been made to categorise the chemistry described, firstly according to the type of sulphone involved (ketosulphone, hydroxy sulphone, etc.), and secondly, by subdividing material according to the type of transformation carried out (e.g. alkylation, acylation, etc., of sulphonyl carbanions). It is hoped that this arrangement will allow for the most rapid retrieval of information and references, although clearly much overlap exists between the chapters, and the placement of some material is somewhat arbitrary.

This book is intended to highlight the synthetic aspects of the sulphone group, with a great emphasis on presenting the widest range of synthetic transformations in schematic form. The space available precludes detailed discussion of the interesting and important theoretical and mechanistic aspects of many of these reactions. Other areas such as the spectroscopic characteristics of sulphones have been omitted entirely. It has also been impossible to be completely comprehensive, and to keep up with the flow of contributions whilst this book was in preparation. Significant new work appearing in the literature up to the end of 1991 has been included, if only as an addendum to an appropriate reference.

Chapter 1 References

1. C. M. Suter, *Organic Chemistry of Sulphur*, Wiley, New York–London, **1944**; S. Oae (Ed.), *Organic Chemistry of Sulphur,* Plenum Press, New York–London **1977**; E. Block, *Organic Chemistry Vol. 37, Reactions of Organosulphur Compounds*, Academic Press, New York, **1978**; see also *Royal Society of Chemistry Specialist Periodical Reports, Organic Compounds of Sulphur Selenium And Tellurium*; L. Field, *Synthesis*, **1978**, 713; L. Field, *Synthesis*, **1972**, 101.

2. T. Durst, in *Comprehensive Organic Chemistry*, Ed. D. H. R. Barton and W. D. Ollis, Pergamon Press, Oxford, **1979**, *3*, 171 and 197; S. Rajappa, in *Comprehensive Heterocyclic Chemistry,* Ed. A. R. Katritzky and C. W. Rees, Pergamon Press, Oxford, **1984**, *4*, 741; A. H. Ingall, in *Comprehensive Heterocyclic Chemistry*, Ed. A. R. Katritzky and C. W. Rees, Pergamon Press, Oxford, **1984**, *3*, 885; see also *Comprehensive Organic Synthesis*, Ed. B. M. Trost and I. Fleming, Pergamon Press, Oxford, **1991**.

3. P. D. Magnus, *Tetrahedron*, **1977**, *33*, 2019.

4. K. Schank, in *Methoden der Organischen Chemie (Houben-Weyl)*, G. Thieme, Stuttgart, **1985**, *E11*, 1132; see also B. M. Trost, *Bull. Chem. Soc. Jpn.*, **1988**, *61*, 107; N. S. Simpkins, *Tetrahedron*, **1990**, *46*, 6951.

5. S. Patai, Z. Rappoport, and C. J. M. Stirling (Eds), *The Chemistry of Sulphones and Sulphoxides,* John Wiley and Sons, Chichester, UK, **1988**.

6. A few sulphones have been reported as natural products, see A. Kjær, *Pure Appl. Chem.*, **1977**, *49*, 137 and references therein; M. K. Jogia, R. J. Andersen, E. K. Mantus, and J. Clardy, *Tetrahedron Lett.*, **1989**, *30*, 4919; H. Nakamura, H. Wu, J. Kobayashi, M. Kobayashi, Y. Ohizumi, and Y. Hirata, *J. Org. Chem.*, **1985**, *50*, 2494; Y. Ichikawa, *Tetrahedron Lett.*, **1988**, *29*, 4957; S. N. Suryawanshi, A. Rani, and D. S. Bhakuni, *Ind. J. Chem.*, **1991**, *30B*, 1098; K. Morita and S. Kobayashi, *Chem. Pharm. Bull.*, **1967**, *15*, 988.

    In addition, sulphone-containing analogues of natural products have been of considerable interest, see for example M. Nakano, S. Atsuumi, Y. Koike, S. Tanaka, H. Funabashi, J. Hashimoto, and H. Morishima, *Tetrahedron Lett.*, **1990**, *31*, 1569; K. C. Schneider and S. A. Benner, *Tetrahedron Lett.*, **1990**, *31*, 335; E. W. Logusch, *Tetrahedron Lett.*, **1988**, *29*, 6055; Y. Girard, M. Larue, T. R. Jones, and J. Rokach, *Tetrahedron Lett.*, **1982**, *23*, 1023; B. Beagley, P. H. Crackett, R. G. Pritchard, R. J. Stoodley, and C. W. Greengrass, *J. Chem. Soc., Perkin Trans. 1*, **1990**, 773; S. Hanessian and M. Alpegiani, *Tetrahedron*, **1989**, *45*, 941.

CHAPTER 2

The Preparation of Sulphones

The ease of preparation of sulphones is central to their utility in organic synthesis. This chapter provides extensive coverage of sulphone preparation, ranging from the simplest dialkyl sulphones to much more complex intermediates combining the sulphone group with other functionality. The chemistry has been divided largely according to the type of sulphone being prepared. Some methods of preparation, such as the oxidation of sulphides, are clearly almost universally applicable, and are here grouped together in the first section. Other general reaction types such as reactions of sulphinate salts can be found in several subsections.

Many sulphones can of course be prepared from simpler ones, primarily through carbon–carbon bond-forming reactions of the derived carbanions. Such sulphonyl carbanion chemistry is dealt with mainly in Chapter 3. Similarly, processes such as rearrangements and cycloadditions useful for preparing sulphones are touched on here and dealt with in more detail in Chapters 5 and 6 respectively.

2.1 Simple Alkyl Aryl and Dialkyl Sulphones

The three most important methods for the preparation of simple sulphones are dealt with in Sections 2.1.1 to 2.1.3, with other miscellaneous methods grouped together in Section 2.1.4.

2.1.1 Oxidation of Sulphides

This is an extremely useful and broadly applicable method for preparing sulphones. A major attraction of the method stems from the variety of high-yielding reactions which may be used to introduce an alkylthio or arylthio grouping. Both nucleophilic and electrophilic reagents can be used, e.g. Scheme 1.

A very large selection of reagents is available for the subsequent oxidation to form sulphones, thus ensuring a highly efficient two-step sequence in very many cases.[1] By comparison, the direct introduction of sulphur at a higher oxidation state can be lower yielding. A classic example is the alkylation of sulphinate salts, which, being much less nucleophilic than thiolates (and ambient), often require forcing conditions or special 'tricks' to obtain satisfactory results (Section 2.1.2).

5

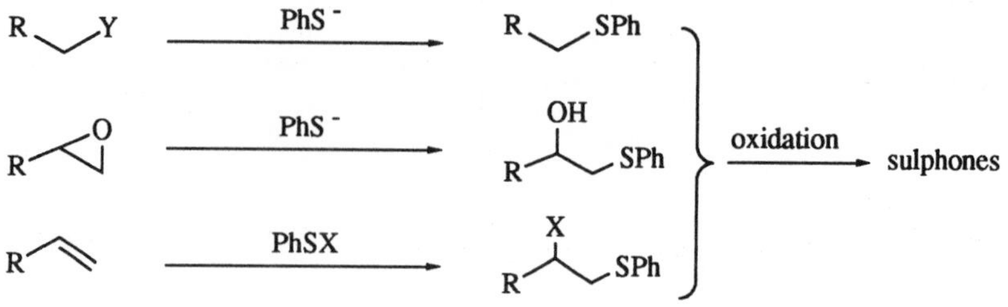

Y = halogen, OTs, etc.

X = halogen, SePh, etc.

Scheme 1

A detailed description of the synthesis of sulphides is not appropriate here, although some of the schemes shown include the sulphur-introducing step for illustration. The oxidation of all types of sulphides is considered in this section to avoid fragmentation and repetition of material between sections of this chapter.

Goheen and Bennett have reported the use of concentrated nitric acid for the preparation of simple dialkyl sulphones from the corresponding sulphides.[2] Although this is one of the cheapest reagents available for this transformation the harsh conditions make it unsuitable for all but the most robust systems.

Peracetic acid, generated *in situ* from hydrogen peroxide, is a very popular choice for this oxidation. Sulphides (1)-(3) were each converted to the corresponding sulphones under typical conditions, i.e. 30% H_2O_2 in glacial acetic acid.[3]

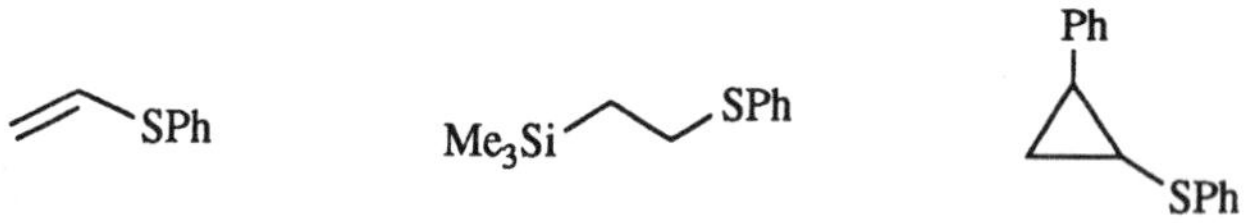

In these reactions the first oxidation to form the sulphoxide is much more rapid than the second oxidation to form the sulphone. Often an excess of H_2O_2 is required, and prolonged reaction times and/or heating may be needed to complete the reaction. Catalysis of the reaction is possible using metal catalysts such as $ZrCl_4$ or tungstic, vanadic or molybdic acid, enabling clean sulphone formation with only stoichiometric amounts of H_2O_2.[4] Other forms of H_2O_2 useful for this reaction include the urea complex[5] and the bis(trimethylsilyl) derivative TMSOOTMS.[6]

Perhaps the most commonly used reagent for sulphide to sulphone conversions is MCPBA. Four examples of sulphone preparation incorporating MCPBA oxidation are highlighted in Scheme 2.[7–10]

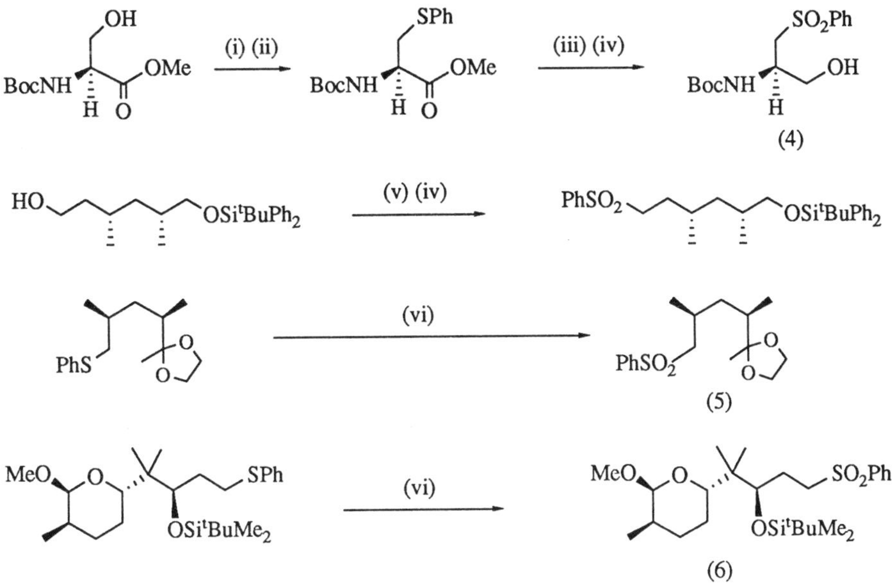

reagents:

(i) TsCl, pyridine (ii) PhSNa, DMF, 0°C
(iii) $NaBH_4$, MeOH (iv) MCPBA, CH_2Cl_2
(v) PhSSPh, $^nBu_3P$, CH_2Cl_2 (vi) MCPBA, $NaHCO_3$, CH_2Cl_2

Scheme 2

Sulphone (**4**) was prepared to allow the homologation of serine using sulphonyl carbanion chemistry.[7] The use of buffered conditions for the oxidation of acid-sensitive systems is illustrated by the preparation of (**5**) and (**6**).

The two oxidations shown in Scheme 3 underline the chemoselectivity possible in sulphide oxidations. Thus the unsaturated sulphide (**7**) is oxidised very cleanly to the mixture of diastereoisomeric sulphoxides (**8**) by the use of MCPBA at low temperature. Treatment of the related sulphoxides (**9**) with MCPBA at 0–20°C provides the sulphone (**10**) in excellent yield.[11] The lack of alkene epoxidation is notable, as is the use of Na_2S in the classical cyclic sulphide preparation which starts the whole sequence.[12]

Other peracids have also been used for this oxidation, including peroxytrifluoroacetic acid, which is highly reactive even at low temperature.[13] As a safer alternative to MCPBA, which is shock sensitive in pure form, the use of magnesium monoperphthalate has been recommended.[14]

Oxone® is a safe, commercially available oxidant which has become widely accepted for the oxidation of sulphides to sulphones. Commercially available Oxone® is a mixture of K_2SO_4, $KHSO_4$ and the active oxidant potassium hydrogen persulphate, $KHSO_5$. The reagent is usually employed in aqueous alcoholic solvent, in which it forms an acidic solution (pH 2–3). Buffering of the solution, for example with borate, enables oxidations to be performed at about pH 5 for acid-sensitive substrates.[15] In the report by Trost and Curran the reagent has been demonstrated to be highly chemoselective, cleanly oxidising substrates in which other functional groups such as ketones, alcohols, and alkenes are unreactive.[16] Examples of compounds oxidised by Trost's group include (11)-(13)[16] and (14).[17]

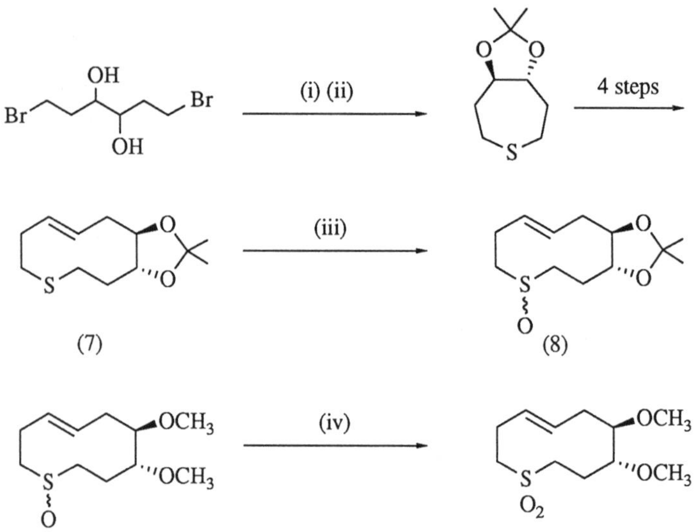

reagents:

(i) Me$_2$C(OMe)$_2$, TsOH, benzene, 96% (ii) Na$_2$S·9H$_2$O, EtOH, 80%
(iii) MCPBA, CH$_2$Cl$_2$, -80°C, 95% (iv) MCPBA, 0–20°C, 96%

Scheme 3

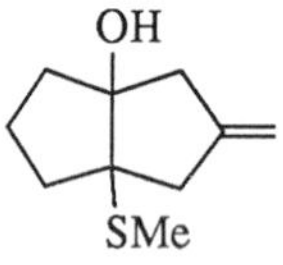

(11) 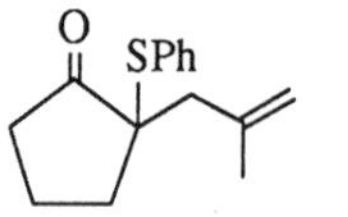 (12) 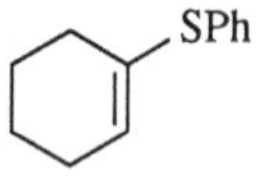 (13) 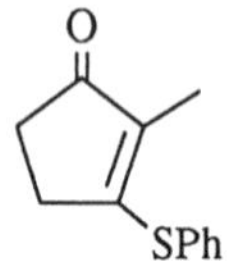 (14)

This method has been quite widely adopted for the oxidation of polyfunctionalised sulphones as demonstrated in a number of total syntheses. Scheme 4 gives three examples of substrates successfully oxidised.

Each of the compounds (15),[18] (16)[19] and (17)[20] is converted to the corresponding sulphone in high yield, giving some idea of the scope of the Oxone® method for the oxidation of chiral functionalised intermediates. Note the use of phenylsulphenyl succinimide for the preparation of sulphide (15) from the corresponding alcohol with inversion of configuration.

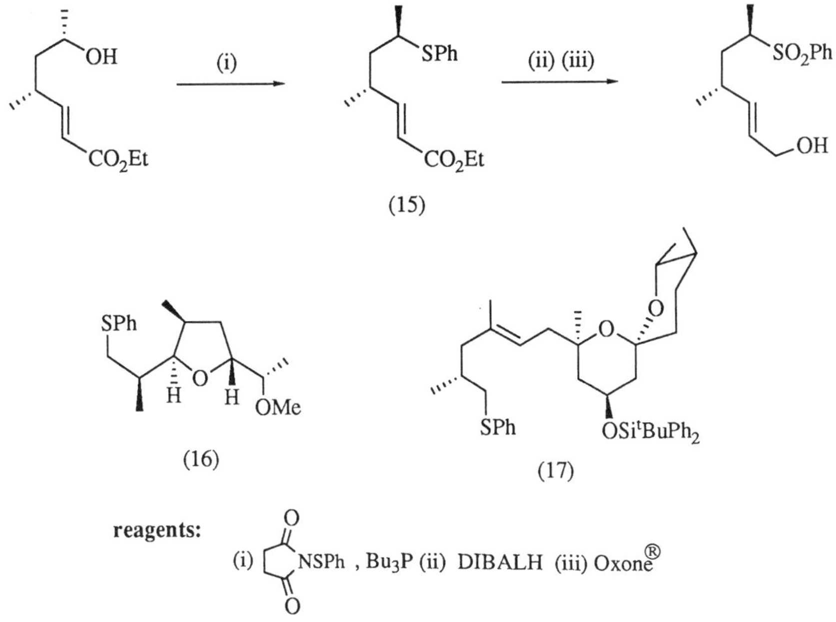

(15)

(16)

(17)

reagents:

(i) NSPh , Bu$_3$P (ii) DIBALH (iii) Oxone$^{\circledR}$

Scheme 4

Another very attractive oxidant for sulphides is sodium perborate.[21] The use of this reagent in glacial acetic acid at 50–55°C results in extremely clean conversion of simple aliphatic or aromatic sulphides to the corresponding sulphones. Anilines are also oxidised to the corresponding nitroarenes, and the two heteroatom oxidations can be carried out concurrently, as in the preparation of the nitrosulphone (**18**), Scheme 5.

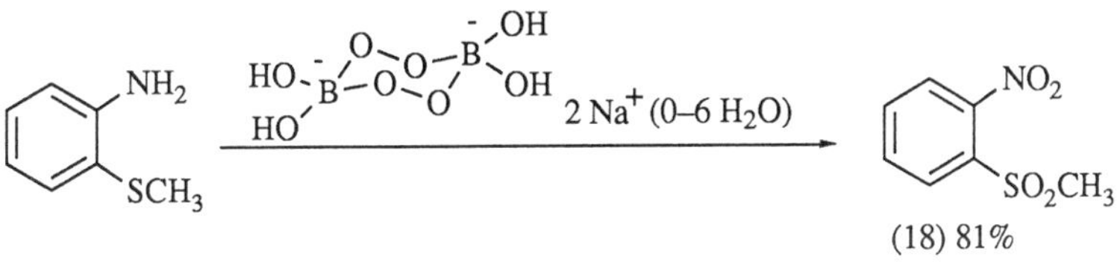

(18) 81%

Scheme 5

A method using sodium periodate and a catalytic amount of ruthenium trichloride in a two-phase system described by Sharpless for the oxidation of alkenes, alcohols, ethers and aromatics has been reported to be highly effective in the conversion of thiophenyl glycosides to the corresponding sulphones.[22] In a recent total synthesis of cytovaricin Evans *et al.* have observed the concomitant oxidation of a sulphide on carrying out the dihydroxylation of an alkene using OsO$_4$, Scheme 6.[23]

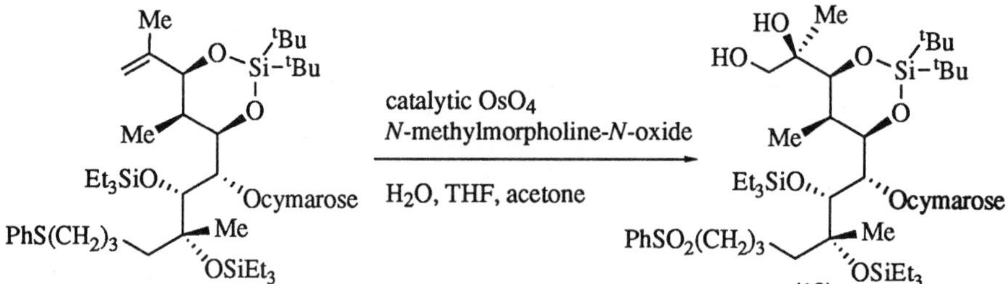

Scheme 6

The dihydroxy sulphone (**19**) is obtained in 96% yield. This result is somewhat unexpected, since earlier reports have indicated that although sulphoxides are rapidly oxidised to sulphones using OsO$_4$, sulphides appear to be quite resistant. Evans *et al.* have shown that in a control experiment without the OsO$_4$ catalyst no oxidation takes place. The ease of sulphone formation was attributed to catalysis of the sulphide oxidation by the tertiary amine (*N*-methylmorpholine) present.

Reich *et al.* have described the use of seleninic acids as catalysts for the oxidation of sulphides using hydrogen peroxide.[24] In this report the use of *ortho*-nitro-benzeneseleninic acid was recommended for the preparation of sulphones. Later reports by Nicolaou *et al.*[25] and Ley and co-workers[26] have revealed that the method can be used chemoselectively with unsaturated substrates, Scheme 7.[26]

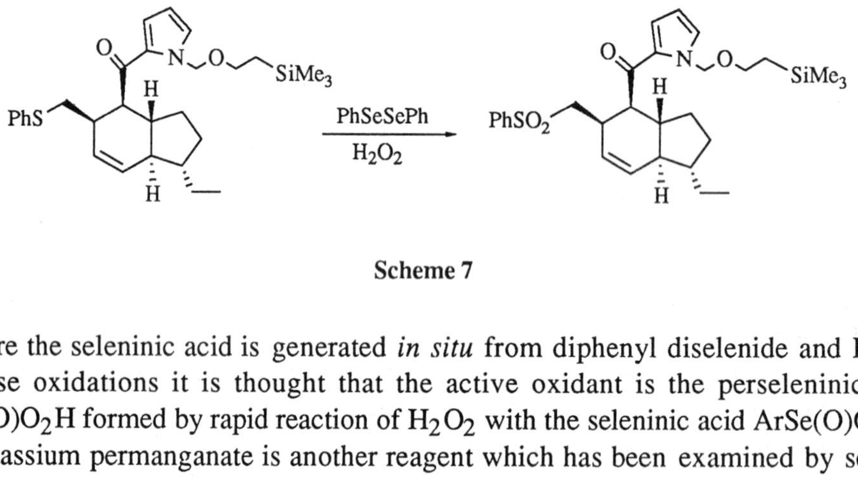

Scheme 7

Here the seleninic acid is generated *in situ* from diphenyl diselenide and H$_2$O$_2$. In these oxidations it is thought that the active oxidant is the perseleninic acid ArSe(O)O$_2$H formed by rapid reaction of H$_2$O$_2$ with the seleninic acid ArSe(O)OH.

Potassium permanganate is another reagent which has been examined by several groups as a useful oxidant for sulphides. Thus KMnO$_4$ has been reported to give high yields of sulphones under heterogeneous (refluxing CH$_2$Cl$_2$ or hexane)[27] or phase-transfer (CH$_2$Cl$_2$, H$_2$O)[28] conditions. Another report describes the reaction of *gem*-disulphides to give monosulphone derivatives using KMnO$_4$ in acetone at 0°C, Scheme 8.[29] The reaction gives good yields but is very slow, taking 8–10 days to reach completion.

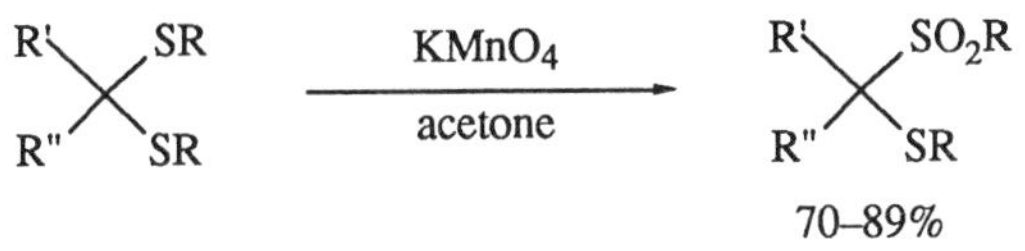

Scheme 8

Many other oxidants for the conversion of either sulphoxides and/or sulphides to sulphones have been described in the literature; however, most of these have not been widely adopted for synthetic work and often offer no advantage over the 'mainstream' methods already discussed. A selection of examples includes superoxide,[30] nitronium salts,[31] iodosylbenzene,[32] [bis(trifluoroacetoxy)iodo]benzene,[33] ozone[34] and microbial methods.[35]

2.1.2 Alkylation of Sulphinate Salts

Sulphones can be obtained from the reaction of sulphinic acids or their derived metal salts with many types of functional groups including alkenes, alcohols, epoxides, alkyl halides and carbonyl compounds.

The additions of sulphinic acids to simple alkenes, i.e. hydrosulphonylation, is not a facile process. With dienes or polyenes this reaction may be accomplished by employing palladium catalysis, e.g. Scheme 9.[36]

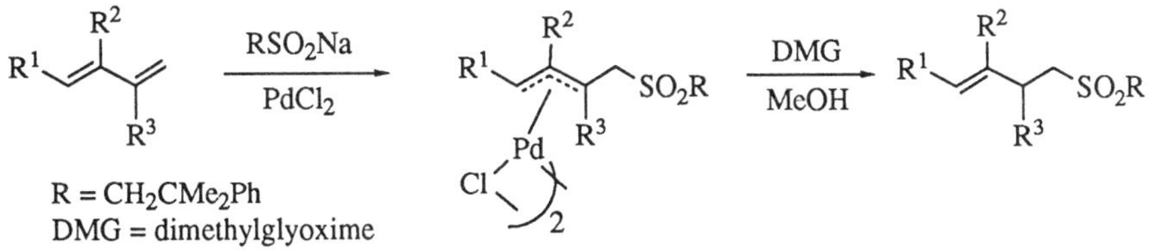

Scheme 9

In the three cases studied anti-Markovnikov products resulting from 1,2-additions were obtained in high yield.

Polarised C=C bonds, either electron rich or electron poor, can react with sulphinic acids with far greater facility. This chemistry leads to functionalised sulphones and is dealt with in Section 2.4 along with sulphinate reactions with carbonyls, amines and epoxides. This section will concentrate on the most widely used reactions of sulphinic acids, those of their metal salts with alkyl halides. At this point it is appropriate to include a brief account of the availability of the sulphinic acids and derived salts themselves, since this will have bearing on the attractiveness of this route to sulphones.

Standard methods for the preparation of sulphinic acids include the reduction of sulphonyl halides (e.g. with metals, or with sodium or potassium sulphite) and the reaction of organometallics with SO_2.[37] Despite the generality of these methods difficulties may be encountered, particularly in the isolation of pure sulphinic acids, which are somewhat unstable (especially the alkanesulphinic acids) and best stored in the form of metal salts.

Other methods which can be used to isolate sulphinic acids include the oxidation of thiols with MCPBA[38] and most recently the reaction of sulphonyl halides with thiols.[39] Of these, the latter appears more straightforward and is particularly suitable for the preparation of certain hydroxyarenesulphinic acids, Scheme 10.

$$RSO_2Cl \xrightarrow{\text{2 eq. TolSH, Et}_3\text{N, CH}_2\text{Cl}_2, -78^\circ\text{C}} RSO_2H$$

R = Cl-, HO-, NO$_2$-substituted aromatic

Scheme 10

Two other reports describe useful modifications of the sulphone cleavage method for the preparation of sulphinate salts. Either benzothiazole derivatives (**20**)[40] or phthalimidomethyl sulphones (**21**)[41] are prepared, starting with thiols, and ultimately cleaved to give the desired sulphinate salts in high yield, Scheme 11.

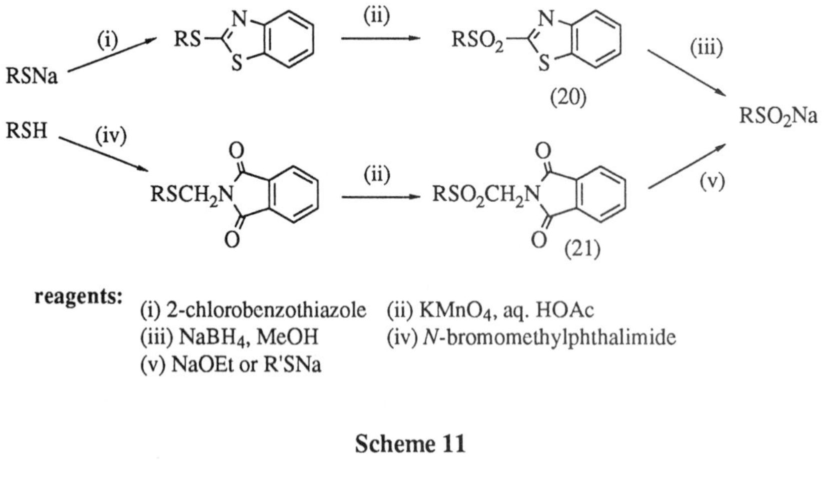

reagents: (i) 2-chlorobenzothiazole   (ii) KMnO$_4$, aq. HOAc
           (iii) NaBH$_4$, MeOH   (iv) N-bromomethylphthalimide
           (v) NaOEt or R'SNa

Scheme 11

For the vast majority of synthetic applications either phenyl sulphones (PhSO$_2$) or *p*-tolyl sulphones (TolSO$_2$) are employed, the aromatic group not usually being involved in carbon–carbon bond formation mediated by the sulphone. Since both benzenesulphinic acid and toluenesulphinic acid are commercially available as their sodium salts, a major preoccupation of the synthetic chemist has been their effective *S*-alkylation, and hence introduction of an arenesulphonyl group into the substrate of interest. The most widely used alkylating agents are alkyl halides, which react according to Scheme 12.

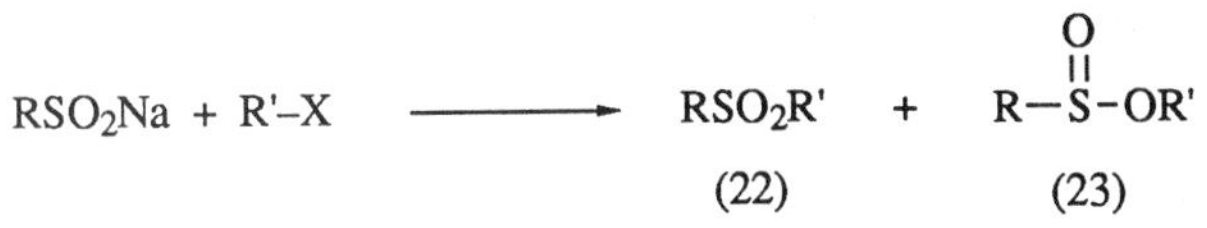

$$RSO_2Na + R'-X \longrightarrow RSO_2R' + R\overset{\overset{\displaystyle O}{\|}}{-}S-OR'$$

$$\qquad\qquad\qquad\qquad\qquad (22) \qquad\qquad (23)$$

Scheme 12

Since sulphinate is an ambident nucleophile the alkylation can occur either on sulphur, giving the sulphone (**22**), or on oxygen, to give the sulphinate ester (**23**). In general, hard alkylating agents react on oxygen, whilst softer electrophiles give *S*-alkylation. The account of Meek and Fowler provides examples of both extremes.[42] Thus alkylation of $TolSO_2Na$ with dimethyl sulphate (or reaction of $TolSO_2H$ with CH_2N_2) gives almost entirely sulphinate ester, whereas the use of MeI gives predominantly the sulphone. Under certain circumstances the competing *O*-alkylations may not cause difficulties, for example, where rearrangement of the sulphinate ester can occur to give a sulphone (as is the case in allylic systems, see Section 5.2). Similarly, if the reaction is conducted under conditions in which hydrolysis of the sulphinate can take place, and excess alkylating agent (e.g. dimethyl sulphate in aqueous bicarbonate[43]) is employed, then it is likely that sulphinate ester is recycled to sulphinate salt by hydrolysis, ultimately leading to a preponderance of stable sulphone product.

The *O/S*-alkylation problem encountered in many sulphinate alkylations under traditional conditions, e.g. hot alcohol or DMF, has stimulated an active search for alternative milder conditions which give high yields of sulphones.

One simple solution is to carry out the alkylation using poly(ethylene glycol)s (PEGs) or their ethers as solvent.[44] These conditions are consistently superior to comparative runs using methanol as solvent, although methanol with a catalytic amount of PEG also gives improved results. In this report the possibility of sulphinate ester hydrolysis under the reaction conditions is again noted.

A topic of considerable interest has been the alkylation of tetraalkylammonium arenesulphinates to give sulphones. A report by Manescalchi *et al.* describes the use of benzenesulphinate anion supported on the anion-exchange resin Amberlyst A-26.[45] The alkylation is performed by stirring a slight excess of the preformed sulphinate resin with an alkyl halide in refluxing benzene for a few hours. The yields of sulphone are excellent, although the necessity of preforming the resin is a minor inconvenience. Two other related reports describe the use of either a preformed tetrabutylammonium arenesulphinate[46] or a catalytic amount of Bu_4NBr to effect superior alkylations.[47] Of these two the second method gives slightly better yields and appears most convenient, whilst the former is conducted under milder conditions. Unfortunately, most of the examples carried out in these studies use only primary and/or otherwise activated alkyl halides and give little indication of the scope or limitations of the methods. A later report by Crandall and Pradat gives a more critical account of a liquid–liquid phase-transfer method, also requiring a catalytic amount of tetrabutylammonium salt.[48] This method is limited to primary bromides and iodides, secondary iodides, and certain activated chlorides. Thus 1-chlorohexane gives no sulphone product whereas, under identical conditions, 1-bromohexane gives the sulphone in 85% yield.

Two recent innovations reported by Biswas and co-workers appear to offer significant advantages for sulphinate alkylations. The first involves treatment of the alkyl halide with $TolSO_2Na$ and DBU in acetonitrile.[49] The reaction occurs rapidly at room temperature to give high yields of sulphone, although, again, a primary chloride (epichlorohydrin) was unreactive. The other paper from this group indicates that remarkable acceleration of sulphinate alkylations is possible using ultrasound.[50] Thus, under ultrasonication, $TolSO_2Na$ in aqueous DMF was alkylated with the usual variety of primary alkyl halides at ambient temperature in times of only 1–15 minutes. The yields are not quite so uniformly high as in some of the other methods and a secondary bromide was found to be unreactive.

The methods described give the organic chemist considerable scope in choosing a method for the preparation of a particular sulphone, although in many cases a compromise between ease of operation and chemical yield may be needed.

Symmetrical sulphones can be prepared in a single operation which forms both C–S bonds. This double alkylation process using sodium formaldehyde sulphoxylate (hydroxymethanesulphinic acid sodium salt) (24) and either a benzylic halide[51] or the salt of a Mannich base[52] gives moderate yields of the desired symmetrical sulphones, Scheme 13.

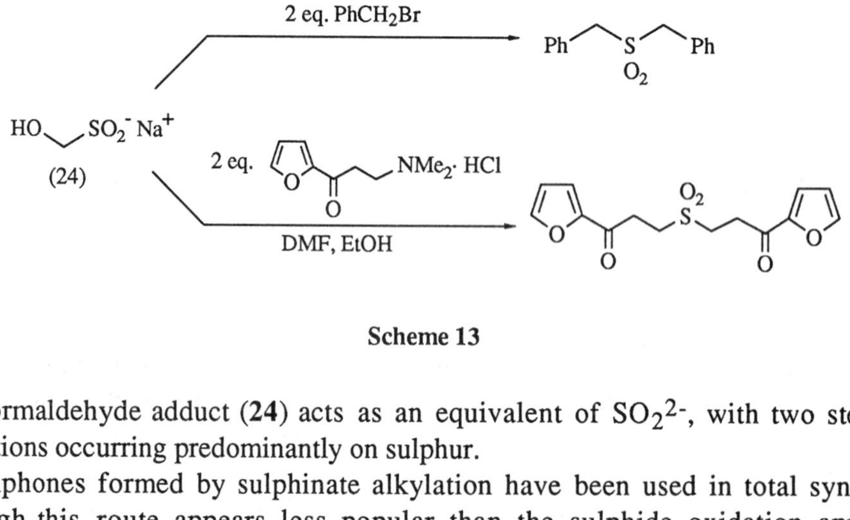

Scheme 13

The formaldehyde adduct (24) acts as an equivalent of SO_2^{2-}, with two stepwise alkylations occurring predominantly on sulphur.

Sulphones formed by sulphinate alkylation have been used in total synthesis, although this route appears less popular than the sulphide oxidation approach described in Section 2.1.1. Sulphones (25)-(27) are each formed by sulphinate alkylation under fairly typical conditions.[53] In the preparation of (25) and (27) the corresponding iodide and chloride respectively are used in the alkylation, whilst the precursor to (26) is, atypically, the corresponding triflate.

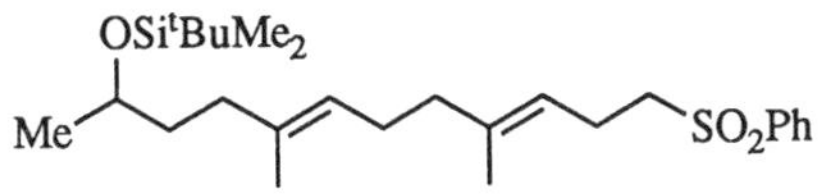

OSi^{t}BuMe$_2$

Me   SO$_2$Ph   (25)

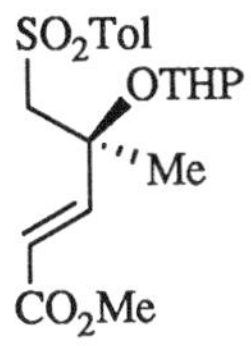

SO$_2$Tol

OTHP

Me

CO$_2$Me

(26)

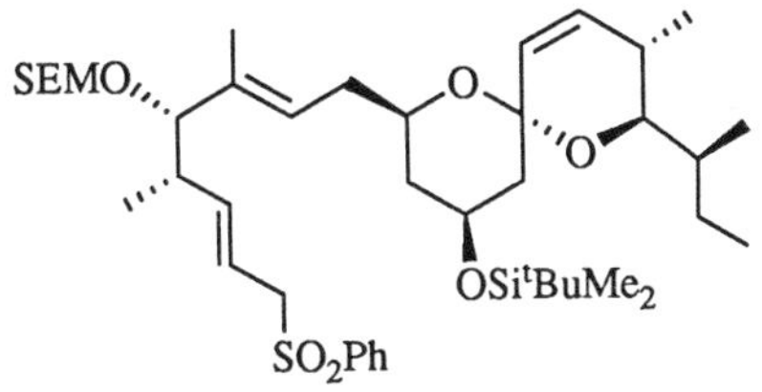

SEMO

O

O

OSi^{t}BuMe$_2$

SO$_2$Ph

(27)

Analogous alkylation procedures can be used to prepare triflones, RSO$_2$CF$_3$, by use of triflinate salts.[54]

Finally, it has proved possible to use certain types of tertiary allylic alcohols to prepare sulphones, e.g. Scheme 14.[55]

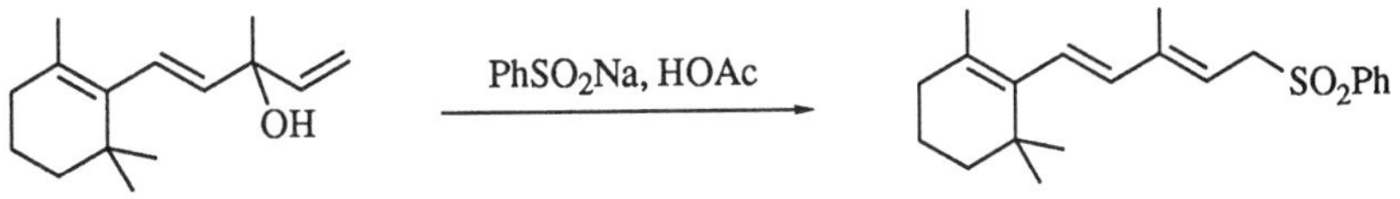

OH   PhSO$_2$Na, HOAc   →   SO$_2$Ph

Scheme 14

This type of reaction is presumably limited to systems capable of generating rather stable carbonium ion intermediates by initial hydroxyl group protonation.

2.1.3 Reactions of Sulphonic Acid Derivatives

The reactions of alkane- or arenesulphonic acid derivatives RSO$_2$X (X = halogen, OR', etc.) with nucleophilic partners represent a large and well-examined group of transformations. The direct substitution of X for a carbon nucleophile can, in principle, lead to sulphones. However, many problems are associated with this seemingly straightforward approach, particularly when using reactive C-nucleophiles such as alkyllithiums (*vide infra*).

Diaryl sulphones can usually be prepared without difficulty by reaction of arenes with arenesulphonyl chlorides in the presence of aluminium chloride.[56] This Friedel–Crafts type of sulphonylation and other diaryl sulphone preparations are dealt with in Section 2.3.4.

The analogous synthesis of alkyl aryl sulphones from alkanesulphonyl chlorides is, however, not general, and is very often complicated by side reactions such as arene chlorination.[57] At least a partial solution to this problem is to substitute chlorine with a more powerfully electronegative group. The use of either alkanesulphonic trifluoromethanesulphonic anhydrides (28),[58] or the corresponding sulphonyl fluorides, e.g. (29) or (30),[59] gives good results, as shown in Scheme 15.

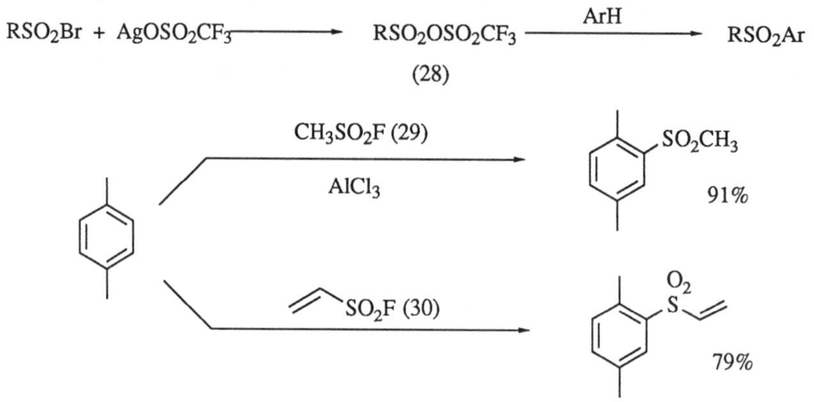

Scheme 15

In the reactions of anhydride (28), where R = Me or Et, reasonable yields of sulphone are obtained with aromatics without the use of a catalyst. However, with (28), where R = ^{i}Pr, the anhydride acts as a source of R$^+$, rather than RSO$_2$$^+$, giving Friedel–Crafts alkylation products in place of sulphones. The sulphonyl fluoride method described by Hyatt and White gives good yields of sulphone product using a variety of substrates, including vinylsulphonyl fluoride (30). Since the sulphonyl fluorides are readily prepared from the corresponding chlorides this method would appear to be the most general and convenient to date. Not surprisingly, regioisomers are obtained from reactions involving monosubstituted aromatics. No secondary sulphonyl fluorides were examined.

Two more recent reports provide alternative solutions to the problem of synthesising alkyl aryl sulphones from arenesulphonyl halides, Scheme 16.[60, 61]

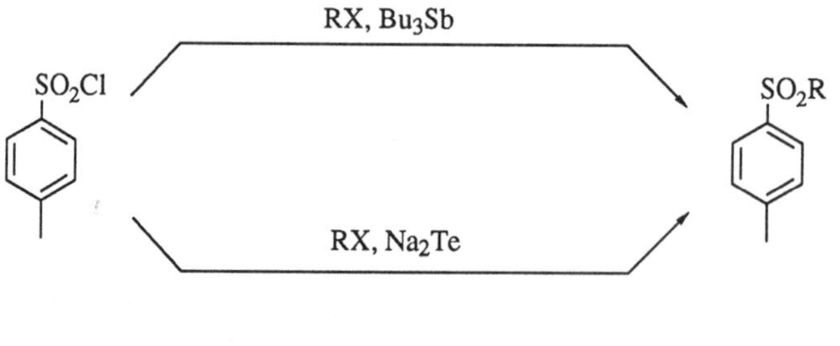

Scheme 16

Both reactions are most likely to involve the alkylation of an intermediate sulphinate formed by a reductive process (and so are variants of the sulphinate alkylations described in Section 2.1.2), and give comparable results. Notably, the method involving sodium telluride could also be applied to the reaction of octanesulphonyl chloride with ethyl iodide to give ethyl octyl sulphone in 74% yield.

As mentioned previously, the reaction between organometallics and sulphonyl halides is rarely a satisfactory route to sulphones. The report of Shirota *et al.* concerning the reaction of benzylsulphonyl halides with phenyllithium adequately illustrates the pitfalls of this chemistry, Scheme 17.[62]

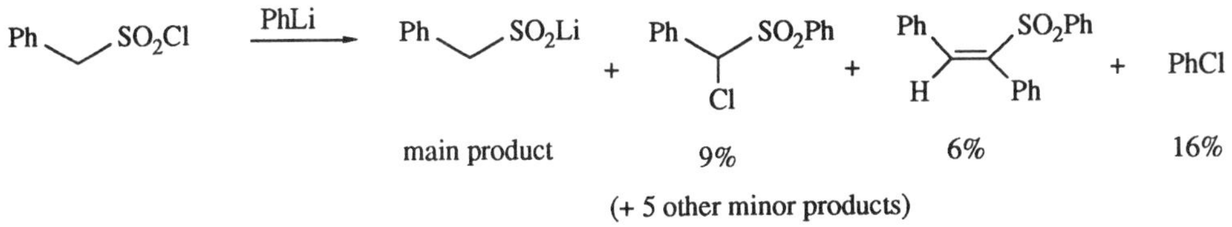

Scheme 17

Thus unwanted halogen–lithium exchange to give chlorobenzene and benzylsulphinic acid lithium salt occurs, accompanied by deprotonation α to the sulphone to give a variety of products, including mono- and dichlorosulphone, di- and trisulphone, and materials which could arise via sulphene intermediates.

When the corresponding sulphonyl fluoride is used in place of the chloride some of these side reactions are virtually eliminated, but the major product is the *gem*-disulphone (**31**), Scheme 18.[62,63]

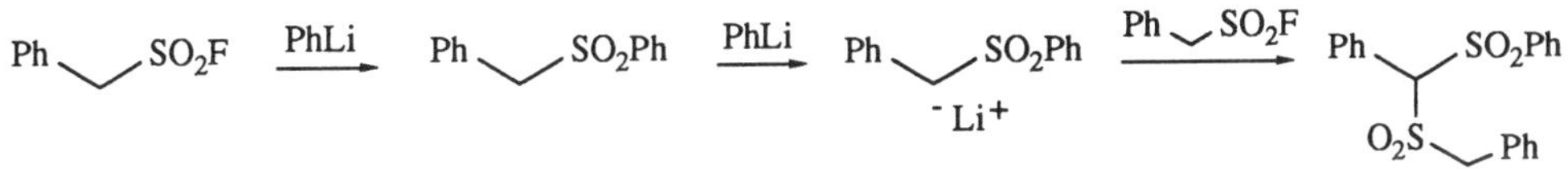

Scheme 18

This is clearly quite a good method for the synthesis of *gem*-disulphones. Further modification using PhMgBr in place of PhLi enables clean formation of the desired benzyl phenyl sulphone in 88% yield.[64]

Esters of aromatic sulphonic acids react efficiently with a range of a organolithiums to furnish diaryl, aryl heteroaryl and alkyl aryl sulphones, Scheme 19.[65]

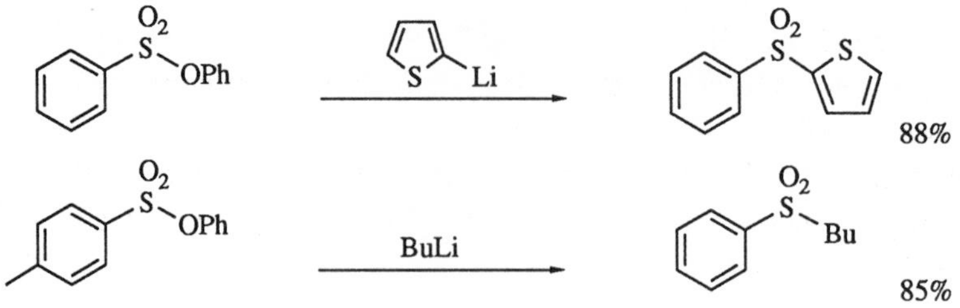

Scheme 19

Attempts to extend this reaction to include mesylate esters have not been successful, presumably due to competing deprotonation. Finally, Hendrickson and Bair have described analogous studies using triflic anhydride in attempts to prepare triflones.[66] Similar difficulties to those described earlier in this section were experienced, with ditriflones being the major products. Some improvement is observed when using $PhN(SO_2CF_3)_2$ as electrophile in place of triflic anhydride. Although some simple triflones can then be obtained in good yield (RSO_2CF_3, R = ^nBu, ^sBu, Et), the reaction still lacks generality.

2.1.4 Miscellaneous Sulphone Preparations

Sulphone radicals RSO_2· are involved in a host of important sulphone chemistry including the additions of RSO_2X (e.g. X = halogen, SePh) reagents to multiple bonds, rearrangements of unsaturated sulphones, and SO_2 extrusions from cyclic sulphones. These reactions are dealt with in later sections concerned with the preparation of functionalised sulphones (Section 2.4), sulphone rearrangements (Section 5.3) and cyclic sulphones (Chapter 8), respectively. Sulphone radicals would appear to be attractive intermediates for sulphone synthesis and yet the preparation of sulphones by the addition of carbon-centred radicals to SO_2 has received rather scant attention.

An early report in this area describes a sulphone preparation involving the passage of SO_2 through a solution of benzoyl peroxide in diphenylmethane.[67] Diphenylmethyl phenyl sulphone was isolated from the reaction in 36% yield, presumably via the sequence of events shown in Scheme 20.

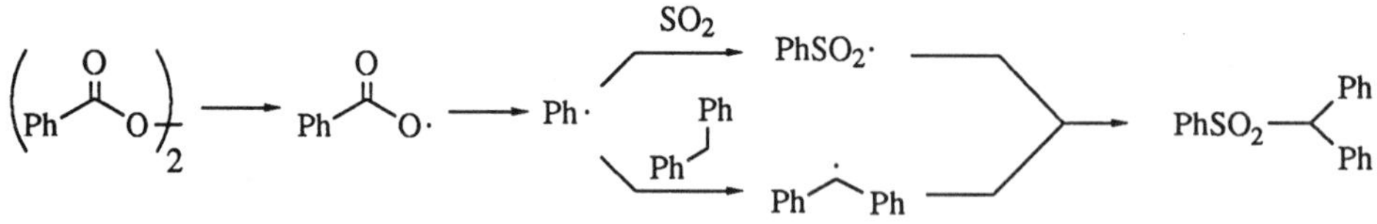

Scheme 20

Such reactions are clearly difficult to control, with by-products arising from unwanted alkyl radical coupling, and hydrogen atom abstraction by sulphonyl radicals leading to sulphinic acids.

Other related reactions in this area which yield sulphones as major products include the preparation of bis(tetrahydrofuranyl) sulphone (**32**),[68] and the phenolic sulphone (**33**),[69] Scheme 21.

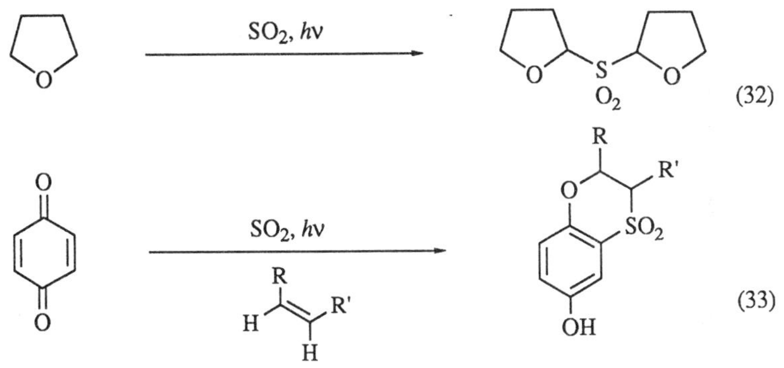

Scheme 21

Whereas sulphone (**32**) is formed only in low yield by an *inter*molecular radical combination (the tetrahydrofuranylsulphinic acid could also be isolated), the synthesis of (**33**) is much more efficient. This is because the latter process proceeds via a series of diradicals, enabling the intermediate sulphonyl radical to complete the cyclisation sequence by *intra*molecular radical combination. The trapping of diradicals by SO_2 may be worth further examination as a route to cyclic sulphones.

More recently other radical trapping reactions have been used to prepare sulphinic acids, which can of course be further transformed into sulphones, e.g. Scheme 22.

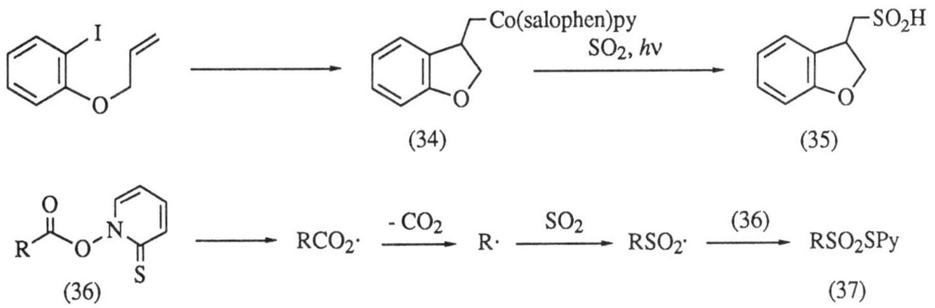

Scheme 22

Irradiation of the cobalt salophen intermediate (34) in the presence of SO_2 gives low yields of the sulphinic acid (35).[70] The Barton procedure, starting with thiohydroxamate (36), is much more efficient, giving yields of thiosulphonates (37) in the range 30–91%.[71] These compounds can be converted directly to sulphones by treatment with KOH followed by an alkylating agent.

Katritzky *et al.* have described a pyrylium salt-mediated conversion of primary amines into the corresponding sulphones.[72] Treatment of intermediate pyridinium salt (38) with $PhSO_2Na$ in refluxing dioxan for 12 h gives good yields of sulphone in two cases, Scheme 23.

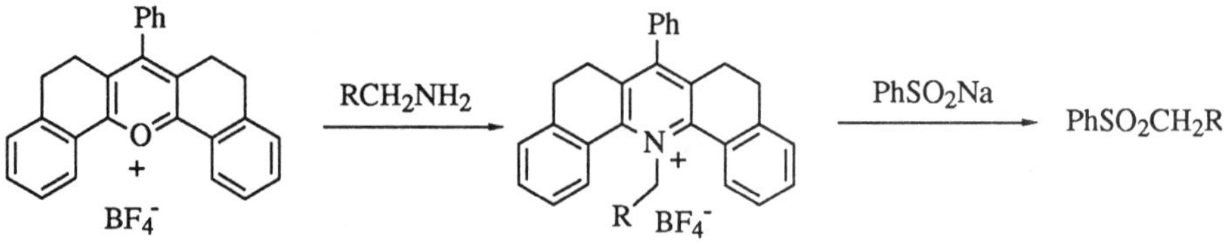

Scheme 23

Both this method and the above-mentioned Barton thiosulphonate preparation rely on sulphinate alkylation to furnish the final sulphone (Section 2.1.2).

Other sulphones prepared by special methods include the cyclopropyl sulphone (39) formed using a phenylsulphonyl carbene[73] and the unusual ferrocenyl sulphone (40).[74]

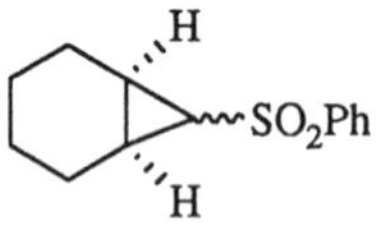

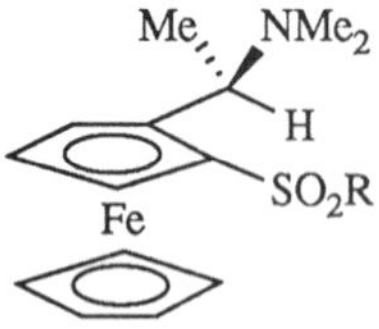

(39)   (40)

Sulphones have been obtained as minor products in the photolysis of tosylhydrazones[75] and in reactions of (aryloxy)oxosulphonium ylides with carbonyl compounds.[76] Neither of these reactions appears to have preparative value.

2.2 Vinyl Sulphones

Vinyl sulphones (α,β-unsaturated sulphones) have now become generally accepted as useful intermediates in organic synthesis. Thus vinyl sulphones serve efficiently as both Michael acceptors and as 2π–partners in cycloaddition reactions. In Michael reactions, a vinyl sulphone can enable complementary chemistry to that available using conventional Michael acceptors such as α,β-unsaturated carbonyl compounds (phenyl vinyl sulphone itself acts as a *two*-carbon acceptor not directly available using carbonyl Michael acceptors). In cycloaddition reactions, vinyl sulphones again serve a useful function as convenient equivalents for ethylene, acetylene, ketene, etc.

Thanks to the pioneering work of Julia and others, many types of vinyl sulphones are readily available, often stereoselectively. The opportunities for synthesis using these intermediates are considerable and are amply demonstrated by the synthetic efforts of Fuchs' group.

Vinyl sulphone chemistry has been reviewed in some detail before[77] and certain aspects have been covered in more general reviews and in a recent book.[78]

2.2.1 *Ionic and Radical Additions to Alkenes, Alkynes and Allenes*

A very broadly applicable strategy for the preparation of vinyl sulphones involves the construction of a β-heterosubstituted sulphone, i.e. (41), which can then undergo elimination, Scheme 24.

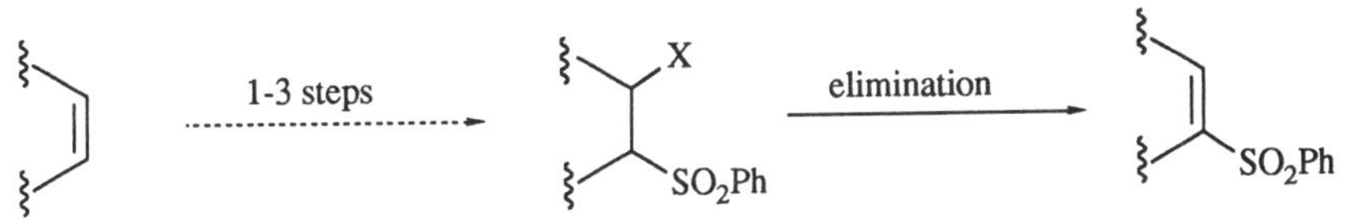

(41) e.g. X = OAc, SePh, halogen

Scheme 24

Whereas intermediates such as (41) can be prepared in a convergent fashion by combination of a sulphonyl carbanion with a carbonyl compound (Section 2.2.2), this section is concerned with the use of unsaturated starting materials, particularly simple alkenes, for the preparation of (41) without changing the carbon skeleton.

An indirect but versatile approach involving chlorosulphenylation–dehydrochlorination has been used by Hopkins and Fuchs to prepare a variety of cyclic vinyl sulphones, e.g. (42)-(44), Scheme 25.[79]

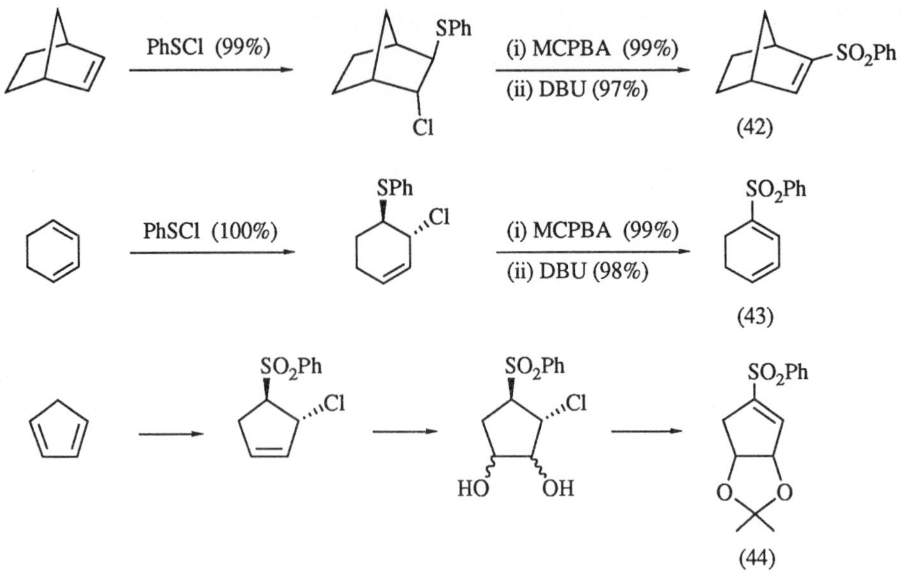

Scheme 25

Combined with an alternative epoxidation–thiolate-opening–oxidation sequence this chemistry allows the preparation of a number of key homochiral intermediates for natural product synthesis.[80]

Several alternative methods are available for the direct synthesis of intermediates (**41**) which involve ionic or radical addition of $PhSO_2X$ (X = Cl, Br, I, SePh, etc.) to an alkene. The reaction of alkenes with $HgCl_2$ in the presence of $PhSO_2Na$ results in the formation of arenesulphonyl mercury intermediates, which can then be eliminated under basic conditions, Scheme 26.[81,82]

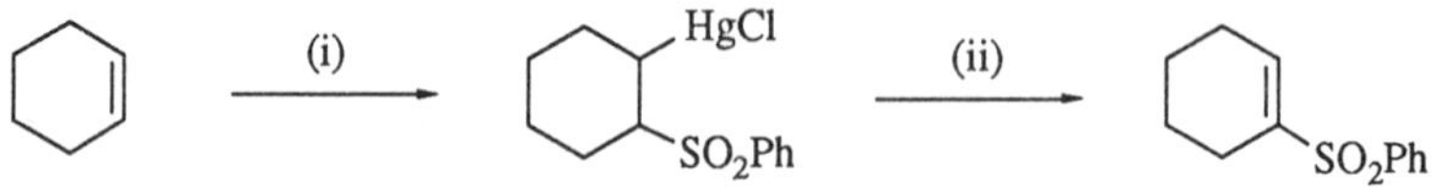

reagents:
(i) $HgCl_2$, $PhSO_2Na$, H_2O (ii) 50% $NaOH_{aq}$/dioxane

Scheme 26

α-Brominated vinyl sulphones can be prepared from simple parent compounds by treatment with bromine, followed by dehydrobromination with base.[83]

Alkene selenosulphonylation can be carried out under ionic or free-radical conditions with complementary regiochemical results. Thus, reaction of $PhSeSO_2Ar$ with alkenes in the presence of catalytic quantities of $BF_3 \cdot OEt_2$ follows a similar course to the sulphonylmercuration already described, namely Markovnikov addition.[84] A subsequent oxidation–*syn*-elimination step then provides vinyl sulphones in excellent yields, Scheme 27.

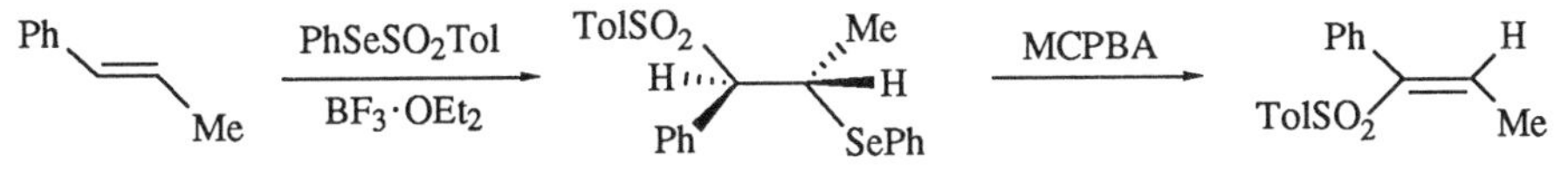

Scheme 27

Since under these conditions both the initial addition step and the *syn*-elimination are stereoselective, the sequence constitutes a stereoselective vinyl sulphone synthesis. Under thermal or photolytic conditions the selenosulphonylation shows all the characteristics of a free-radical chain reaction, giving anti-Markovnikov addition and ring opening in certain cases, e.g. (45), Scheme 28.[84,85]

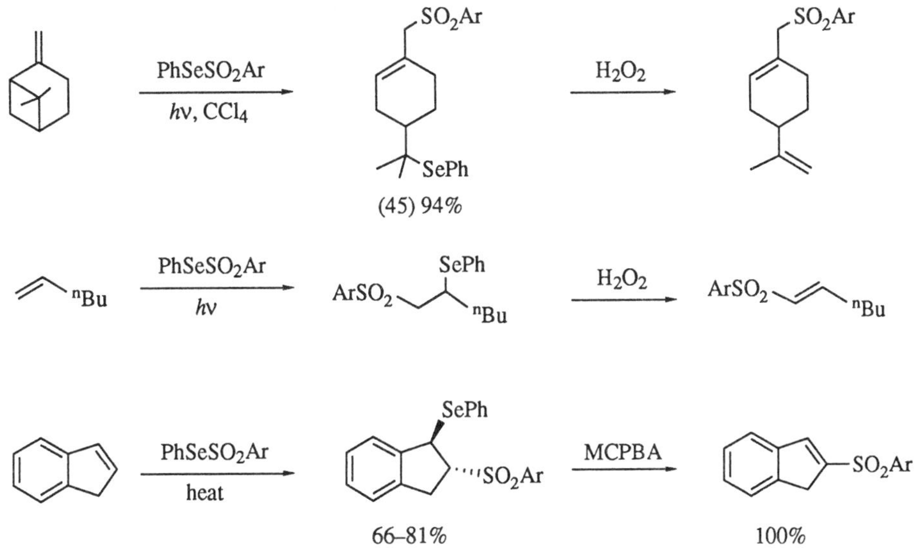

Scheme 28

Analogous chemistry can be carried out using sulphonyl iodides in place of selenides. Thus, Liu *et al.* carried out the radical addition of preformed $PhSO_2I$ or $TolSO_2I$ to alkenes in the presence of catalytic quantities of $CuCl_2$.[86] Similar results can be obtained using the convenient $TolSO_2Na/I_2$ mixture in MeOH, which presumably forms the somewhat unstable $TolSO_2I$ *in situ*, Scheme 29.[87]

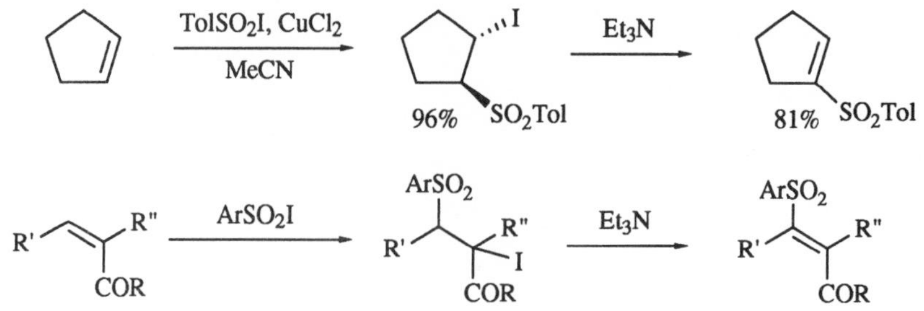

Scheme 29

This approach avoids the use of selenium, which is both costly and toxic, and allows the second elimination step to be carried out under very mild, non-oxidative conditions. The instability of the intermediate iodosulphones is, however, one potential drawback in some cases. As shown, this type of iodosulphonylation can also be applied to the preparation of functionalised vinyl sulphones, starting from α,β-unsaturated carbonyl compounds.[88] Both the seleno- and iodosulphonylation procedures can also be conducted with alkynes. In these cases the intermediate product is a β-seleno-[89] or β-iodovinyl sulphone,[90] both of which can then be modified to give other products, Scheme 30.

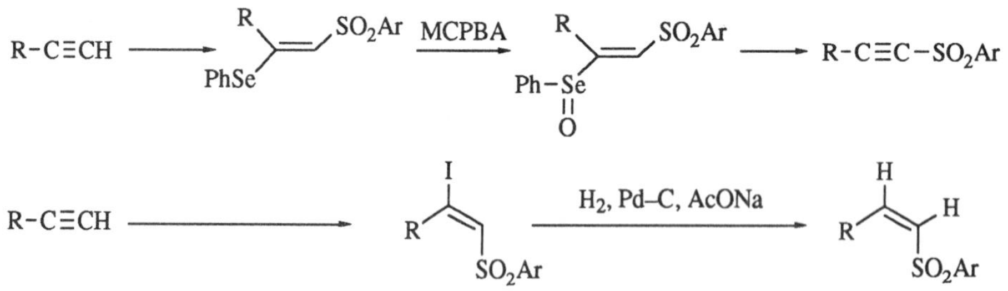

Scheme 30

In both cases the addition process is anti-Markovnikov and gives the *trans*-disubstituted product. The high level of stereoselectivity is somewhat surprising in view of the known rapid isomerisation of the presumed intermediate vinyl radical and is apparently a consequence of an even more rapid chain-transfer reaction. Back and Muralidharan have recently studied the rate of this process using the classical free-radical clock approach.[91] Alkene or alkyne substrates, e.g. (**46**) or (**47**), having an appropriately placed cyclopropane group, are exposed to typical selenosulphonylation conditions and products arising from simple 1,2-selenosulphonylation or the alternative ring opening to give a 1,5-adduct are obtained, e.g. Scheme 31.

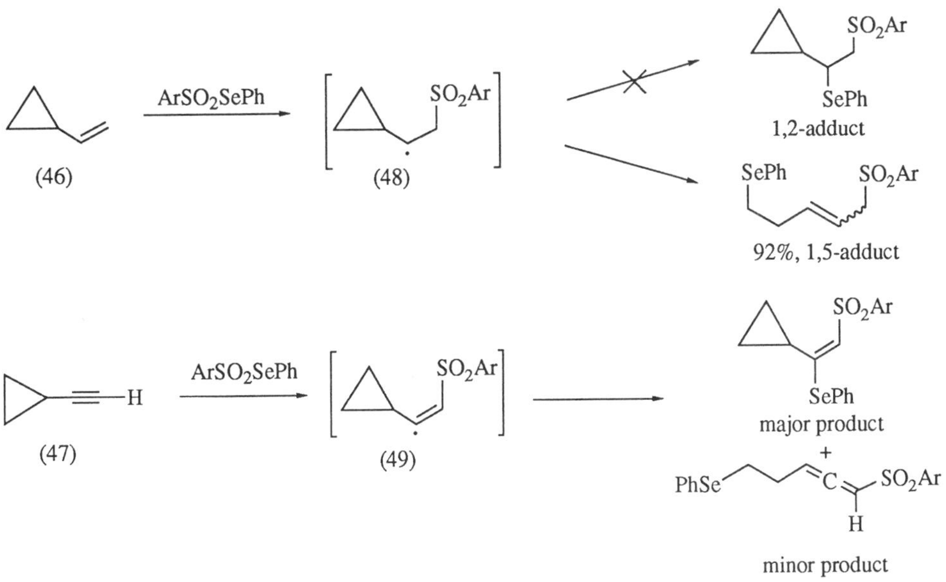

Scheme 31

In the case of vinylcyclopropanes such as (46) none of the simple 1,2-adduct is obtained; the radical chain-transfer step is clearly not competing with the more rapid ring opening of the intermediate cyclopropylcarbinyl radical (48). In contrast, vinyl radical (49) gives the 1,2-adduct as the major product, particularly when an excess of selenosulphonate is employed. If it is assumed that ring opening occurs at comparable rates for the radicals (48) and (49), then clearly selenosulphonates must be substantially more reactive towards (49) than (48). The rate of inversion of β-sulphonylvinyl radicals was deduced to be $\leq 2 \times 10^6$ s^{-1}.

The use of β-functionalised vinyl sulphones, such as those in Scheme 30, in alkylative processes leading to substituted products is described in Scheme 49.

Further work by Truce's group has extended the scope of the iodosulphonylation reaction to the use of allene substrates, e.g. Scheme 32.[92]

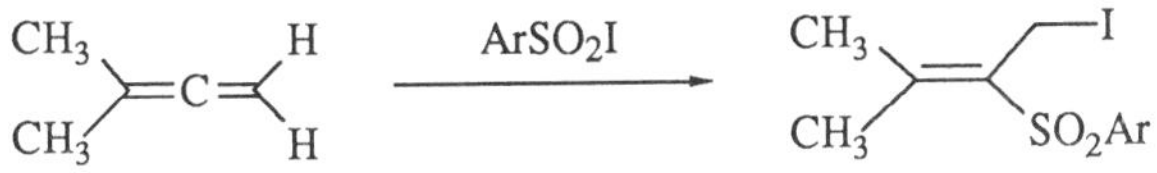

Scheme 32

The major products obtained result from addition of the $ArSO_2$ radical to the central position, followed by chain-transfer iodine abstraction at the terminal position.

Recently $ArSO_2$ radicals have been generated from sulphonate esters; however, the synthetic aspects of this process remain to be explored.[93]

Many of the procedures described can be extended to give dienyl sulphones, e.g. Scheme 33.[94–96] Whilst iodosulphonylation gives products of exclusive 1,4-addition, the other two processes shown involve predominant 1,2-addition. Various elimination procedures can then be applied to these intermediates to furnish the final dienyl sulphones, which have utility in reverse electron demand Diels–Alder reactions, see Chapter 6.

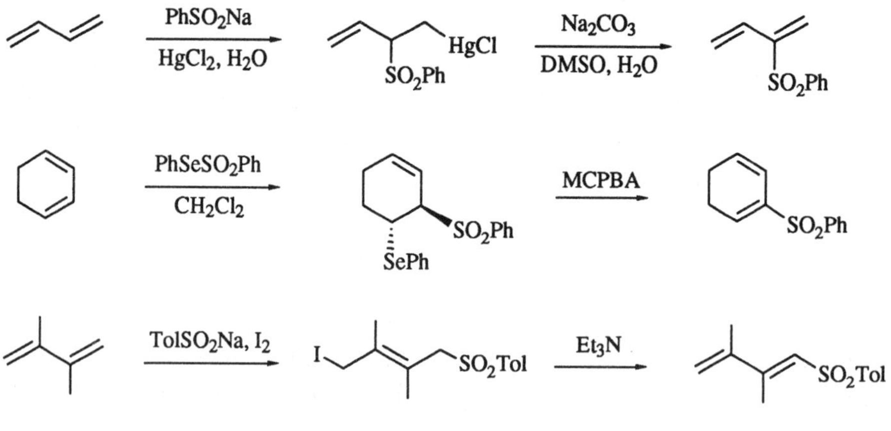

Scheme 33

Some other related reactions of sulphonyl chlorides which have been reported and which rely on transition metal catalysis are shown in Scheme 34.[97–100]

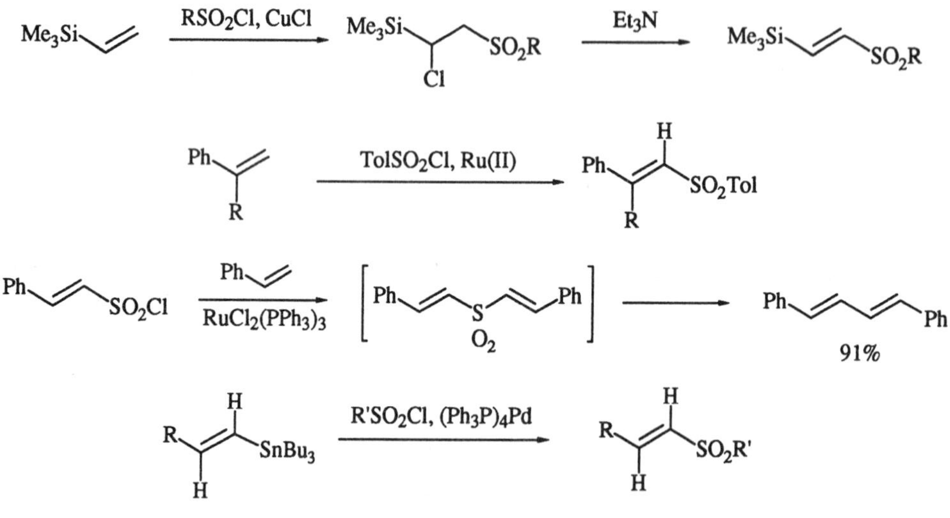

Scheme 34

The chlorosulphonylation of unsaturated silanes, including trimethylvinylsilane, occurs using RSO_2Cl and a copper catalyst.[97] In this case a separate step is required for the elimination of HCl using triethylamine.

The two related ruthenium-catalysed reactions both involve the regioselective chlorosulphonylation of styrene derivatives.[98,99] When a simple aromatic sulphonyl halide is employed ($ArSO_2Cl$) the expected simple vinyl sulphone product is obtained directly.[98] The homologous version of this reaction, in which both partners are styrene derivatives, does not stop at the divinyl sulphone stage, but proceeds further by way of SO_2 extrusion to give 1,3-dienes.[99]

The palladium-catalysed cross-coupling of simple and substituted vinyl (or allyl) stannanes with sulphonyl halides appears to be rather more versatile than the ruthenium-catalysed reactions, allowing a wider range of products to be accessed. However, this reaction presumably also suffers from the inherent limitation of using partners in which β-elimination from organometallic intermediates is not possible and only (*E*)-products are available due to isomerisation of (Z)-intermediates.[100]

2.2.2 *Aldol, Wittig, Peterson and Related Reactions*

Many variants for the synthesis of vinyl sulphones rely on the addition of a sulphonyl carbanion to a carbonyl compound, followed by a subsequent elimination step, Scheme 35.

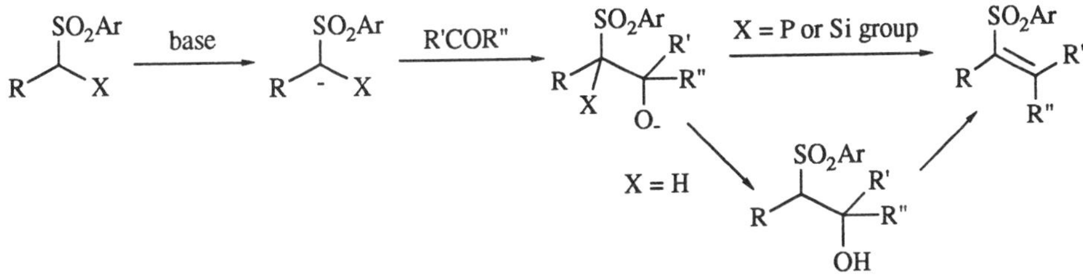

Scheme 35

In the simplest case, where X = H, dehydration of the intermediate hydroxy sulphone (or some derivative thereof) is necessary, usually in a separate step,[101] whereas the use of phosphorus (e.g. X = P(O)(OR)$_2$) or silicon (X = $SiMe_3$) groups allows direct *in situ* elimination to give the vinyl sulphone product. A variety of examples illustrating the first approach (X = H) is shown in Scheme 36. Thus, certain methyl styryl sulphones, i.e. (50), are very simply prepared by a classical Knoevenagel method involving condensation of an aromatic aldehyde with an ester of methanesulphonylacetic acid, followed by dealkylation–decarboxylation using LiI in DMF.[102]

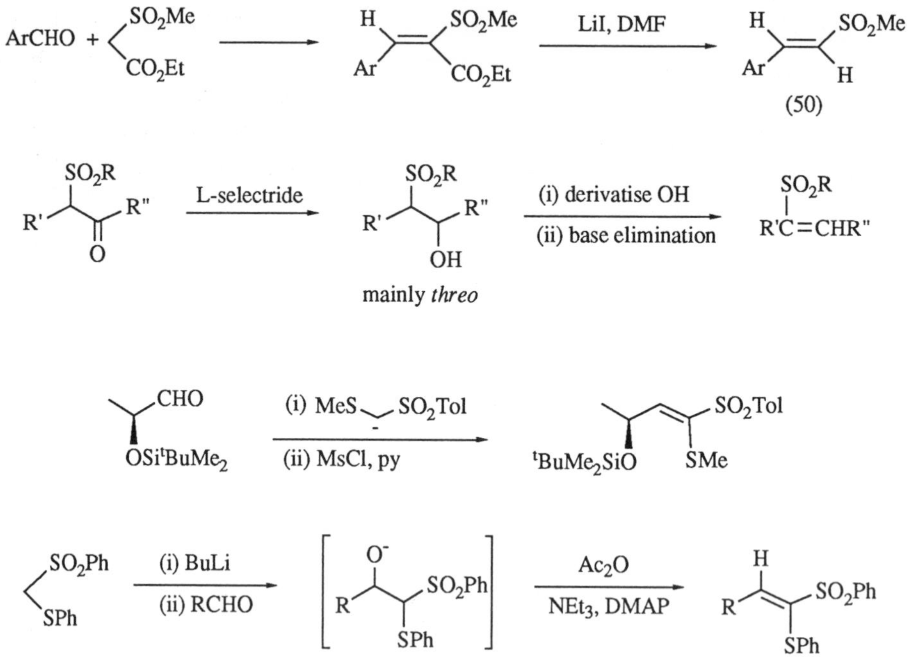

Scheme 36

The *threo*-selective reduction of β-ketosulphones to β-hydroxy sulphones using L-selectride, reported by Julia *et al.*, provides the basis for a stereoselective synthesis of vinyl sulphones.[103] Elimination via the corresponding tosylates is stereospecific (*anti*), whereas both diastereoisomers of the derived acetates give predominantly the (E)-vinyl sulphone product. Coupled to stereoselective desulphonylation procedures, also developed by Julia, this chemistry has allowed the development of a new stereodefined alkene synthesis which is described in Chapter 9.

The Julia strategy for stereoselective vinyl sulphone synthesis is easily extended to include the preparation of dienyl sulphones[104] and, as illustrated, is also applicable to the synthesis of alkylthio- and arylthiovinyl sulphone derivatives.[105,106] If sulphinyl sulphone (51) is employed in such aldol-type condensations then hydroxy-substituted vinyl sulphones are cleanly produced. This overall transformation, along with the proposed intermediates, is shown in Scheme 37.[107]

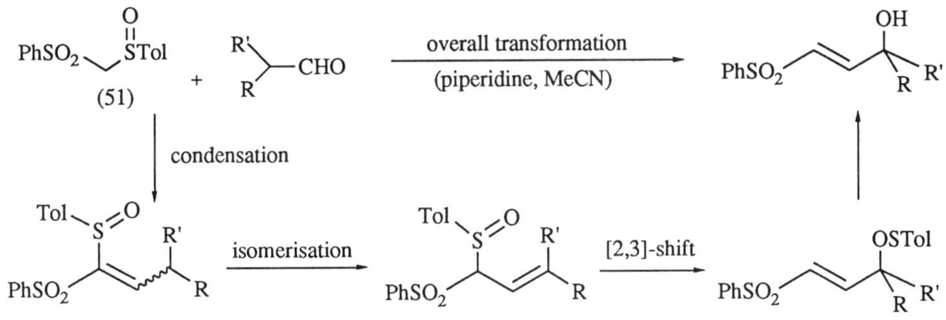

Scheme 37

Fluorinated vinyl sulphones can also be prepared using the aldol–elimination approach by using fluoromethyl phenyl sulphone as the starting material.[108]

α-Methylene sulphones can be prepared by the reaction of certain additionally stabilised sulphonyl carbanions with formaldehyde, Scheme 38.[109,110]

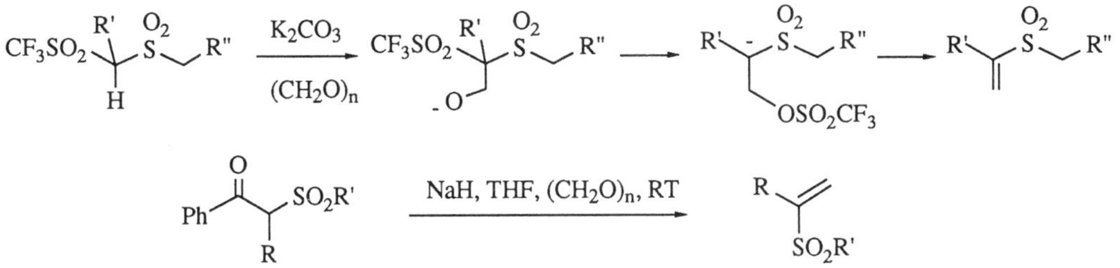

Scheme 38

In both cases vinyl sulphone formation occurs following the migration of the additional stabilising group (either triflyl or phenylcarbonyl) from carbon to oxygen. Yields for either process, with a few examples of R' and R", are generally very high, although no reactions could be obtained with other aldehydes.

The sulphonyl phosphonates (**52**) and related derivatives have proven very useful in Horner–Emmons reactions, giving substituted vinyl sulphones, Scheme 39.[111]

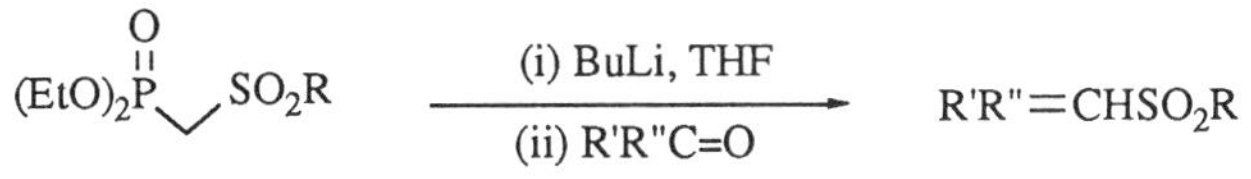

Scheme 39

Using this method with aldehydes the product vinyl sulphones are invariably formed as the *trans*-isomers. Since the original report by Posner and Brunelle[111] a number of modifications have been reported, including the use of a two-phase reaction system,[112] the use of alkylated homologues of phosphonate **(52)**,[113] and an *in situ* preparation of the anion of **(52)** (R = Ph) by phosphorylation of the dianion of methyl phenyl sulphone.[114]   Some other recent advances and applications of this chemistry are highlighted in Scheme 40.[115–118]

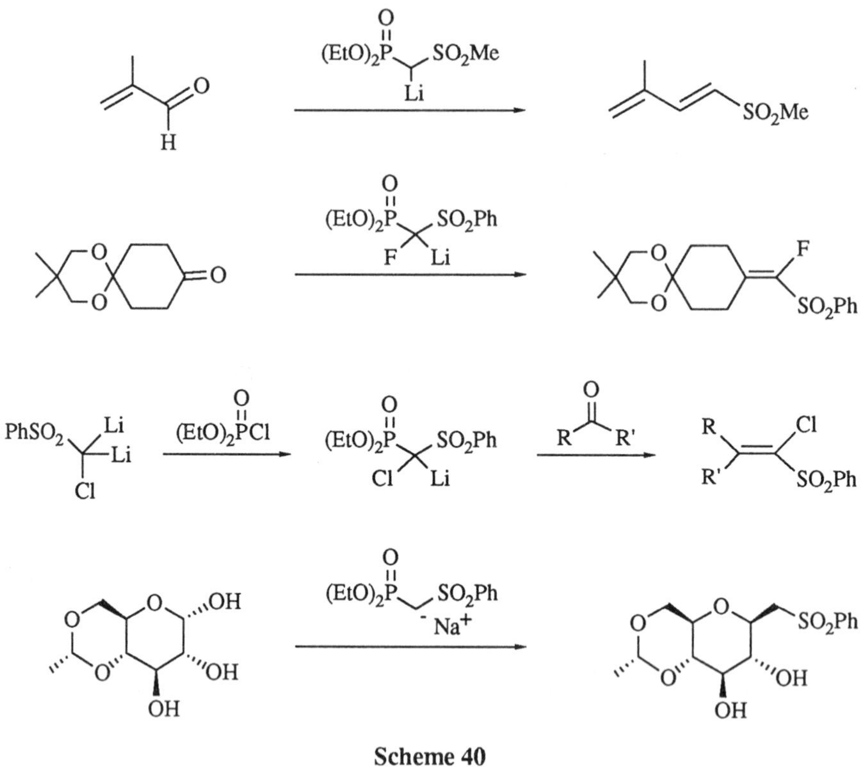

Scheme 40

Thus reagents incorporating an α-halogen in addition to the sulphone group appear to give good results. Notably, enolisable ketones and α,β-unsaturated aldehydes give good yields of vinyl sulphone and, even more significantly, certain lactols (i.e. sugars) appear to react smoothly even without hydroxyl protection.[118]

Similar chemistry can be conducted by Peterson olefination using the silyl sulphone **(53)**, Scheme 41.[119] This reagent gives good results with aldehydes, although competing carbonyl deprotonation can occur when ketones are used. In this case the use of DME as solvent in place of THF usually gives improved results. The vinyl sulphone products are usually obtained as mixtures of stereoisomers, one notable exception being compound **(54)**, which is obtained as the (*E*)-isomer exclusively.

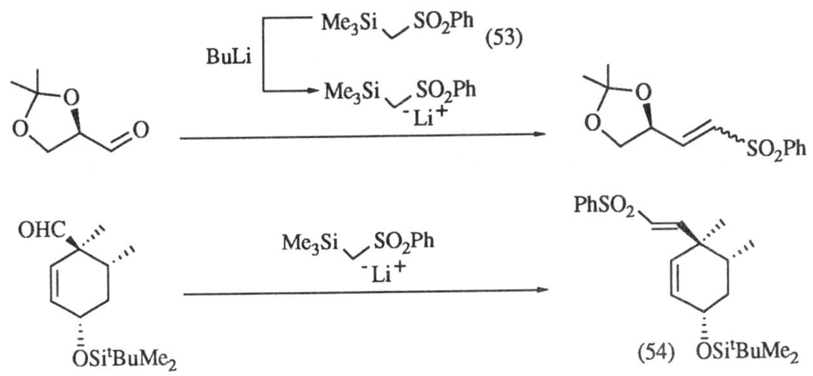

Scheme 41

A further innovation is the reaction of the dianion derived from (**53**) with diisopropoxytitanium dichloride prior to reaction with the carbonyl compound.[120] Under these conditions, using benzaldehyde as the carbonyl partner, the α-silylvinyl sulphone (**55**) is formed in good yield as a 2:1 (*E*):(*Z*) mixture, Scheme 42.

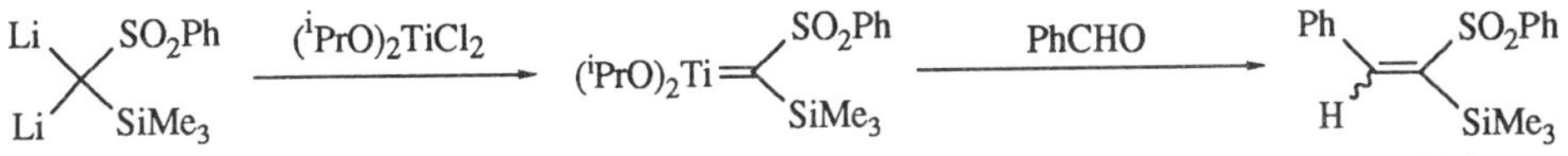

Scheme 42

2.2.3 Substitution of Simple Vinyl Sulphones and their Synthetic Equivalents

The discovery that simple vinyl sulphones can be deprotonated to give the corresponding vinylic anions has opened up new and attractive possibilities for the direct synthesis of more highly substituted vinyl sulphones, Scheme 43.[121]

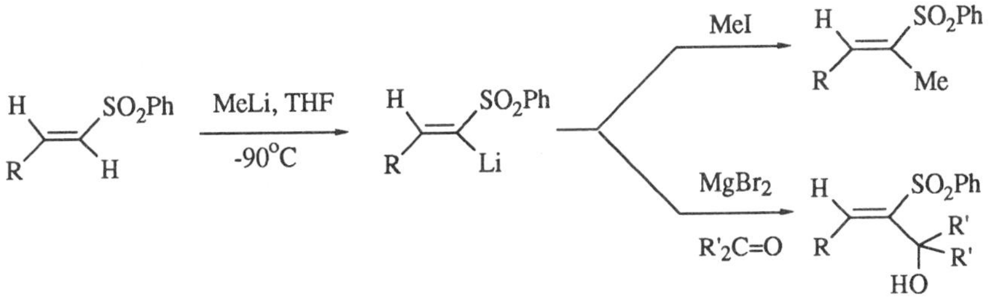

Scheme 43

The α-vinylic hydrogen is abstracted highly selectively under such kinetically controlled conditions, despite the possibility of forming the much more stable allylic α-sulphonyl carbanion through γ-deprotonation. However, it is not possible to form such a vinylic anion from the parent unsubstituted vinyl sulphone, due to rapid polymerisation, presumably initiated by Michael addition. This problem can be overcome by using suitably substituted sulphones as masked vinyl sulphones, for example β-amino[122] or β-silyl sulphones,[123] Scheme 44.

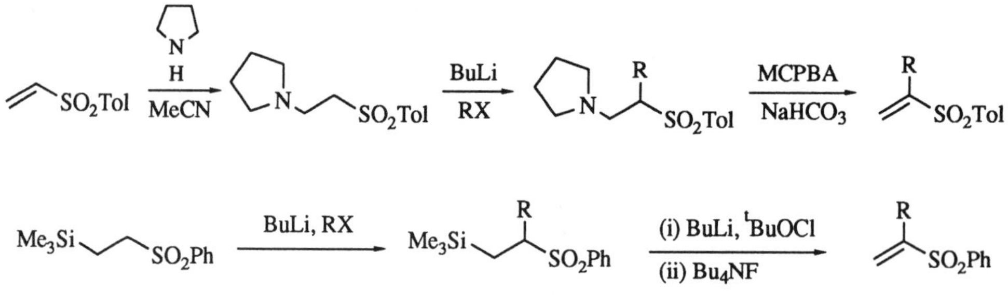

Scheme 44

The preparation of β-pyrrolidinyl sulphones has also proved possible using allyl sulphones as starting materials, presumably due to double-bond isomerisation under the reaction conditions.[124] In the second sequence the Bu_4NF-induced elimination of Me_3SiCl (from an intermediate α-chlorosulphone) is observed, underlining the relatively poor leaving group ability of the sulphone (the SO_2Ph group is lost in the analogous reaction of the non-chlorinated β-silyl sulphone).

In a series of papers, Knochel's group has shown that vinyl sulphones such as **(56)** and **(57)** react in a variety of ways to give substituted vinyl sulphone products, Scheme 45.[125]

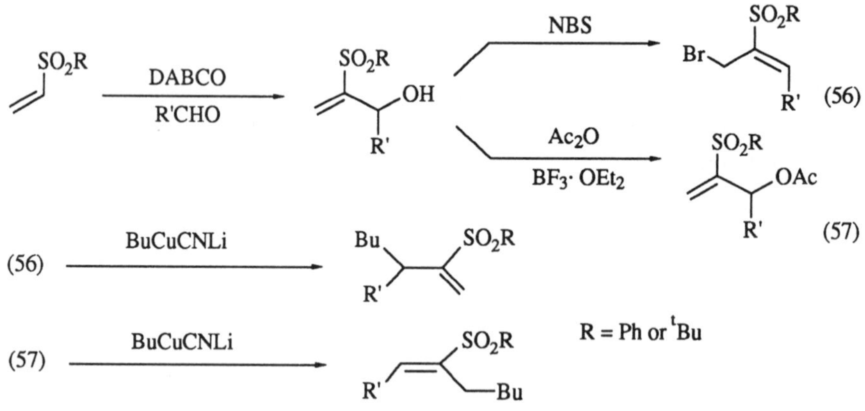

Scheme 45

Thus reaction of (**56**) and (**57**) with carbon nucleophiles such as organocuprates takes place with complementary regiochemistry. The bromides (**56**) have also proved amenable to zinc-promoted coupling with aldehydes to give hydroxy sulphone products such as (**58**), Scheme 46.[126]

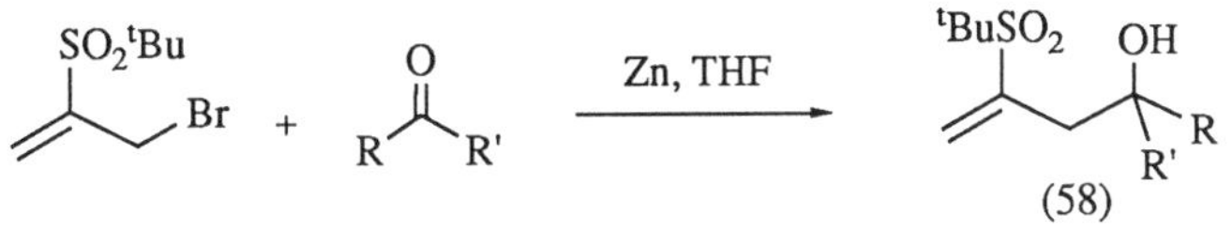

Scheme 46

An interesting extension of the chemistry of β-silyl sulphones enables the formation of vinylsulphinic acid salts which can be alkylated with MeI, Scheme 47.[127]

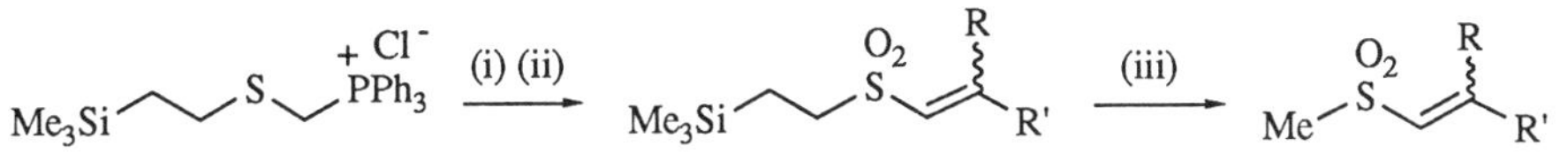

reagents:

    (i) base (KO^{t}Bu or LDA), RR"C=O   (ii) MCPBA   (iii) Bu$_4$NF, MeI.

Scheme 47

The method affords roughly 1:1 mixtures of stereoisomers in most cases, although facile separation can be achieved by virtue of the more rapid oxidation of the (*E*)-vinyl sulphide to the vinyl sulphone (under the conditions used the (*Z*)-vinyl sulphide oxidises only as far as the sulphoxide). Unfortunately, in the final step, fluoride proves basic enough to promote isomerisation of some of the vinyl sulphone products to the corresponding allylic sulphones.

    The β-stannylvinyl sulphone (**59**) undergoes quite efficient intermolecular radical addition–fragmentation reactions to give substituted vinyl sulphone products, Scheme 48.[128]

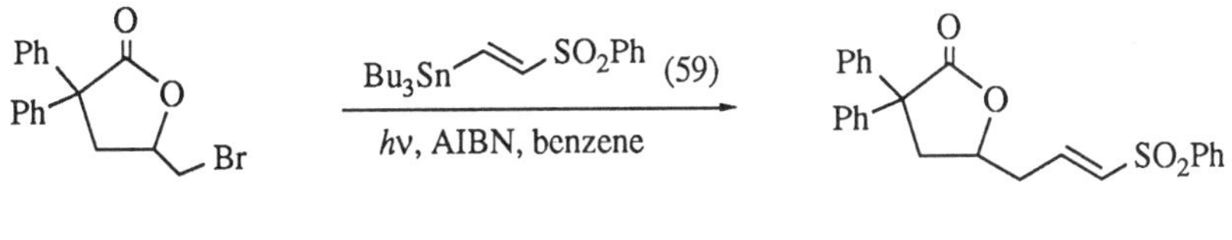

Scheme 48

The β-phenylselenovinyl sulphones prepared by addition of $ArSO_2SePh$ to alkynes (e.g. Scheme 30) are useful precursors to other alkylated vinyl sulphones.[129] Thus, reaction with the unconventional cuprate reagents $R'Cu(SePh)Li$ gives products arising from substitution of the SePh group, e.g. Scheme 49.

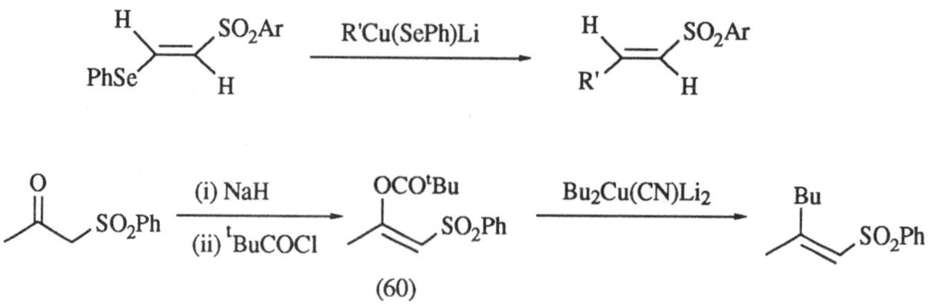

H₂C=C(SO₂Ar) with PhSe and H substituents   ── R'Cu(SePh)Li →   product with R' replacing SePh, SO₂Ar

O=C–CH₃ ... SO₂Ph ──(i) NaH (ii) ^{t}BuCOCl──→ OCO^{t}Bu / SO₂Ph (60) ── Bu₂Cu(CN)Li₂ → Bu / SO₂Ph

Scheme 49

Although the substitution reaction is efficient and stereoselective, the necessity of using the selenocuprate, which requires prior preparation of CuSePh from PhSeH, is a significant drawback. Difficulties presumably also arise in many cases in controlling the regioselective construction of the phenylselenovinyl sulphone starting materials.

A similar but complementary approach involves the preparation of the enol pivalate derivatives (60) which undergo clean substitution on reaction with standard higher-order cyanocuprates.[130] Here the construction of the β-substituted vinyl sulphone is clearly always regiospecific, although the stereochemical outcome usually favours the more stable (E)-vinyl sulphone product. Somewhat surprisingly, neither of these methods suffers from overreaction of the vinyl sulphone products, even with excess cuprate reagent.

2.2.4 Reactions of Alkynyl Sulphones

Not surprisingly vinyl sulphones have been prepared by hydrogenation of the corresponding alkynyl sulphones. For example, Paquette obtained either (E)- or (Z)-vinyl sulphones on hydrogenation of silylethynyl sulphone (61), Scheme 50.[131]

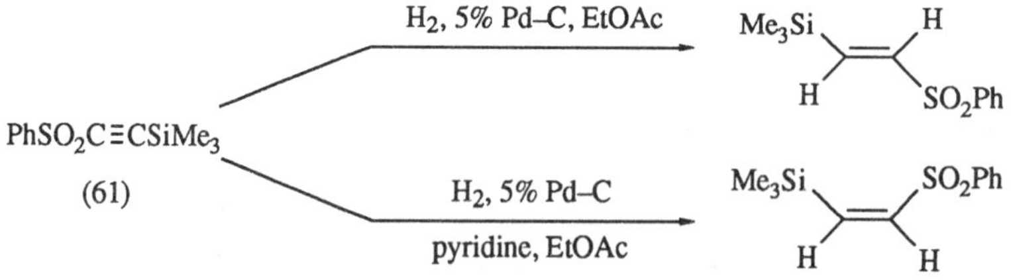

PhSO₂C≡CSiMe₃ (61)

H₂, 5% Pd–C, EtOAc → Me₃Si / H ... H / SO₂Ph

H₂, 5% Pd–C, pyridine, EtOAc → Me₃Si / SO₂Ph ... H / H

Scheme 50

A more recent method which provides (Z)-vinyl sulphones stereoselectively involves the use of a $Cu(BF_4)_2/HSiEt_2Me$ reagent combination, Scheme 51.[132]

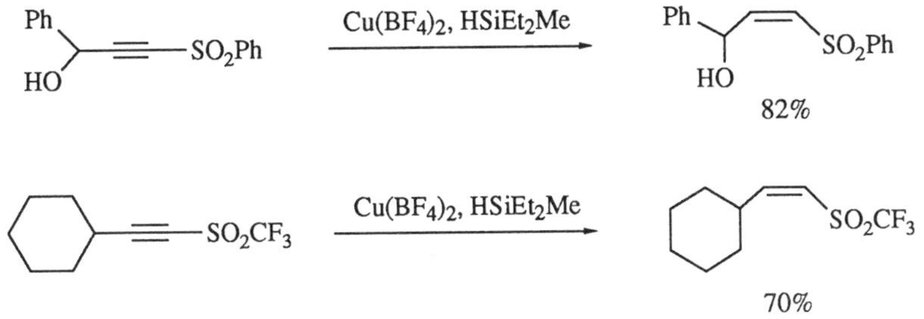

Scheme 51

The reaction presumably involves some kind of Cu(II) hydride species, perhaps $HCu^+BF_4^-$. A competition reaction using an alkynyl sulphone in the presence of cyclohexenone and 2-phenylethyl bromide results in reduction of only the sulphone, attesting to the chemoselectivity of the method. Treatment of (61) with $^iBu_2AlH$ results in clean formation of (E)-2-trimethylsilylvinyl sulphone.[133]

Both single and double addition reactions are possible in reactions of organocuprates with alkynyl sulphones.[134] The stereochemical course of the mono-addition reaction to give the vinyl sulphone products involves mainly *syn*-addition, Scheme 52.

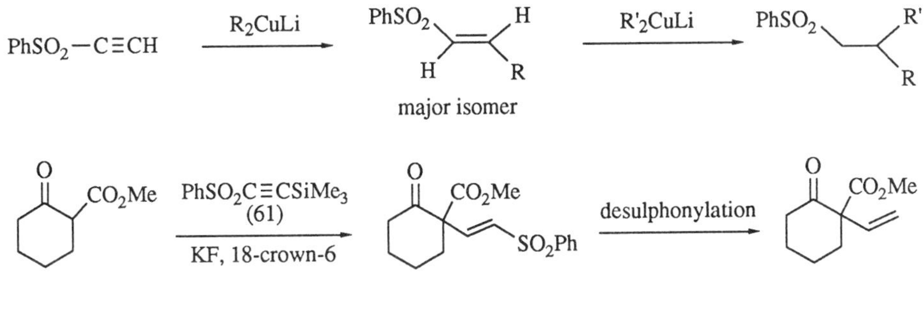

Scheme 52

If desired, the double addition reaction, i.e. reaction with a further cuprate reagent, can be carried out without isolating the initial adduct by quenching the first reaction with thiophenol and then adding to another preformed cuprate mixture. More stable anions such as those derived from malonates and β-keto esters add smoothly to the alkynyl sulphone (61) to give vinyl sulphone intermediates.[135] Desulphonylation is effected directly using aluminium amalgam, or in some cases indirectly via cyanohydrin intermediates, thus allowing overall vinylation of the carbonyl starting materials.

A number of substituted alkynyl sulphones including (61) can be employed in Diels–Alder cycloadditions to give cyclic β-silylvinyl sulphone products, e.g. Scheme 53.[136,137]

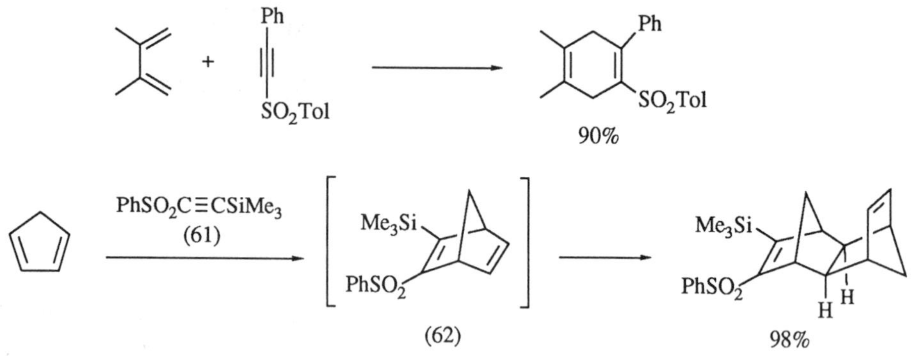

Scheme 53

Somewhat surprisingly, intermediate (62) reacts regioselectively with a further equivalent of cyclopentadiene at the unsubstituted alkene.[137] Cycloaddition chemistry of unsaturated sulphones is dealt with in detail in Chapter 6.

With suitable alkenyl substrates ethynyl *p*-tolyl sulphone undergoes smooth EtAlCl$_2$-catalysed ene-reaction, Scheme 54.[138]

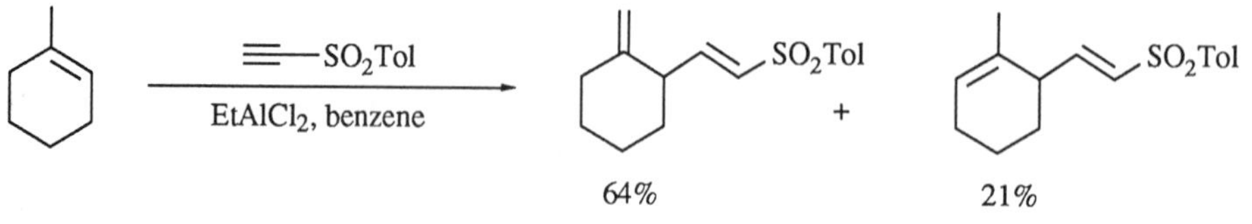

Scheme 54

The reaction is very sensitive to both the reaction conditions (the use of an aromatic solvent is important) and the substitution pattern of the starting alkene.

Both β-sulphur- and β-nitrogen-substituted vinyl sulphones can be prepared stereoselectively by the Michael addition of appropriate nucleophiles to an alkynyl sulphone, e.g. Scheme 55.[139,140] Such addition reactions occur stereoselectively in an *anti*-fashion, thus resulting in the formation of the (Z)-isomer in the cases shown. However, isomerisation of the kinetically formed products to give more stable isomers can compete under some conditions.

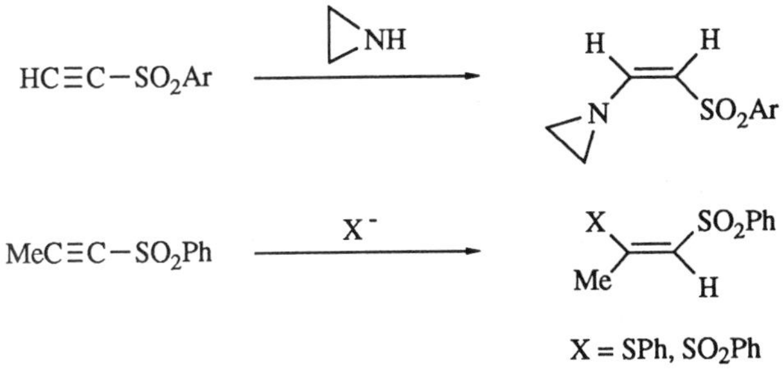

Scheme 55

2.2.5 Miscellaneous Vinyl Sulphone Preparations

A standard approach to the synthesis of sulphones involves oxidation of the corresponding sulphides. Vinyl sulphones can be prepared in this way, for example using the economical Oxone® reagent as described in Section 2.1.1.

A palladium(II)-mediated process allows access to acetoxy- and chloro-substituted vinyl sulphones, Scheme 56.[141]

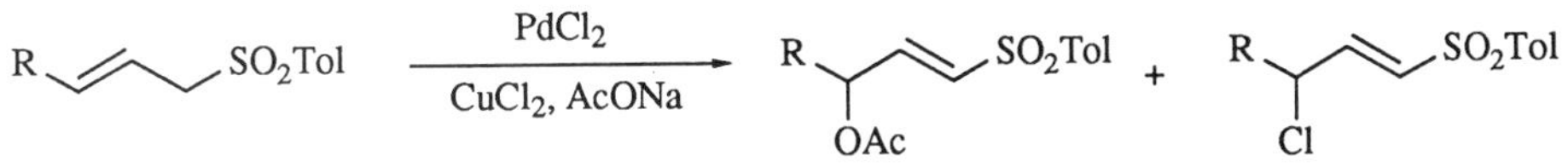

Scheme 56

The two products are formed regio- and stereoselectively, the ratio of the two depending on the amounts of additives such as $CuCl_2$ in the reaction. Similarly functionalised vinyl sulphones are available from allylic sulphones by bromination–dehydrobromination or from base-mediated opening of derived epoxides, Scheme 57.[142]

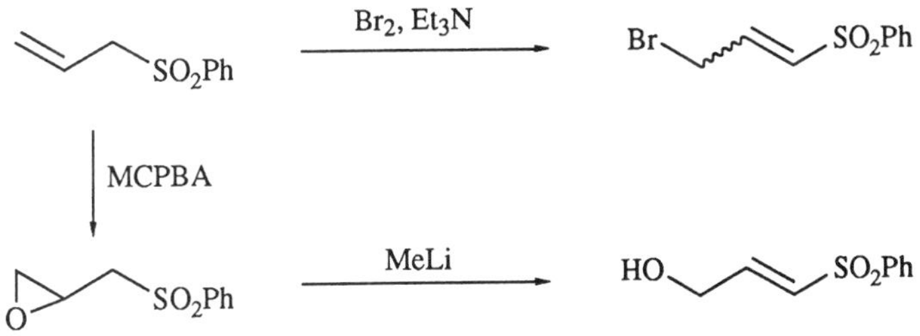

Scheme 57

The addition of carbanions derived from allylic sulphones to an enone can, in principle, give 1,2- or 1,4-adducts resulting from either α- or γ-alkylation with respect to the sulphone, Scheme 58.

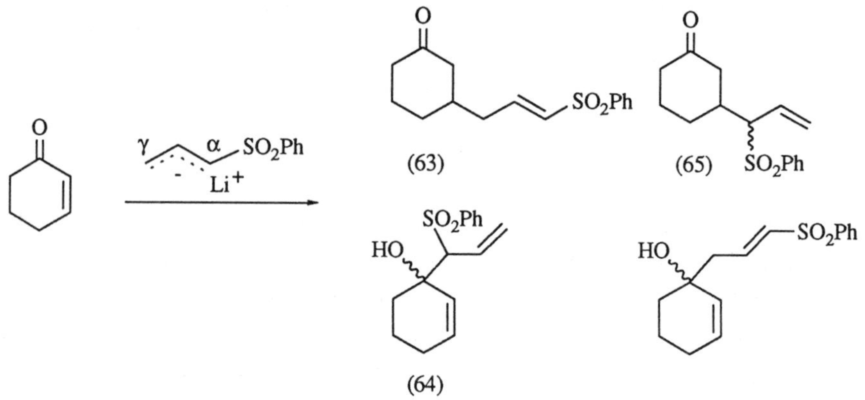

Scheme 58

Haynes and co-workers have studied this reaction (and that of the corresponding allylic sulphoxides) in some detail.[143] Vinyl sulphones such as (**63**) can be formed directly in a kinetically controlled 1,4-addition reaction; alternatively they can arise through initial 1,2-addition, to give (**64**), followed by rearrangement, depending on the enone involved. By contrast, reactions conducted in the presence of HMPA give predominantly allylic sulphone products such as (**65**).

Treatment of lithiated sulphones with two equivalents of $Cu(OAc)_2$ results in the formation of the corresponding vinyl sulphones, Scheme 59.[144]

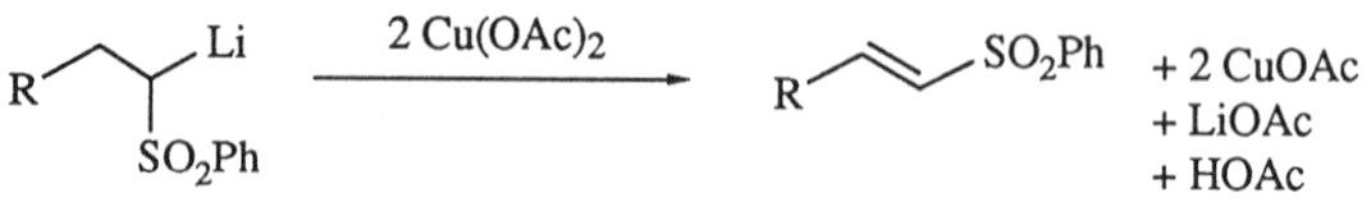

Scheme 59

This result contrasts with the dimerisation reaction obtained using $Cu(OTf)_2$ in place of $Cu(OAc)_2$.

Vinyl sulphones are formed in good yields from the corresponding alkenyl mercurials by treatment with the sodium salts of sulphinic acids under photochemical conditions.[145]

Optically active vinyl sulphones (**66**) were prepared by elimination from the corresponding mannose derivatives (**67**), Scheme 60.[146]

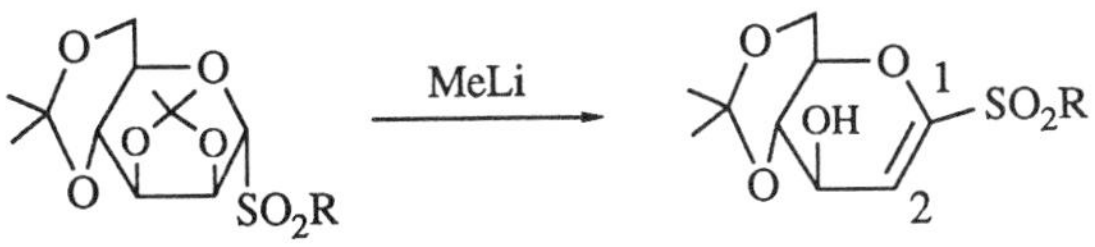

(67) R = Ph, ^{t}Bu   (66)

Scheme 60

Compound (66) does not participate in Michael additions, but instead can be metallated at C-2 to form a vinyl anion which reacts with MeI.

It has been reported that β,γ-unsaturated (allyl) sulphones are more thermodynamically stable than the corresponding α,β-unsaturated (vinyl) isomers.[147] However, depending on the substitution pattern of the sulphone in question, it is possible to isomerise in either direction (i.e. allyl→vinyl or vinyl→allyl) given the appropriate choice of reaction conditions, Scheme 61.[148,149]

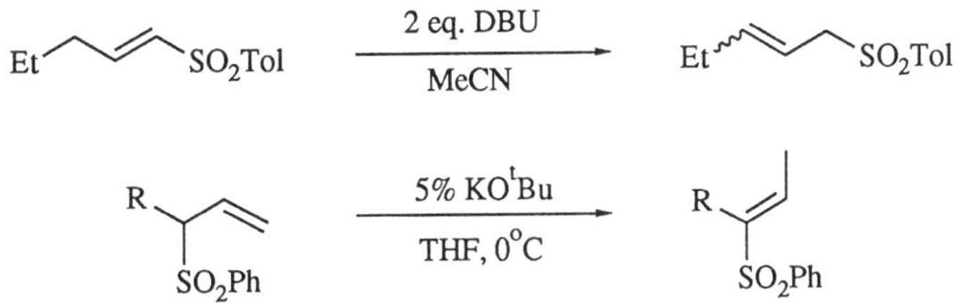

(69)   (68)

Scheme 61

Thus the use of two equivalents of DBU in acetonitrile facilitates the isomerisation of a variety of vinyl sulphones to the corresponding allyl isomers.[148]   A more limited selection of vinyl sulphones of general formula (68) could be obtained by treatment of allyl sulphones (69) with catalytic quantities of potassium *tert*-butoxide in THF at 0°C.[149]   It seems likely that equilibration of the isomers occurs in most cases, the favoured product having the more highly alkyl-substituted alkene, although steric effects can also be important.[150]

The selenylation–selenoxide *syn*-elimination protocol, so popular for dehydrogenation of carbonyl compounds, has been rarely used for converting saturated sulphones to vinyl sulphones. One notable example is the transformation of the highly functionalised sulphone (70) to the corresponding unsaturated compound (71) carried out by Ley's group in a synthesis of (+)-milbemycin β1, Scheme 62.[151]

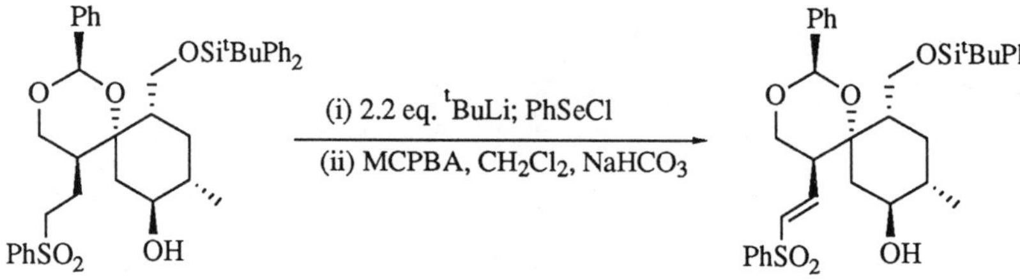

(70) (71)

Scheme 62

Finally, Julia has shown that the substitution of a sulphonyl group in allylidene *gem*-disulphones can be carried out using a variety of heteronucleophiles to give γ-functionalised (*E*)-vinyl sulphones, Scheme 63.[152]

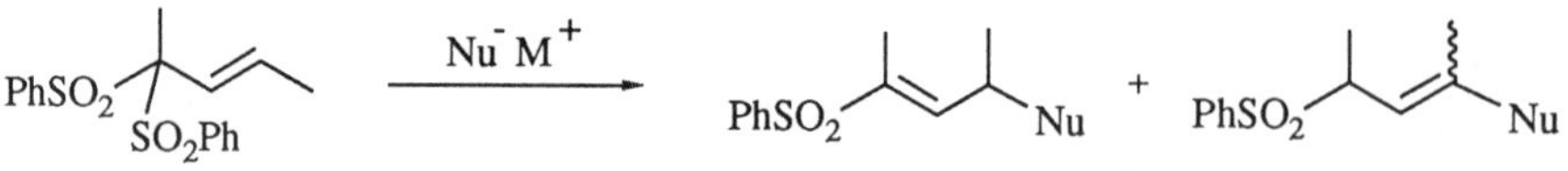

Nu = SPh, SO_2Ph, OH, NR_2

Scheme 63

2.3 Allylic, Allenic, Alkynyl and Aryl Sulphones

2.3.1 Palladium-catalysed Synthesis of Allylic Sulphones

An important method specifically for the preparation of allylic sulphones involves the treatment of an allylic acetate or nitro compound (**72**) with a sulphinate salt in the presence of a catalyst such as $Pd(PPh_3)_4$, Scheme 64.

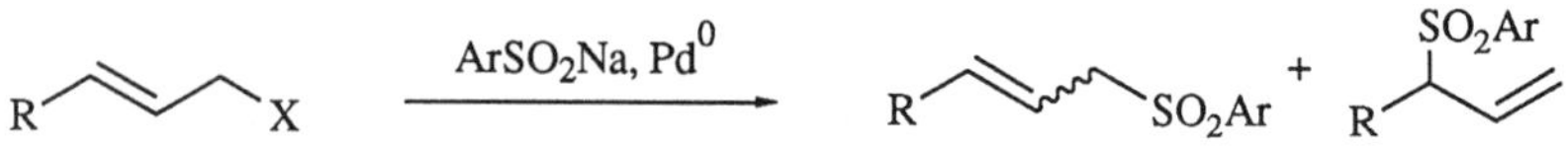

(72) X = OAc or NO_2 (73) (74)

Scheme 64

Of prime importance in such a transformation are, firstly, the regiochemical outcome, i.e. (**73**) versus (**74**), and, secondly, the stereochemistry of the products, e.g. of (**73**). An example of the type of regiocontrol which can be achieved in such reactions is illustrated in Scheme 65.

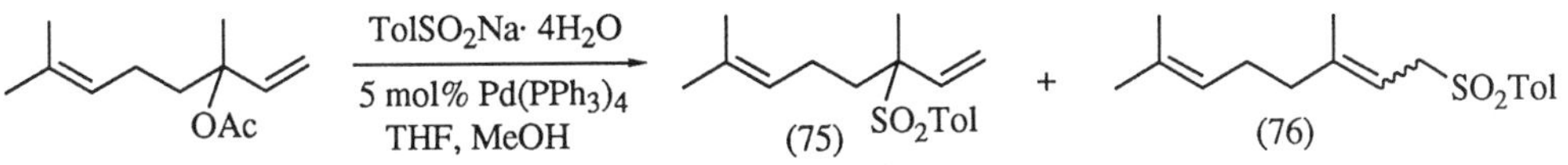

	(75)		(76)	
0°C, 3.5h	78%	:	9% (*E*:*Z* 81:19)	
RT overnight	–	:	84% (*E*:*Z* 87:13)	

Scheme 65

It appears that (**75**) is the kinetically controlled product, whilst (**76**) is the thermodynamically favoured isomer.[153] Furthermore, (**75**) was found to be isomerised to (**76**) in the presence of a catalytic amount of $Pd(PPh_3)_4$, and in a related example an initially formed (*Z*)-allylic sulphone could be isomerised to the corresponding (*E*)-form. This suggests the reversible formation of π-allyl complexes, which can eventually allow 'pumping' of kinetically formed products in the direction of the more stable isomer. Julia and co-workers have further examined the role of the ligands on palladium in controlling this reaction and observed a range of product distributions (e.g. between (**75**) and (**76**)) using various catalyst systems, over extended reaction times, e.g. 65 h.[154] The report also describes isomerisation of allylic sulphones, this time using sulphinic acid catalysts, which can be carried out most rapidly in acidic solvents such as acetic acid.

Even more remarkable results have been obtained by utilising allylic nitro compounds in place of acetates in such palladium-catalysed reactions.[155] Initial studies using compound (**77**) as starting material appear to mirror the kind of kinetic product distribution described above, Scheme 66.

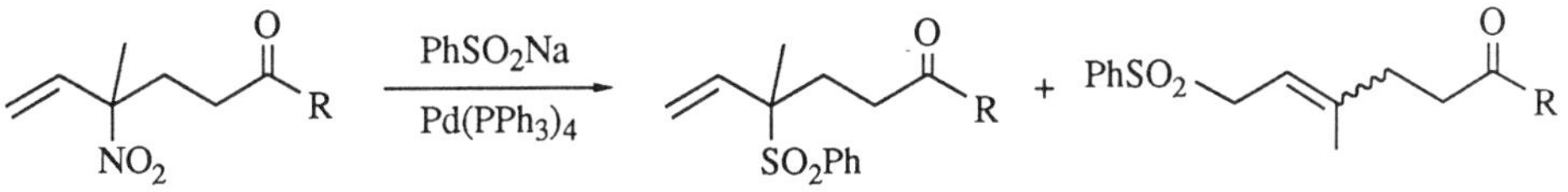

(77) (78) major (79) minor

Scheme 66

However, in contrast to the experiments using allylic acetates, in this case the product distribution does *not* change under the reaction conditions. It is found that running the reaction in the presence of added $NaNO_2$ results in exclusive formation of the kinetic product (**78**). These results indicate that the $NaNO_2$ liberated from the nitroalkene starting material as the reaction progresses inhibits the isomerisation of kinetic products, e.g. (**78**), to the thermodynamic ones, e.g. (**79**). Presumably the $NaNO_2$ modifies the catalyst so that π-allyl complexes can be formed from an allylic acetate or nitro compound but *not* from an allylic sulphone. This method provides considerable scope for obtaining kinetic-type products, even from allylic acetates, simply by including NaNO$_2$ in the reaction mixture. A later full account of this work describes, in addition, competition reactions between allylic nitro compounds, acetates and sulphones, and gives further details and rationalisations of the product control possible.[156]

A further variant on this chemistry allows the use of vinyl nitro compounds (available by the Henry reaction) in place of allylic nitro compounds, Scheme 67.[157]

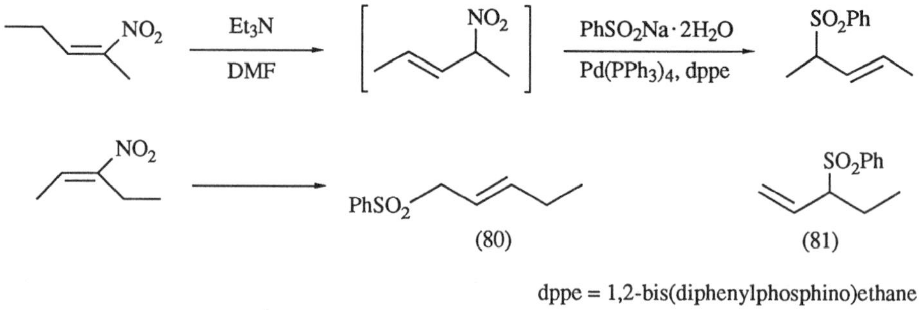

dppe = 1,2-bis(diphenylphosphino)ethane

Scheme 67

Somewhat surprisingly, in view of the usual kinetic-type of product obtained, this method gives one product (**80**) which appears to arise by isomerisation of the expected kinetic product (**81**). That isolated sulphones are unreactive under such reaction conditions is demonstrated by a related reductive denitration reaction using sulphone (**82**), Scheme 68.[158]

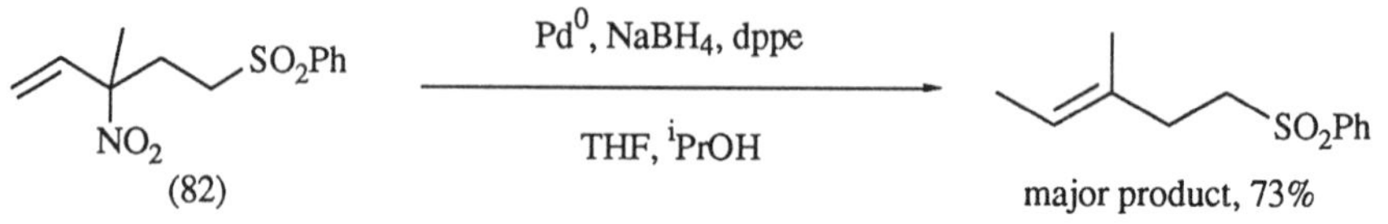

Scheme 68

If desired, allylic sulphones can be prepared from esters of allylic alcohols by reaction with sulphinate salts in the presence of heterogeneous catalysts such as palladium on graphite.[159]

2.3.2 Unsaturated Sulphones through Sulphinate Rearrangements

The rearrangement of sulphinate esters to the corresponding sulphones has been known for many years. However, the reactions involving saturated compounds, which proceed by ion-pair formation–recombination, are not generally synthetically useful.[160] In contrast, the rearrangement of allylic and propargylic systems is a very useful method of preparing unsaturated sulphones, Scheme 69.

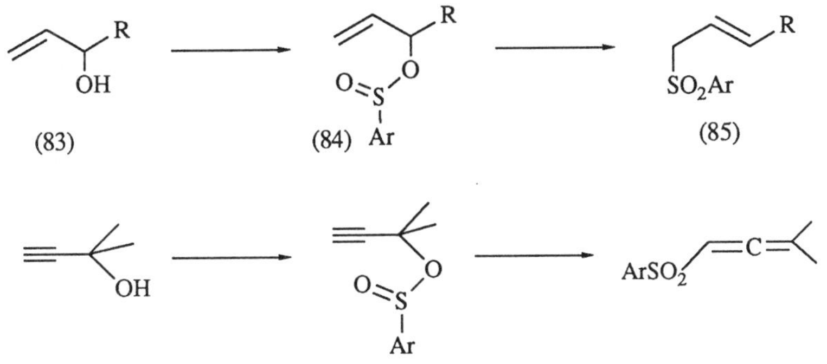

Scheme 69

As shown, the starting sulphinate esters are usually prepared from the corresponding alcohol, e.g. (83), by treatment with a sulphinyl halide, very often PhS(O)Cl. Rearrangement can then be effected by heating in various solvents. Two important observations are that firstly, in simple systems, the rearrangement of, for example, (84) to (85) occurs with clean allylic 'inversion' (i.e. a 1,3-transposition of the allylic functionality); and secondly, optically active (S)-(86) rearranges to give (S)-(87) with 87% stereospecificity (i.e. 87% fidelity in the transfer of chirality from sulphur to carbon), Scheme 70.[161]

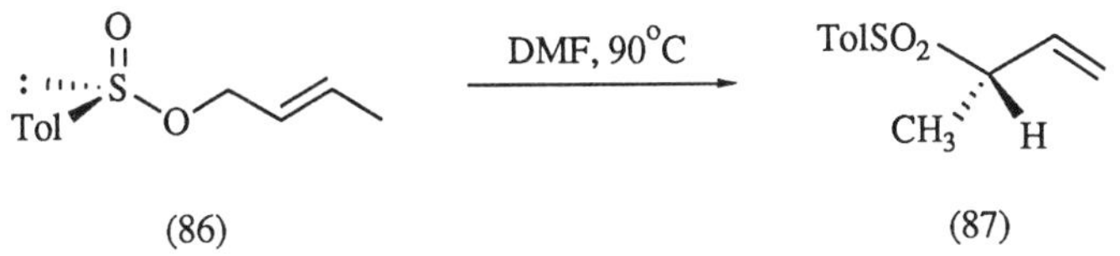

Scheme 70

Thus in many such rearrangements a [2,3] sigmatropic shift mechanism operates (although ion-pair formation–recombination appears important in cyclic systems). This rearrangement is dealt with in much more detail in Chapter 5; here it is appropriate just to include three more examples to illustrate the process, Scheme 71.[162–164]

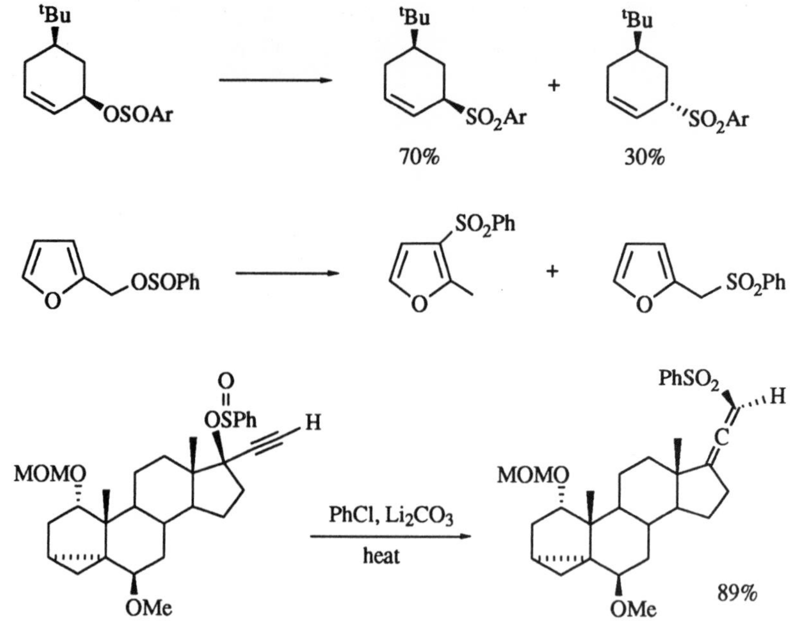

70%   30%

Scheme 71

2.3.3   Radical-,   Selenium-,   and   Phosphorous-mediated   Methods   (and Miscellaneous Preparations)

As mentioned in Section 2.1.4 sulphonyl radicals can be useful intermediates for the synthesis of sulphones, although the method has not attracted widespread use. In a series of useful contributions to this area the groups of Johnson[165] and Gupta[166] have utilised the reaction between certain unsaturated cobaloximes and sulphonyl halides to prepare a range of unsaturated sulphones. Thus the use of allyl or allenyl cobaloximes gives rise to the corresponding sulphones in regioselective fashion, Scheme 72.

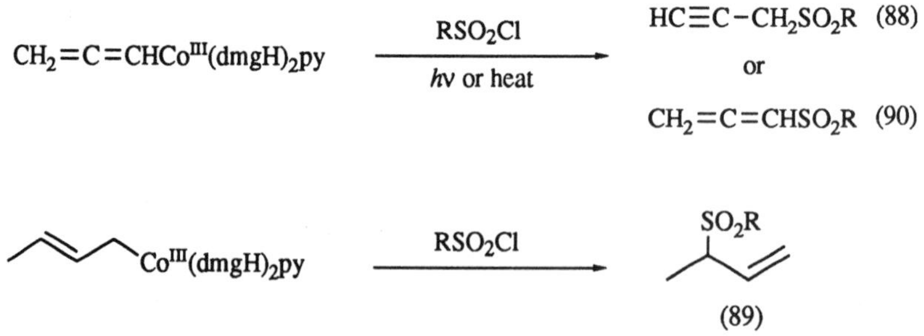

Scheme 72

Attack of an $RSO_2\cdot$ radical at the distal position of the unsaturated group of the Co(III) organometallic is the usual pathway, e.g. to give (**88**) or (**89**). However, depending on the precise conditions used for the reaction, it appears that products arising from α-attack, e.g. (**90**), can also be formed with good selectivity.[167]

The predominant γ-attack can be used to great effect in the cyclisation reactions of butenyl cobaloximes, such as (**91**), to give cyclopropyl products, e.g. Scheme 73.[168]

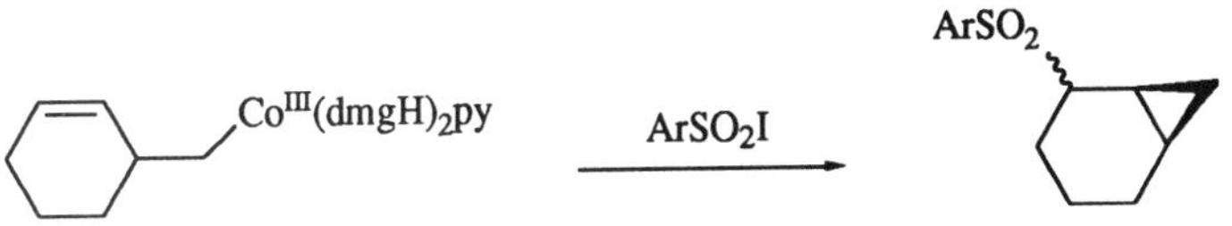

Scheme 73

This is an attractive route to such compounds, since the alternative 3-*exo-trig* radical ring closure is almost never viable due to rapid ring opening of the cyclopropylcarbinyl radical.

These homolytic displacement reactions are chain processes involving homolysis of some X–Y bond (here X = $ArSO_2$, Y = halogen, usually Cl), according to the general Scheme 74.

$$\dot{C}o^{II}(dmgH)_2py \;+\; XY \longrightarrow YCo^{III}(dmgH)_2py \;+\; X\cdot$$

$$RCo^{III}(dmgH)_2py \;+\; X\cdot \longrightarrow \dot{C}o^{II}(dmgH)_2py \;+\; RX$$

Scheme 74

The chain-propagating species, Co(II), can be formed initially by trace homolysis of the starting organocobaloxime.

In some of this chemistry products arising from halosulphonylation of C=C bonds can be formed. As already discussed in Section 2.2 the addition of $PhSO_2X$ to multiply bonded systems is a good route to a range of vinyl sulphone products. The studies of Back and co-workers have extended the scope of their selenosulphonylation chemistry to enable the facile preparation of allylic, allenic and alkynyl sulphones.[169–171] This chemistry proceeds mainly by oxidation of the seleno-sulphone intermediates to the corresponding selenoxides, followed by thermal *syn*-elimination, for example, Scheme 75.

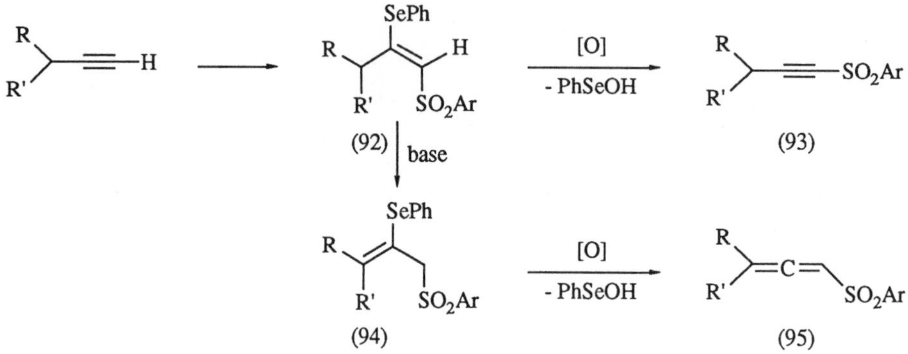

Scheme 75

Selenoxides derived from the usual selenosulphonylation product from a terminal alkyne, e.g. (92), undergo *syn*-elimination towards the SO_2Ar group to give alkynyl sulphone products, e.g. (93). However, if isomerisation of vinyl sulphones (92) to their allylic counterparts (94) is carried out (base = LDA, Et_3N or ^tBuOK) prior to oxidative elimination, then allenyl sulphones, i.e. (95), are obtained in high yield. An alternative [2,3] sigmatropic rearrangement pathway is followed by selenoxides derived by selenosulphonylation of conjugated enynes, Scheme 76.[170]

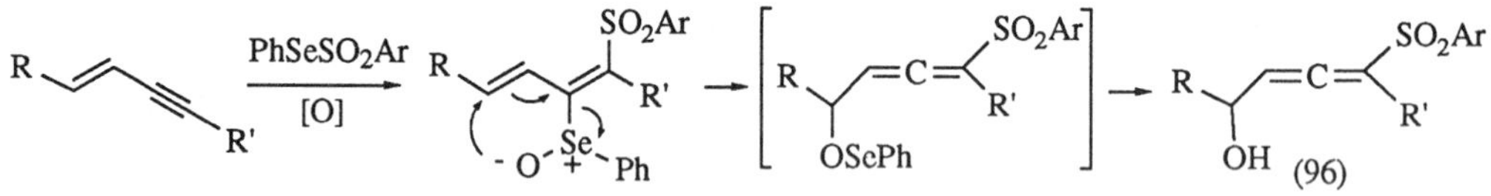

Scheme 76

This appears to be a most versatile and elegant route to compounds such as (96), which in turn are useful for further elaboration, for example by elimination, or cuprate conjugate addition reactions. The selenoxides, which are intermediates in these eliminations, can be made to undergo substitution reactions with suitable nucleophiles, including propargylic carbanions and malonate anions, e.g. Scheme 77.[171] The scope and limitations of the reaction were explored and the reaction mechanism was deduced to involve initial elimination to give an alkynyl sulphone.

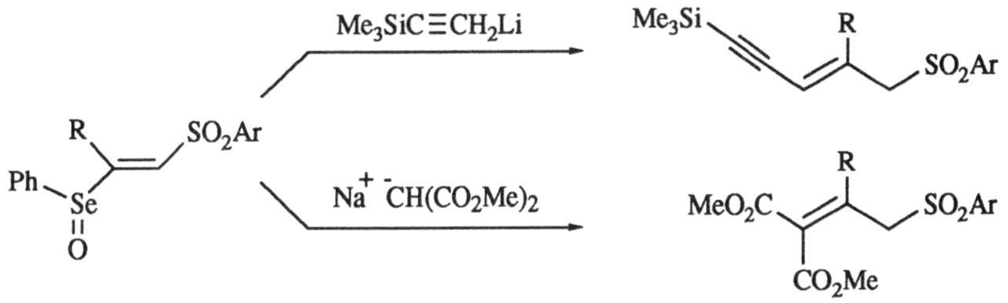

Scheme 77

The simplest method of preparing an alkynyl sulphone is undoubtedly via the corresponding sulphide. This oxidation route was used in several early studies of the reactions of unsaturated sulphones, notably by Truce and Markley[139] and Stirling.[140] Dialkynyl sulphones can also be prepared in this way simply by employing SCl_2 as the sulphenylating agent to access the required dialkynyl sulphide.[172] Stirling also reported the first example of the very useful isomerisation of a propargyl sulphone into an allenic sulphone.[140]

Both alkynyl and allenic sulphones have been accessed by use of the Horner–Emmons reaction. In a report describing the synthesis of alkynyl sulphones in this way the chlorinated sulphonyl phosphonate (**97**) is used to generate an intermediate chlorovinyl sulphone which is then immediately dehydrochlorinated using ^tBuOK, Scheme 78.[173]

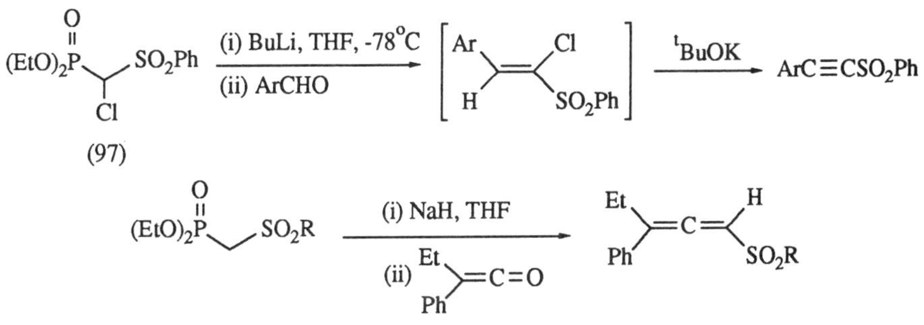

Scheme 78

Only terminal aromatic-substituted compounds were prepared. Even more severe limitations are apparent in the synthesis of allenic sulphones using this approach, starting from ketenes.[174] Here the main problem is the inherent instability of the ketene starting materials, for example towards polymerisation. A modification involving ultrasonication apparently gives improved results with some substrates.[175]

Two further preparations of alkynyl sulphones involving substitution and elimination respectively are highlighted in Scheme 79.

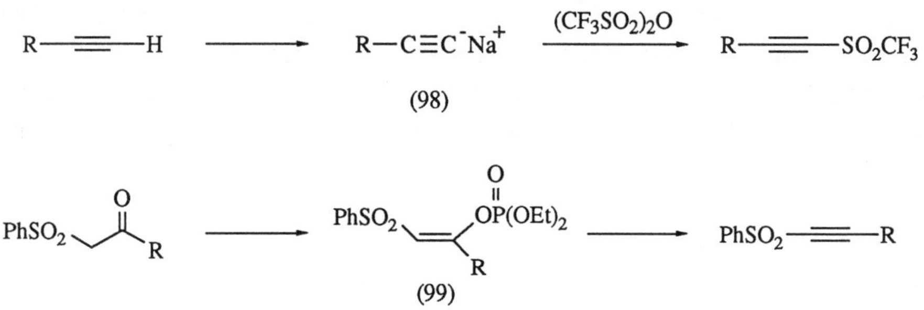

$$R\text{---}\!\!\equiv\!\!\text{---}H \quad\longrightarrow\quad R\text{---}C\equiv C^-Na^+ \quad\xrightarrow{(CF_3SO_2)_2O}\quad R\text{---}\!\!\equiv\!\!\text{---}SO_2CF_3$$

(98)

$$PhSO_2\overset{O}{\overbrace{\quad\quad}}R \quad\longrightarrow\quad PhSO_2\overset{\overset{O}{\underset{||}{OP(OEt)_2}}}{\underset{R}{\diagup\!\!\diagdown}} \quad\longrightarrow\quad PhSO_2\text{---}\!\!\equiv\!\!\text{---}R$$

(99)

Scheme 79

As discussed in Section 2.1.3, the reaction of organometallics with sulphonic acid derivatives is often troublesome. Here the use of sodium acetylides (**98**) for reaction with triflic anhydride overcomes some of the problems due to undesirable side reactions experienced when lithium acetylides are employed.[66,176] The elimination reaction of the enol phosphates (**99**) occurs smoothly on treatment with ^{t}BuOK in THF at low temperature.[177]

Julia *et al.* have described the direct conversion of conjugated dienes into allylic sulphones, using only a catalytic amount of palladium in the presence of an arenesulphinic acid, Scheme 80.[178]

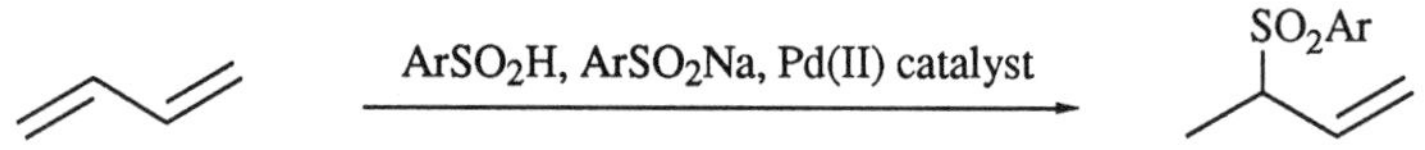

$$\diagup\!\!\diagdown\!\!\diagup\quad\xrightarrow[\quad\quad\quad\quad\quad]{ArSO_2H,\ ArSO_2Na,\ Pd(II)\ catalyst}\quad\overset{SO_2Ar}{\diagup\!\!\diagdown\!\!\diagup}$$

Scheme 80

This chemistry contrasts to other reports in which loss of SO_2 from the palladium intermediate occurs, resulting in the formation of *C*-coupled products.[179]   Finally, two other direct routes to allylic sulphones from either allylic halides or allylic alcohols are highlighted in Scheme 81.[180, 181]

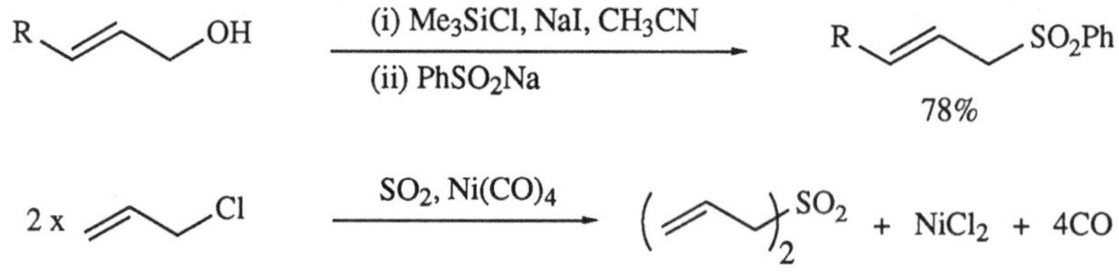

$$R\diagdown\!\!\diagup\!\!\diagdown OH \quad\xrightarrow[\text{(ii) }PhSO_2Na]{\text{(i) }Me_3SiCl,\ NaI,\ CH_3CN}\quad R\diagdown\!\!\diagup\!\!\diagdown SO_2Ph$$

78%

$$2\times\ \diagup\!\!\diagdown\!\!\diagup Cl \quad\xrightarrow{SO_2,\ Ni(CO)_4}\quad \left(\diagup\!\!\diagdown\!\!\diagup\right)_2\!SO_2\ +\ NiCl_2\ +\ 4CO$$

Scheme 81

The first route appears to simply involve *in situ* HI generation to form an intermediate allylic iodide, which then reacts in the same pot with $PhSO_2Na$, as described in Section 2.1.2.[180] The second method presumably involves insertion of SO_2 into nickel π-allyl intermediates and is limited to the preparation of symmetrical diallyl sulphones.[181]

2.3.4 Diaryl Sulphones

Diaryl sulphones have been traditionally prepared by the oxidation of diaryl sulphides and by the reaction of aromatic sulphonyl halides with either aromatic partners (in Friedel–Crafts-type sulphonylations) or aromatic organometallics such as Grignard reagents. Both of these topics have been discussed in earlier sections of this chapter and so here it is appropriate only to mention variants on these themes before reviewing alternative routes to diaryl sulphone products.

Firstly, the use of aryl silanes or stannanes in Friedel–Crafts-type reactions with arenesulphonyl halides gives reasonable yields of diaryl sulphones, Scheme 82.[182]

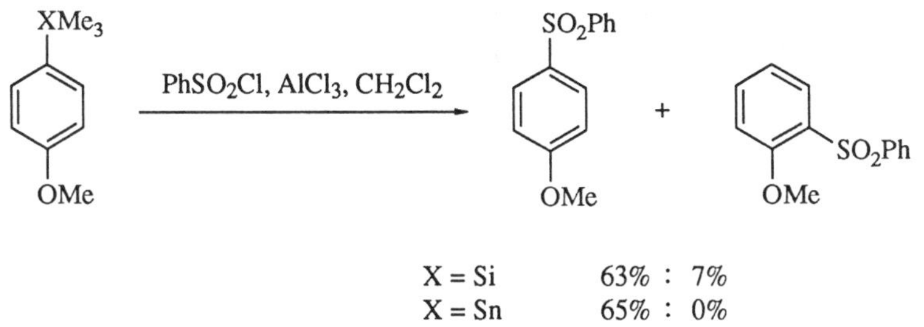

X = Si	63% :	7%
X = Sn	65% :	0%

Scheme 82

As shown, mixtures of regioisomers are sometimes obtained using silicon compounds, due to competing 'normal' sulphonylation. However, unwanted *ortho*-isomers are not obtained using aryl stannanes as starting materials, which was attributed to the more facile C–M bond cleavage in the case of stannanes. Mention is given in the paper of previous isolated reports of the use of organomercury and organocadmium reagents in analogous reactions.

A sulphonylating mixture comprising sulphuric acid and trifluoroacetic anhydride gives good to excellent yields of diaryl sulphones in a direct process from the parent aromatics, Scheme 83.

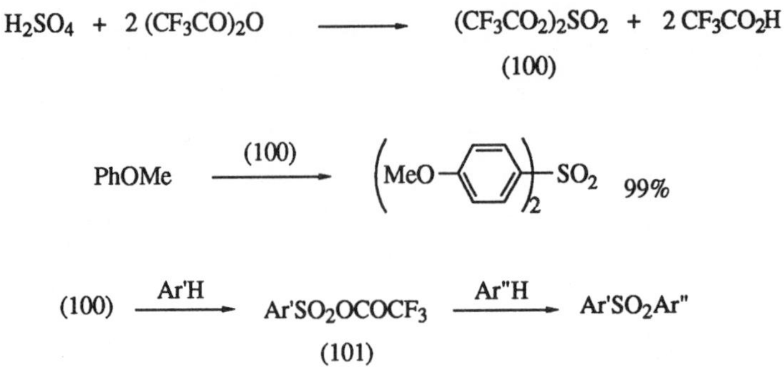

$$\text{H}_2\text{SO}_4 \ + \ 2\,(\text{CF}_3\text{CO})_2\text{O} \quad \longrightarrow \quad (\text{CF}_3\text{CO}_2)_2\text{SO}_2 \ + \ 2\,\text{CF}_3\text{CO}_2\text{H}$$
$$(\mathbf{100})$$

PhOMe $\quad \xrightarrow{\ (\mathbf{100})\ } \quad \left(\text{MeO}-\!\!\left\langle\!\!\bigcirc\!\!\right\rangle\!\!-\right)_{\!2}\!\!\text{SO}_2 \quad$ 99%

$(\mathbf{100}) \quad \xrightarrow{\ \text{Ar'H}\ } \quad \text{Ar'SO}_2\text{OCOCF}_3 \quad \xrightarrow{\ \text{Ar''H}\ } \quad \text{Ar'SO}_2\text{Ar''}$
$$(\mathbf{101})$$

Scheme 83

The method was first applied to the formation of symmetrical sulphones, with (**100**) being formed *in situ* and best results being obtained by using nitromethane as solvent.[183] By using a stoichiometric amount of the reagent unsymmetrical sulphones can also be obtained, since the initial reaction of Ar'H with (**100**) is faster than the subsequent reaction with intermediate (**101**).[184]

Another variant on this type of electrophilic aromatic substitution route involves the direct use of sulphonic acids in an acidic medium such as polyphosphoric acid (PPA). This method was first explored in detail by Graybill[185] and subsequently modified by Sipe *et al.*[186] The modified procedure gives improved yields and involves both the addition of P_2O_5 to the reaction mixture, and the pretreatment of the sulphonic acid with P_2O_5, Scheme 84.

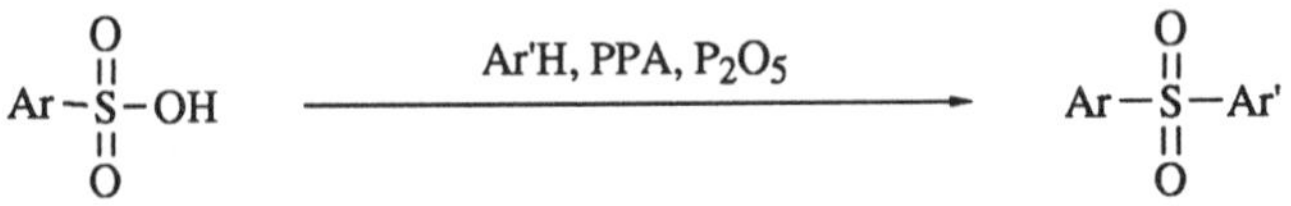

$$\underset{\underset{\text{O}}{\overset{\text{O}}{\|}}{\overset{\|}{\text{Ar}-\text{S}-\text{OH}}} \quad \xrightarrow{\quad \text{Ar'H, PPA, P}_2\text{O}_5 \quad} \quad \underset{\underset{\text{O}}{\overset{\text{O}}{\|}}{\overset{\|}{\text{Ar}-\text{S}-\text{Ar'}}}$$

Scheme 84

Not surprisingly, more substituted, electron-rich Ar'H partners facilitate the reaction, whilst excessive substitution on the ArSO_3H partner can lead to some competing desulphonation.[185] Other types of heteropolyacid such as $\text{H}_4[\text{SiMo}_{12}\text{O}_{40}]$, $\text{H}_3[\text{PMo}_{12}\text{O}_{40}]$ and $\text{H}_4[\text{SiN}_{12}\text{O}_{40}]$ have also been employed in such sulphonylation reactions.[187]

Finally, Hancock and Orszulik have reported the formation of sulphones by reaction of aryl thallium compounds with a mixture of sodium benzenesulphinate and cupric sulphate.[188]

2.4 Functionalised Sulphones

Many sulphones having additional functional groups can be prepared by methods already described in this chapter, such as oxidation of sulphides (Section 2.1.1) or alkylation of sulphinate salts (Section 2.1.2). Other reactions of use in the preparation of functionalised sulphones involve the interplay between two or more functional groups, for example, transformation of an epoxy sulphone to a substituted hydroxy sulphone by nucleophilic epoxide opening. These methods are dealt with in this section, with emphasis on the latter category of special methods. Many syntheses of complex sulphones rely on reactions of simpler sulphones via their derived anions. Some of these reactions are mentioned briefly here, and dealt with in much more detail in Chapter 3.

2.4.1 Hydroxy and Alkoxy Sulphones

Perhaps the most widely utilised route to hydroxy sulphones is the condensation of a sulphonyl carbanion with a carbonyl compound. The most common use for the resulting hydroxy sulphone is reductive elimination to form an alkene. The two-stage process is known as the Julia olefination and is dealt with in Chapter 7.

A more detailed examination of the type of epoxide-opening process outlined in Scheme 57 combines a subsequent lithiation of the resulting alkoxide-substituted vinyl sulphone to furnish hydroxy sulphones of type (102), Scheme 85.[189]

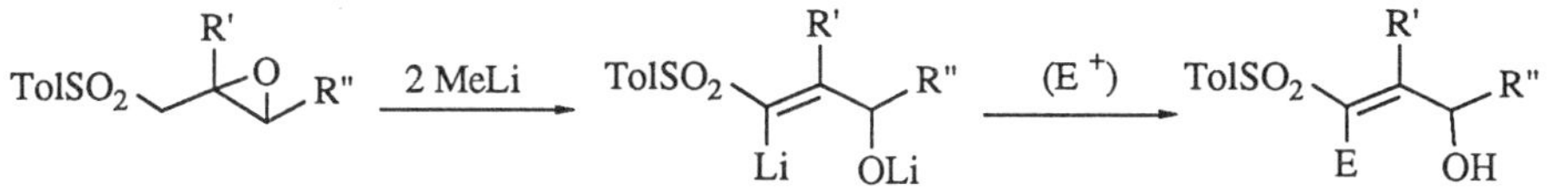

Scheme 85

If Grignard reagents are used, then the eliminative epoxide opening can be followed by conjugate addition to the vinyl sulphone group, Scheme 86.[190]

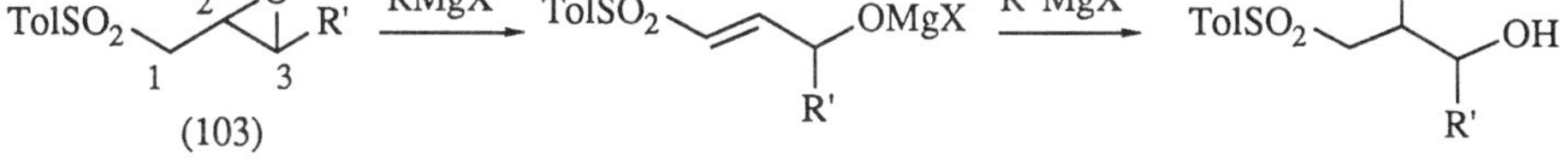

Scheme 86

The overall result is that of nucleophilic opening of the epoxide in (103) by the Grignard R"MgX at C-2. The alternative direct nucleophilic opening of such epoxides at C-3 can also be carried out in some instances by the use of either Grignard reagents in THF in the presence of $CuBr$[190] or in Et_2O in the presence of CuI, Scheme 87.[191]

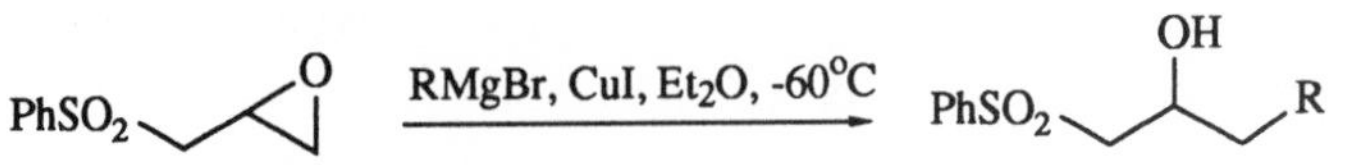

Scheme 87

The opening of epoxides by the rather weakly nucleophilic sulphinate anion is a viable route to β-hydroxy sulphones, Scheme 88.[192]

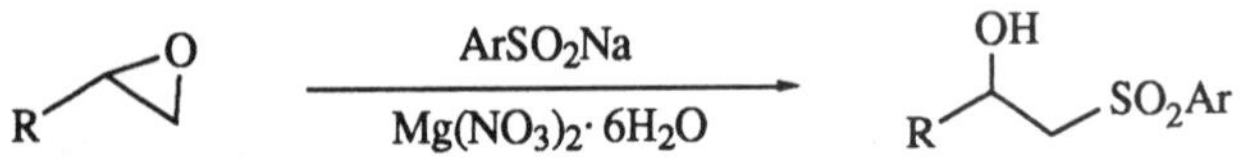

Scheme 88

The reaction appears to be rather limited in scope, as only terminal epoxides give reasonable results. The use of magnesium nitrate as Lewis acid catalyst is found to give the best results, whereas magnesium bromide gives unwanted bromohydrin products.

The *threo*-selective reduction of β-ketosulphones to the corresponding hydroxy sulphones has been mentioned in Scheme 36 as part of the Julia vinyl sulphone synthesis. This interesting reduction has been studied in some detail by Grossert *et al.*[193] The stereochemical outcome of the reaction, e.g. (104)→(105), can be explained by invoking a Felkin–Anh type model in the nucleophilic attack of hydride, Scheme 89.

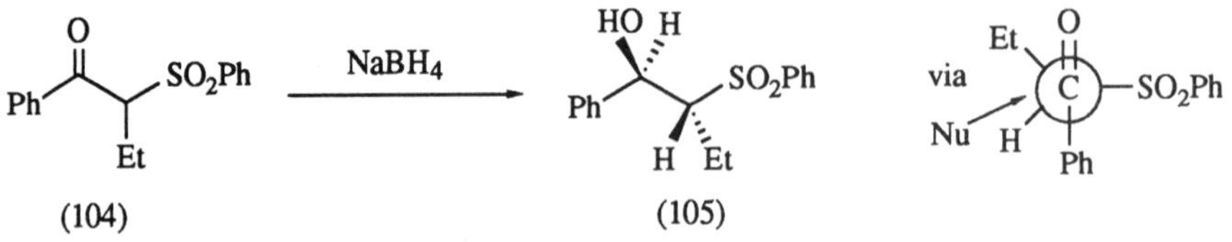

Scheme 89

High levels of stereoselection are obtained using $NaBH_4$, whereas in Julia's report[103] this reagent gives inferior selectivity to both $LiAlH_4$ and $LiHB^SBu_3$.

Because of the potential usefulness of optically active hydroxy sulphones as building blocks in asymmetric synthesis (particularly for linking of homochiral fragments via sulphonyl carbanions), their preparation by enantioselective reduction of ketosulphones has attracted considerable attention. Two reports which describe such transformations mediated by yeast are highlighted in Scheme 90. The preparation of sulphone (**106**) in greater than 95% ee has been described by Crumbie *et al.*[194] In this study, a range of β-ketosulphones, as well as the corresponding sulphoxides and sulphides, are reduced, mainly with good results.

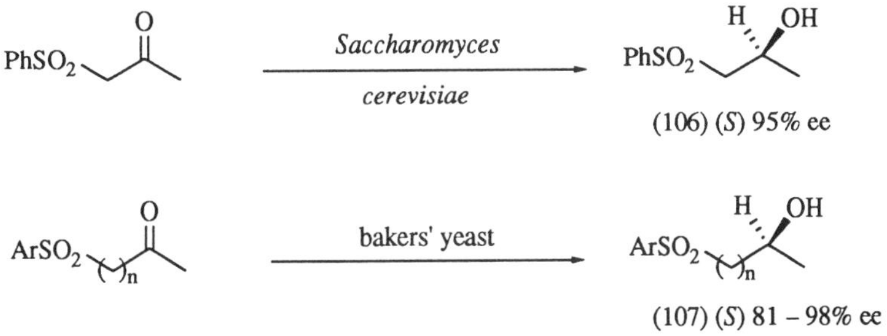

Scheme 90

The utilisation of a γ- or δ-ketosulphone in this chemistry considerably broadens the potential for the products in organic synthesis.[195] Unfortunately (**107**) (n = 4) can be obtained only in a very low chemical yield (6%). In both studies it is clear that variations in substrate structure may have a highly detrimental effect on the ee of the product alcohol.

Chemical methods for such reductions have also been examined, for example the use of a tartaric acid-modified Raney nickel reagent gives products in up to 70% ee in simple cases.[196] More recently, Corey and Bakshi have introduced a new catalytic procedure for the asymmetric reduction of achiral ketones.[197] This method, which requires only 0.1 equivalents of the catalyst (**108**), has been applied to sulphone (**109**), yielding the corresponding alcohol (**110**) in 91% ee, Scheme 91.

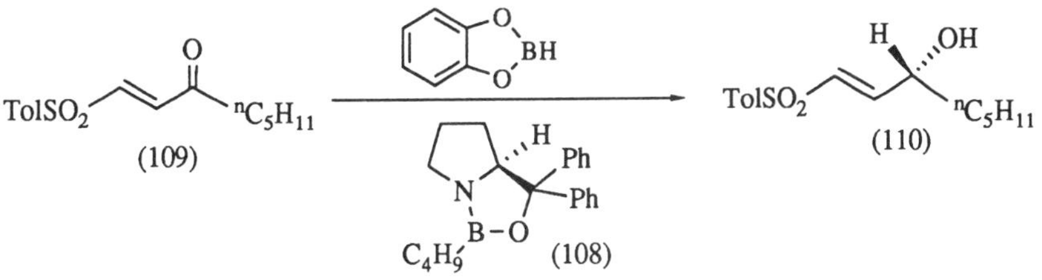

Scheme 91

The use of (**108**) in combination with catecholborane is a modification of the original procedure which used a slightly different catalyst and BH_3 as reductant. Hydroxy sulphone (**110**) could be enriched to very high optical purity (>98% ee) by a single recrystallisation, the absolute configuration of the product being established by a correlation sequence involving oxidative cleavage of the C=C bond. Similar optically active sulphones can also be obtained by a lipase-catalysed enantioselective esterification.[198] This kinetic resolution procedure, although giving rather variable optical purities, can easily be used to obtain either enantiomer of a particular hydroxy sulphone.

A recent radical cyclisation reaction coincidentally allows easy incorporation of an hydroxy or ether functionality into the sulphone product, e.g. Scheme 92.[199]

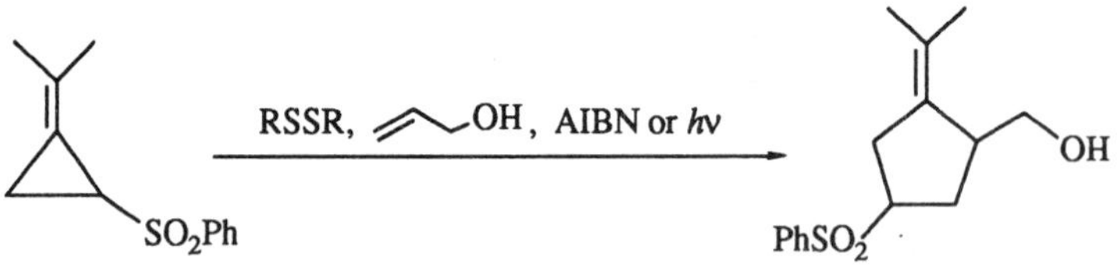

Scheme 92

The process employs a catalytic quantity of disulphide which initiates cyclopropane cleavage of the starting sulphone by attack of RS· formed by disulphide cleavage. The reaction is remarkable in that unactivated or electron-rich alkenes give good results (due to the electrophilic nature of the α-sulphonyl radical intermediate) and that changing the bulk of the disulphide catalyst can influence the stereochemical outcome of the cyclisation.

Schank's group has conducted extensive studies on the synthesis and chemistry of oxygenated sulphones, including sulphonyl ethers.[200] The synthesis of simple α-sulphonyl ethers by three complementary routes nicely illustrates some of the general methods for sulphone synthesis outlined in earlier sections of this chapter, Scheme 93.[201]

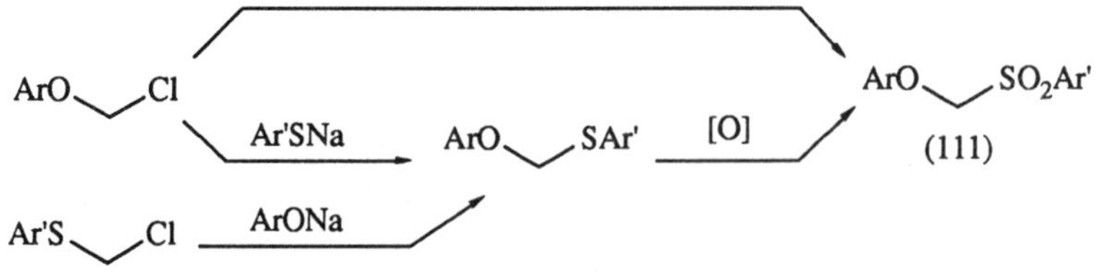

Scheme 93

Derivatives of α-alkoxy sulphones such as (111) are available simply through alkylation or acylation of the corresponding anions.[202] α-Sulphonyl cyclic ethers such as 2-phenylsulphonyltetrahydropyran (112) are very easily prepared by the reaction of either dihydropyran or 2-methoxytetrahydropyran with benzenesulphinic acid, Scheme 94.[203]

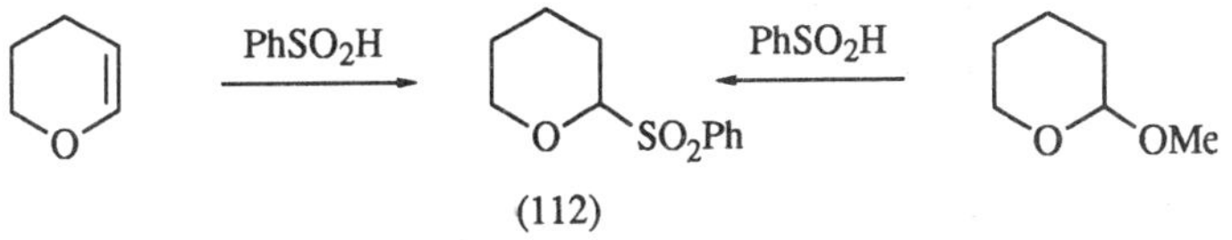

(112)

Scheme 94

Most of the interest in these systems involves utilising the sulphone group to generate an anion which then reacts to give products in which the sulphonyl group is lost or is easily replaced in a subsequent transformation. The anionic and desulphonylation aspects of this chemistry are dealt with in Chapters 3 and 9, respectively.

An interesting *gem*-disulphonyl ether (113) was prepared by Trost via the corresponding *gem*-disulphide (114) starting with anthranilic acid, carbon disulphide and β-trimethylsilylethanol, Scheme 95.[204] The oxidation step requires the use of $MoO_5 \cdot HMPA \cdot H_2O$ in CH_2Cl_2 at 0°C.

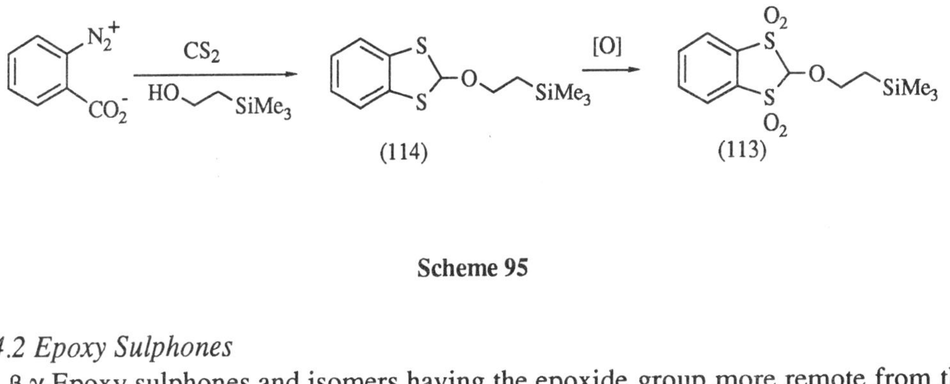

(114)   (113)

Scheme 95

2.4.2 Epoxy Sulphones

β,γ-Epoxy sulphones and isomers having the epoxide group more remote from the sulphone can be prepared by conventional epoxidation (e.g. using peracids) of the corresponding alkenyl sulphones. α,β-Epoxy sulphones represent a special sub-group which is most usually prepared by nucleophilic epoxidation of α,β-unsaturated sulphones. Whereas epoxidation of *cis*-sulphone (115) with H_2O_2/HO^- gives the corresponding *trans*-epoxy sulphone (116),[205] the use of ClO^- results in the highly stereoselective production of the *cis*-isomer (117), Scheme 96.[206]

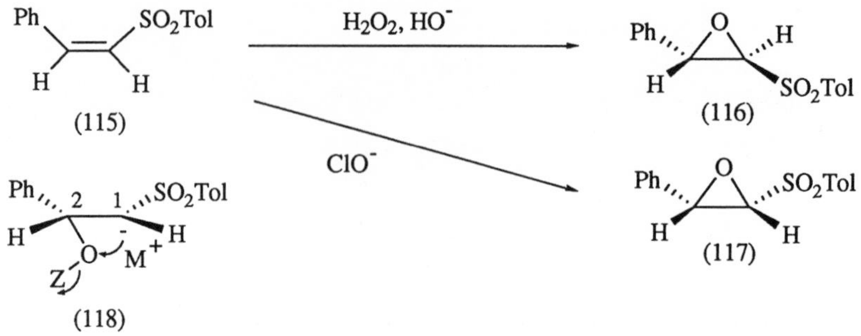

Scheme 96

These results can be interpreted in terms of the rate of expulsion of group Z (Z = OH or Cl) compared to the rate of rotation about the C-1–C-2 bond in an intermediate carbanion (**118**). Thus, in the first case, the second step is sufficiently slow to allow sulphonyl carbanion equilibration and hence formation of the more stable epoxide product (also obtained from the *trans* starting material). With ClO⁻ the reaction is either highly stereoselective or stereospecific.

In the report by Curci and DiFuria[206] the use of ᵗBuOO⁻ generated under aqueous conditions gives similar non-stereospecific results to those obtained with HOO⁻. In contrast, a more recent method from Meth-Cohn's laboratory involves generating ᵗBuOOLi under anhydrous conditions and allows stereospecific epoxide formation.[207] A report by Ashwell and Jackson describes the use of this method for the stereospecific preparation of either (**119**) or (**120**), Scheme 97.[208]

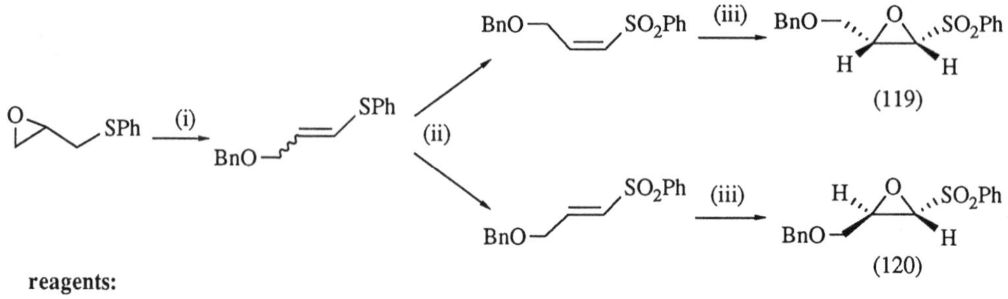

reagents:

(i) NaH, BnBr   (ii) MCPBA, separate isomers   (iii) ᵗBuOOH, BuLi, THF, -20°C

Scheme 97

Ashwell and Jackson have also examined the homologation of such epoxy sulphones via their derived anions, along with the well-established transformation of the products by reaction with MgBr₂, e.g. Scheme 98.[209]

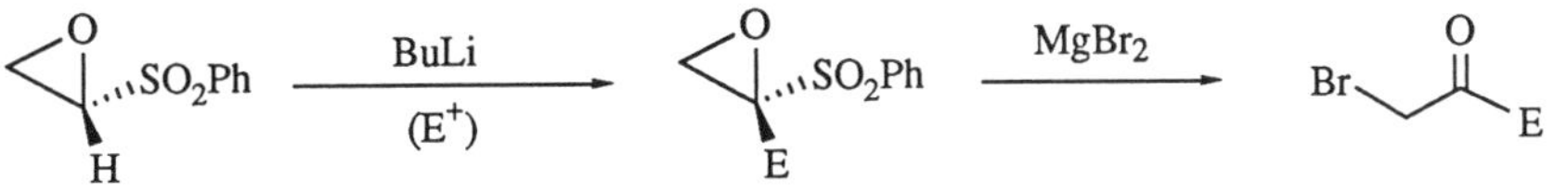

Scheme 98

The intermediate anions involved in this homologation are rather unstable and are best formed and reacted at very low temperature (-102°C).

The contrasting reactivity of simple alkenes and vinyl sulphones in epoxidation reactions is nicely illustrated by the highly selective epoxidation of dienyl sulphones such as (**121**), Scheme 99.[210]

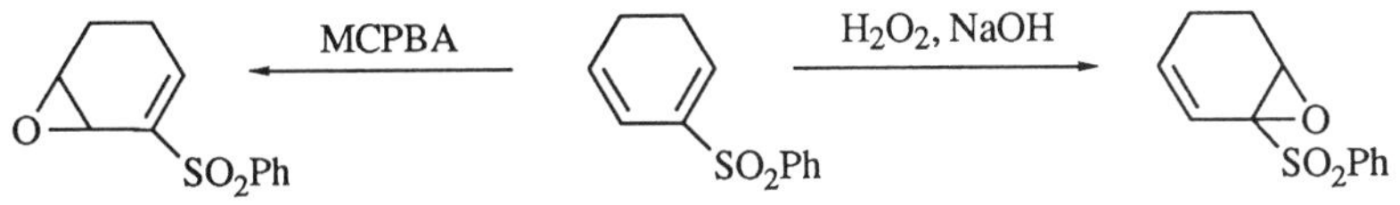

Scheme 99

α,β-Epoxy sulphones have also been prepared by oxidation of the corresponding epoxy sulphides, using either monoperphthalic acid or MCPBA as oxidant.[211]

The Darzens approach to α,β-epoxy sulphones reported by Makosza and co-workers involves combination of the anion of an α-chlorosulphone, generated under phase-transfer conditions, with a carbonyl partner.[212] The reaction has been widely used, notably by Durst and co-workers[213] and more recently by Hewkin and Jackson,[214] e.g. Scheme 100.

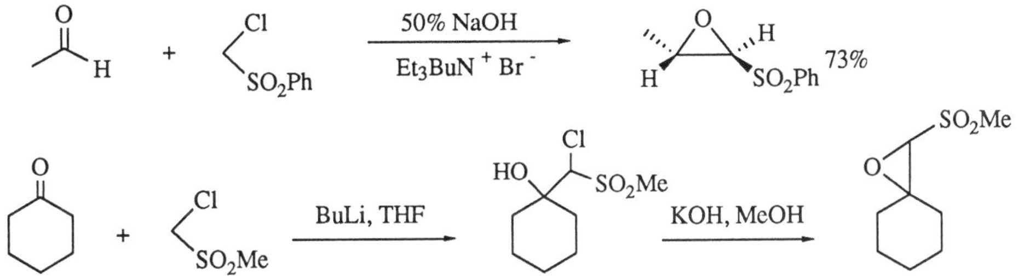

Scheme 100

Durst has also examined a two-stage variant of this method involving initial condensation of a chlorosulphonyl anion with a carbonyl compound to give isolable chlorohydrins such as (122) followed by a separate cyclisation. In general, this method appears less convenient and lower-yielding than the direct phase-transfer method, in part due to the propensity of the chlorohydrin intermediates to undergo competing retro-aldol reaction, thus regenerating the two starting materials. Both methods are capable of furnishing epoxy sulphone products using starting chloro-sulphones which might be expected to undergo competing Ramberg–Bäcklund reactions.

The phase-transfer method produces *trans*-epoxy sulphones from aldehydes, and usually gives mixtures of stereoisomers from ketones. The studies of Grossert *et al.* have shown that the stereochemistry of the epoxy sulphones does not necessarily reflect the stereochemistry of the intermediate chlorohydrin intermediates.[215]

2.4.3 Carbonyl-substituted Sulphones

Sulphones of general structure (123)-(125), i.e. α-, β- and γ-acyl sulphones, are all well known. Of these, β-ketosulphones (124) (R = alkyl) have undoubtedly received the most attention, largely due to their interesting and useful anion and dianion chemistry. The chemistry of homologues (125) as well as unsaturated analogues (which could also be regarded as vinyl sulphones) has also been well explored.

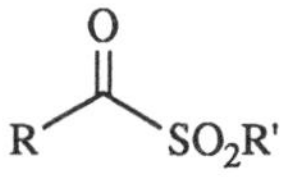

(123) 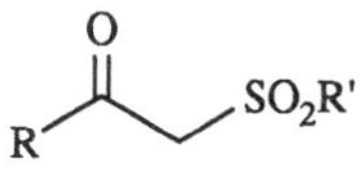   (124) 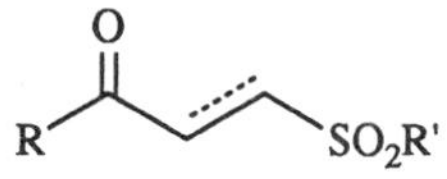   (125)

By contrast, α-acyl derivatives, e.g. α-oxosulphones R'SO$_2$COR (123), are much less familiar as synthetic intermediates. Oxidative routes are broadly applicable to the preparation of such compounds, for example α-sulphonyl amides, Scheme 101.[216–218]

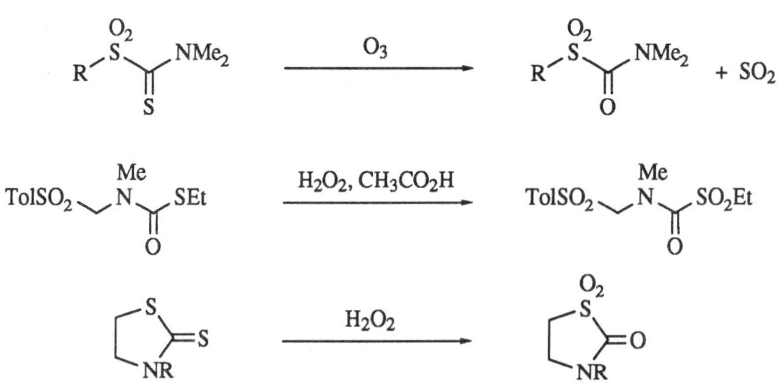

Scheme 101

The corresponding ketone and ester derivatives can be similarly prepared, Scheme 102.[219–221] α-Ketosulphones such as (126) have been examined in detail by Schank's group.[219] They can be prepared by ozonolysis of α-diazosulphones or α-methylene sulphones.

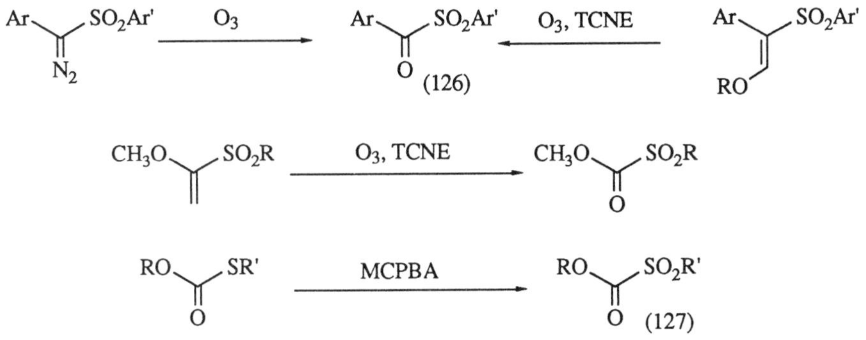

Scheme 102

These compounds, as well as the acyl sulphones such as (127), are sensitive compounds which generally react as powerful acylating agents, e.g. Scheme 103.[219–222]

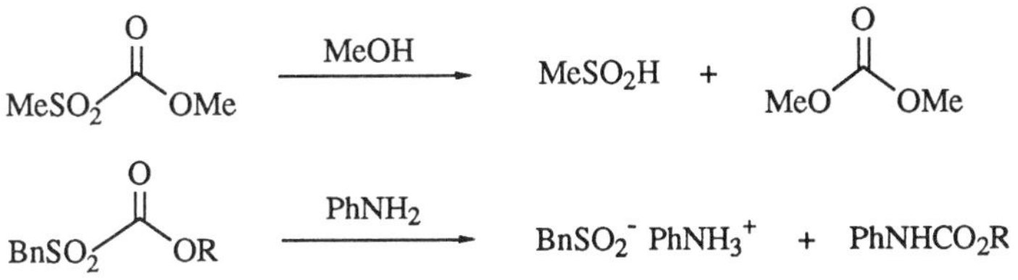

Scheme 103

This type of acylation chemistry can likewise be carried out by using *gem*-disulphonyl ethers derived from (113), mentioned in Scheme 95.[204]   A notable additional feature of this approach is that aromatic systems undergo intramolecular *C*-acylation, Scheme 104.

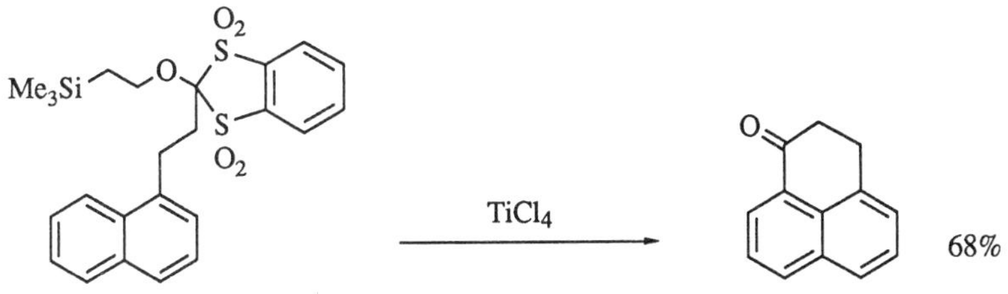

Scheme 104

In this case Lewis acids such as BCl_3 or $TiCl_4$ are far superior to nucleophilic fluoride sources for initiating acyl sulphone reactivity.

Schank and Werner have also ascertained that an α-oxosulphone (**128**) is capable of typical ketone-like chemistry, rather than the acylation chemistry already described. One such reaction utilises diazomethane to give an epoxy sulphone (**129**), Scheme 105.[223]

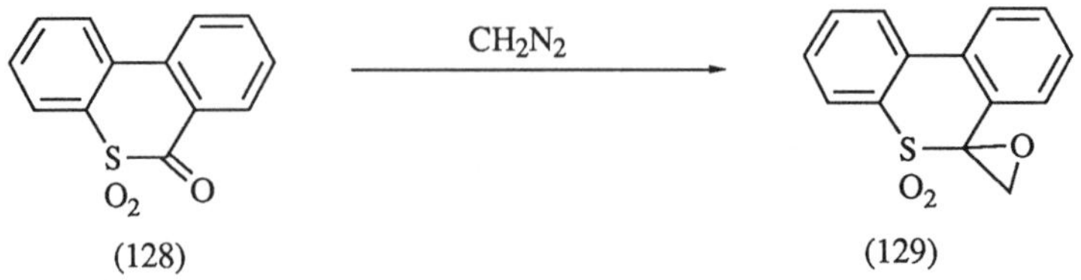

Scheme 105

β-Acyl sulphones can be prepared by reaction of either nucleophilic or electrophilic sulphur reagents at the α-position of a carbonyl starting material. The chemistry follows the protocols outlined in previous sections, for example sulphinate alkylation[224] or sulphide oxidation, Scheme 106.[225]

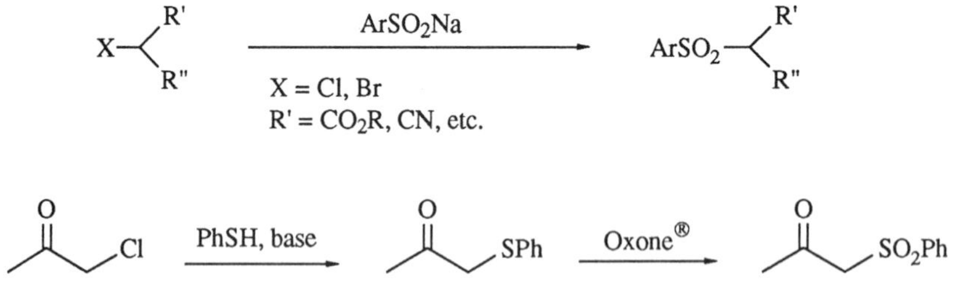

Scheme 106

Sulphonylation α to a carbonyl group can also be achieved indirectly by reaction of an enamine with an arenesulphonyl chloride, Scheme 107.[226,227] However, under some conditions, sulphene intermediates can be formed from alkanesulphonyl halides, in which case [2+2] cycloadducts are the principle products.[228]

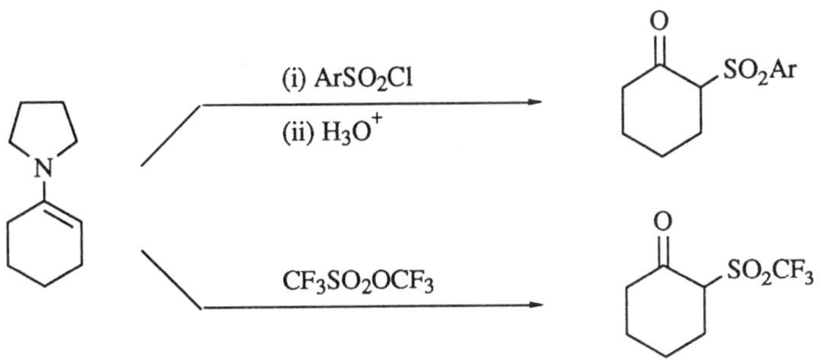

Scheme 107

Direct sulphonylation of carbonyl compounds via their derived enolates can also be achieved, for example using SO_2 or $ArSO_2F$ as the electrophilic quenching agents, Scheme 108.

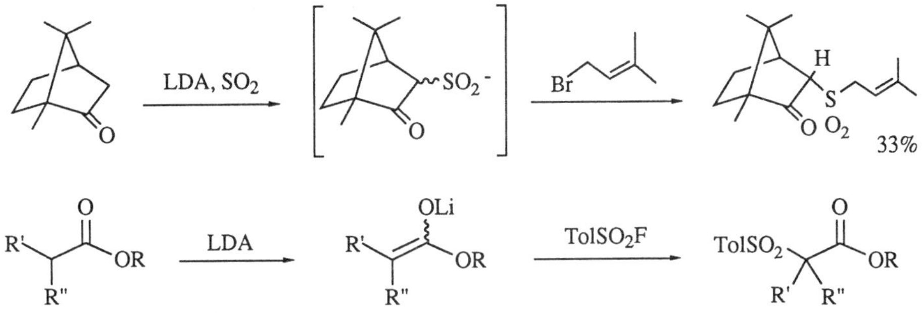

Scheme 108

The first sequence is rather low-yielding, presumably due to the reversibility of the sulphonylation step.[229]   The reaction of $TolSO_2F$ with either ester enolates or their derived silyl ketene acetals as described by Kende and Mendoza proceeds well with α-disubstituted esters ($R_1 = R_2 = $ alkyl).[230]   However, monosubstituted systems give low yields, due to rapid proton transfer between the starting enolate and the rather acidic ester sulphone product. Similar effects have been observed by Hunig and co-workers with the enolate derived from *tert*-butyl methyl ketone, Scheme 109.[231]

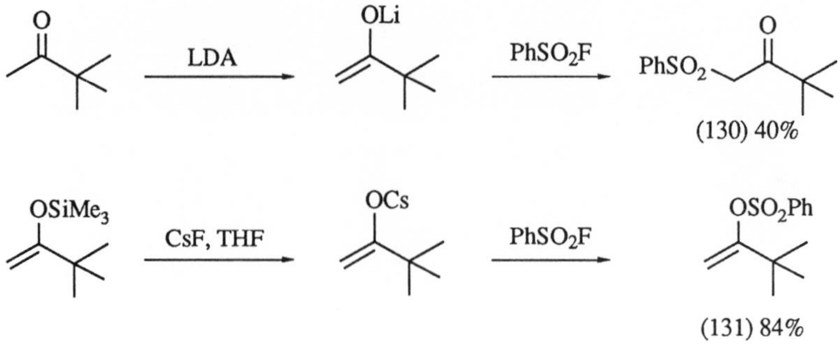

Scheme 109

In this study the effects of different solvents and enolate counterions were examined. With Li^+ or Na^+ the β-ketosulphone (**130**) is the sole or major product, whilst by using Cs^+ or R_4N^+ counterions the product of *O*-sulphonylation (**131**) is obtained exclusively in high yield.

Another well-developed strategy for the synthesis of β-ketosulphones and related derivatives is the acylation of sulphonyl carbanions. This topic is dealt with in detail in Section 3.2.5; here just a few simple examples will serve to illustrate this approach. Thus, simple β-ketosulphones have been obtained by Becker and Russell simply by treating esters with dimethyl sulphone and ^tBuOK in hot DMSO.[232] Truce and co-workers also carried out significant early studies on the synthesis of various β-ketosulphones, as well as reporting on their acidities and spectroscopic characteristics.[233] One particular area of interest involves the use of Thorpe cyclisations of cyanosulphones to form cyclic β-ketosulphones, Scheme 110.

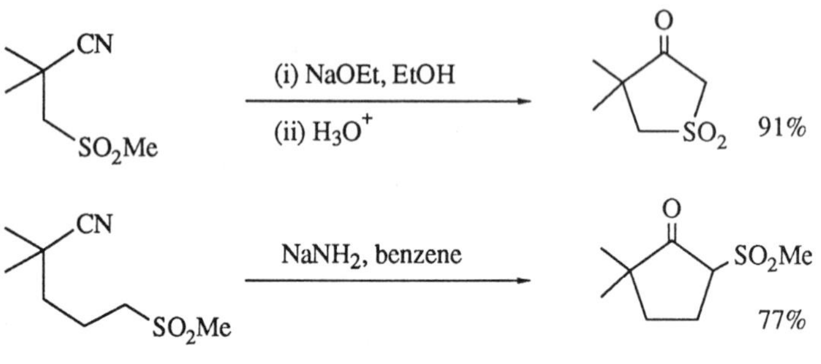

Scheme 110

Compounds without the blocking *gem*-dimethyl group α to the nitrile were found not to react, due to preferential deprotonation at this site. β-Formyl sulphones can easily be isolated as their metal salts following condensation of a sulphonyl carbanion with formate esters, e.g. Scheme 111.[234]

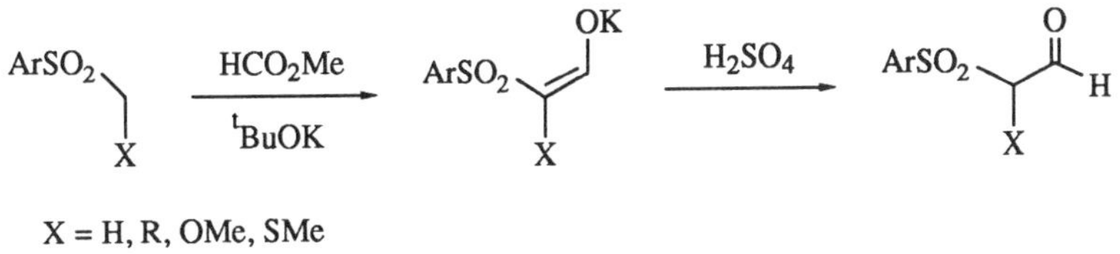

X = H, R, OMe, SMe

Scheme 111

A problem that occurs when a sulphone and ester are combined using a stoichiometric amount of base is consumption of the starting sulphonyl carbanion by proton transfer from the ketosulphone product. This problem is often overcome by the use of excess base or sulphonyl carbanion, e.g. Scheme 112.[235]

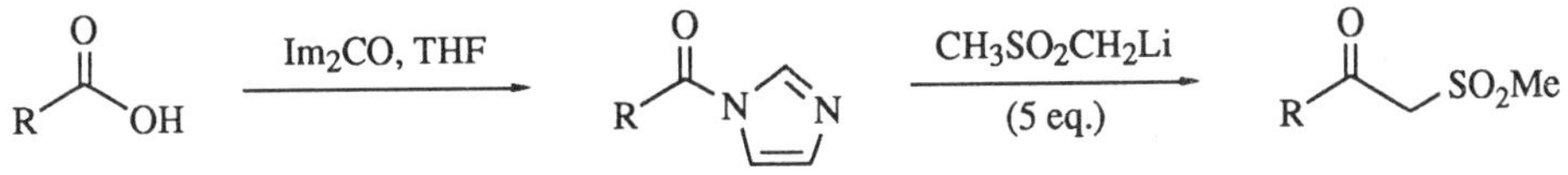

Scheme 112

An alternative solution is to use a *gem*-dilithiosulphone in the condensation. This approach was first described by Kaiser *et al.*,[236] but has since been used quite often, e.g. in the synthesis of precursors to *cis*-jasmone, Scheme 113.[237]

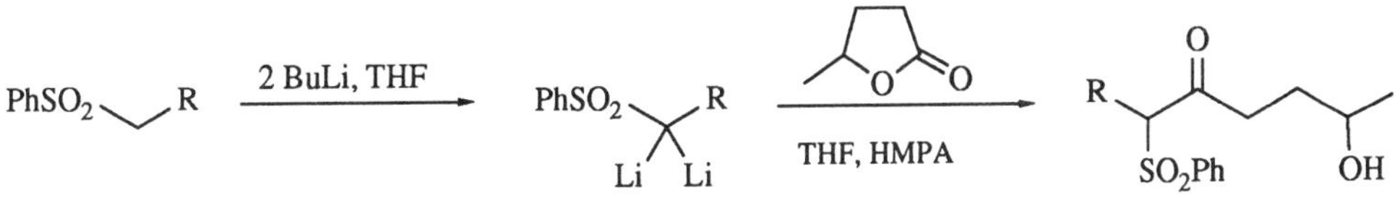

Scheme 113

The reaction of dilithiated sulphones with acid chlorides is reported to be even faster and cleaner than that with esters.[238]

γ-Acyl sulphones can be prepared by Michael addition of thiolate to a suitable α,β-unsaturated compound followed by an oxidation step.[239] The same types of products are also available more directly by treatment of an α,β-unsaturated ketone with a sulphinic acid, Scheme 114.[240]

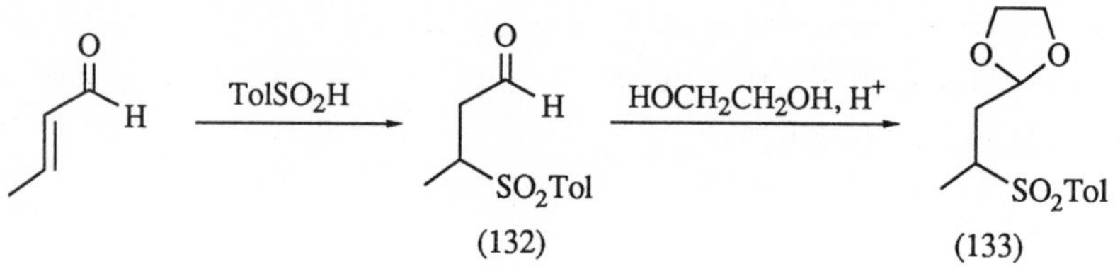

Scheme 114

It is usually most convenient to generate the sulphinic acid *in situ*, for example by use of a sulphinate salt and an equivalent of acetic acid.[241] The γ-ketosulphone products such as (132) can be protected by ketalisation to give compounds such as (133) which can then be elaborated using sulphonyl carbanion chemistry. Sulphones related to (132), and homologues, have also been prepared by alkylation of sulphinate intermediates formed in dithionite reductions of α,β-unsaturated carbonyl compounds.[242]

Unsaturated acyl sulphones such as (134) or (135) can be prepared by the iodo-sulphonylation methods outlined in Section 2.2.1, Scheme 115.[88]

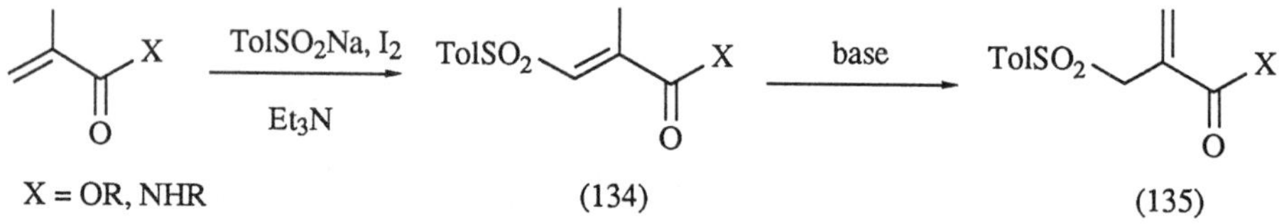

Scheme 115

Both these series of compounds undergo sulphone substitution reactions with nucleophiles such as Grignard reagents.[243,244] Particularly interesting is the use of the parent acrylamide sulphone (136) in reactions with two equivalents of Grignard reagent, which allows overall substitution of the acrylamide at the α-position, Scheme 116.[244]

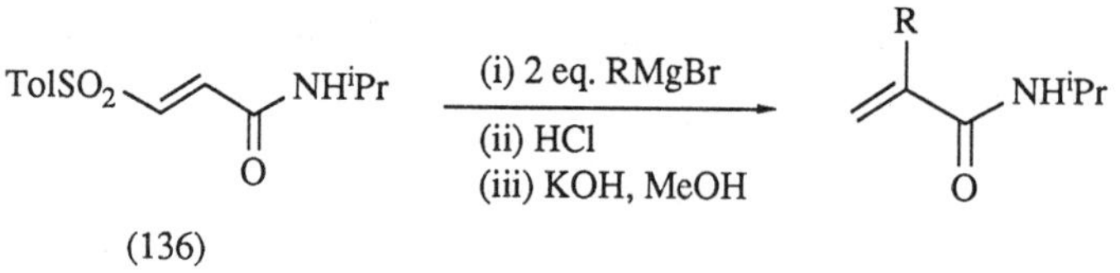

Scheme 116

The reaction relies on addition to the vinyl sulphone group, the amide being deactivated by deprotonation by the first equivalent of RMgBr.

Certain 1,3-diketones undergo β-sulphonylmethylation when treated with a mixture of sulphinic acid and paraformaldehyde.[245]   A recent example of a synthetic application of this type of reaction comes from the laboratory of Zwanenburg, Scheme 117.[246]

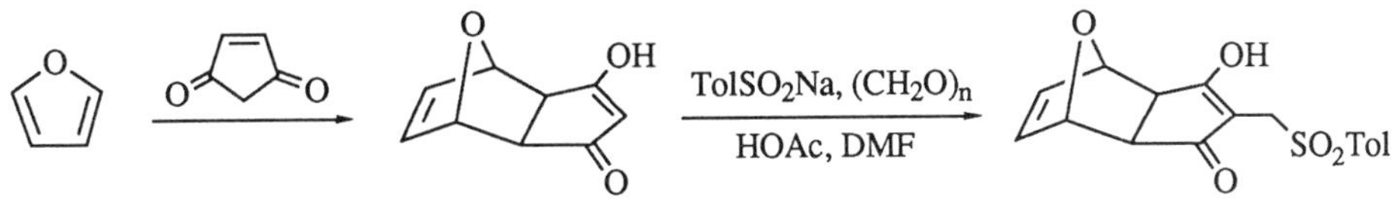

Scheme 117

Sulphones such as (**137**) were found useful for further elaboration, followed by cycloelimination (i.e. the reverse of the first step in Scheme 117) to eventually furnish *epi*-pentenomycin products.

β-Ketosulphones can be converted to cyclic derivatives by a thermal or copper-catalysed C–H insertion reaction of their derived α-diazo compounds.[247]   Monteiro has described improvements, both in the preparation[248] and cyclisation[249] of such α-diazo-β-ketosulphones, Scheme 118.

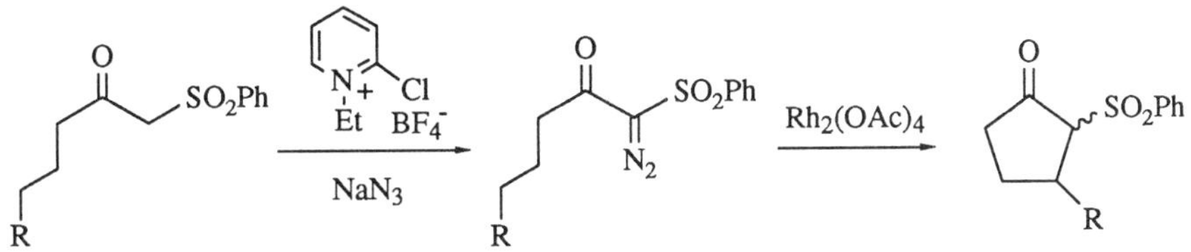

Scheme 118

The rhodium-catalysed carbene insertion sequence gives good yields (typically *ca.* 65%), and can also be used for internal cyclopropanation, i.e. if R is a vinyl group.

A variety of unsaturated ketosulphones have been prepared by the Mn(OAc)₃-mediated addition of acetone to isopentenyl phenyl sulphone.[250]   This reaction usually gives mixtures of products, but can give useful amounts of methylene compounds such as (**138**) which, if desired, can be further transformed to a diketo-sulphone, e.g. (**139**), by ozonolysis, Scheme 119.

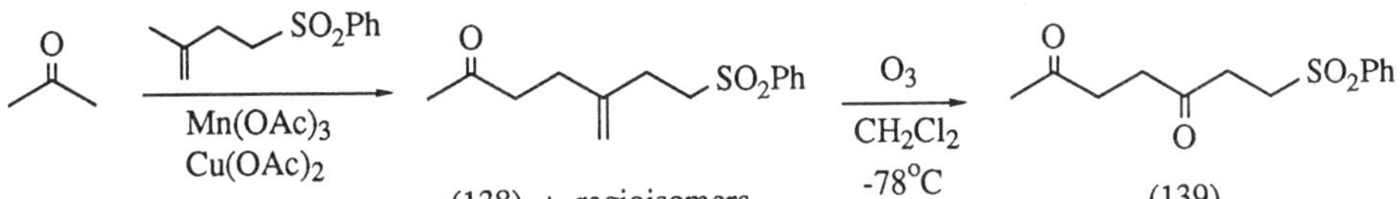

Scheme 119

A range of substituted ketosulphones can be prepared by the Michael addition of enolates (or enolate equivalents) to unsaturated sulphones. Two typical examples are shown in Scheme 120.[251,252]

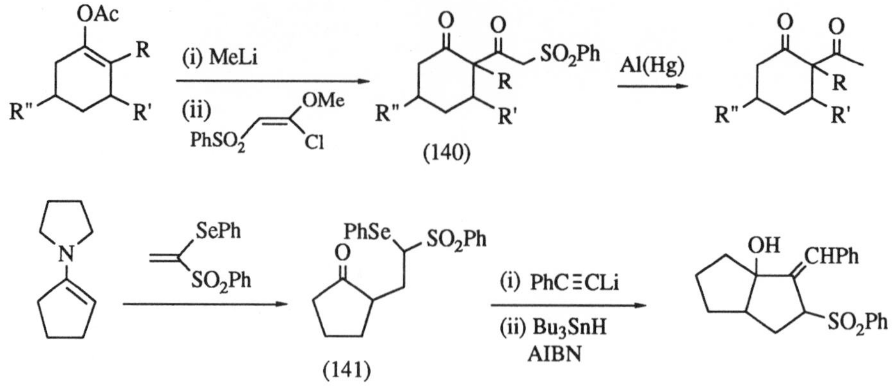

Scheme 120

In each case the intermediate ketosulphones formed in the Michael addition are then subjected to further transformation: in the case of (**140**) a simple desulphonylation[251] and in the case of (**141**) elaboration and free-radical cyclisation.[252] The use of such α-sulphonyl carbon-centred radicals in cyclisations has received rather sparse attention despite obvious attractions for the preparation of cyclic (especially cyclopentane) sulphones, e.g. Scheme 121.[253]

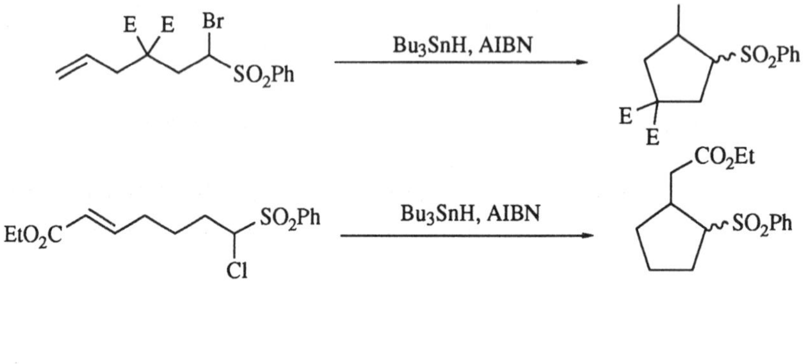

Scheme 121

Both of these systems undergo efficient 5-*exo-trig* cyclisation of intermediate α-sulphonyl radicals to give sulphonyl ester products.

Sulphonyl nitriles as well as sulphonyl esters and ketones are of synthetic interest and may often be prepared analogously. Three alternative preparations resulting in various sulphonyl-substituted nitriles are illustrated in Scheme 122.[254–256]

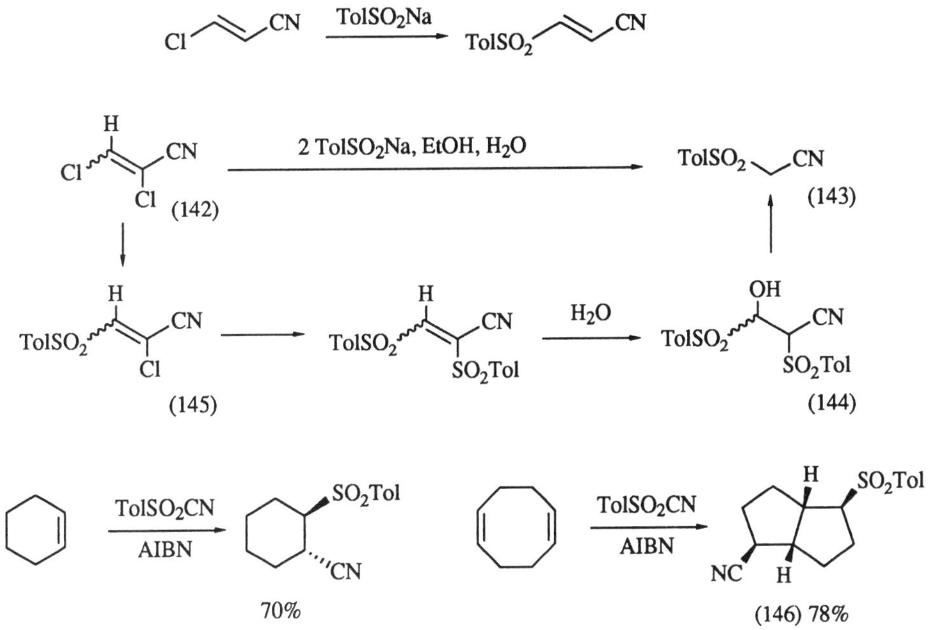

Scheme 122

The unexpected chain-shortening conversion of (**142**) to α-cyanosulphone (**143**) occurs through sequential Michael additions to give intermediate (**144**) which can then decompose by expulsion of sulphinate and deformylation. Adjustment of the reaction conditions also allows the isolation of (**1 4 5**).[255] Free-radical cyanosulphonylation occurs in an analogous fashion to iodosulphonylation, etc., described previously.[256] Particularly appealing is the formation of bicyclic cyano-sulphone (**146**) in good yield using a transannular cyclisation.

The preparation and reactions of α-cyanosulphones, RSO₂CN, have been reported by van Leusen *et al.*[257] These compounds react rather like cyanogen halides with a range of nucleophiles.[258]

2.4.4 Nitrogen-substituted Sulphones

Aminosulphones can be prepared by special methods which are largely dealt with in other chapters, such as the conjugate addition of amines to α,β-unsaturated sulphones described in Chapter 4 (see also Scheme 44). Other nitrogen-containing sulphones can be found earlier in this chapter, e.g. Schemes 2, 5 and 11.

A very widely applicable synthesis of α-aminosulphone derivatives is the Mannich-type condensation of a sulphinic acid with an aldehyde in the presence of a nucleophilic amine component.[259] The method can be used with amines and also with less nucleophilic amino components such as amides,[260] sulphonamides[261] and related compounds, e.g. Scheme 123.

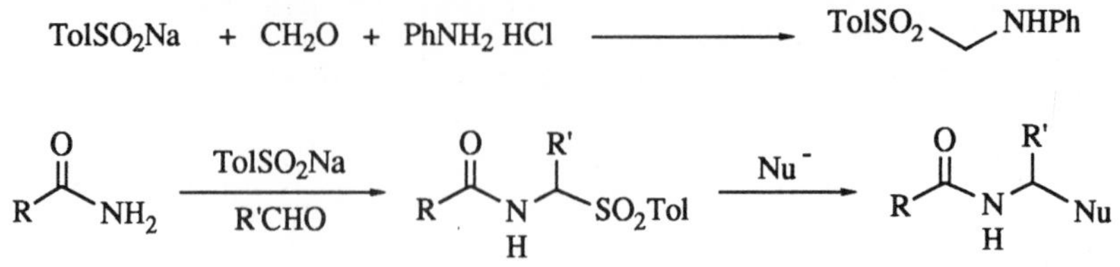

Scheme 123

Products such as (147) can act as useful amidoalkylating agents in reactions with sulphur, nitrogen and carbon nucleophiles.[262]

Sulphonyl enamines (see also Scheme 55) can be prepared by treatment of α,β-epoxy sulphones with $BF_3 \cdot OEt_2$ followed by reaction with a secondary amine.[263]

An interesting route to aminosulphones involves addition of the carbon-centred radical $TolSO_2CH_2\cdot$ to enamines, Scheme 124.[264] With cyclic enamines the *cis*-product is formed selectively, and if a suitable homochiral amine is used to make the starting enamine then the radical addition also shows good face selectivity.[265]

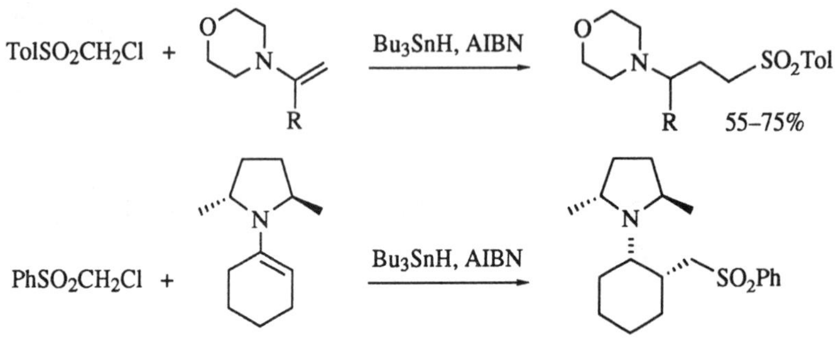

Scheme 124

Three- and four-membered rings having vicinal amine and sulphone groups can be prepared starting with enamines, Scheme 125.[266]

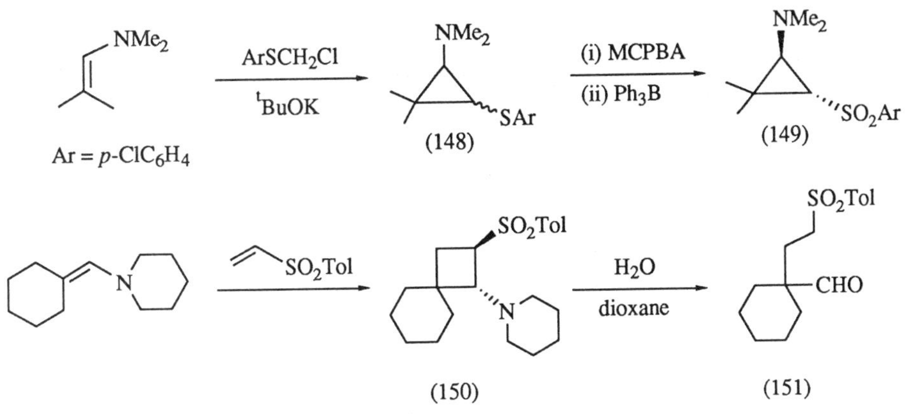

Scheme 125

The preparation of the three-membered systems such as (**149**) by oxidation of (**148**) was found to be possible only after some experimentation. An indirect route was eventually found which converted the *trans*-isomer of (**148**) to (**149**) via the intermediate *N*-oxide which was reduced using Ph_3B. Both the three- and four-membered ring compounds are prone to ring opening; for example, treatment of (**150**) with aqueous dioxane gives aldehyde (**151**). Saturated nitrogen heterocycles of various ring sizes having sulphone substitution have been prepared, Scheme 126.[267–270]

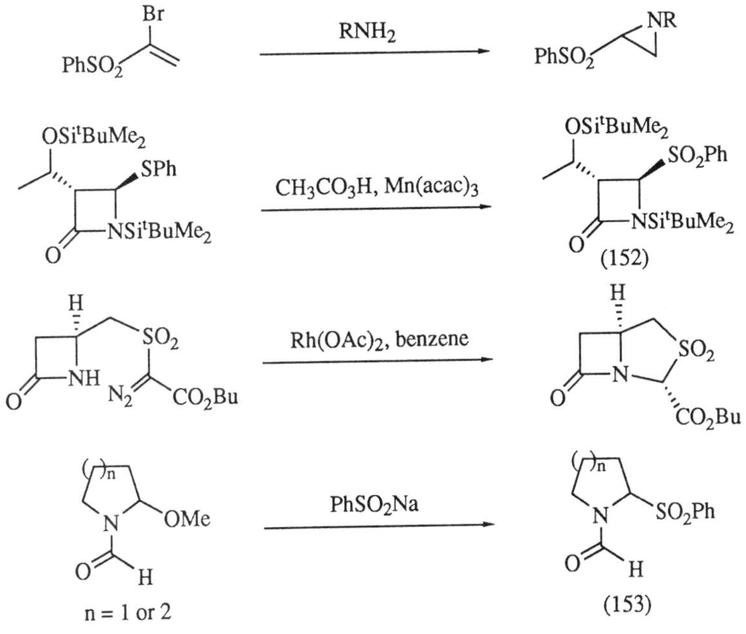

Scheme 126

Compounds such as (152) and (153) are useful for forming new C–C bonds α to nitrogen by a nucleophilic sulphone substitution reaction as indicated in Scheme 123 for the acyclic analogue (147).

The most widely utilised nitrogen-substituted sulphone reagent is undoubtedly tosylmethylisocyanide (TosMIC). This compound (154) and its relatives are most easily prepared by reaction of a metallated isocyanide with an arenesulphonyl fluoride,[271,272] although alternative access by dehydration of sulphonyl formamides is also possible,[272] Scheme 127.

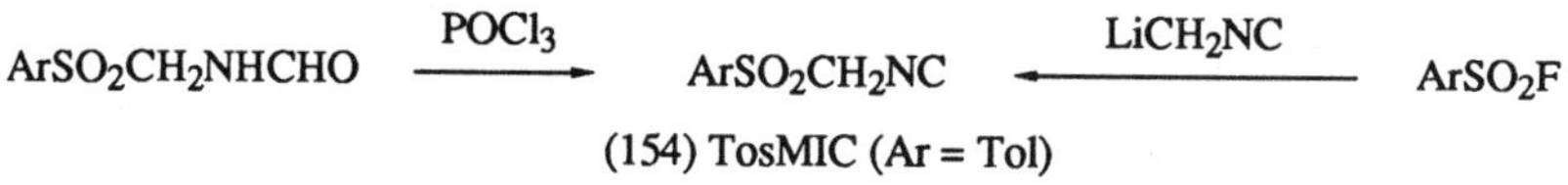

$$ArSO_2CH_2NHCHO \xrightarrow{\text{POCl}_3} ArSO_2CH_2NC \xleftarrow{\text{LiCH}_2NC} ArSO_2F$$

(154) TosMIC (Ar = Tol)

Scheme 127

Substituted derivatives can also be prepared by alkylation of the parent compound. This chemistry, along with the use of such versatile derivatives, for example in heterocyclic synthesis, is dealt with in Chapter 3. Related *N*-(tosylmethyl)-carbodiimides (155) have also been prepared via the corresponding thiourea derivatives, Scheme 128.[273]

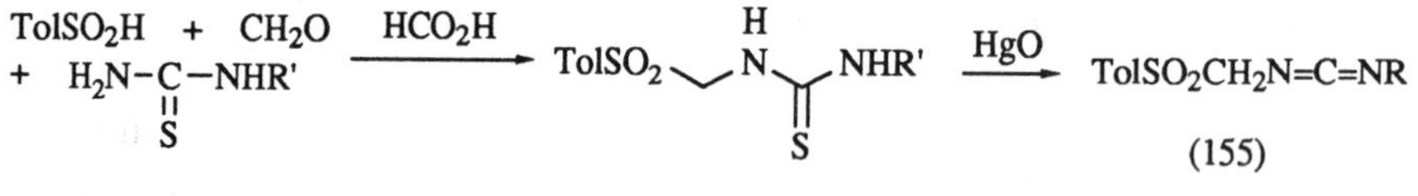

$$TolSO_2H + CH_2O + H_2N\text{-}\underset{\underset{S}{\|}}{C}\text{-}NHR' \xrightarrow{\text{HCO}_2H} TolSO_2\diagdown N \diagup \underset{\underset{S}{\|}}{\overset{H}{N}} NHR' \xrightarrow{\text{HgO}} TolSO_2CH_2N\text{=}C\text{=}NR$$

(155)

Scheme 128

These compounds react analogously to TosMIC, forming heterocyclic products in base-catalysed condensation with electrophilic partners such as aldehydes.

α-Nitrosulphones can be prepared by a number of straightforward procedures starting with either a sulphone or a nitro compound, Scheme 129.[274–277] In the first sequence an α-iodo nitroalkane is treated with a sulphinate salt to give the expected substitution product.[274] Direct conversion of a nitroalkane to a nitro- sulphone is also possible by reaction of the derived nitronate anion with either $TolSO_2I$ or, as shown, a mixture of $TolSO_2Na$ and iodine.[275]

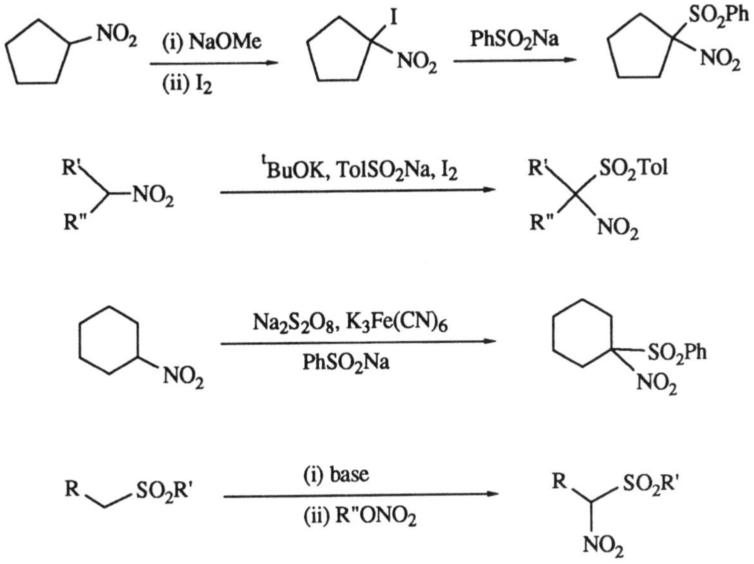

Scheme 129

Again the α-iodo nitroalkane is an intermediate, and the replacement of iodine for the ArSO$_2$ group occurs, not by simple substitution, but by a free-radical chain reaction. A related oxidative sulphonylation, e.g. of nitrocyclohexane, also gives good results using sodium persulphate with a catalytic quantity of ferricyanide.[276]

The reaction of a sulphonyl carbanion with an organic nitrate also gives nitro-sulphones.[277] Depending on the nature of the starting sulphone either a ^{t}BuOK/THF or a KNH$_2$/NH$_3$ system gives good results. Halogenations of the products were also examined. Homologation of the parent nitrosulphone PhSO$_2$CH$_2$NO$_2$ is possible simply by alkylation of the derived carbanion.[278]

Perhaps the most important route to α-diazosulphones is the diazo transfer reaction of a suitably activated sulphone with tosyl azide, Scheme 130.[279] A major shortcoming of the reaction is the need for an additional activating group, here an ester, besides the sulphone.

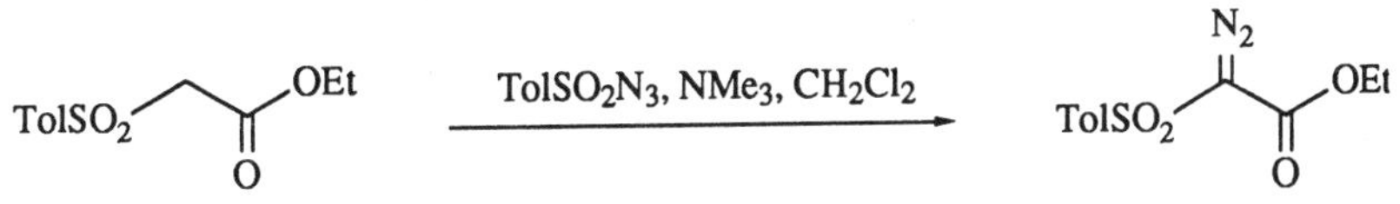

Scheme 130

Other routes to these compounds include the original preparation involving the cleavage of nitrosocarbamates reported by van Leusen and Strating[280] and the acylation of sulphonyl diazomethanes.[281] Other variants have also been described by van Leusen *et al.* which are especially suitable for some classes of diazo-sulphone, such as those having α-aryl groups.[282]

α-Azidosulphones can easily be prepared by oxidation of the corresponding sulphides,[283] or by reaction of nitronates with TolSO$_2$N$_3$,[284] Scheme 131.

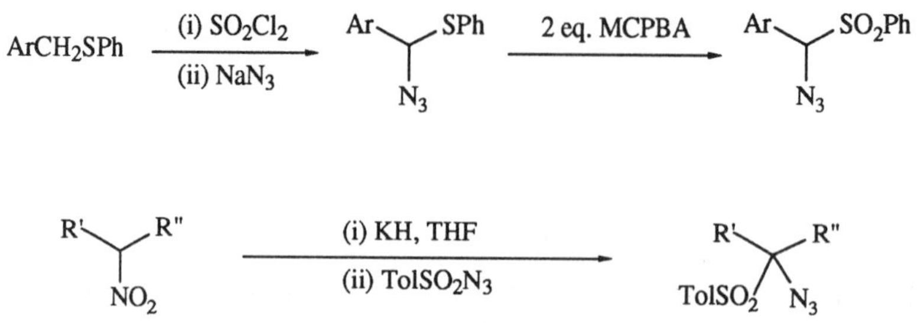

Scheme showing:

ArCH$_2$SPh →(i) SO$_2$Cl$_2$ (ii) NaN$_3$→ Ar—CH(N$_3$)—SPh →2 eq. MCPBA→ Ar—CH(N$_3$)—SO$_2$Ph

R'—CH(NO$_2$)—R" →(i) KH, THF (ii) TolSO$_2$N$_3$→ R'—C(SO$_2$Tol)(N$_3$)—R"

Scheme 131

α-Azidosulphones are sensitive to basic conditions, under which they undergo loss of nitrogen and benzenesulphinate to yield a nitrile.

2.4.5 Halogenosulphones

α-Halogenosulphones are very important intermediates in the Ramberg– Bäcklund reaction which converts dialkyl sulphones to alkenes with extrusion of SO$_2$ (see Chapter 7). Other applications of these compounds include the Darzens-type reactions of their derived carbanions with carbonyl compounds to give epoxy sulphones as described in Section 2.4.2. Not surprisingly, then, the preparation of α-halogenosulphones has attracted considerable attention. The most important route to such compounds involves α-halogenation (especially chlorination) of sulphones via their derived carbanions. Meyers has described the use of a CCl$_4$/KOH/^{t}BuOH system for such chlorinations.[285] In these reactions the sulphonyl carbanion is chlorinated *in situ* by reaction with CCl$_4$. In general this method gives Ramberg–Bäcklund products when applied to sulphones possessing α- and α'-hydrogens. De Waard and co-workers have found that the use of hexachloroethane (HCE) as chlorinating agent with preformed sulphonyl carbanions can give monohalogenated products from certain cyclic sulphones, e.g. (156), Scheme 132.[286]

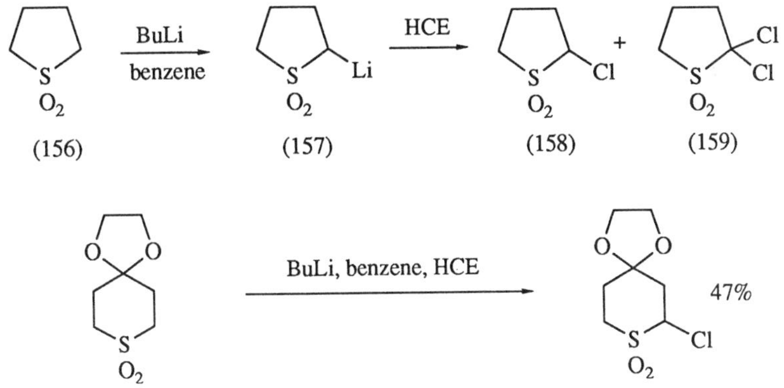

Scheme 132

The reaction proceeds better in benzene than in THF, but always gives some recovered starting sulphone and usually dichlorinated products. This behaviour may be ascribed to the sulphonyl carbanion, e.g. (157), deprotonating the initially formed product (158) to give a new anion which is again chlorinated to give (159). Very clean polychlorination can be achieved this way, for example using BuLi in THF followed by addition of a tenfold quantity of CCl_4, Scheme 133.[287]

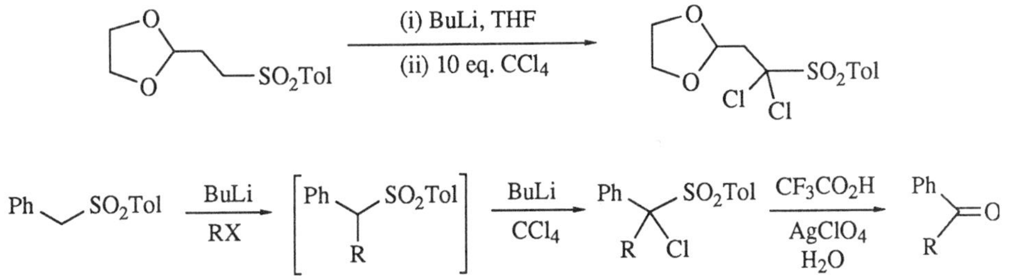

Scheme 133

As shown, the chlorination can also be applied to an *in situ* formed secondary sulphone to give cleanly a monochlorinated product which can then be converted to the corresponding aromatic ketone. Phase-transfer conditions have also been used for such chlorinations, for example a $CH_2Cl_2/NaOH_{aq}$/benzyltriethylammonium chloride (TEBA) system.[288]

These sulphone halogenations could conceivably proceed either by direct S_N2-type attack, or alternatively by a single electron transfer (SET) mechanism. Attempts to trap, by intramolecular cyclisation, the carbon-centred radical intermediates which could be formed in a SET halogenation of (160) failed, despite the cyclisation reaction of **(161)** to **(162)** being efficient under alternative radical-generating conditions, Scheme 134.[289]

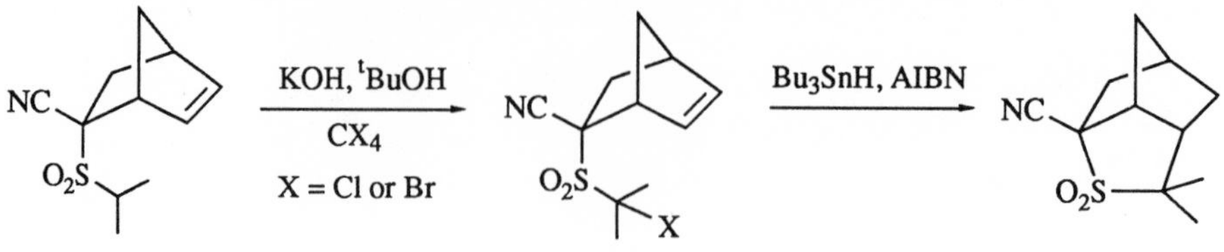

Scheme 134

Thus either an SN2 mechanism operates in such reactions, or the second step in the SET process, shown in Scheme 135, is very rapid indeed.

$$RSO_2\diagdown R' \;+\; CX_4 \;\longrightarrow\; \left[RSO_2\dot{C}HR'\;\; C\dot{X}_4\right] \;\longrightarrow\; RSO_2CHXR' \;+\; CX_3{}^{\cdot}$$

Scheme 135

Of course, halogens other than chlorine can be introduced analogously, most notably at bridgehead positions, Scheme 136.[290,291]

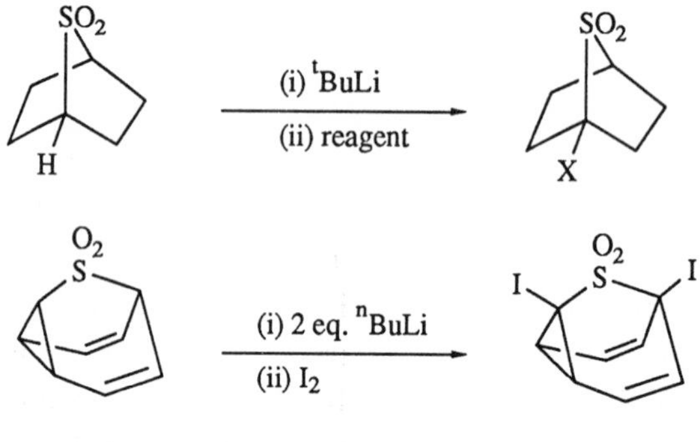

reagent	X
Cl_3CSO_2Cl	Cl
Br_2, BrCN	Br
I_2	I

Scheme 136

The simple α-fluorosulphone **(163)** is prepared by oxidation of the corresponding α-fluorosulphide, which itself is available either by fluoride substitution of the α-chlorosulphide,[292] or by a diethylaminosulphur trifluoride (DAST) reaction using methyl phenyl sulphoxide as starting material, Scheme 137.[108,116]

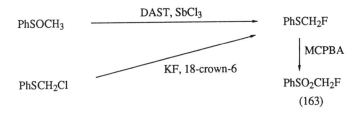

Scheme 137

Various other α-halogenosulphones have been prepared by classical procedures involving sulphide oxidation,[293] sulphinate alkylation,[294] and also by alkylation of iodomethyl aryl sulphones,[295] Scheme 138.

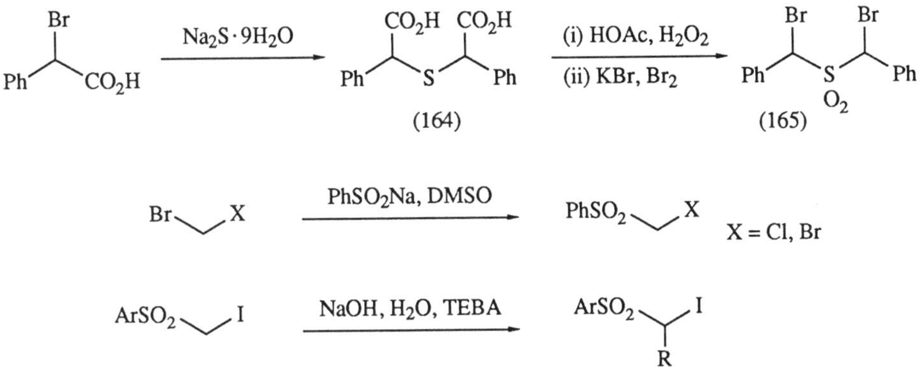

Scheme 138

The conversion of (**164**) to (**165**) using bromine is an example of halogenative decarboxylation.[293] Bordwell *et al.* have examined this process in some detail, and shown that the reaction can usually be performed simply by heating a mixture of starting sulphonyl carboxylic acid and *N*-halosuccinimide in CCl_4.[296] The reaction most likely proceeds by enolisation, halogenation (the *N*-halosuccinimide acts as a convenient source of low concentrations of X_2), and finally decarboxylation. Similar halogenative deacylation has also been achieved using ketosulphones.[297] The formation of β-halogenosulphones by reaction of sulphenyl halides with alkenes and subsequent oxidation has been discussed in Section 2.2.1 in the context of vinyl sulphone preparation.[79] Paquette and Houser have observed an interesting intramolecular variant of this process when treating some bridged sulphides with sulphuryl chloride, Scheme 139.[298]

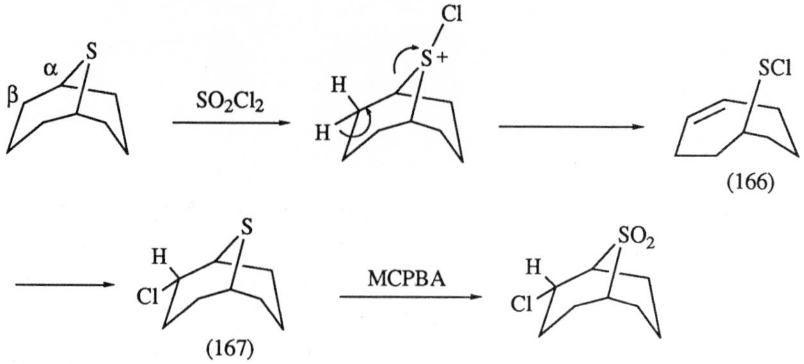

Scheme 139

Proton removal occurs from the β-position rather than as expected from the α-position (which would normally lead to the α-chloro sulphide) to give (**166**), which then adds in a *trans* fashion to the internal double bond to give (**167**). Oxidation then gives the corresponding β-chloro-sulphone. The desired α-chloro- sulphone is obtained by direct sulphonyl carbanion chlorination using SO_2Cl_2.

Free-radical halosulphonylation can be carried out with concomitant cyclisation, especially with systems which can undergo the particularly favourable 5-*exo-trig* cyclisation, Scheme 140.[299]

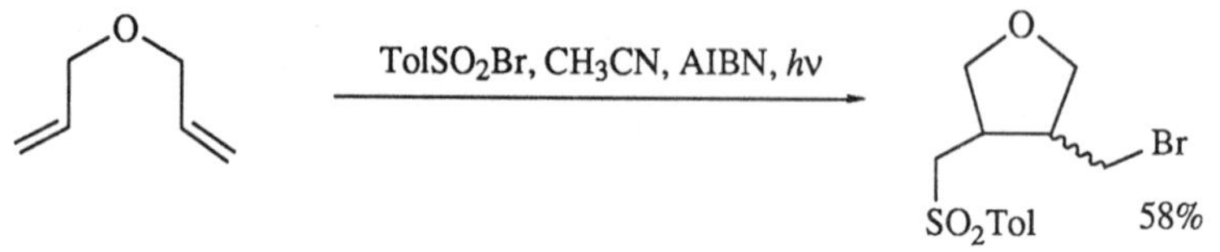

Scheme 140

This little-used variant looks especially attractive for assembling functionalised cyclic sulphones. The *cis*-isomers are formed predominantly.

Durst *et al.* have discovered that treatment of certain β-, γ- and δ-hydroxy sulphoxides with sulphuryl chloride results in the formation of the corresponding chlorosulphones.[300] The reaction occurs using SO_2Cl_2, $PhICl_2$ and ^tBuOCl, and the products are thought to be formed by ring opening of an intermediate alkoxyoxosulphonium ion (**168**), Scheme 141.

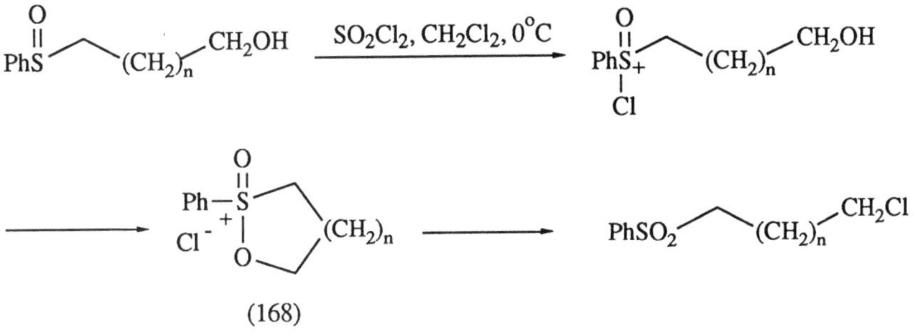

Scheme 141

Similar neighbouring group participation is also seen when sulphoxides having additional acid or amide functionality in place of an alcohol are reacted in the same way.

Familiar reactions which have been used to obtain unsaturated halogenated sulphones include the conversion of hydroxy sulphone (**169**) to bromide (**170**) using NBS/Me$_2$S,[125] and the acid-catalysed opening of cyclopropylcarbinols such as (**171**), Scheme 142.[301]

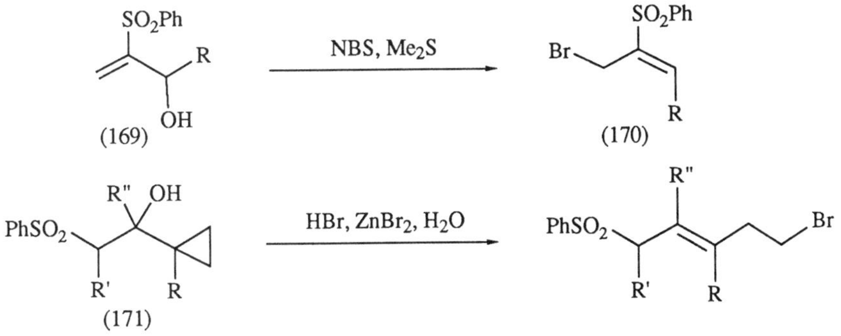

Scheme 142

Finally, trichloromethyl phenyl sulphone reacts with terminal alkenes in the presence of a ruthenium(II) catalyst to give 1:1 adducts, Scheme 143.[302]

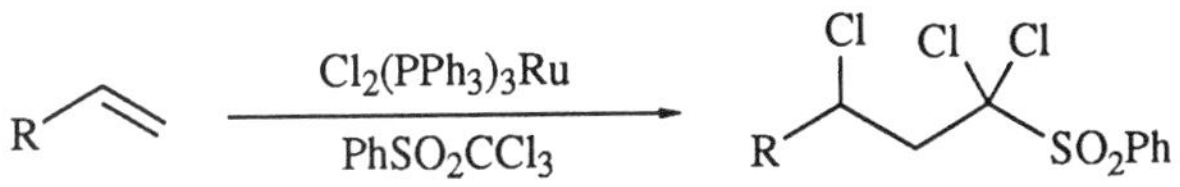

Scheme 143

$PhSO_2CCl_3$ is less reactive than either $TolSO_2Cl$ or CCl_4 in this reaction. The absence of reaction in the presence of a radical inhibitor suggests that radicals are involved.

2.4.6 Sulphur- and Selenium-substituted Sulphones

The selective oxidation of one sulphur atom in a thioacetal to give a monosulphone derivative has been mentioned in Section 2.1.1.[29] Ogura *et al.* have used such oxidations to obtain selectively either the bis-sulphoxide (**172**) or the sulphide sulphone (**173**), Scheme 144.[303]

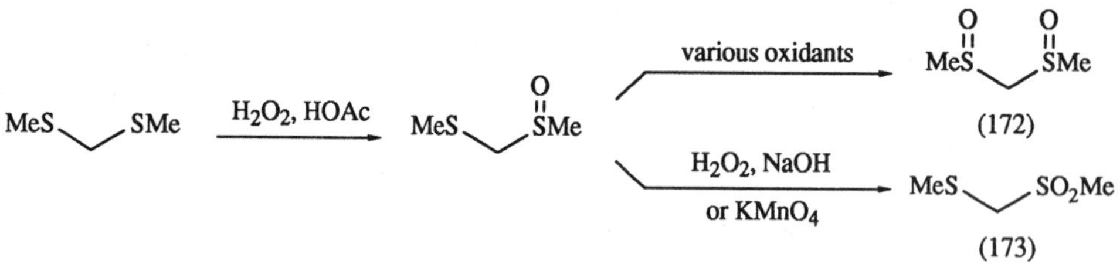

Scheme 144

A switch from electrophilic oxidation which furnishes (**172**) (e.g. with MCPBA) to nucleophilic oxidation with HOONa or $KMnO_4$ explains the selectivity. A one-pot procedure also described by Ogura *et al.* allows easy preparation of sulphone (**174**), Scheme 145.[304]

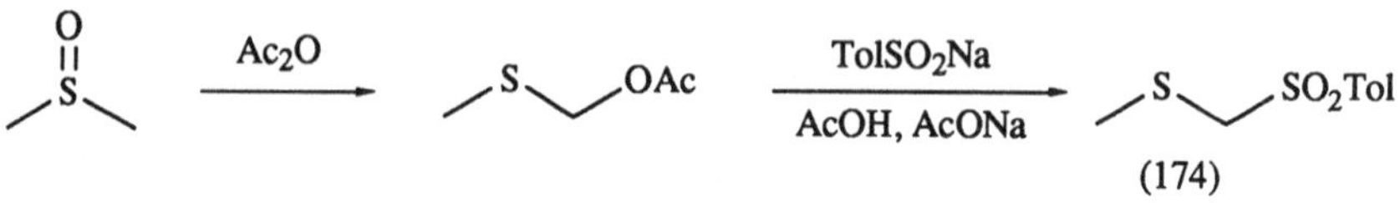

Scheme 145

The sequence relies on a sulphinate displacement on a Pummerer product derived from dimethyl sulphoxide. Here acetic acid was found superior to DMF as solvent, and the addition of AcONa was found further to improve the yield. Ogura has established sulphone (**174**) as a versatile intermediate for the synthesis of many types of carbonyl compound by alkylations and aldol-type reactions, followed by conversion of the sulphur-substituted carbon centre into a carbonyl group, e.g. Scheme 146.[305]

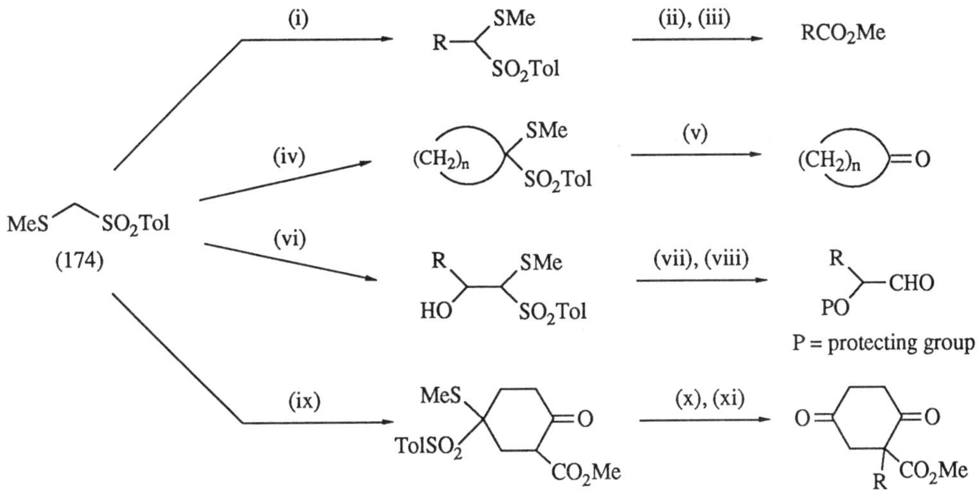

reagents:
(i) trioctylmethylammonium chloride (TOMAC), toluene/50% NaOH, RX   (ii) MCPBA
(iii) MeOH, H$^+$ (iv) Br–(CH$_2$)$_n$–Br, TOMAC, toluene/50% NaOH   (v) HCl, H$_2$O
(vi) BuLi, THF, RCHO   (vii) OH protection   (viii) $h\nu$, (254nm), H$_2$O/dioxane
(ix) 2 eq. CH$_2$=CHCO$_2$Me, NaH, DMF   (x) RX, K$_2$CO$_3$ (xi) H$^+$

Scheme 146

Thus, following alkylation, aldol or Michael reaction of the anion of (**174**) the masked carbonyl functionality can be revealed by treatment with acid or by photolysis.

Elimination reactions of thioacetal 1,1-dioxides occur to give thiocarbonyl compounds according to Scheme 147.[306]

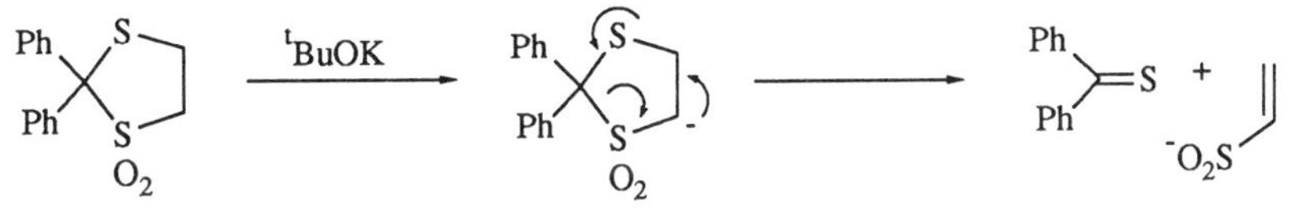

Scheme 147

Analogous reactions using 2-alkylidene-1,3-dithiolane 1,1-dioxides furnish thio-ketenes.

Sulphenylation of sulphonyl carbanions is another well-established route to a range of sulphur-substituted sulphones. Wladislaw and co-workers have obtained some sulphenylated benzyl phenyl sulphones in this way by using NaH as base and MeSSMe as the sulphenylating agent.[307] Similarly, anions derived from secondary allylic sulphones using BuLi can be simply sulphenylated using MeSSMe, and subsequently converted to carbonyl compounds in a similar way to that shown in Scheme 146.[308] Such sulphenylations (and also selenenylations as mentioned in Section 2.2.5) can be difficult to control, with over-sulphenylation occurring. Grossert and Dubey have examined this point with a range of carbonyl- and sulphone-containing substrates, and have obtained high yields of either mono- or bis-sulphenylated products, Scheme 148.[309]

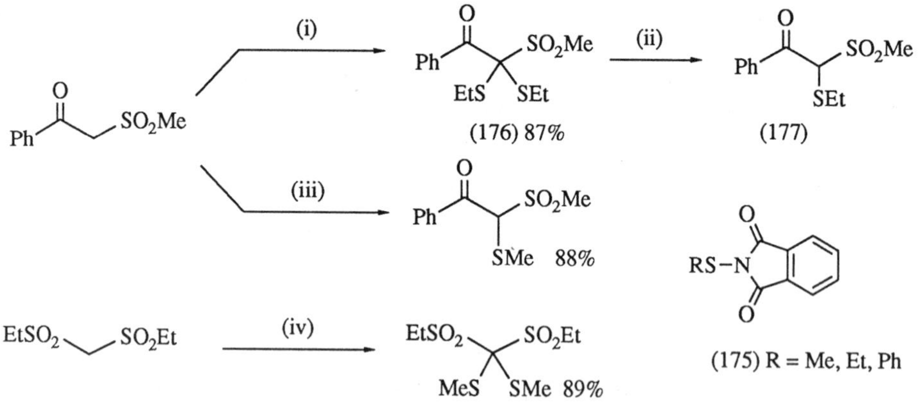

reagents:
(i) 2 eq. (175) R = Et, base (ii) NaSEt, NaH
(iii) 1 eq. (175) R = Me, base (iv) 2 eq. (175) R = Me, base

Scheme 148

Thus in some cases the desired degree of sulphenylation can be effected simply by adjusting the stoichiometry of the reagents (**175**). However, in other cases, only doubly sulphenylated compounds can be obtained cleanly. Significantly, a method of converting polysulphenylated products back to monosulphenylated derivatives has been found, e.g. (**176**)→(**177**), using thiolate in the presence of excess base.

A good way of obtaining sulphinyl sulphones is to react a sulphonyl carbanion with a sulphinate ester, Scheme 149.[310,311]

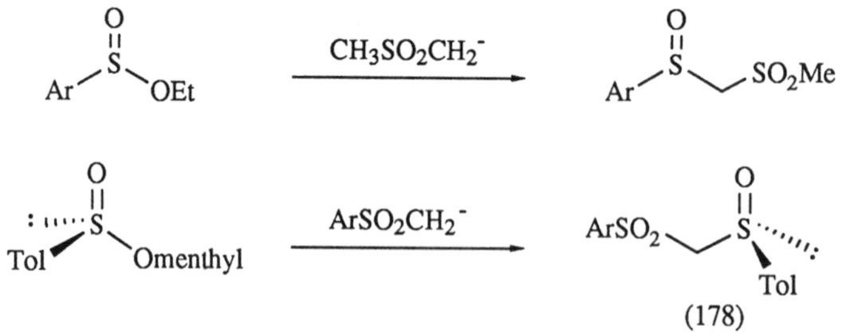

Scheme 149

Of particular interest is the generation of optically active sulphinyl sulphone (**178**) by an Andersen-type approach starting with commercially available (-)-menthyl *p*-toluenesulphinate.[311]

Barton *et al.* have obtained α-thiopyridyl sulphones (**179**) by irradiation of the *O*-acyl derivatives of *N*-hydroxy-2-thiopyridone (**180**) in the presence of phenyl vinyl sulphone, Scheme 150.[312]

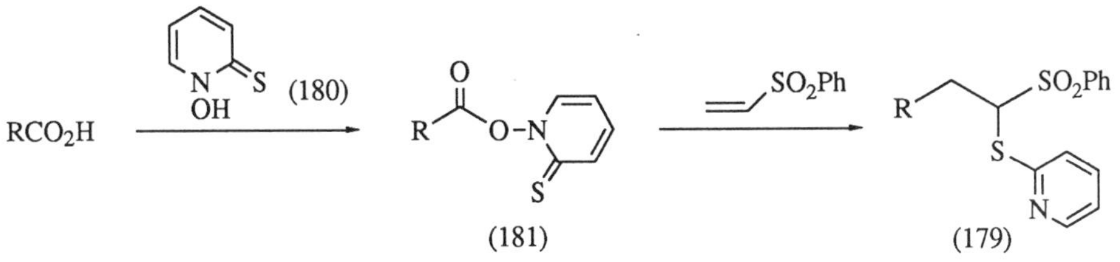

Scheme 150

This conversion involves homolysis of the N–O bond in (**181**) followed by loss of CO_2 to give a radical R·, which adds to the vinyl sulphone, giving a product radical which continues the chain reaction by attacking another molecule of (**181**). The report includes further transformations of adducts such as (**179**) along with intramolecular variants which give interesting polycyclic thiosulphones.

Some examples of reactions which furnish unsaturated sulphur-substituted sulphones are shown in Scheme 151.[313,314]

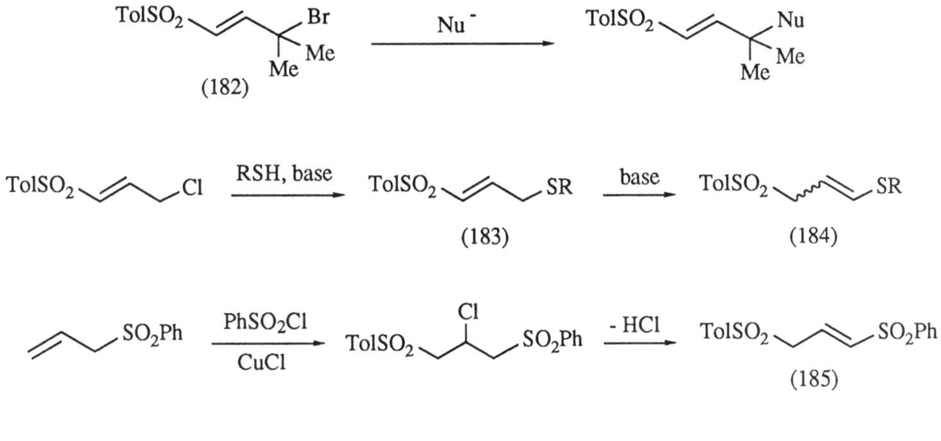

Scheme 151

Surprisingly, Bordwell and Pagani have shown that bromide (**182**) gives S_N2-type products with nucleophiles including thiolate.[313] This reaction is, however, complicated by other reaction pathways leading to different substitution products (e.g. S_N2') as well as elimination products. A later report by Ogura *et al.* is concerned with synthetic access to isomeric sulphones (**183**) and (**184**), either of which can be obtained in high yield depending on the reaction conditions.[314] Again complications can arise, for example through Michael addition to the vinyl sulphone and subsequent cyclisation to form cyclopropanes. The formation of (**185**) utilises the same chlorosulphonylation procedure described for silane substrates in Scheme 34. In fact a double sulphonylation can give (**185**) directly from allyltrimethylsilane by use of excess sulphonyl chloride.

Bis(arenesulphonyl)cyclopropanes can be prepared by making use of the carbenoid-like behaviour of phenyliodonium bis(arenesulphonyl)methylides towards alkenes and alkynes, e.g. Scheme 152.[315]

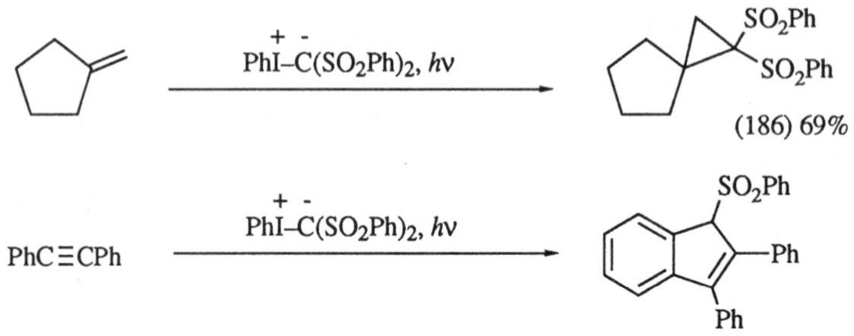

Scheme 152

Besides bis(arenesulphonyl)cyclopropanes such as (**186**), products arising by loss of SO_2 and subsequent rearrangement also result from reactions of alkenes and alkynes, e.g. (**187**). The ready availability of the iodonium reagents from bis(arenesulphonyl)methane and diacetoxyiodosylbenzene makes this an attractive route to such products. Additional examples of the reactivity of these reagents include additions to sulphur heterocycles such as thiophene[316] and iodination with *N*-iodosuccinimide to give α-iodo-bis-sulphones.[317]

Some of the procedures described in this section are undoubtedly applicable to the preparation of selenosulphones, although, notwithstanding the studies of Back and co-workers already described, these compounds have not been examined in great detail. An additional report from Back describes the reaction of a seleno-sulphonate with diazomethane, Scheme 153.[318]

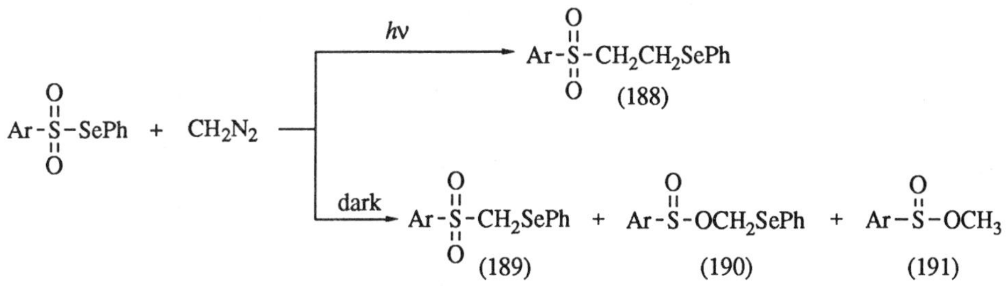

Scheme 153

Either β-selenosulphone (**188**) or α-selenosulphone (**189**) are obtained, depending on whether the reaction is carried out under photochemical conditions or in the dark. The formation of (**188**) proceeds by a radical process in which the sequence $ArSO_2 \cdot$ → $ArSO_2CH_2 \cdot$ → $ArSO_2CH_2CH_2 \cdot$ → (**188**) is brought about by two successive reactions with CH_2N_2 followed by further reaction with $ArSO_2SePh$. The reaction in the dark proceeds by ionic attack of diazomethane on $ArSO_2SePh$ to give $[ArSO_2^- \ PhSeCH_2N_2^+]$, which then reacts further by sulphinate displacement of nitrogen to give a mixture of (**189**)-(**191**).

Finally, both alkylthio- and alkylselenosulphones can be obtained by alkylation of the parent methylene compounds.[319] The alkylated derivatives, i.e. (**192**), undergo alkylative desulphonylation when treated with allyltrimethylsilane in the presence of $EtAlCl_2$, Scheme 154.

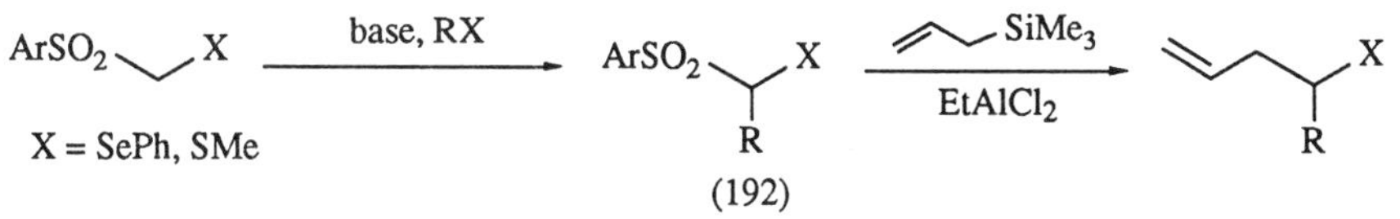

Scheme 154

2.4.7 Phosphorus- and Silicon-substituted Sulphones

Many sulphones in these categories have been discussed in the preceding sections, particularly in the context of vinyl sulphone preparation using Wittig and Peterson approaches. Simple α-silyl sulphones such as (**53**) and phosphonates such as (**52**) used in such reactions are usually prepared by oxidation of the corresponding sulphides or by reactions of sulphonyl carbanions with suitable silicon or phosphorus halides, Scheme 155.[111,119]

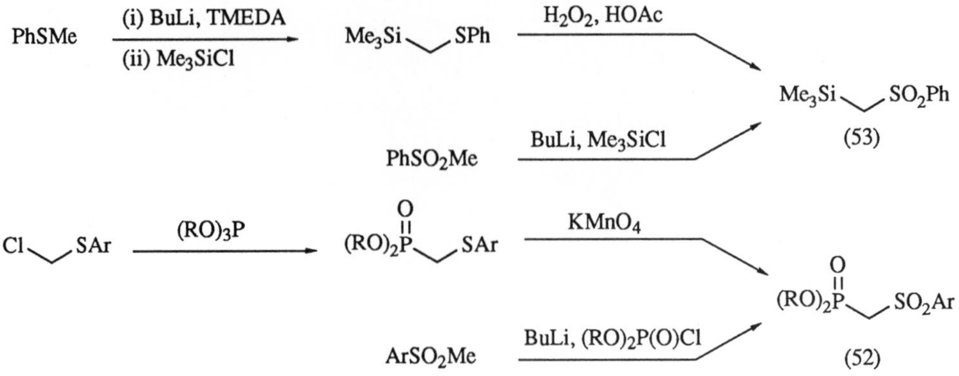

PhSMe $\xrightarrow[\text{(ii) Me}_3\text{SiCl}]{\text{(i) BuLi, TMEDA}}$ Me$_3$Si⌵SPh $\xrightarrow{\text{H}_2\text{O}_2, \text{HOAc}}$

Me$_3$Si⌵SO$_2$Ph

(53)

PhSO$_2$Me $\xrightarrow{\text{BuLi, Me}_3\text{SiCl}}$

Cl⌵SAr $\xrightarrow{\text{(RO)}_3\text{P}}$ (RO)$_2\overset{\overset{\text{O}}{\|}}{\text{P}}$⌵SAr $\xrightarrow{\text{KMnO}_4}$

(RO)$_2\overset{\overset{\text{O}}{\|}}{\text{P}}$⌵SO$_2$Ar

(52)

ArSO$_2$Me $\xrightarrow{\text{BuLi, (RO)}_2\text{P(O)Cl}}$

Scheme 155

The anion of phosphonate (**52**) and derivatives are also available directly by reaction of the dianion of methyl phenyl sulphone with (RO)$_2$P(O)Cl. [114,117]

The less well-known sulphonyl phosphoranes (**193**) can be prepared by reaction of a sulphonyl fluoride with a phosphorane, or by reaction of an α-halogeno-sulphone with triphenylphosphine, Scheme 156. [320]

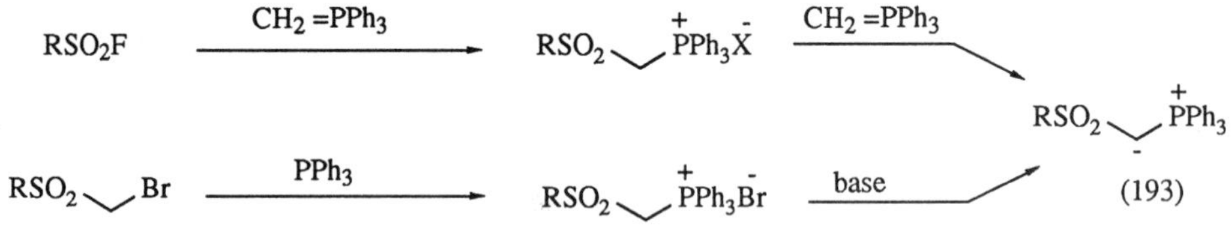

RSO$_2$F $\xrightarrow{\text{CH}_2=\text{PPh}_3}$ RSO$_2$⌵$\overset{+}{\text{PPh}_3}\overset{-}{\text{X}}$ $\xrightarrow{\text{CH}_2=\text{PPh}_3}$

RSO$_2$⌵$\overset{+}{\text{PPh}_3}$

RSO$_2$⌵Br $\xrightarrow{\text{PPh}_3}$ RSO$_2$⌵$\overset{+}{\text{PPh}_3}\overset{-}{\text{Br}}$ $\xrightarrow{\text{base}}$

(193)

Scheme 156

In some cases using the first method unexpected rearranged products were obtained, due to the intervention of sulphenes. [321]

The topic of mixed sulphur and silicon reagents has recently been reviewed and contains much relevant chemistry of silyl sulphones. [322] A variety of silicon-containing sulphones, mainly vinyl sulphones, has received mention in Section 2.2, but a final report worthy of note describes the use of α-silyl sulphones as latent α-sulphonyl anions. [323] Here, Anderson and Fuchs were able to maintain an α-silyl sulphone, formed initially by Michael addition to a vinyl sulphone, through several subsequent manipulations. Finally, the use of fluoride to regenerate the α-sulphonyl carbanion enables an intramolecular aldol-type reaction to occur, e.g. Scheme 157.

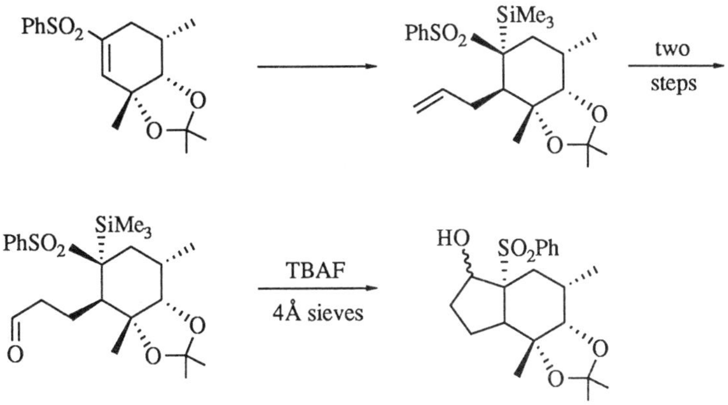

Scheme 157

α-Stannyl sulphones have also been reported, although no synthetic applications have so far appeared.[324]

Chapter 2 References

1. W. S. Trakanovsky, *Oxidation in Organic Chemistry Part C*, Academic Press, **1978**; S. Uemura, in *Comprehensive Organic Synthesis,* Ed. B. M. Trost and I. Fleming, Pergamon Press, Oxford, **1991**, 7, 757. For more general references which include accounts of sulphide oxidation, see K. Schank, in *The Chemistry of Sulphones and Sulphoxides,* Ed. S. Patai, Z. Rappoport, and C. J. M. Stirling, John Wiley and Sons, **1988**, p.165; G. Solladie, in *Comprehensive Organic Synthesis*, Ed. B. M. Trost and I. Fleming, Pergamon Press, Oxford, **1991**, *6*, 133.
2. D. W. Goheen and C. F. Bennett, *J. Org. Chem.,* **1961**, *26*, 1331.
3. C-N. Hsiao and H. Shechter, *Tetrahedron Lett.,* **1982**, *23,* 1963; L.A. Paquette and R.V.C. Carr, *Organic Syntheses*, **1986**, *64*, 157; T. Balaji and D. B. Reddy, *Bull. Chem. Soc. Jpn.,* **1979**, *52*, 3434.
4. H. S. Schultz, H. B. Freyermuth, and S. R. Buc, *J. Org. Chem.,* **1963**, *28*, 1140; N.W. Connon, *Org. Chem. Bull.,* **1972**, *44*, 1.
5. M. S. Cooper, H. Heaney, A. J. Newbold, and W. R. Sanderson, *Synlett,* **1990**, 533.
6. P. Kocienski and M. Todd, *J. Chem. Soc., Chem. Commun.,* **1982**, 1078.
7. N. A. Sasaki, C. Hashimoto, and P. Potier, *Tetrahedron Lett.,* **1987**, *28*, 6069.
8. D. A. Evans, R. L. Dow, T. L. Shih, J. M. Takacs, and R. Zahler, *J. Am. Chem. Soc.,* **1990**, *112*, 5290.
9. D. V. Patel, F. VanMiddlesworth, J. Donaubauer, P. Gannett, and C. J. Sih, *J. Am. Chem. Soc.,* **1986**, *108*, 4603.
10. F. Matsuda, N. Tomiyosi, N. Yanagiya, and T. Matsumoto, *Chem. Lett.,* **1987**, 2097.
11. V. Cere, C. Paolucci, S. Pollicino, E. Sandri, and A. Fava, *J. Org. Chem.,* **1988**, *53*, 5689.
12. M. Sander, *Chem. Rev.,* **1966**, 341; K. Nagasawa and A. Yoneta, *Chem. Pharm. Bull.,* **1985**, *33*, 5048.
13. C. G. Venier, T. G. Squires, Y-Y. Chen, G. P. Hussmann, J. C. Shei, and B. F. Smith, *J. Org. Chem.,* **1982**, *47*, 3773.
14. P. Brougham, M. S. Cooper, D. A. Cummerson, H. Heaney, and N. Thompson, *Synthesis,* **1987**, 1015.
15. B. M. Trost and M. Acemoglu, *Tetrahedron Lett.,* **1989**, *30*, 1495.
16. B. M. Trost and D. P. Curran, *Tetrahedron Lett.,* **1981**, *22*, 1287; B. M. Trost and R. Braslau, *J. Org. Chem.,* **1988**, *53*, 532. See also R. Bloch, J. Abecassis, and D. Hassan, *J. Org. Chem.,* **1985**, *50,* 1544.
17. B. M. Trost, P. Seoane, S. Mignani, and M. Acemoglu, *J. Am. Chem. Soc.,* **1989**, *111*, 7487.
18. A. G. M. Barrett, R. A. E. Carr, S. V. Attwood, G. Richardson, and N. D. A. Walshe, *J. Org. Chem.,* **1986**, *51*, 4840.

19. S. E. de Laszlo, M. J. Ford, S. V. Ley, and G. N. Maw, *Tetrahedron Lett.,* **1990,** *31*, 5525.
20. R. Baker, M. J. O'Mahony, and C. J. Swain, *J. Chem. Soc., Perkin Trans. 1,* **1987,** 1623.
21. A. McKillop and J. A. Tarbin, *Tetrahedron Lett.,* **1983,** *24*, 1505.
22. A. Fernandez-Mayoralas, A. Marra, M. Trumtel, A. Veyrieres, and P. Sinay, *Tetrahedron Lett.,* **1989,** *30*, 2537.
23. D. A. Evans, S. W. Kaldor, T. K. Jones, J. Clardy, and T. J. Stout, *J. Am. Chem. Soc.,* **1990,** *112*, 7001.
24. H. J. Reich, F. Chow, and S. L. Peake, *Synthesis,* **1978,** 299.
25. K. C. Nicolaou, R. L. Magolda, W. L. Sipio, W. E. Barnette, Z. Lysenko, and M. M. Joullie, *J. Am. Chem. Soc.,* **1980,** *102*, 3784.
26. M. P. Edwards, S. V. Ley, S. G. Lister, B. D. Palmer, and D. J. Williams, *J. Org. Chem.,* **1984,** *49*, 3503.
27. N. A. Noureldin, W. B. McConnell, and D. G. Lee, *Can. J. Chem.,* **1984,** *62*, 2113; F. M. Menger and C. Lee, *Tetrahedron Lett.,* **1981,** *22*, 1655; S. T. Purrington and A. G. Glenn, *OPPI Briefs*, **1985,** *17*, 227.
28. G. W. Gokel, H. M. Gerdes, and D. M. Dishong, *J. Org. Chem.,* **1980,** *45*, 3634; D. G. Lee and N. S. Srinivasan, *Sulphur Lett.,* **1982,** *1*, 1.
29. M. Poje and K. Balenovic, *Tetrahedron Lett.,* **1978,** 1231; see also T. P. Ahern, H. O. Fong, R. F. Langler, and P. M. Mason, *Can. J. Chem.,* **1980,** *58*, 878.
30. Y. H. Kim and H. K. Lee, *Chem. Lett.,* **1987,** 1499; H. K. Lee and Y. H. Kim, *Sulphur Lett.,* **1987,** *7*, 1; M. Miura, M. Nojima, and S. Kusabayashi, *J. Chem. Soc., Chem. Commun.,* **1982,** 1352.
31. G. A. Olah and B. G. B. Gupta, *J. Org. Chem.,* **1983,** *48*, 3585.
32. P. Müller and J. Godoy, *Helv. Chim. Acta,* **1983,** *66*, 1790; see also A. R. Humffray and H. E. Imberger, *J. Chem. Soc., Perkin Trans. 2,* **1981,** 382.
33. D. Barbas, S. Spyroudis, and A. Varvoglis, *J. Chem. Res. (S),* **1985,** 186.
34. P. S. Bailey, *Ozonation in Organic Chemistry, Vol II,* Academic Press, **1982,** p. 207.
35. H. Ohta, Y. Okamoto, and G. Tsuchihashi, *Agric. Biol. Chem.,* **1985,** *49*, 2229; H. Ohta, Y. Okamoto, and G. Tsuchihashi, *Chem. Lett.,* **1984,** 205.
36. Y. Tamaru, M. Kagotani, and Z. Yoshida, *J. Chem. Soc., Chem. Commun.,* **1978,** 367; see also Y. Tamaru and Z. Yoshida, *J. Org. Chem.,* **1979,** *44*, 1188.
37. W. E. Truce and A. M. Murphy, *Chem. Rev.,* **1951,** *48*, 69.
38. W. G. Filby, K. Günther, and R. D. Penzhorn, *J. Org. Chem.,* **1973,** *38*, 4070.
39. C. Lee and L. Field, *Synthesis,* **1990,** 391.
40. Y. Ueno, A. Kojima, and M. Okawara, *Chem. Lett.,* **1984,** 2125.
41. M. Uchino, K. Suzuki, and M. Sekiya, *Synthesis,* **1977,** 794; M. Uchino, K. Suzuki, and M. Sekiya, *Chem. Pharm. Bull.,* **1978,** *26*, 1837. See also J. M. Blanco, O. Caamano, A. Eirin, F. Fernandez, and L. Medina, *Synthesis,* **1990,** 584.

42. J. S. Meek and J. S. Fowler, *J. Org. Chem.,* **1968**, *33*, 3422; see also P. Kielbasinski, R. Zurawinski, J. Drabowicz, and M. Mikolajczyk, *Tetrahedron,* **1988,** *44,* 6687.
43. L. Field and R. D. Clark, *Organic Syntheses, Coll.Vol. IV,* **1963,** 674.
44. K. Sukata, *Bull. Chem. Soc. Jpn.,* **1984,** *57,* 613.
45. F. Manescalchi, M. Orena, and D. Savoia, *Synthesis,* **1979,** 445.
46. G. E. Vennstra and B. Zwaneburg, *Synthesis,* **1975,** 519.
47. J. Wildeman and A. M. van Leusen, *Synthesis,* **1979,** 733.
48. J. K. Crandall and C. Pradat, *J. Org. Chem.,* **1985,** *50,* 1327.
49. G. Biswas and D. Mal, *J. Chem. Res. (S),* **1988,** 308.
50. G. K. Biswas, S. S. Jash, and P. Bhattacharyya, *Ind. J. Chem.,* **1990,** *29B,* 491.
51. A. R. Harris, *Synth. Commun.,* **1988,** *18,* 659; see also K. Matsuo, M. Kobayashi, and H. Minato, *Bull. Chem. Soc. Jpn.,* **1970,** *43,* 260. For an electrochemical approach see D. Knittel and B. Kastening, *J. App. Electrochem.,* **1973,** *3,* 291.
52. P. Messinger and H. Greve, *Synthesis,* **1977,** 259.
53. U. C. Dyer and J. A. Robinson, *J. Chem. Soc., Perkin Trans. 1,* **1988,** 53; D. Tanner and P. Somfai, *Tetrahedron,* **1987,** *43,* 4395; J. D. White and G. L. Bolton, *J. Am. Chem. Soc.,* **1990,** *112,* 1626.
54. J. B. Hendrickson, D. D. Sternbach, and K. W. Bair, *Acc. Chem. Res.,* **1977,** *10,* 306; J. B. Hendrickson, D. A. Judelson, and T. Chancellor, *Synthesis,* **1984,** 320.
55. D. Arnould, P. Chabardes, G. Farge, and M. Julia, *Bull. Chem. Soc. Fr.,* **1985,** 130.
56. F. Effenberger, *Angew. Chem., Int. Ed. Engl.,* **1980,** *19,* 151; F. R. Jensen and G. Goldman, in *Friedel–Crafts and Related Reactions,* Ed. G. Olah, Wiley-Interscience, New York, **1964.**
57. J. Boeseken and H. VanOckenburg, *Recl. Trav. Chim. Pays-Bas,* **1914,** *33,* 320; W. Truce and C. Vriesen, *J. Am. Chem. Soc.,* **1955,** *75,* 5032.
58. K. Huthmacher, G. Konig, and F. Effenberger, *Chem. Ber.,* **1975,** *108,* 2947.
59. J. A. Hyatt and A. W. White, *Synthesis,* **1984,** 214.
60. C. Chen, F. Zhu, and Y-Z. Huang, *J. Chem. Res. (S),* **1989,** 381.
61. H. Suzuki, Y. Nishioka, S. I. Padmanabhan, and T. Ogawa, *Chem. Lett.,* **1988,** 727; see also X. Huang and J-H. Pi, *Synth. Commun.,* **1990,** *20,* 2291.
62. Y. Shirota, T. Nagai, and N. Tokura, *Tetrahedron,* **1967,** *23,* 639.
63. H. Fukuda, F. J. Frank, and W. E. Truce, *J. Org. Chem.,* **1963,** *28,* 1420; Y. Shirota, T. Nagai and N. Tokura, *Bull. Chem. Soc. Jpn.,* **1966,** *39,* 405.
64. Y. Shirota, T. Nagai, and N. Tokura, *Tetrahedron,* **1969,** *25,* 3193.
65. W. H. Baarschers, *Can. J. Chem.,* **1976,** *54,* 3056.
66. J. B. Hendrickson and K. W. Bair, *J. Org. Chem.,* **1977,** *42,* 3875; see also X. Creary, *J. Org. Chem.,* **1980,** *45,* 2727.

67. C. M. M. da Silva Correa, A. S. Lindsay, and W. A. Waters, *J. Chem. Soc. (C)*, **1968**, 1872.

68. H. Takeuchi, T. Nagai, and N. Tokura, *Bull. Chem. Soc. Jpn.*, **1973**, *46*, 695.

69. R. M. Wilson and S. W. Wunderly, *J. Am. Chem. Soc.*, **1974**, *96*, 7350.

70. V. F. Patel and G. Pattenden, *Tetrahedron Lett.*, **1987**, *28*, 1451; V. F. Patel and G. Pattenden, *J. Chem. Soc., Perkin Trans. 1*, **1990**, 2703.

71. D. H. R. Barton, B. Lacher, B. Misterkiewicz, and S. Z. Zard, *Tetrahedron*, **1988**, *44*, 1153.

72. A. R. Katritzky, A. Saba, and R. C. Patel, *J. Chem. Soc., Perkin Trans. 1*, **1981**, 1492.

73. R. A. Abramovitch and J. Roy, *J. Chem. Soc., Chem. Commun.*, **1965**, 542.

74. R. Herrmann, G. Hubener, and I. Ugi, *Tetrahedron*, **1985**, *41*, 941.

75. F. Bellesia, R. Grandi, U. M. Pagnoni, and R. Trave, *J. Chem. Res. (S)*, **1981**, 112.

76. K. Okuma, K. Nakanishi, and H. Ohta, *J. Org. Chem.*, **1984**, *49*, 1402.

77. N. S. Simpkins, *Tetrahedron*, **1990**, *46*, 6951; H. Zhang and X. Huang, *Huaxue Tongbao*, **1989**, *9*, 7.

78. See for example: a) K. Schank, in *Methoden der Organischen Chemie (Houben-Weyl)*, Vol. E11, 4th Edn, Ed. D. Klamann, Thieme, Stuttgart, **1985**, p. 1129; b) K. Schank, in *The Chemistry of Sulphones and Sulphoxides*, Ed. S. Patai, Z. Rappoport and C. Stirling, John Wiley and Sons Ltd, **1988**, p. 165.

79. P. B. Hopkins and P. L. Fuchs, *J. Org. Chem.*, **1978**, *43*, 1208; see also W. Böll, *Liebigs Ann. Chem.*, **1979**, 1665.

80. J. C. Carretero, J. L. Garcia Ruano, M. C. Martinez, and J. H. Rodriguez, *Tetrahedron*, **1987**, *43*, 4417; D. R. Marshall, P. J. Thomas, and C. J. M. Stirling, *J. Chem. Soc., Perkin Trans. 2*, **1977**, 1914; K. N. Barlow, D. R. Marshall, and C. J. M. Stirling, *J. Chem. Soc., Perkin Trans. 2*, **1977**, 1920.

81. W. Sas, *J. Chem. Soc., Chem. Commun.*, **1984**, 862.

82. P. Rajakumar and A. Kannan, *J. Chem. Soc., Chem. Commun.*, **1989**, 154.

83. J. C. Philips, M. Aregullin, M. Oku, and A. Sierra, *Tetrahedron Lett.*, **1974**, 4157; J. C. Philips and L. C. Hernandez, *Tetrahedron Lett.*, **1977**, 4461; see also M. S. R. Naidu and R. Prabhakara, *Ind. J. Chem.*, **1984**, *23B*, 140.

84. T. G. Back and S. Collins, *J. Org. Chem.*, **1981**, *46*, 3249.

85. Y-H. Kang and J. L. Kice, *J. Org. Chem.*, **1984**, *49*, 1507; R. A. Gancarz and J. L. Kice, *J. Org. Chem.*, **1981**, *46*, 4899.

86. L. K. Liu, Y. Chi, and K-Y. Jen, *J. Org. Chem.*, **1980**, *45*, 406.

87. K. Inomata, T. Kobayashi, S. Sasaoka, H. Kinoshita, and H. Kotake, *Chem. Lett.*, **1986**, 289; K. Inomata, S. Sasaoka, T. Kobayashi, Y. Tanaka, S. Igarashi, T. Ohtani, H. Kinoshita, and H. Kotake, *Bull. Chem. Soc. Jpn.*, **1987**, *60*, 1767.

88. C. Najera, B. Baldo, and M. Yus, *J. Chem. Soc., Perkin Trans. 1*, **1988**, 1029.

89. T. G. Back, S. Collins, and R. G. Kerr, *J. Org. Chem.*, **1983**, *48*, 3077.

90. T. Kobayashi, Y. Tanaka, T. Ohtani, H. Kinoshita, K. Inomata, and H. Kotake, *Chem. Lett.*, **1987**, 1209; W. E. Truce and G. C. Wolf, *J. Org. Chem.*, **1971**, *36*, 1727; see also M. Ozawa, N. Iwata, H. Kinoshita, and K. Inomata, *Chem. Lett.*, **1990**, 1689.

91. T. G. Back and K. R. Muralidharan, *J. Org. Chem.*, **1989**, *54*, 121; see also M. C. M. de C. Alpoim, A. D. Morris, W. B. Motherwell, and D. M. O'Shea, *Tetrahedron Lett.*, **1988**, *29*, 4173.

92. W. E. Truce, D. L. Heuring, and G. C. Wolf, *J. Org. Chem.*, **1974**, *39*, 238.

93. P. N. Culshaw and J. C. Walton, *Tetrahedron Lett.*, **1990**, *31*, 2457.

94. O. S. Andell and J-E. Bäckvall, *Tetrahedron Lett.*, **1985**, *26*, 4555.

95. J-E. Bäckvall, C. Najera, and M. Yus, *Tetrahedron Lett.*, **1988**, *29*, 1445; see also T. G. Back, E. K. Y. Lai, and K. R. Muralidharan, *Tetrahedron Lett.*, **1989**, *30*, 6481.

96. J. Barluenga, J. M. Martinez-Gallo, C. Najera, F. J. Fananas, and M. Yus, *J. Chem. Soc., Perkin Trans. 1*, **1987**, 2605.

97. J-P. Pillot, J. Dunogues, and R. Calas, *Synthesis*, **1977**, 469.

98. N. Kamigata, H. Sawada, and M. Kobayashi, *Chem. Lett.*, **1979**, 159.

99. N. Kamigata, J. Ozaki, and M. Kobayashi, *J. Org. Chem.*, **1985**, *50*, 5045; M. Kameyama, H. Shimezawa, T. Satoh, and N. Kamigata, *Bull. Chem. Soc. Jpn.*, **1988**, *61*, 1231.

100. S. S. Labadie, *J. Org. Chem.*, **1989**, *54*, 2496.

101. A. Bongini, D. Savoia, and A. Umani-Ronchi, *J. Organomet. Chem.*, **1976**, *112*, 1; V. Pascali, N. Tangari, and A. Umani-Ronchi, *J. Chem. Soc., Perkin Trans. 1*, **1973**, 1166.

102. D. A. R. Happer and B. E. Steenson, *Synthesis*, **1980**, 806; R. Tanikaga, T. Tamura, Y. Nozaki, and A. Kaji, *J. Chem. Soc., Chem. Commun.*, **1984**, 87; H. Dressler and J. E. Graham, *J. Org. Chem.*, **1967**, *32*, 985; M. V. R. Reddy, S. Reddy, D. B. Reddy, and V. Padmavathi, *Synth. Commun.*, **1989**, *19*, 1101; M. V. R. Reddy, S. Balasubramanyam, D. B. Reddy, S. Reddy, and B. Seenaiah, *Sulphur Lett.*, **1988**, *8*, 237.

103. M. Julia, M. Launay, J-P. Stacino, and J-N. Verpeaux, **1982**, *23*, 2465.

104. T. Cuvigny, C. Herve du Penhoat, and M. Julia, *Tetrahedron*, **1986**, *42*, 5329.

105. M. Hirama, H. Hioki, and S. Ito, *Tetrahedron Lett.*, **1988**, *29*, 3125.

106. C. T. Hewkin, R. F. W. Jackson, and W. Clegg, *Tetrahedron Lett.*, **1988**, *29*, 4889.

107. E. Dominguez and J. C. Carretero, *Tetrahedron Lett.*, **1990**, *31*, 2487; E. Dominguez and J. C. Carretero, *Tetrahedron*, **1990**, *46*, 7197; B. M. Trost and T. A. Grese, *J. Org. Chem.*, **1991**, *56*, 3189.

108. M. Inbasekaran, N. P. Peet, J. R. McCarthy, and M. E. LeTourneau, *J. Chem. Soc., Chem. Commun.*, **1985**, 678.

109. J. B. Hendrickson and P. S. Palumbo, *Tetrahedron Lett.*, **1985**, *26*, 2849.

110. Y. Ueno, H. Setoi, and M. Okawara, *Chem. Lett.*, **1979**, 47; see also G. Ferdinand and K. Schank, *Synthesis*, **1976**, 404; G. Ferdinand, K. Schank, and A. Weber, *Liebigs Ann. Chem.*, **1975**, 1484.

111. G. H. Posner and D. J. Brunelle, *J. Org. Chem.*, **1972**, *37*, 3547.

112. M. Mikolajczyk, S. Grzejszczak, W. Midura, and A. Zatorski, *Synthesis*, **1975**, 278.

113. P. O. Ellingsen and K. Undheim, *Acta Chem. Scand.*, **1979**, *B33*, 528.

114. J. W. Lee and D. Y. Oh, *Synth. Commun.*, **1989**, *19*, 2209.

115. B. E. de Jong, H. de Koning, and H. O. Huisman, *Recl. Trav. Chim. Pays-Bas*, **1981**, *100*, 410.

116. J. R. McCarthy, D. P. Matthews, M. L. Edwards, D. M. Stemerick, and E. T. Jarvi, *Tetrahedron Lett.*, **1990**, *31*, 5449.

117. J. W. Lee and D. Y. Oh, *Synth. Commun.*, **1990**, *20*, 273.

118. A. H. Davidson, L. R. Hughes, S. S. Qureshi, and B. Wright, *Tetrahedron Lett.*, **1988**, *29*, 693.

119. D. Craig, S. V. Ley, N. S. Simpkins, G. H. Whitham, and M. J. Prior, *J. Chem. Soc., Perkin Trans. 1*, **1985**, 1949; D. J. Ager, *J. Chem. Soc., Perkin Trans. 1*, **1986**, 183; K. Schank and F. Schroeder, *Liebigs Ann. Chem.*, **1977**, 1676; see also C. Palomo, J. M. Aizpurua, J. M. Garcia, I. Ganboa, F. P. Cossio, B. Lecea, and C. Lopez, *J. Org. Chem.*, **1990**, *55*, 2498.

120. J. Vollhardt, H-J. Gais, and K. L. Lukas, *Angew. Chem., Int. Ed. Engl.*, **1985**, *24*, 696.

121. J. J. Eisch and J. E. Galle, *J. Org. Chem.*, **1979**, *44*, 3279.

122. J. J. Eisch and J. E. Galle, *J. Org. Chem.*, **1980**, *45*, 4534.

123. C-N. Hsiao and H. Shechter, *J. Org. Chem.*, **1988**, *53*, 2688.

124. K. Inomata, Y. Tanaka, S. Sasaoka, H. Kinoshita, and H. Kotake, *Chem. Lett.*, **1986**, 341.

125. P. Auvray, P. Knochel, and J. F. Normant, *Tetrahedron*, **1988**, *44*, 4509 and 6095.

126. P. Auvray, P. Knochel, and J. F. Normant, *Tetrahedron Lett.*, **1985**, *26*, 2329.

127. J. T. Palmer and P. L. Fuchs, *Synth. Commun.*, **1988**, 233.

128. G. E. Keck, J. H. Byers, and A. M. Tafesh, *J. Org. Chem.*, **1988**, *53*, 1127; see also V. Farina and S. I. Hauck, *J. Org. Chem.*, **1991**, *56*, 4317.

129. T. G. Back, S. Collins, M. V. Krishna, and K-W. Law, *J. Org. Chem.*, **1987**, *52*, 4258; see also T. G. Back, S. Collins, and K-W. Law, *Can. J. Chem.*, **1985**, *63*, 2313.

130. G. M. P. Giblin and N. S. Simpkins, *J. Chem. Soc., Chem. Commun.*, **1987**, 207.

131. R. V. C. Carr, R. V. Williams, and L. A. Paquette, *J. Org. Chem.*, **1983**, *48*, 4976.

132. I. Ryu, N. Kusumoto, A. Ogawa, N. Kambe, and N. Sonoda, *Organometallics*, **1989**, *8*, 2279.

133. J. J. Eisch, M. Behrooz, and S. K. Dua, *J. Organomet. Chem.*, **1985**, *285*, 121.

134. V. Fiandanese, G. Marchese, and F. Naso, *Tetrahedron Lett.*, **1978**, 5131.

135. T. Ohnuma, N. Hata, H. Fujiwara, and Y. Ban, *J. Org. Chem.*, **1982**, *47*, 4713.
136. O. De Lucchi, G. Licini, L. Pasquato, and M. Senta, *Tetrahedron Lett.*, **1988**, *29*, 831.
137. R. V. Williams and C-L. A. Sung, *J. Chem. Soc., Chem. Commun.*, **1987**, 590.
138. B. B. Snider, T. C. Kirk, D. M. Roush, and D. Gonzalez, *J. Org. Chem.*, **1980**, *45*, 5015.
139. W. E. Truce and L. D. Markley, *J. Org. Chem.*, **1970**, *35*, 3275.
140. C. J. M. Stirling, *J. Chem. Soc.*, **1964**, 5856.
141. K. Ogura, N. Shibuya, K. Takahashi, and H. Iida, *Bull. Chem. Soc. Jpn.*, **1984**, *57*, 1092; see also J. Muzart, A. Riahi, and J. P. Pete, *J. Organomet. Chem.*, **1985**, *280*, 269.
142. J. J. Eisch and J. E. Galle, *J. Org. Chem.*, **1979**, *44*, 3277; see also C. C. J. Culvenor, W. Davies, and W. E. Savige, *J. Chem. Soc.*, **1949**, 2198.
143. M. R. Binns, R. K. Haynes, A. G. Katsifis, P. A. Schober, and S. C. Vonwiller, *J. Org. Chem.*, **1989**, *54*, 1960.
144. J-B. Baudin, M. Julia, C. Rolando, and J-N. Verpeaux, *Tetrahedron Lett.*, **1984**, *25*, 3203; M. Julia, H. Lauron, J-N. Verpeaux, Y. Jeannin, and C. Bois, *J. Organomet. Chem.*, **1988**, *358*, C11.
145. J. Hershberger and G. A. Russell, *Synthesis*, **1980**, 475.
146. J. F. Cassidy and J. M. Williams, *Tetrahedron Lett.*, **1986**, *27*, 4355.
147. D. E. O'Connor and W. I. Lyness, *J. Am. Chem. Soc.*, **1964**, *86*, 3840; J. Hine and M. J. Skoglund, *J. Org. Chem.*, **1982**, *47*, 4766; V. Svata, M. Prochazka, and V. Bakos, *Collect. Czech. Chem. Commun.*, **1978**, *43*, 2619.
148. K. Inomata, T. Hirata, H. Suhara, H. Kinoshita, H. Kotake, and H. Senda, *Chem. Lett.*, **1988**, 2009, and references therein.
149. D. Savoia, C. Trombini, and A. Umani-Ronchi, *J. Chem. Soc., Perkin Trans. 1*, **1977**, 123.
150. I. Sataty and C. Y. Meyers, *Tetrahedron Lett.*, **1974**, 4161.
151. S. V. Ley, N. J. Anthony, A. Armstrong, M. G. Brasca, T. Clarke, D. Culshaw, C. Greck, P. Grice, A. B. Jones, B. Lygo, A. Madin, R. N. Sheppard, A. M. Z. Slawin, and D. J. Williams, *Tetrahedron*, **1989**, *45*, 7161; see also references 87 and 148.
152. T. Cuvigny, C. Herve du Penhoat, and M. Julia, *Recl. Trav. Chim. Pays-Bas*, **1986**, *105*, 409.
153. K. Inomata, T. Yamamoto, and H. Kotake, *Chem. Lett.*, **1981**, 1357.
154. M. Julia, M. Nel, A. Righini, and D. Uguen, *J. Organomet. Chem.*, **1982**, *235*, 113.
155. R. Tamura, K. Hayashi, M. Kakihana, M. Tsuji, and D. Oda, *Tetrahedron Lett.*, **1985**, *26*, 851; N. Ono, I. Hamamoto, T. Yanai, and A. Kaji, *J. Chem. Soc., Chem. Commun.*, **1985**, 523.
156. N. Ono, I. Hamamoto, T. Kawai, A. Kaji, R. Tamura, and M. Kakihana, *Bull. Chem. Soc., Jpn.*, **1986**, *59*, 405.

157. R. Tamura, K. Hayashi, M. Kakihana, M. Tsuji, and D. Oda, *Chem. Lett.*, **1985**, 229.
158. N. Ono, I. Hamamoto, A. Kamimura, and A. Kaji, *J. Org. Chem.*, **1986**, *51*, 3734.
159. G. P. Boldrini, D. Savoia, E. Tagliavini, C. Trombini, and A. Umani-Ronchi, *J. Organomet. Chem.*, **1984**, *268*, 97.
160. A. H. Wragg, J. S. McFadyen, and T. S. Stevens, *J. Chem. Soc.*, **1958**, 3603.
161. K. Hiroi, R. Kitayama, and S. Sato, *Chem. Pharm. Bull.*, **1984**, *32*, 2628.
162. D. J. Knight, P. Lin, S. T. Russell, and G. H. Whitham, *J. Chem. Soc., Perkin Trans. 1*, **1987**, 2701; D. J. Knight, G. H. Whitham, and J. G. Williams, *J. Chem. Soc., Perkin Trans. 1*, **1987**, 2149.
163. S. Braverman and T. Globerman, *Tetrahedron*, **1974**, *30*, 3873.
164. M. Ohmori, Y. Takano, S. Yamada, and H. Takayama, *Tetrahedron Lett.*, **1986**, *27*, 71.
165. A. E. Crease, B. D. Gupta, M. D. Johnson, E. Bialkowska, K. N. V. Duong, and A. Gaudemer, *J. Chem. Soc., Perkin Trans. 1*, **1979**, 2611.
166. B. D. Gupta, S. Roy, and S. Sen, *Ind. J. Chem.*, **1985**, *24B*, 1032.
167. B. D. Gupta and S. Roy, *Tetrahedron Lett.*, **1986**, *27*, 4905; B. D. Gupta and S. Roy, *J. Chem. Soc., Perkin Trans. 2*, **1988**, 1377.
168. M. R. Ashcroft, R. Bougeard, A. Bury, C. J. Cooksey, and M. D. Johnson, *J. Organomet. Chem.*, **1985**, *289*, 403.
169. T. G. Back, M. V. Krishna, and K. R. Muralidharan, *J. Org. Chem*, **1989**, *54*, 4146; T. Miura and M. Kobayashi, *J. Chem. Soc., Chem. Commun.*, **1982**, 438.
170. T. G. Back, E. K. Y. Lai, and K. R. Muralidharan, *J. Org. Chem.*, **1990**, *55*, 4595.
171. T. G. Back and M. V. Krishna, *J. Org. Chem.*, **1987**, *52*, 4265.
172. W. Verboom, M. Schoufs, J. Meijer, H. D. Verkruijsse, and L. Brandsma, *Recl. Trav. Chim. Pays-Bas*, **1978**, *97*, 244.
173. J. W. Lee, T. H. Kim, and D. Y. Oh, *Synth. Commun.*, **1989**, *19*, 2633.
174. H. Fillion, A. Hseine, M-H. Pera, V. Dufaud, and B. Refouvelet, *Synthesis*, **1987**, 708.
175. H. Fillion, B. Refouvelet, M. H. Pera, V. Dufaud, and J. L. Luche, *Synth. Commun.*, **1989**, *19*, 3343.
176. M. Hanack, B. Wilhelm, and L. R. Subramanian, *Synthesis*, **1988**, 592.
177. J. W. Lee and D. Y. Oh, *Synlett*, **1990**, 290.
178. M. Julia, M. Nel, and L. Saussine, *J. Organomet. Chem.*, **1979**, *181*, C17.
179. Y. Tamura and Z. Yoshida, *Tetrahedron Lett.*, **1978**, 4527; see also reference 36.
180. T. Kanai, Y. Kanagawa, and Y. Ishii, *J. Org. Chem.*, **1990**, *55*, 3274.
181. M. Foa and M. T. Venturi, *Gazz. Chim. Ital.*, **1975**, *105*, 1199.
182. S. N. Bhattacharya, C. Eaborn, and D. R. M. Walton, *J. Org. Chem. (C)*, **1969**, 1367.

183. T. E. Tyobeka, R. A. Hancock, and H. Weigel, *J. Chem. Soc., Chem. Commun.,* **1980**, 114.
184. R. A. Hancock, T. E. Tyobeka, and H. Weigel, *J. Chem. Res. (S),* **1980**, 270.
185. B. M. Graybill, *J. Org. Chem.,* **1967**, *32*, 2931.
186. H. J. Sipe, Jr., D. W. Clary, and S. B. White, *Synthesis,* **1984**, 283.
187. K. Nomiya, Y. Sugaya, and M. Miwa, *Bull. Chem. Soc. Jpn.,* **1980**, *53*, 3389.
188. R. A. Hancock and S. T. Orszulik, *Tetrahedron Lett.,* **1979**, 3789.
189. C. Najera and M. Yus, *J. Org. Chem.,* **1989**, *54*, 1491.
190. C. Najera and J. M. Sansano, *Tetrahedron,* **1990**, *46*, 3993.
191. R. Tanikaga, K. Hosoya, and A. Kaji, *J. Chem. Soc., Chem. Commun.,* **1986**, 836.
192. I. W. J. Still and F. J. Ablenas, *Synth. Commun.,* **1982**, *12*, 1103.
193. J. S. Grossert, H. R. W. Dharmaratne, T. S. Cameron, and B. R. Vincent, *Can. J. Chem.,* **1988**, *66*, 2860.
194. R. L. Crumbie, B. S. Deol, J. E. Nemorin, and D. D. Ridley, *Aust. J. Chem.,* **1978**, *31*, 1965.
195. A. S. Gopalan and H. K. Jacobs, *Tetrahedron Lett.,* **1990**, *31*, 5575. For an alternative approach see P. Ferraboschi, P. Grisenti, A. Manzocchi, and E. Santaniello, *J. Org. Chem.,* **1990**, *55*, 6214.
196. Y. Hiraki, K. Ito, T. Harada, and A. Tai, *Chem. Lett.,* **1981**, 131.
197. E. J. Corey and R. K. Bakshi, *Tetrahedron Lett.,* **1990**, *31*, 611.
198. R. Chinchilla, C. Najera, J. Pardo, and M. Yus, *Tetrahedron:Asymmetry,* **1990**, *1*, 575.
199. D. A. Singleton, K. M. Church, and M. J. Lucero, *Tetrahedron Lett.,* **1990**, *31*, 5551; D. A. Singleton and K. M. Church, *J. Org. Chem.,* **1990**, *55*, 4780.
200. K. Schank, in *Methoden der Organischen Chemie (Houben-Weyl),* Vol. E11, 4th Edn, Ed. D. Klamann, Thieme, Stuttgart, **1985**, p. 1129.
201. K. Schank and H-G. Schmitt, *Chem. Ber.,* **1978**, *111*, 3497; see also R. J. Mulder, A. M. van Leusen, and J. Strating, *Tetrahedron Lett.,* **1967**, 3057; R. J. Mulder, A. M. van Leusen, and J. Strating, *Tetrahedron Lett.,* **1967**, 3061.
202. K. Schank, H-G. Schmitt, F. Schroeder, and A. Weber, *Liebigs Ann. Chem.,* **1977**, 1116.
203. S. V. Ley, B. Lygo, and A. Wonnacott, *Tetrahedron Lett.,* **1985**, *26*, 535.
204. B. M. Trost and P. Quayle, *J. Am. Chem. Soc.,* **1984**, *106*, 2469.
205. B. Zwanenburg and J. ter Wiel, *Tetrahedron Lett.,* **1970**, 935.
206. R. Curci and F. DiFuria, *Tetrahedron Lett.,* **1974**, 4085.
207. C. Clark, P. Hermans, O. Meth-Cohn, C. Moore, H. C. Taljaard, and G. van Vuuren, *J. Chem. Soc., Chem. Commun.,* **1986**, 1378; O. Meth-Cohn, C. Moore, and H. C. Taljaard, *J. Chem. Soc., Perkin Trans. 1,* **1988**, 2663.
208. M. Ashwell and R. F. W. Jackson, *J. Chem. Soc., Perkin Trans. 1,* **1989**, 835.

209. M. Ashwell and R. F. W. Jackson, *J. Chem. Soc., Chem. Commun.,* **1988**, 645; M. Ashwell, W. Clegg, and R. F. W. Jackson, *J. Chem. Soc., Perkin Trans. 1,* **1991**, 897; C. T. Hewkin, R. F. W. Jackson, and W. Clegg, *J. Chem. Soc., Perkin Trans. 1,* **1991**, 3091; C. T. Hewkin and R. F. W. Jackson, *J. Chem. Soc., Perkin Trans. 1,* **1991**, 3103. Two earlier reports deal with the lithiation of a range of α-heterosubstituted epoxides, including epoxy sulphones, see J. J. Eisch and J. E. Galle, *J. Organomet. Chem.,* **1976**, *121*, C10; J. J. Eisch and J. E. Galle, *J. Organomet. Chem.,* **1988**, *341*, 293.

210. J. E. Bäckvall and S. K. Juntunen, *J. Org. Chem.,* **1988**, *53*, 2396.

211. K. El-Berembally, M. El-Kersh, and H. El-Fatatry, *Sulphur Lett.,* **1990**, *11*, 157.

212. A. Jonczyk, K. Banko, and M. Makosza, *J. Org. Chem.,* **1975**, *40*, 266.

213. T. Durst, in *Topics in Organic Sulphur Chemistry* (Plenary Lectures of the 8th International Symposium on Organic Sulphur Chemistry), Ed. M. Tisler, University Press, Ljubljana, **1978**, p. 29; T. Durst, K-C. Tin, F. De Reinach-Hirtzbach, J. M. Decesare, and M. D. Ryan, *Can. J. Chem.,* **1979**, *57*, 258.

214. C. T. Hewkin and R. F. W. Jackson, *Tetrahedron Lett.,* **1990**, *31*, 1877.

215. J. S. Grossert, S. Sotheeswaran, H. R. W. Dharmaratne, and T. S. Cameron, *Can. J. Chem.,* **1988**, *66*, 2870.

216. A. Senning, O. N. Sorensen, and C. Jacobsen, *Angew. Chem., Int. Ed. Engl.,* **1968**, *7*, 734.

217. T. Olijnsma, J. B. F. N. Engberts, and J. Strating, *Recl. Trav. Chim. Pays-Bas,* **1970**, *89*, 897.

218. R. J. Gaul and W. J. Fremuth, *J. Org. Chem.,* **1961**, *26*, 5103.

219. K. Schank and F. Werner, *Liebigs Ann. Chem.,* **1979**, 1977; K. Schank, A. Frisch, and B. Zwanenburg, *J. Org. Chem.,* **1983**, *48*, 4580.

220. G. Ferdinand and K. Schank, *Synthesis,* **1976**, 408.

221. D. H. R. Barton, D. P. Manly, and D. A. Widdowson, *J. Chem. Soc., Perkin Trans.1,* **1975**, 1568.

222. The corresponding acyl sulphoxides, formed *in situ*, react similarly, see for example H. Minato, H. Kodama, T. Miura, and M. Kobayashi, *Chem. Lett.,* **1977**, 413.

223. K. Schank and F. Werner, *Liebigs Ann. Chem.,* **1983**, 1739.

224. G. Bram, A. Loupy, M. C. Roux-Schmitt, J. Sansoulet, T. Strzalko, and J. Seyden-Penne, *Synthesis,* **1987**, 56.

225. R. Davis, *Synth. Commun.,* **1987**, *17*, 823.

226. M. E. Kuehne, *J. Org. Chem.,* **1963**, *28*, 2124.

227. Y. Kobayashi, T, Yoshida, and I. Kumadaki, *Tetrahedron Lett.,* **1979**, 3865.

228. G. Stork and I. J. Borowitz, *J. Am. Chem. Soc.,* **1962**, *84*, 313; see also G. Opitz and H. Adolph, *Angew. Chem., Int. Ed. Engl.,* **1962**, *1*, 113.

229. N. P. Singh and J-F. Biellmann, *Synth. Commun.,* **1988**, 1061.

230. A. S. Kende and J. S. Mendoza, *J. Org. Chem.*, **1990**, *55*, 1125; see also A. I. Meyers, A. Nabeya, H. W. Adickes, and I. R. Politzer, *J. Am. Chem. Soc.*, **1969**, *91*, 763.
231. E. Hirsch, S. Hünig, and H-U. Reisig, *Chem. Ber.*, **1982**, *115*, 399 and 3687.
232. H-D. Becker and G. A. Russell, *J. Org. Chem.*, **1963**, *28*, 1896.
233. W. E. Truce, W. W. Bannister, and R. H. Knospe, *J. Org. Chem.*, **1962**, *27*, 2821; W. E. Truce and R. H. Knospe, *J. Am. Chem. Soc.*, **1955**, *77*, 5063.
234. G. Ferdinand and K. Schank, *Synthesis,* **1976**, 406.
235. C. A. Ibarra, R. C. Rodriguez, M. C. F. Monreal, F. J. G. Navarro, and J. M. Tesorero, *J. Org. Chem.*, **1989**, *54*, 5620.
236. E. M. Kaiser, L. E. Solter, R. A. Schwarz, R. D. Beard, and C. R. Hauser, *J. Am. Chem. Soc.*, **1971**, *93*, 4237.
237. M. C. Mussatto, D. Savoia, C. Trombini, and A. Umani-Ronchi, *J. Org. Chem.*, **1980**, *45*, 4002.
238. M. W. Thomsen, B. M. Handwerker, S. A. Katz, and R. B. Belser, *J. Org. Chem.*, **1988**, *53*, 906.
239. K. Kondo and D. Tunemoto, *Tetrahedron Lett.*, **1975**, 1007; J. Otera, T. Mandai, M. Shiba, T. Saito, K. Shimohata, K. Takemori, and Y. Kawasaki, *Organometallics*, **1983**, *2*, 332.
240. G. K. Cooper and L. J. Dolby, *J. Org. Chem.* **1979**, *44*, 3414; G. K. Cooper and L. J. Dolby, *Tetrahedron Lett.*, **1976**, 4675.
241. J. Fayos, J. Clardy, L. J. Dolby, and T. Farnham, *J. Org. Chem.*, **1977**, *42*, 1349.
242. S-K. Chung, *J. Org. Chem.*, **1981**, *46*, 5457; F. Camps, J. Coll, A. Guerrero, J. Guitart, and M. Riba, *Chem. Lett.*, **1982**, 715.
243. C. Najera, B. Mancheno, and M. Yus, *Tetrahedron Lett.*, **1989**, *30*, 3837.
244. C. Najera, B. Mancheno, and M. Yus, *Tetrahedron Lett.*, **1989**, *30*, 6085.
245. H. Hellman and K. Müller, *Chem. Ber.*, **1965**, *98*, 638.
246. A. J. H. Klunder, A. A. M. Houwen-Claassen, M. G. Kooy, and B. Zwanenburg, *Tetrahedron Lett.*, **1987**, *28*, 1329; A. J. H. Klunder, M. J. F. M. Crul, A. A. M. Houwen-Claassen, M. G. Kooy, and B. Zwanenburg, *Tetrahedron Lett.*, **1987**, *28*, 3147.
247. I. Kuwajima, Y. Higuchi, H. Iwasawa, and T. Sato, *Chem. Lett.*, **1976**, 1271.
248. H. J. Monteiro, *Synth. Commun.*, **1987**, *17*, 983.
249. H. J. Monteiro, *Tetrahedron Lett.*, **1987**, *28*, 3459.
250. P. Breuilles and D. Uguen, *Bull. Soc. Chim. Fr.*, **1988**, 705.
251. H. Kinoshita, I. Hori, T. Oishi, and Y. Ban, *Chem. Lett.*, **1984**, 1517.
252. D. L. J. Clive, T. L. B. Boivin, and A. G. Angoh, *J. Org. Chem.*, **1987**, *52*, 4943.

253. Y-M. Tsai, B-W. Ke, and C-H. Lin, *Tetrahedron Lett.*, **1990**, *31*, 6047; see also V. Reutrakul, C. Poolsanong, and M. Pohmakotr, *Tetrahedron Lett.*, **1989**, *30*, 6913; M. Julia, C. Rolando, and J. N. Verpeaux, *Tetrahedron Lett.*, **1982**, *23*, 4319.

254. F. Scotti and F. J. Frazza, *J. Org. Chem.*, **1964**, *29*, 1800.

255. B. Miller and M. V. Kalnins, *Tetrahedron*, **1967**, *23*, 1145.

256. J-M. Fang and M-Y. Chen, *Tetrahedron Lett.*, **1987**, *28*, 2853.

257. A. M. van Leusen, A. J. W. Iedema, and J. Strating, *J. Chem. Soc., Chem. Commun.*, **1968**, 440.

258. A. M. van Leusen and J. C. Jagt, , **1970**, 967.

259. H. Meijer, R. M. T. J. Strating, and J. B. F. N. Engberts, *Recl. Trav. Chim. Pays-Bas*, **1973**, *92*, 72; E. Bäder and H. D. Hermann, *Chem. Ber.*, **1955**, *88*, 41.

260. T. Olijnsma, J. B. F. N. Engberts, and J. Strating, *Recl. Trav. Chim. Pays-Bas*, **1972**, *91*, 209.

261. T. Olijnsma, J. B. F. N. Engberts, and J. Strating, *Recl. Trav. Chim. Pays-Bas*, **1967**, *86*, 463.

262. J. Morton, A. Rahim, and E. R. H. Walker, *Tetrahedron Lett.*, **1982**, *23*, 4123.

263. A. A. M. Houwen-Claassen, J. W. McFarland, B. H. M. Lammerink, L. Thijs, and B. Zwanenburg, *Synthesis*, **1983**, 628.

264. P. Renaud and S. Schubert, *Angew. Chem., Int. Ed. Engl.*, **1990**, *29*, 433; P. Renaud, *Tetrahedron Lett.*, **1990**, *31*, 4601.

265. P. Renaud and S. Schubert, *Synlett*, **1990**, 624.

266. R. H. Rynbrandt and F. E. Dutton, *J. Org. Chem.*, **1975**, *40*, 2282.

267. P. Carlier, Y. Gelas-Mialhe, and R. Vessiere, *Can. J. Chem.*, **1977**, *55*, 3190; J. Golinski, M. Makosza, and A. Rykowski, *Tetrahedron Lett.*, **1983**, *24*, 3279.

268. A. Yoshida, T. Hayashi, N. Takeda, S. Oida, and E. Ohki, *Chem. Pharm. Bull.*, **1981**, *29*, 2899.

269. J. Brennan and I. L Pinto, *Tetrahedron Lett.*, **1983**, *24*, 4731.

270. D. S. Brown, T. Hansson, and S. V. Ley, *Synlett*, **1990**, 48.

271. U. Schöllkopf and R. Schroder, *Angew. Chem., Int. Ed. Engl.*, **1972**, *11*, 311.

272. A. M. van Leusen, G. J. M. Boerma, R. B. Helmholdt, H. Siderius, and J. Strating, *Tetrahedron Lett.*, **1972**, 2367.

273. A. M. van Leusen, H. J. Jeuring, J. Wildeman, and S. P. J. M. van Nispen, *J. Org. Chem.*, **1981**, *46*, 2069.

274. N. Kornblum, M. M. Kestner, S. D. Boyd, and L. C. Cattran, *J. Am. Chem. Soc.*, **1973**, *95*, 3356.

275 J. J. Zeilstra and J. B. F. N. Engberts, *Recl. Trav. Chim. Pays-Bas*, **1974**, *93*, 11.

276. L. C. Garver, V. Grakauskas, and K. Baum, *J. Org. Chem.*, **1985**, *50*, 1699

277. W. E. Truce, T. C. Klingler, J. E. Paar, H. Feuer, and D. K. Wu, *J. Org. Chem.*, **1969**, *34*, 3104.

278. P. A. Wade, H. R. Hinney, N. V. Amin, P. D. Vail, S. D. Morrow, S. A. Hardinger, and M. S. Saft, *J. Org. Chem.*, **1981**, *46*, 765.
279. M. Regitz, *Synthesis*, **1972**, 351; M. Regitz and W. Bartz, *Chem. Ber.*, **1970**, *103*, 1477; see also references 247–249.
280. A. M. van Leusen and J. Strating, *Recl. Trav. Chim. Pays-Bas,* **1965**, *84*, 151; A. M. van Leusen and J. Strating, *Quart. Rep. Sulphur Chem.*, **1970**, *5*, 67.
281. Y-C. Kuo, T. Aoyama, and T. Shiori, *Chem. Pharm. Bull.*, **1982**, *30*, 526.
282. A. M. van Leusen, B. A. Reith, and D. van Leusen, *Tetrahedron*, **1975**, *31*, 597.
283. B. B. Jarvis and P. E. Nicholas, *J. Org. Chem.*, **1979**, *44*, 2951; B. B. Jarvis and P. E. Nicholas, *J. Org. Chem.*, **1980**, *45*, 2265.
284. E. R. Koft, *J. Org. Chem.*, **1987**, *52*, 3466.
285. C. Y. Meyers, A. M. Malte, and W. S. Matthews, *J. Am. Chem. Soc.*, **1969**, *91*, 7510; C. Y. Meyers, in *Topics in Organic Sulphur Chemistry*, Ed. M. Tislev, University Press, Ljubljana, **1978**, p. 207; C. Y. Meyers, A. M. Malte, and W. S. Matthews, *Quart. Rep. Sulphur Chem.*, **1970**, *5*, 229; N. S. Zefirov and D. I. Makhon'kov, *Chem. Rev.*, **1982**, *82*, 615.
286. J. Kattenberg, E. R. De Waard, and H. O. Huisman, *Tetrahedron*, **1973**, *29*, 4149.
287. H. Kotake, K. Inomata, H. Kinoshita, Y. Sakamoto, and Y. Kaneto, *Bull. Chem. Soc. Jpn.*, **1980**, *53*, 3027.
288. S. E. Lauritzen, C. Romming, and L. Skattebol, *Acta Chem. Scand.*, **1981**, *B35*, 263; R. R. Regis and A. M. Doweyko, *Tetrahedron Lett.*, **1982**, *23*, 2539.
289. B. Vacher, A. Samat, and M. Chanon, *Tetrahedron Lett.*, **1985**, *26*, 5129.
290. E. J. Corey and E. Block, *J. Org. Chem.*, **1969**, *34*, 1233; see also W. M. Ziegler and R. Connor, *J. Am. Chem. Soc.*, **1940**, *62*, 2596.
291. L. A. Paquette and U. Jacobsson, unpublished results, see L. A. Paquette, *Org. React.*, **1977**, *25*, 1.
292. K. M. More and J. Wemple, *Synthesis*, **1977**, 791.
293. F. G. Bordwell and B. B. Jarvis, *J. Am. Chem. Soc.*, **1973**, *95*, 3585; L. A. Carpino, L. V. McAdams III, R. H. Rynbrandt, and J. W. Spiewak, *J. Am. Chem. Soc.*, **1971**, *93*, 476.
294. M. Makosza, J. Golinski, and J. Baran, *J. Org. Chem.*, **1984**, *49*, 1488.
295. A. Jonczyk and T. Pytlewski, *Synthesis*, **1978**, 883.
296. F. G. Bordwell, M. D. Wolfinger, and J. B. O'Dwyer, *J. Org. Chem.*, **1974**, *39*, 2516.
297. G. Mignani, D. Morel, and F. Grass, *Tetrahedron Lett.*, **1987**, *28*, 5505; J. S. Grossert, P. K. Dubey, and T. Elwood, *Can. J. Chem.*, **1985**, *63*, 1263.
298. L. A. Paquette and R. W. Houser, *J. Am. Chem. Soc.*, **1969**, *91*, 3870.
299. I. De Riggi, J-M. Surzur, and M. P. Bertrand, *Tetrahedron*, **1988**, *44*, 7119; R. Nouguier, C. Lesueur, E. De Riggi, M. P. Bertrand, and A. Virgili, *Tetrahedron Lett.*, **1990**, *31*, 3541.

300. T. Durst, K-C. Tin, and M. J. V. Marcil, *Can. J. Chem.*, **1973**, *51*, 1704.
301. M. Julia and J-M. Paris, *Tetrahedron Lett.*, **1974**, 3445.
302. N. Kamigata, S. Kodate, J. Ozaki, M. Shyono, and M. Kobayashi, *Sulphur Lett.*, **1984**, *2*, 255.
303. K. Ogura, M. Suzuki, and G. Tsuchihashi, *Bull. Chem. Soc. Jpn.*, **1980**, *53*, 1414.
304. K. Ogura, N. Yahata, J. Watanabe, K. Takahashi, and H. Iida, *Bull. Chem. Soc. Jpn.*, **1983**, *56*, 3543.
305. K. Ogura, *Pure Appl. Chem.*, **1987**, *59*, 1033.
306. E. Schaumann, in *Perspectives in the Organic Chemistry of Sulphur*, Ed. B. Zwanenburg and A. J. H. Klunder, Elsevier, Amsterdam, **1987**, p. 251.
307. B. Wladislaw, L. Marzorati, and G. Ebeling, *Phosphorus, Sulphur and Silicon*, **1990**, *48*, 163; B. Wladislaw, L. Marzorati, R. B. Uchoa, and H. Viertler, *Synthesis*, **1985**, 553.
308. Y. Murata, K. Inomata, H. Kinoshita, and H. Kotake, *Bull. Chem. Soc. Jpn.*, **1983**, *56*, 2539.
309. J. S. Grossert and P. K. Dubey, *J. Chem. Soc., Chem. Commun.*, **1982**, 1183.
310. H. Böhme and B. Clement, *Tetrahedron Lett.*, **1979**, 1737.
311. R. Annunziata, M. Cinquini, and F. Cozzi, *Synthesis*, **1979**, 535.
312. D. H. R. Barton, J. Boivin, J. Sarma, E. da Silva, and S. Z. Zard, *Tetrahedron Lett.*, **1989**, *30*, 4237. Further examples can be found in Chapter 4.
313. F. G. Bordwell and G. A. Pagani, *J. Am. Chem. Soc.*, **1975**, *97*, 118.
314. K. Ogura, T. Iihama, K. Takahashi, and H. Iida, *Bull. Chem. Soc. Jpn.*, **1984**, *57*, 3347.
315. L. Hatjiarapoglou, A. Varvoglis, N. W. Alcock, and G. A. Pike, *J. Chem. Soc., Perkin Trans. 1*, **1988**, 2839.
316. L. Hatjiarapoglou and A. Varvoglis, *J. Heterocyclic Chem.*, **1988**, *25*, 1599.
317. L. Hatjiarapoglou and A. Varvoglis, *J. Chem. Res. (S)*, **1988**, 306.
318. T. G. Back, *J. Org. Chem.*, **1981**, *46*, 5443.
319. N. S. Simpkins, *Tetrahedron Lett.*, **1988**, *29*, 6787.
320. A. M. van Leusen, B. A. Reith, A. J. W. Iedema, and J. Strating, *Recl. Trav. Chim., Pays-Bas*, **1972**, *91*, 37.
321. B. A. Reith, J. Strating, and A. M. van Leusen, *J. Org. Chem.*, **1974**, *39*, 2728.
322. E. Block and M. Aslam, *Tetrahedron*, **1988**, *44*, 281; see also H-H. Otto, in *Perspectives in the Organic Chemistry of Sulphur*, Ed. B. Zwanenburg and A. J. H. Klunder, Elsevier, Amsterdam, **1987**, p. 231.
323. M. B. Anderson and P. L. Fuchs, *J. Org. Chem.*, **1990**, *55*, 337.
324. D. J. Peterson, *J. Organomet. Chem.*, **1971**, *26*, 215.

CHAPTER 3

Sulphonyl Carbanions

The chemistry of sulphones is dominated by the reactions of their derived carbanions. The sulphone group is unique in its ability to enable deprotonation of attached alkyl, alkenyl and aryl groups, and also in its ability to allow multiple deprotonation to form a range of polyanions. Combined with the relative inertness of the sulphone group to nucleophilic attack, these properties have elevated the sulphone to a premier position amongst carbanion-stabilising groups, and have resulted in the widespread use of sulphones in synthesis.

3.1 Introduction to Sulphonyl Carbanions

The synthetic importance of sulphonyl carbanions, combined with their intriguing stereochemical characteristics, has made them targets for intensive study.[1] This section describes important general features of sulphonyl carbanions, including solution and solid-state structure, stereoelectronics of carbanion formation and electrophilic quench, and the nature of sulphone acidity.

3.1.1 The Structure of Sulphonyl Carbanions

The question of the structure of sulphonyl carbanions arose out of early observations that optically active acyclic sulphones having an asymmetric α-carbon centre undergo deuteration (i.e. H–D exchange) faster than racemisation ($k_{exchange}/k_{racemise}$ up to about 40), e.g. Scheme 1.[2,3]

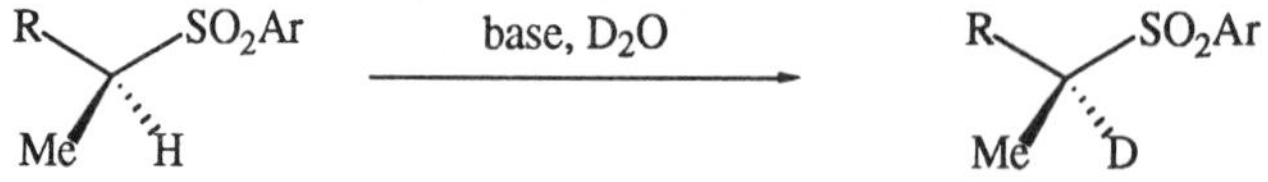

Scheme 1

Similar results are also obtained in decarboxylation reactions of α-sulphonyl carboxylic acids, which occur with retention of configuration, e.g. Scheme 2.[4,5]

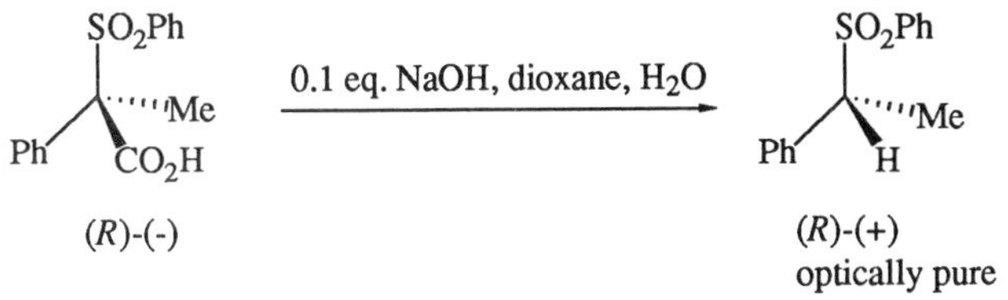

Scheme 2

This evidence clearly points to the formation of a chiral intermediate carbanion. Two extreme representations which would fit the data are the pyramidal sp^3 structure (**1**), or the alternative structure (**2**) having an essentially planar sp^2 carbanionic centre.

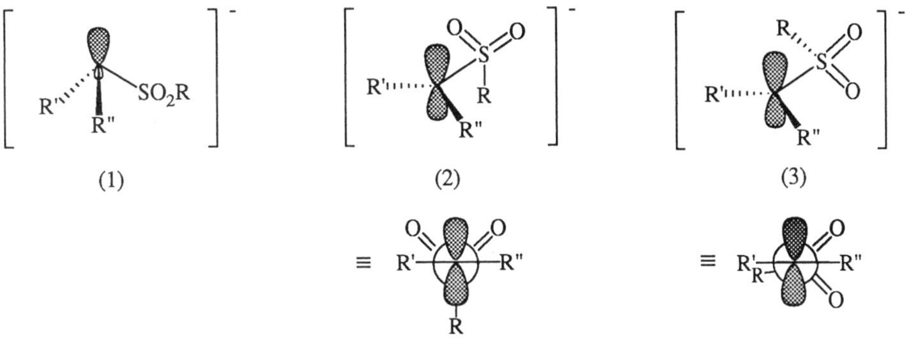

The retention of configuration observed in reactions such as those in Schemes 1 and 2 could then be explained by invoking a significant energy barrier, either to inversion in (**1**), or to rotation about the α-C–S bond in (**2**). Achiral structures such as (**3**) (a rotamer of (**2**)) are clearly ruled out. Corey *et al.* presented a variety of additional evidence which indicates that a pyramidal structure (**1**) is unlikely, at least in the cases examined. Thus, in H–D exchange experiments of three related sulphones (**4**)-(**6**), approximately similar $k_{exchange}/k_{racemise}$ ratios were obtained.[6]

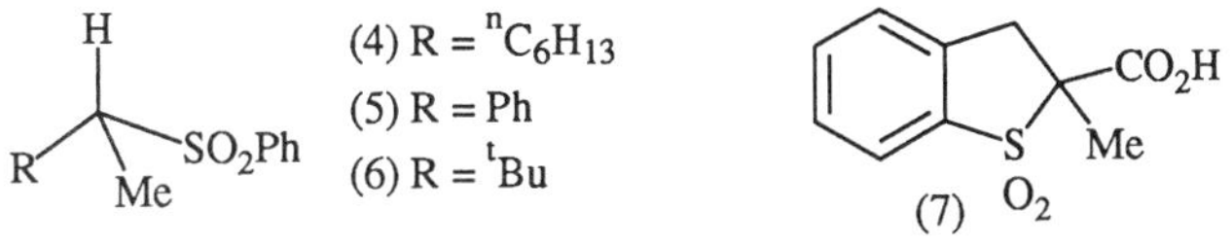

(4) R = $^nC_6H_{13}$
(5) R = Ph
(6) R = ^tBu

(7) O_2

If a pyramidal intermediate were responsible for the chirality of the intermediate carbanions then the rates of racemisation of (5) and (6) would be expected to be greater than that of (4), due to the electronic (in (5)) and steric (in (6)) driving forces towards a planar structure. In addition, decarboxylation of optically active (7) is found to give a racemic sulphone product, in contrast to the result in Scheme 2. In this case a pyramidal intermediate could retain chirality, whereas, because of the constraint of the ring, a planar carbanion would necessarily adopt the achiral arrangement represented by (3). The formation of a racemic product has therefore been presented as evidence for the formation of planar chiral sulphonyl carbanions, although, as other authors have pointed out, this inference assumes similar behaviour of sulphonyl carbanions in cyclic and acyclic systems.[5,7]

The question as to whether the sulphonyl carbanion is planar or pyramidal has become even more contentious following further experiments using the cyclopropane effect.[8] Thus both the kinetic and equilibrium acidities of cyclopropyl sulphones, e.g. (8) and (9), have been measured and compared with the acyclic isopropyl analogues (10) and (11).

(8) R = Ph   (10) R = Ph   (12)   (13)   (14)
(9) R = CF$_3$ (11) R = CF$_3$

If removal of the α-hydrogen to form a carbanion involves rehybridisation at the α-carbon from sp^3 to sp^2 (i.e. the sulphone demands substantial *p*-character at the carbanion centre), then (8) and (9) would be expected to be substantially less acidic than their counterparts, (10) and (11) respectively, due to the strain involved in such a rehybridisation in a cyclopropane ring. However, if pyramidal sulphonyl carbanions are involved, then the intrinsically greater acidity of cyclopropyl hydrogens should make (8) and (9) *more* acidic than their acyclic counterparts. Most of the evidence indicates that the phenyl sulphones (8) and (10) are of very similar acidity.[9,10] In fact, H–D exchange experiments, including intramolecular competitions using compounds (12) and (13), indicate that the cyclopropyl hydrogen is kinetically more acidic.[8,11]

Thus the idea that a planar configuration of the sulphonyl carbanion is a prerequisite for sulphone α-hydrogen acidity is groundless. This is similarly supported by the finding that sulphonyl carbanion formation is facile, even at bridgehead positions, e.g. in (14).[12]

Surprisingly, Bordwell *et al.* describe contradictory results with the triflones (9) ($pK_a = 27$) and (11) ($pK_a = 22$) which more closely resemble results obtained with the corresponding nitro compounds.[13]

That the configuration of the carbanionic centre is actually dependent on the structure of the sulphone concerned has been demonstrated by recent NMR and X-ray results. Thus in ^{13}C NMR experiments the magnitude of the C–H coupling constant at the α-carbon centre can be used to probe the hybridisation state. In benzylic anions such as $PhSO_2CHPhLi$ the α-carbon appears to be rather flat; by contrast, other non-benzylic examples appear to have carbanion hybridisation between sp^2 and sp^3.[14] Such trends are illustrated even more dramatically by X-ray structure determinations of lithiosulphones carried out by the groups of Boche and Gais.[15,16] The structures of each of the lithiosulphones (15)-(21) have been determined; each of these forms a dimer linked in an eight-membered ring with lithium atoms bridging the two SO_2 groups as indicated in (22).

(15) $[PhSO_2CHPhLi \cdot TMEDA]_2$
(16) $[PhSO_2CH_2Li \cdot TMEDA]_2$
(17) $[PhSO_2CH(CH=CH_2)Li \cdot Diglyme]_2$
(18) $[PhSO_2CH(SiMe_3)Li \cdot TMEDA]_2$
(19) $[PhSO_2C(CH_3)_2Li \cdot Diglyme]_2$
(20) $[PhSO_2C(CH_3)PhLi \cdot TMEDA]_2$
(21) $[CH_3SO_2CHPhLi \cdot TMEDA]_2$

(22)

(23)

Flattening of the configuration at the α-carbon atom to an approximately planar arrangement is evident for the benzylic systems, particularly (15) and (21). By contrast, (19) is very strongly pyramidalised as, to a lesser extent, are the other structures including, perhaps surprisingly, the allylic carbanion system (17).

The above-mentioned results, indicating facile formation of a cyclopropyl sulphonyl carbanion, and therefore little rehybridisation, are also reinforced by the crystal structure of the sulphonyl carbanion (23), which shows the $PhSO_2$ group very significantly bent (61.7°) out of the plane of the cyclopropane ring.[15] As expected, in these structures the α-C–S bond is shortened considerably and the S=O bonds slightly lengthened, compared with the parent sulphone. Most notably the crystal structures verify the approximately *gauche* arrangement of the carbanion lone pair with respect to the sulphone oxygens inferred previously, i.e. (2), and required for chirality. Also, no bond is evident between the α-carbon atom and the lithium atom in these structures, with the exception of (23). Additional significant crystal structure determinations have been carried out on triflyl carbanions,[16] on a *C*-titanated cyclopropyl sulphonyl carbanion[17] and on bis(trimethylsilyl)methyl phenyl sulphonyl carbanion.[18] The latter two reports are particularly interesting, since the titanated species is a true organotitanium compound,[17] and the bis-silylated anion forms a monomeric potassium 18-crown-6 salt.[18] Grossert *et al.* describe similar characteristics for crystal structures of carbanions stabilised by more than one sulphonyl group.[19] Finally, the above findings are also largely consistent with quantum mechanical calculations carried out by Wolfe and co-workers[20] and by Bors and Streitwieser.[21]

Following the experiments described by the groups of Corey and Cram in which H–D exchange α to a homochiral sulphone occurs with a high degree of retention of configuration,[3] it might be expected that chiral and indeed homochiral sulphonyl carbanions would have found widespread use in synthesis. This has not been the case for the simple reason that although H–D exchange can be carried out with retention of configuration (under protic conditions where an incipient carbanion is very rapidly reprotonated or deuterated) it has not been possible to use strong bases in aprotic solvents, even at low temperature, to generate stoichiometric amounts of a configurationally stable sulphonyl carbanion from acyclic sulphones. Thus treatment of optically active sulphone $PhSO_2CHMePh$ with BuLi in THF at -80°C results in rapid racemisation, as monitored by polarimetry.[16] Interestingly the corresponding lithiotriflones CF_3SO_2CLiRR' appear to show improved configurational stability, being formed and reprotonated with retention of configuration.

Further attesting to the transient nature of the chirality of phenyl sulphone-derived carbanions generated under aprotic conditions is an experiment involving quenching of the carbanion from sulphone (24) with Me_3SiCl, Scheme 3.[22]

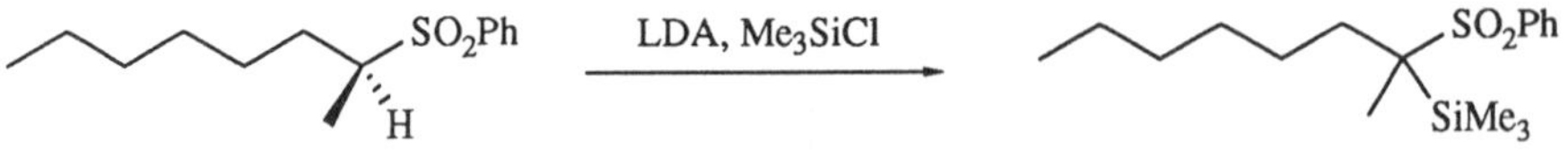

(24) : $[\alpha]_D$ -11.5

LDA, Me_3SiCl

(25)
external quench : racemic
internal quench : $[\alpha]_D$ +22.7

Scheme 3

When sulphone (24) is added to LDA and then Me_3SiCl immediately added, only racemic (25) is formed. Alternatively, when sulphone (24) is added to a mixture of LDA and Me_3SiCl (Corey and Gross's internal quench procedure[23]) optically active (25) is formed. Enantiomer interconversion appears to be rapid on the NMR timescale even for cyclopropyl sulphonyl carbanions.[10]

Finally, it should be noted that clean inversion of configuration on forming and quenching a sulphonyl carbanion occurs in the Ramberg–Bäcklund reaction, see Chapter 7.[24] Inversion is also observed in protonation of a carbanion derived from a cyclic sulphone by a retro-aldol process.[25]

3.1.2 Stereochemistry of Formation and Quenching of Sulphonyl Carbanions

From the findings described in Section 3.1.1 it can be surmised that sulphonyl carbanions exist in a chiral form corresponding to (2), or to pyramidalised versions of (2) in which the lone pair on carbon is approximately *gauche* to the two sulphone oxygens. Both in the solid state and in solution it appears that the lithium atom is associated primarily with the two sulphone oxygen atoms, with little or no lithium–carbon contact.[16] The question arises as to whether the preferred carbanion configuration is formed as a result of kinetic or thermodynamic control. This question has been addressed by studying the relative rates of H–D exchange of diastereotopic hydrogens in conformationally anchored cyclic sulphones. The preference for kinetic deprotonation involves removal of an α-hydrogen which lies along the internal bisector of the O–S–O angle (i.e. antiperiplanar to the α'-C–S bond), H[1] in (26).[26-28]

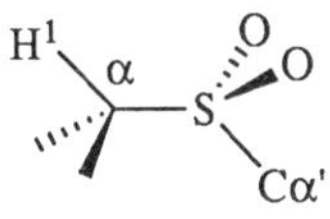

(26) 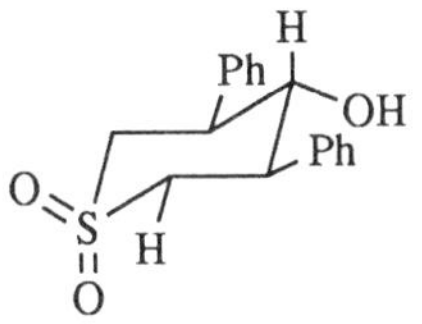　　　(27) 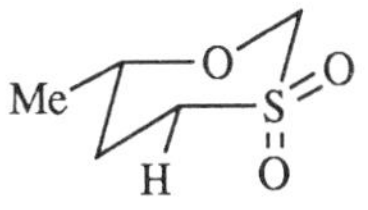　　　(28) 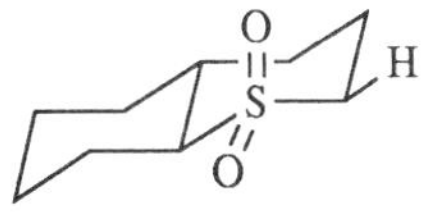　　　(29)

Thus in compounds (27)-(29) exchange of the equatorial hydrogen for deuterium was significantly faster than exchange of the axial hydrogen. Thus the stereoelectronically preferred conformation for deprotonation resembles the preferred carbanion conformation found in the crystal structures already described. In cyclic sulphones such as (28) and (30) electrophilic quenching, to give (31) and (32) respectively, reflects the equatorial preference of the carbanion lone pair, Scheme 4.[27,29]

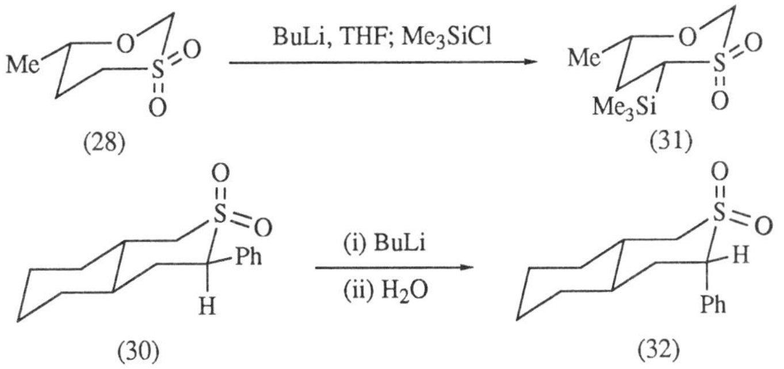

(28)　　　BuLi, THF; Me₃SiCl　　　(31)

(30)　　　(i) BuLi　　(ii) H₂O　　　(32)

Scheme 4

Notice in the conversion of (30) (which initially must undergo either axial deprotonation at the α-carbon, or a change in conformation to a chair–boat form) to

(**32**) the phenyl group ends up axial, again attesting to the strong preference for an equatorial carbanion.

Carbocyclic systems having $ArSO_2$ substitution also give interesting stereochemical results in electrophilic quenches of derived carbanions. The early work of Zimmerman and Thyagarajan contrasted the stereochemical outcome of protonation of the anion formed from sulphone (**33**) with that from nitro compound (**34**), Scheme 5.[30]

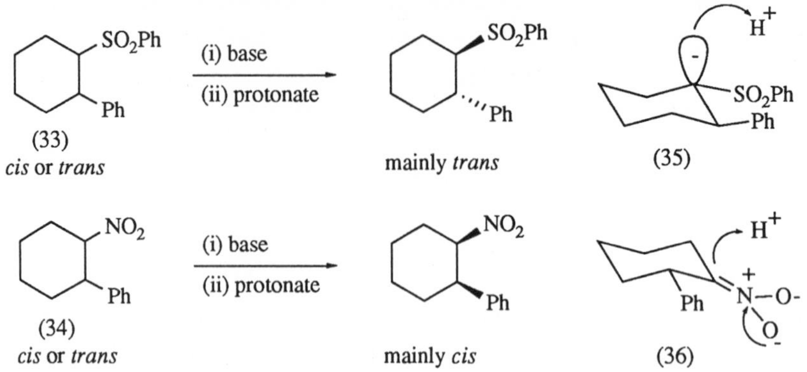

Scheme 5

This result suggests that the transition states for protonation of the two systems have contrasting geometries. Since the nitronate is expected to be sp^2 as in (**36**), formation of the *trans*-sulphone product is best explained by invoking an sp^3 hybridised arrangement for anion protonation such as (**35**).

Further results pointing to a tetrahedral sulphonyl carbanion have also been found by Trost and Schmuff in methylations of bicyclic sulphones (**37**) and (**38**) to give the pairs of diastereoisomeric products (**39**) and (**40**), and (**41**) and (**42**), respectively, Scheme 6.[31]

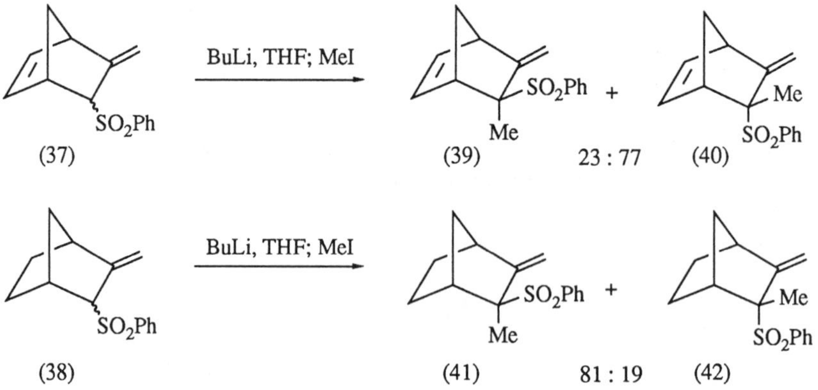

Scheme 6

Alkylation of (**37**) gives predominantly the *exo*-product (**40**) regardless of the stereochemistry of the starting sulphone. By contrast, similar methylation of (**38**) gives mainly (**41**), the product of *endo*-attack. Another group reports similar results using the two sulphones (**43**) and (**44**), Scheme 7.[32]

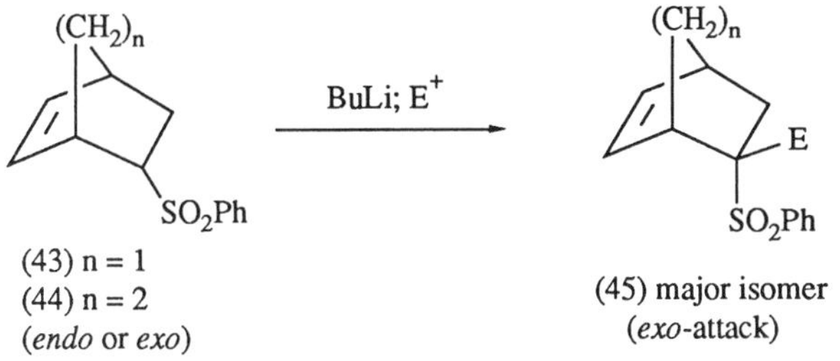

(**43**) n = 1
(**44**) n = 2
(*endo* or *exo*)

(**45**) major isomer
(*exo*-attack)

Scheme 7

In this study a range of electrophiles gives quite consistent stereochemical results, although rather surprisingly the precise *endo/exo* ratios of the product (**45**) are reported to depend on the stereochemistry of the starting sulphone.

The above results appear consistent with the idea that the intermediate sulphonyl carbanions have a relatively low intrinsic preference for either a planar or a pyramidal configuration. Hence steric or electronic influences imposed by the rest of the molecule may push the carbanion into one arrangement or the other (or somewhere in between!). Thus *exo*-alkylation in (**37**), (**43**) and (**44**) may be explained by invoking a somewhat pyramidalised carbanion with an *endo*-orientated $PhSO_2$ group, as in (**46**).

(**46**)   (**47**)

Trost suggests that this arrangement could be stabilised by some $\sigma-\pi^*$ interaction between the C–S bond and the neighbouring π system. On the other hand, in the reaction of (**38**), a flat or pyramidalised carbanion such as (**47**) with the phenyl group of the SO_2Ph in the least sterically encumbered orientation may explain the result.

Intramolecular chelation of the lithium atom in a lithiosulphone to another heteroatom functionality in a substrate molecule can also play a part in determining sulphone configuration. Thus Padwa *et al.* have reported the quantitative conversion of *cis*-(**48**) to the corresponding *trans*-isomer, Scheme 8.[33]

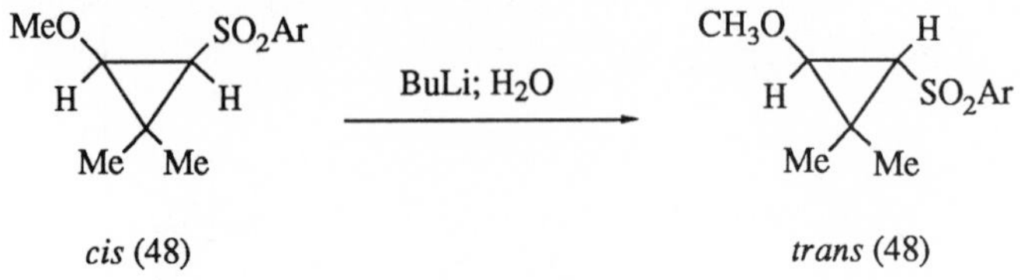

cis (48)   trans (48)

Scheme 8

The implied carbanion interconversion is explained by invoking intramolecular coordination of the lithium to the neighbouring oxygen atom, which is possible only in the *trans*-carbanion. Similar effects have been observed in acyclic systems by Tanikaga *et al*.[34]

  Both intramolecular and intermolecular proton-exchange reactions have been noted in reactions of sulphonyl carbanions.[35,36] The intriguing report of De Waard and co-workers describes the behaviour of carbanions derived from sulphones such as **(49)**.[35]   Treatment of **(49)** with NaOD in refluxing D_2O/dioxane leads to the corresponding *cis*-trideuterated sulphone **(50)**. Remarkably, however, treatment with BuLi in benzene at room temperature followed by protonation leads to the *trans*-fused isomer **(51)**, Scheme 9.

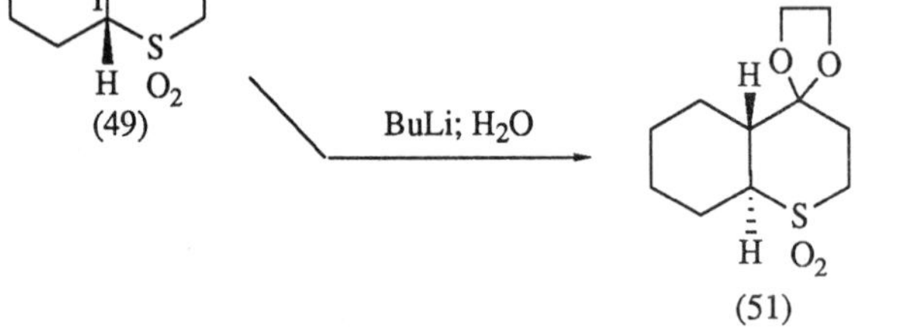

Scheme 9

The first process clearly fits with expectations from earlier work, the sulphonyl carbanion formed at C-1 being deuterated with retention of configuration. As described in Scheme 3 the behaviour of carbanions formed in aprotic media is quite different, and leads here to epimerisation at C-1. De Waard has found that the *cis–trans* isomerisation is an intramolecular process and has proposed the sequence of events shown in Scheme 10 to account for the result.

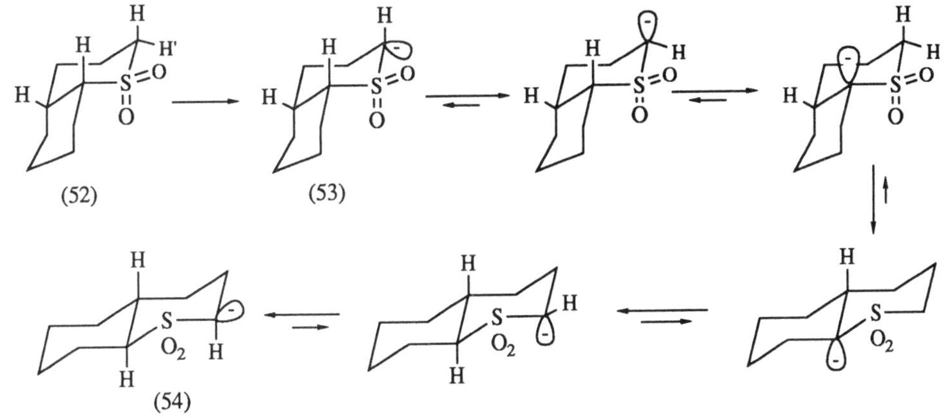

(52)   (53)

(54)

Scheme 10

Kinetically preferred removal of H' from (52) should initially occur to give (53). This anion can then equilibrate with other (less stable) forms by inversion and intramolecular H-transfer processes, ultimately leading to the most stable isomer (54). The finding that no intermolecular equilibration occurs is in contrast to both the earlier findings of Zimmerman and Thyagarajan,[9] and to the recent report of Pine *et al.* describing the polyalkylation of methyl phenyl sulphone which arises due to intermolecular sulphonyl carbanion equilibration.[36] Thus treatment of $PhSO_2Me$ with BuLi in THF at 0°C followed by addition of alkylating agents RX always gives dialkylated by-products $PhSO_2CHR_2$. This was determined to be due to the enhanced nucleophilicity of the more substituted carbanion (55) (compared to (56)) formed as shown in Scheme 11.

$$PhSO_2CH_2^- \ Li^+ + \ PhSO_2CH_2R \ \rightleftharpoons \ PhSO_2CH_3 \ + \ PhSO_2^-CHR \ Li^+$$

(56)   (55)

Scheme 11

Clearly some of the stereochemical results and the proton-exchange phenomena described require further study before a full picture of sulphonyl carbanion behaviour in solution is to be gained.

3.1.3 Equilibrium Acidities, Bases, Counterions and Solvents

Extensive studies have been made of sulphone equilibrium acidities (pK_a values), most notably by Bordwell et al.[37] A selection of values from these studies (in DMSO) is given in Table 1.

(1)	$MeSO_2Me$	31.1	(6)	$PhSO_2CH_2SMe$	23.4
(2)	$PhSO_2Me$	29.0	(7)	$PhSO_2CH_2SO_2Ph$	12.2
(3)	$PhSO_2Et$	31.0	(8)	$PhSO_2CH_2CN$	12.0
(4)	$MeSO_2CF_3$	18.8	(9)	$PhSO_2CH_2COPh$	11.4
(5)	$PhSO_2CH_2OMe$	30.7	(10)	$PhSO_2CH_2NO_2$	7.1

Table 1

Examination of these figures allows some generalisations to be made. Firstly, in broad terms the sulphone is somewhat less acidifying than a carbonyl group (i.e. $PhSO_2Me$, pK_a 29.0, vs PhCOMe, pK_a 24.7) and considerably less so than a nitro group ($MeNO_2$, pK_a 17.2). Secondly, carbon substitution at the carbanion centre reduces the acidity (entry (2) vs (3)) whereas in a nitro compound the reverse is true. Examination of entries (7) and (8) reveals a close comparison between the additive effect of a SO_2Ph and a CN group. Finally, a triflone group, SO_2CF_3, entry (4), is a much more powerful acidifier than an ordinary sulphone, giving a pK_a value similar to a nitro group.

The acidifying effect of the sulphone group has traditionally been ascribed to a $p\pi$–$d\pi$ interaction, effectively analogous to the $p\pi$–$p\pi$ overlap found in enolates and nitronates. The current view, however, is that $p\pi$–$d\pi$ interactions are less significant than n_C–σ^*_{S-R} interactions in stabilisation of negative charge.[38]

Sulphonyl carbanions can be generated and reacted under a wide range of conditions, ranging from aqueous phase-transfer conditions using NaOH as base, to the use of alkyllithium as base in aprotic solvents. The choice of conditions will depend on the structure of the sulphone (especially the approximate pK_a value) under consideration and the nature of the transformation being carried out. The reactivity of the sulphonyl carbanion is dependent on the nature of metal counterions present (Li, Na, K and Mg being the most important), the solvent (e.g. benzene, THF, DME, DMSO) and the presence of 'additives' such as TMEDA, HMPA and Lewis acids. Such possibilities are illustrated in the following sections of this chapter.

3.2 Reactions of Simple Sulphonyl Carbanions

This section deals with typical reactions of simple sulphonyl carbanions, i.e. those lacking additional carbanion-stabilising groups such as carbonyls or heteroatoms, including alkylation (e.g. with halogenoalkanes and epoxides), aldol-type reaction with carbonyls, Michael additions and acylations. Similar chemistry of systems possessing functionality at the carbanion centre, particularly heteroatoms such as oxygen, nitrogen and sulphur, and also the carbanion chemistry of β-ketosulphones and related compounds, is dealt with in Section 3.3.

3.2.1 Alkylations

A very early example of the type of alkylation chemistry typical of sulphonyl carbanions is the methylation of sulphone (**57**), Scheme 12.[39]

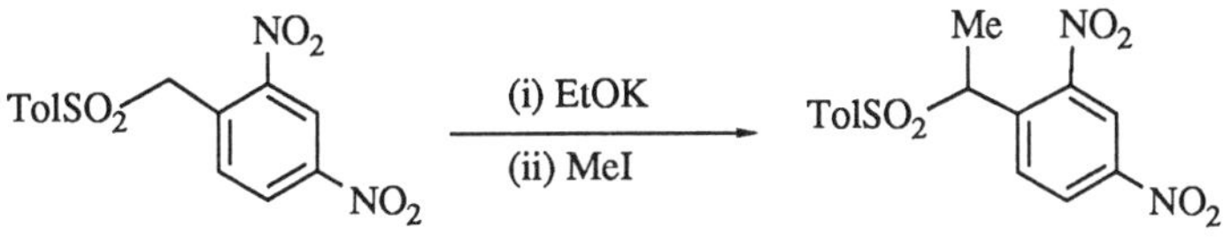

(57)

Scheme 12

In this study the dinitrobenzyl sulphone was reported to give a satisfactory methylation reaction, whereas analogues with only one or no nitro groups could not be alkylated. Other now familiar reactions such as sulphonyl carbanion bromination and dimerisation were also noted.

More recent studies have shown that, with modern techniques, and particularly the ready availability of alkyllithium and lithium amide bases, the alkylation of sulphonyl carbanions is a very general and high-yielding process. The following examples serve to illustrate the structural variation possible in both the nucleophilic and electrophilic partners in typical sulphone alkylations.

Julia and Blasioli have used a variety of allylic halides, having additional vinylic halogen substitution, in reactions with sulphonyl carbanions generated using BuLi, e.g. Scheme 13.[40]

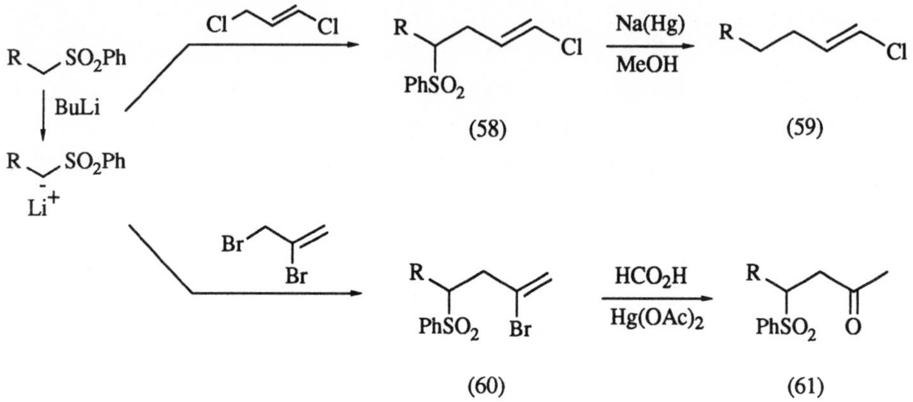

Scheme 13

The sulphone products can be transformed in a variety of ways, for example by desulphonylation to give substituted vinyl halides (**58**) → (**59**), or by vinyl halide hydrolysis to give carbonyls, e.g. (**60**) → (**61**). Alkylation of a sulphonyl carbanion can be an excellent way of linking optically active partners to form a chiral product having distant asymmetric centres. An example of this approach is the preparation of sulphone (**62**), which is an intermediate in the synthesis of the pheromones of the pine sawfly, Scheme 14.[41]

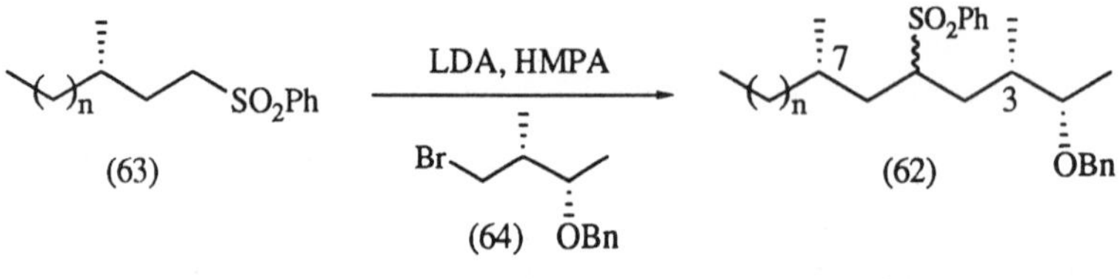

Scheme 14

This strategy ensures the correct relative (and absolute) stereochemistry at the remote carbons, C-3 and C-7, which would otherwise be hard to achieve through relative asymmetric induction. The ready availability of suitable homochiral fragments such as (**63**) and (**64**), the high yields in the sulphone alkylations and the ease of desulphonylation make this a very effective route to long-chain homochiral targets. Three more examples serve to illustrate the scope of this strategy for the synthesis of complex targets, Scheme 15.[42–44]

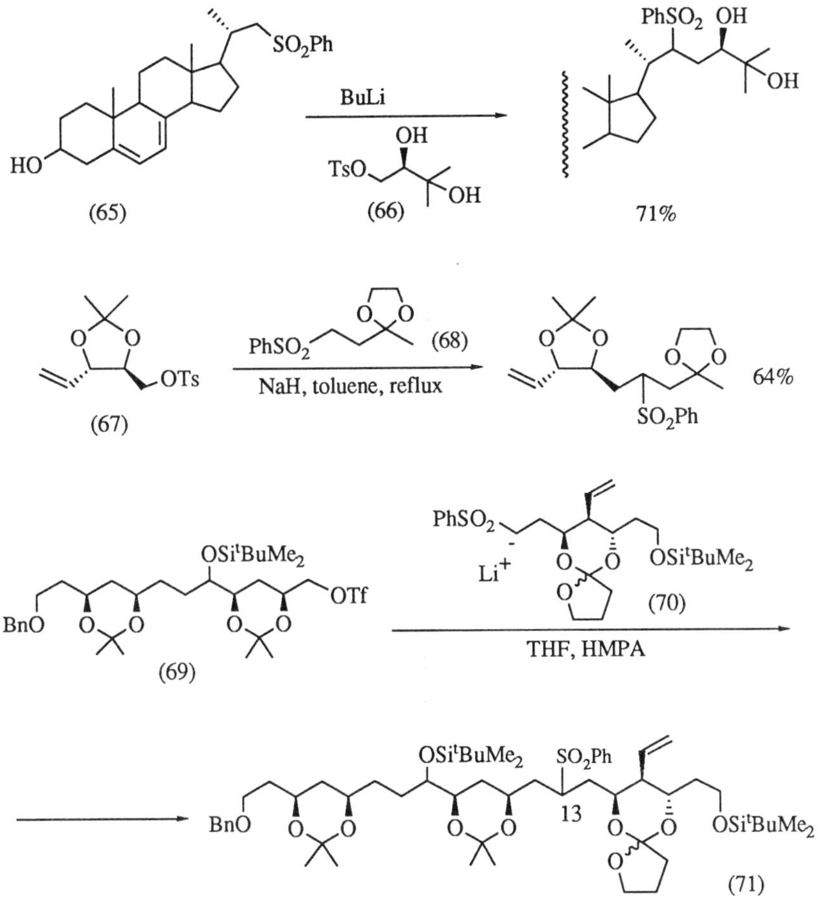

Scheme 15

Each example shows a variation on the general theme, using a different base, solvent or leaving group. Notably the use of a large excess of BuLi in the coupling of (**65**) and (**66**) avoids the need for diol protection (an epoxide derived from (**66**) is probably the actual reaction partner).[42] The linking of (**67**) with (**68**), and (**69**) with (**70**), demonstrates the utility of this approach for preparing rather complex synthetic intermediates. Only in the case of (**71**) does the sulphone perform an additional useful function, being subsequently converted to a carbonyl at C-13 by reaction of the sulphonyl carbanion with MoOPH.[44]

The alkylation of cyclopropyl sulphone (**72**), using either BuLi or LDA as base, occurs smoothly to give a wide range of substituted derivatives (**73**).[45] The products can then be transformed into α,β-unsaturated aldehydes by removal of the MEM protecting group followed by base-mediated ring opening, Scheme 16.

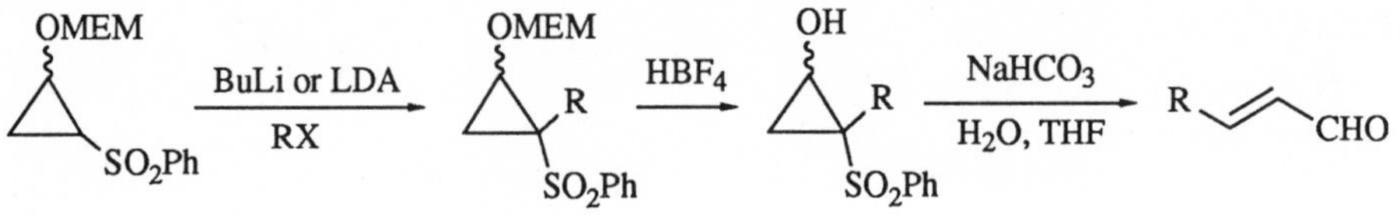

Scheme 16

Sulphones having additional ketal or acetal functionality have likewise been alkylated and then deprotected to give α,β-unsaturated carbonyl products, Scheme 17.[46]

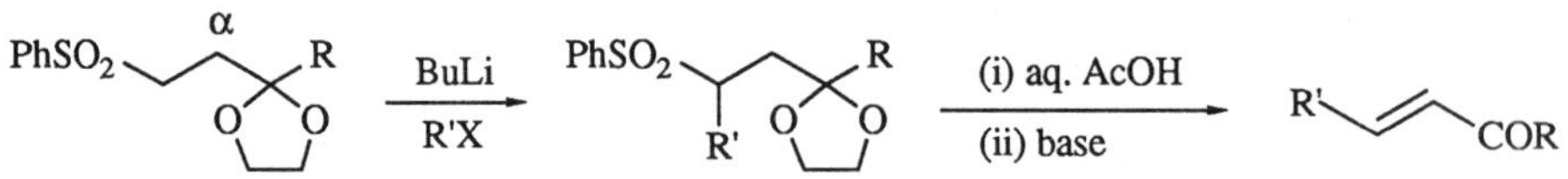

Scheme 17

This method has been extended to starting materials having additional substitution at the α-carbon and to the introduction of a second substituent by a subsequent alkylation of a carbanion derived from (**74**). Analogous chemistry involving sulphone acylation rather than alkylation has also been carried out by the same group.[47]

Intramolecular alkylations, i.e. cyclisation, of ω-halogenoalkyl sulphones and related derivatives is a good way of making cyclic sulphones. Chang and Pinnick used this approach to effect the overall conversion of an α,β-unsaturated carbonyl compound to a cyclopropane, i.e. (**75**) → (**76**).[48]

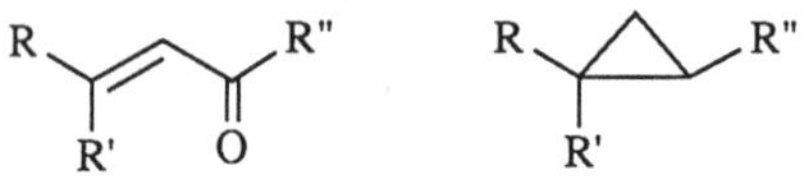

The best sequence involves conjugate addition of thiophenol to the unsaturated carbonyl partner, followed by sulphur oxidation and conversion of the carbonyl group into a leaving group by reduction and tosylation. Cyclisation using LDA in THF occurs very cleanly to give the desired cyclopropyl sulphone, e.g. (**77**)→(**78**), Scheme 18.

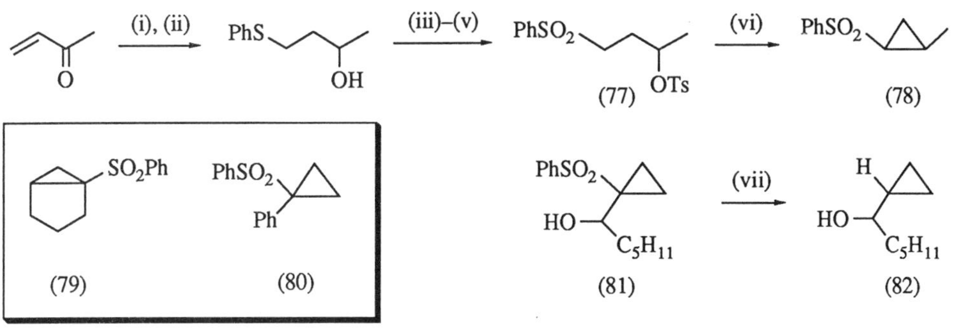

reagents:
(i) PhSNa (ii) $NaBH_4$ (iii) H_2O_2, HOAc
(iv) K_2CO_3, MeOH (v) TsCl, py (vi) LDA, THF (vii) 6% Na(Hg), EtOH

Scheme 18

The overall yields for this cyclopropane synthesis are very good, with the ring-forming reaction being practically quantitative. Examples of the sulphones formed in this way include the bicyclic product (**79**) and phenyl-substituted compound (**80**). The report also includes details of the alkylation and aldol reactions of the anion of cyclopropyl phenyl sulphone. Hydroxy sulphone (**81**) is one example prepared in this way, subsequent desulphonylation (without Julia-type β-elimination) giving cyclopropyl carbinol (**82**) in 90% yield.

Stirling and co-workers have carried out extensive studies of this type of cyclisation, including examples using sulphones[49] and bis-sulphones.[50] These studies showed that three-, four- and five-membered ring cyclisation can be carried out in excellent yield and that the rate of cyclisation to form a three-membered ring is extremely fast compared to four- and five-membered cases. This is thought to be primarily due to favourable entropy factors, i.e. the lack of any need for orientation of the chain connecting nucleophile and electrophile.

Large rings can also be formed in the same way, as indicated by the formation of (**83**), which has the cembrene skeleton, Scheme 19.[51]

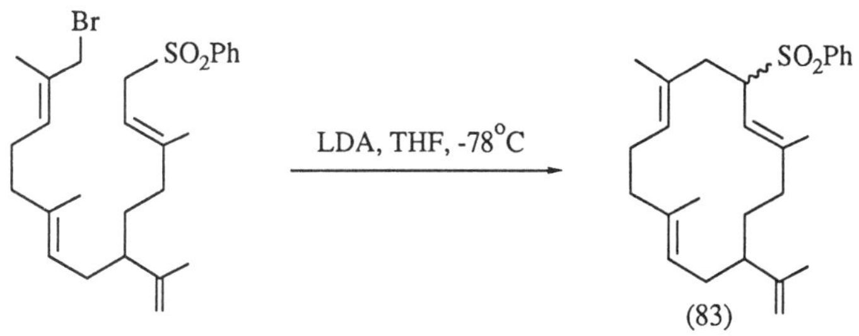

Scheme 19

An unusual cyclisation of the lithiosulphone (**84**) to give the cyclic sulphonyl anion (**85**) occurs on warming from -78°C to about -55°C, Scheme 20.[52]

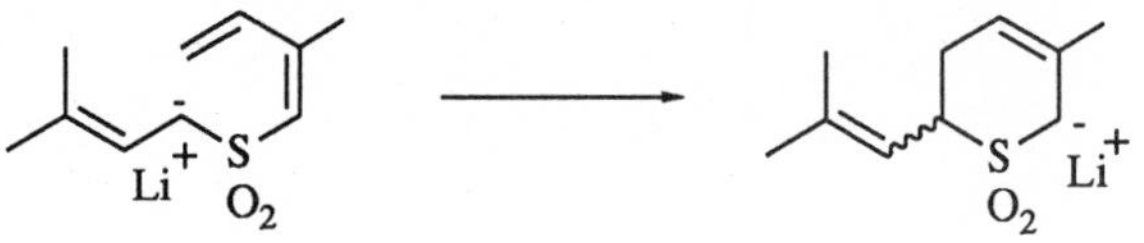

(84)   (85)

Scheme 20

These reactions also illustrate the general tendency of carbanions derived from allylic sulphones to undergo α- rather than γ-alkylation (i.e. reaction occurs α to the SO_2R group). Such alkylation can be carried out regioselectively on both acyclic and cyclic sulphones, using either anhydrous conditions (typically LDA or [n]BuLi in THF) or phase-transfer conditions, Scheme 21.[31,53,54]

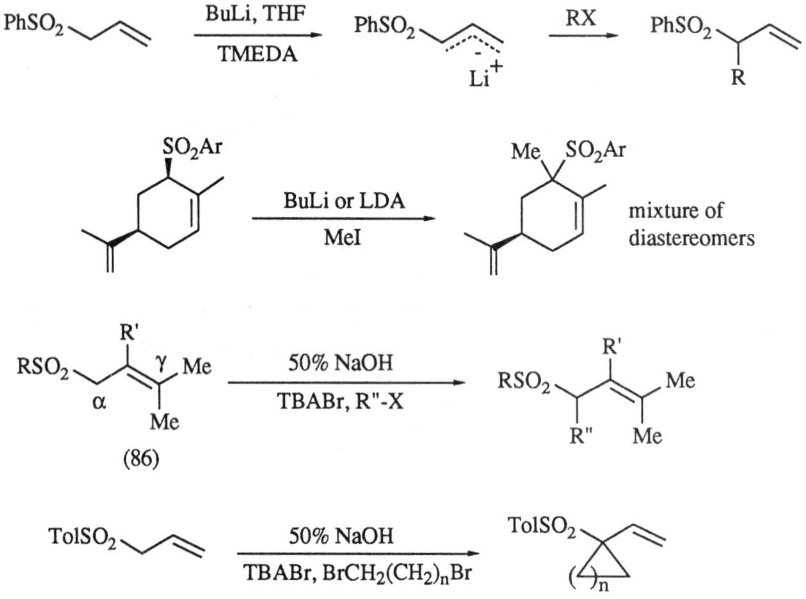

Scheme 21

Although effective with sulphone (**86**), the phase-transfer conditions do not always permit such selective alkylations of allylic sulphones not having the two γ-methyl groups, e.g. (**87**). These compounds can, however, be used in reactions involving ω-dibromoalkanes to give cyclic sulphones such as (**88**). Another report from the same group describes analogous chemistry using benzyl sulphones.[55]

The alkylation of allylic sulphones, and homologues, is very well suited to the assembly of various isoprenoid structures, Scheme 22.[56,57]

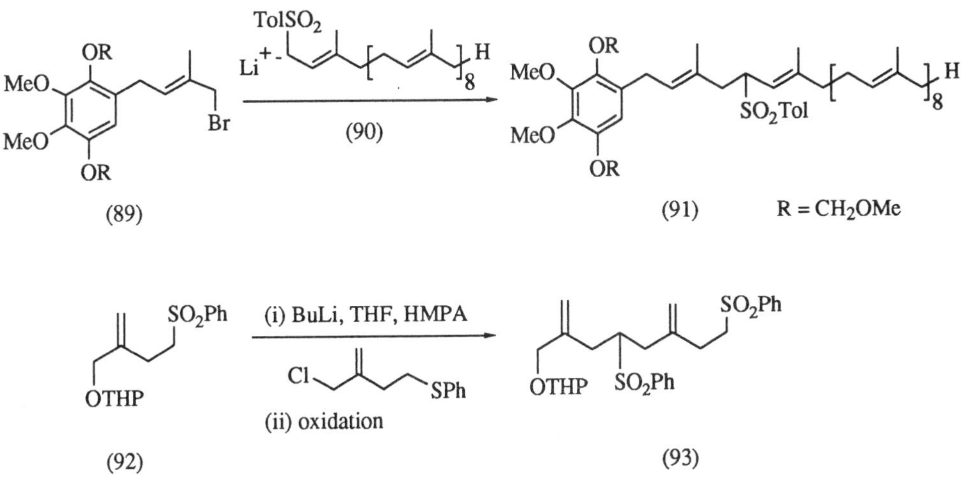

Scheme 22

Thus alkylation of allylic bromide (**91**) with the anion of solanesyl-*p*-tolyl sulphone (**90**) gives (**91**) which is subsequently converted to co-enzyme Q_{10}.[56] A similar alkylation of C_5-unit (**92**) gives the C_{10}-unit (**93**) after a subsequent oxidation.[57] This process can be repeated to produce C_{15}- and C_{20}-units, for example by selective terminal alkylation of the dianion of (**93**).

The anion of cyclopentadienyl phenyl sulphone (**94**) can be generated by treatment of the disulphone (**95**) with two equivalents of BuLi. The anion is rather unreactive but can be methylated regioselectively in good yield to give (**96**) by addition of MeI in HMPA, Scheme 23.[58]

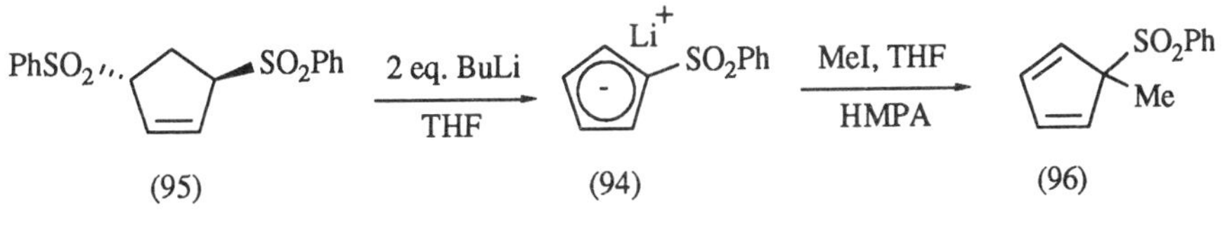

Scheme 23

The sulphone product may be further manipulated, for example by stereoselective epoxidation, without problems due to cyclopentadiene dimerisation.

In contrast to ordinary alkyl sulphones, triflones can be alkylated using mildly basic conditions, e.g. Scheme 24.[59]

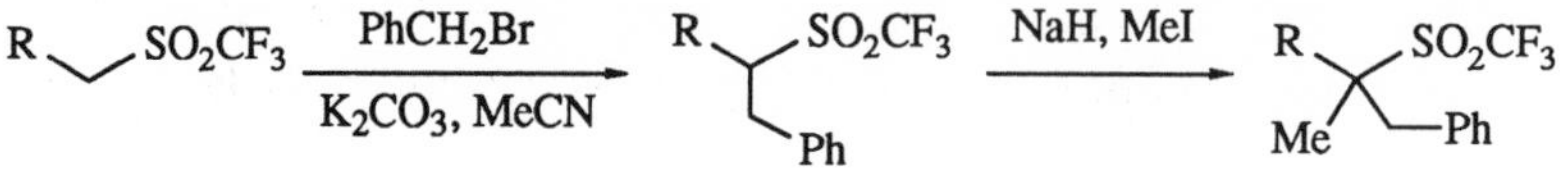

Scheme 24

Monoalkylation requires only the use of K_2CO_3 in refluxing CH_3CN, whereas the use of a preformed triflyl anion, generated using NaH, is preferred for introduction of a second alkyl group.

3.2.2 Reaction with Epoxides

The epoxide-opening reactions of sulphonyl carbanions were first studied in detail by Julia and Uguen.[60] They obtained good yields of products from the reaction of a variety of sulphonyl carbanions, generated in THF using ^{n}BuLi, with terminal epoxides. This study focused on the combination of isoprenoid partners and the subsequent desulphonylation of the hydroxy sulphone products, Scheme 25.

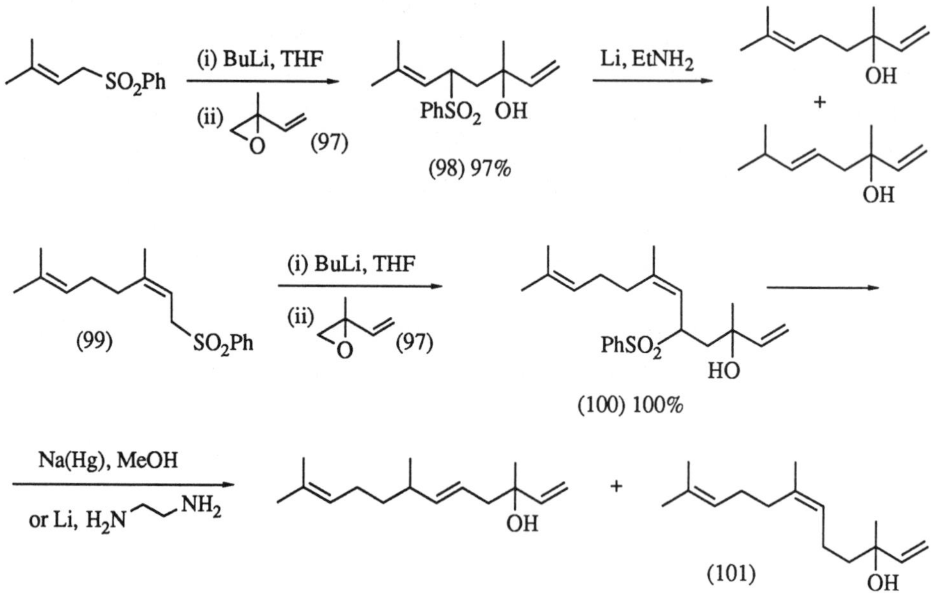

Scheme 25

In the reactions of isoprene monoepoxide (**97**) exclusive attack at the terminal position to give (**98**) is observed, with no evidence of competing SN2' epoxide opening. In reactions of (*Z*)-(**99**) and the corresponding (*E*)-isomer the double-bond geometry of the starting sulphone is preserved in the products, e.g. (**1 0 0**). Subsequent desulphonylations, however, give mixtures of double-bond regioisomers, although double-bond stereochemistry is retained in the non-transposed products, e.g. (**101**).

Another substantial study probed the reaction of variously substituted sulphonyl carbanions with a range of epoxides, including cyclic epoxides, Scheme 26.[61]

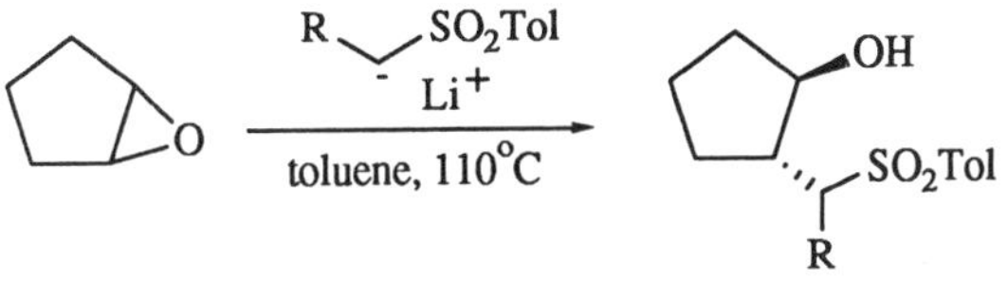

(102) R = H, 72%
(103) R = Me, 98%

Scheme 26

These reactions are carried out by heating a mixture of epoxide and sulphonyl carbanion in toluene at 110°C for 10–24 h. With the simple examples examined, these rather drastic conditions gave quite satisfactory results, yields typically in the range 50–80%. Notably, product (**103**), formed using the anion from ethyl phenyl sulphone, is formed more efficiently than (**102**), from methyl phenyl sulphone, presumably due to the enhanced nucleophilicity of the more substituted carbanion. Not unexpectedly, however, additional substitution on the sulphonyl carbanion very effectively inhibits the epoxide opening. In examples such as (**102**) and (**103**) the expected *trans*-epoxide opening is observed, and attack at the less hindered end of the epoxide is seen in unsymmetrically substituted epoxides. However, analogous reactions using triflyl carbanions are much less successful. Although clearly acceptable for simple systems, the rather vigorous reaction conditions employed appear unattractive for applications involving more highly functionalised sensitive compounds.

A recent application of sulphonyl carbanion epoxide opening described by Carretero and Ghosez utilises reaction conditions similar to those described by Julia (ⁿBuLi, THF, -78°C to room temperature), Scheme 27.[62]

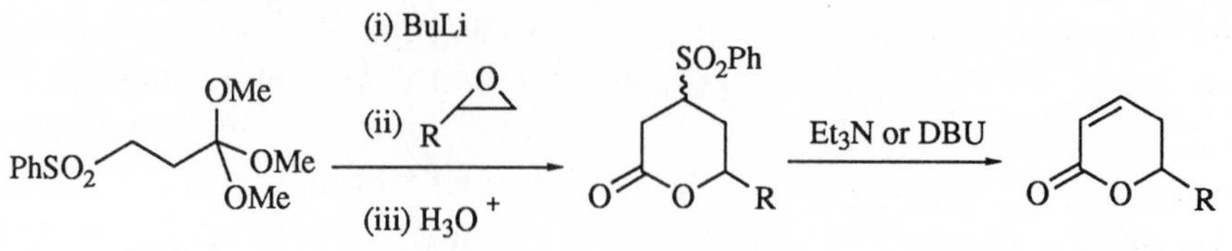

Scheme 27

In this case a carbanion derived from functionalised sulphone (**104**) reacts with a range of terminal epoxides. Deprotection of the orthoester functionality and lactone formation furnishes (**105**) and subsequently the unsaturated analogues (**106**). Two reactions involving disubstituted epoxides also give the desired products, although in very reduced yields.

The rather sluggish reactivity of sulphonyl carbanions towards epoxides (especially non-terminal examples) has stimulated a search for modified, mild conditions that allow smoother reaction. A very interesting report by Marshall and Andrews describes the results of coupling reactions involving lithiosulphone (**107**) and a range of derivatives (**108**)-(**111**), Scheme 28.[63]

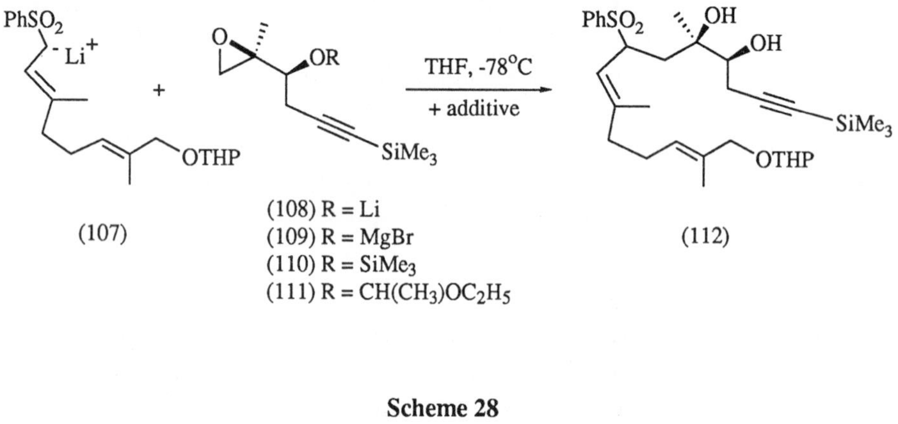

(108) R = Li
(109) R = MgBr
(110) R = SiMe₃
(111) R = CH(CH₃)OC₂H₅

Scheme 28

The best yield of the desired product (**112**) (70%) is obtained when the magnesio-derivative (**109**) is allowed to react with the sulphonyl carbanion at -78°C in a mixture of THF and HMPA. HMPA gives slightly superior results in comparison to other additives such as TMEDA or DABCO. Most interesting, however, is the finding that the presence of the neighbouring magnesium alkoxide is critical to the success of the epoxide opening. Thus with the lithio-derivative (**108**) a very low yield of product is obtained, whilst the protected compounds (**110**) and (**111**) give no product at all. The effect is presumably due to Lewis-acid neighbouring group assistance in the epoxide opening, although MgBr$_2$ added to reactions involving (**110**) or (**111**) does not have the same effect. The generality of this method is further indicated by its use in a total synthesis of (+)-ionomycin by Hanessian *et al.*[64]

Other reports have described the smooth coupling of sulphonyl carbanions with epoxides in the presence of mild Lewis acids such as BF$_3$·OEt$_2$, Scheme 29.[65,66]

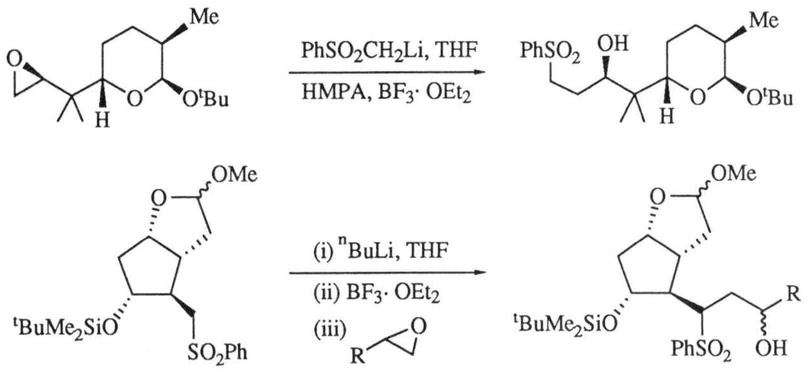

Scheme 29

Thus the difficulties experienced in conducting this type of condensation under mild conditions, especially with hindered partners,[67] may often be overcome by the use of solvent additives (especially HMPA) and/or Lewis acids such as BF$_3$·OEt$_2$ or Ti(O^{i}Pr)$_4$.[68] A recent report from Craig and Smith shows how useful this reaction can be once the optimum reaction conditions have been determined, Scheme 30.[69]

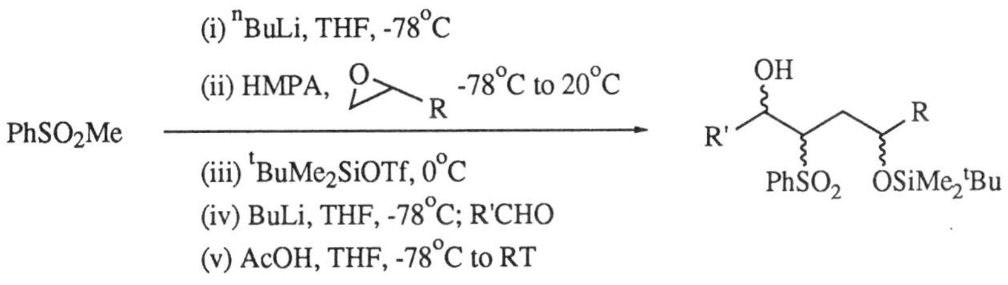

Scheme 30

Thus two successive couplings, involving firstly an epoxide and secondly a carbonyl component, are achieved in very good overall yield.

Cyclisations of epoxy sulphones to yield cyclic γ-hydroxy sulphones of various ring sizes have been carried out, most notably by Durst's group, Scheme 31.[70]

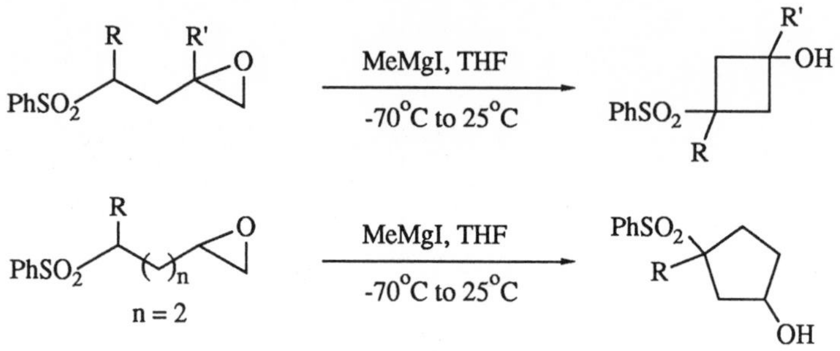

Scheme 31

Both cyclobutanols and cyclopentanols are readily prepared, and may be further transformed to cyclobutenones and cyclopentenones respectively by oxidation and sulphinate elimination. Versatility is increased by the finding that differently sized ring products can be accessed from the same starting materials, simply by changing the base used from CH_3MgI to LDA, Scheme 32.[71]

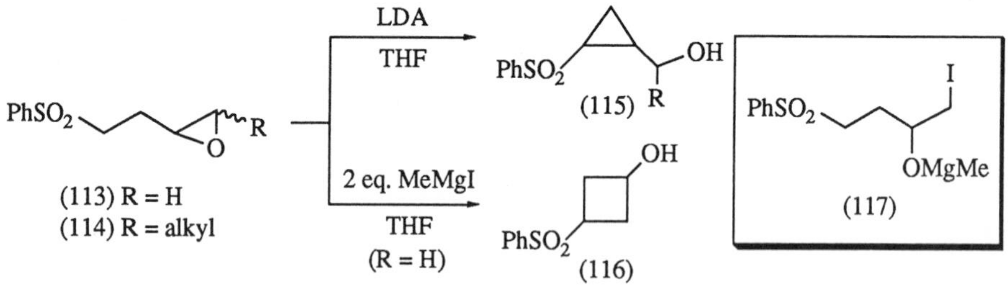

Scheme 32

The formation of the smaller-ring product (**115**) from (**113**) is to be expected, based on previous findings by Stork in analogous closures of epoxy nitriles. The formation of cyclobutanols (**116**) is explained by the intermediacy of the iodide (**117**), which also accounts for the requirement of two equivalents of Grignard reagent. The study includes discussion of the stereochemical aspects of these cyclisations, as well as giving results for higher homologues such as (**114**).

Similar studies have been conducted using epoxides having remote geminal disulphone substitution.[72] Here the relative rates of cyclisation to form various ring sizes are rather different to those expected based on the previous results of Stirling.

As expected, three-membered rings are formed in preference to four-membered rings but, surprisingly, cyclopentane formation is more rapid than cyclopropane formation. One substrate capable of forming a six-membered ring forms only the alternative seven-membered ring product, Scheme 33.

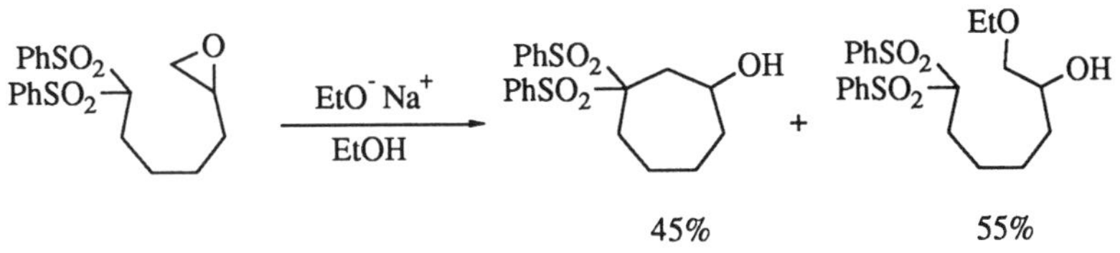

Scheme 33

Other applications of this type of sulphone-mediated cyclisation include the preparation of bicyclobutanes described by Gaoni,[73] and more recently the synthesis of functionalised cyclohexanes such as (**118**), Scheme 34.[74]

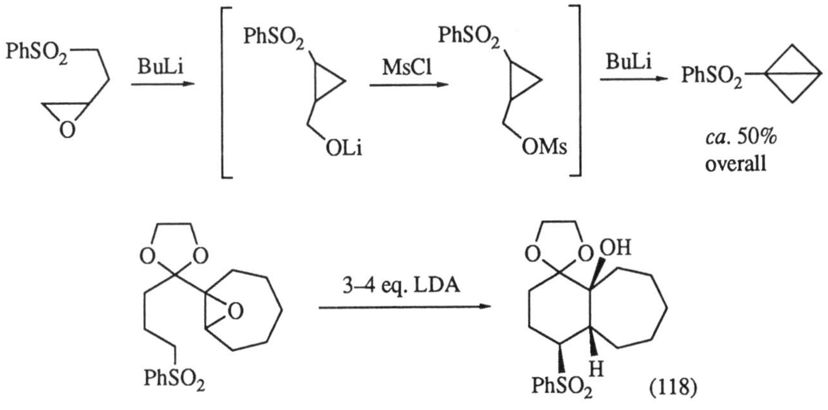

Scheme 34

3.2.3 Aldol-type Reactions with Carbonyl Compounds

The addition of sulphonyl carbanions to aldehydes and ketones is a reaction of great significance, due in great part to the utility of the β-elimination reactions of the so-formed β-hydroxy sulphones to form alkenes – the Julia reaction. However, this addition process has a substantial history stretching back long before the discovery of the Julia reaction. Pioneers in this area, including Kohler, Field and Truce, examined varied reactions of sulphonyl carbanions from as early as the mid-1930s, and focused particularly on additions to carbonyl compounds. Most of these early reports used

'Grignard reagents of sulphones', i.e. the magnesio-derivatives formed by reaction of a sulphone with an alkyl Grignard reagent. For example, Field reported the formation of hydroxy sulphone (**119**) in 90% yield by this method, along with transformations of (**119**) to β-ketosulphone (**120**) and vinyl sulphone (**121**), Scheme 35.[75]

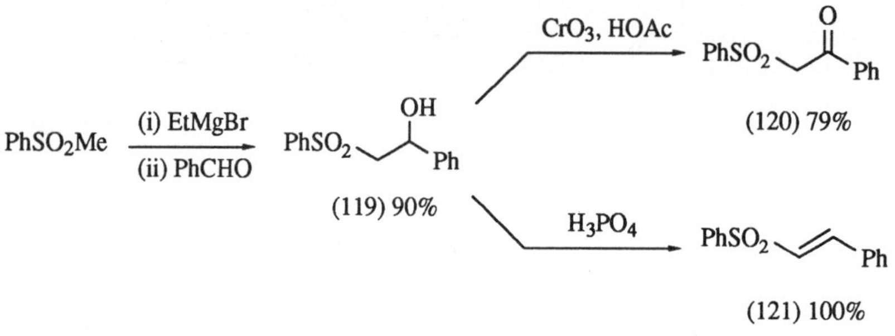

Scheme 35

Field reported that treatment of (**119**) with potassium hydroxide effects a retro-aldol reaction to give methyl phenyl sulphone and benzaldehyde. Hydroxy sulphones are also obtained as the exclusive products in reactions of magnesiosulphones with α,β-unsaturated aldehydes and ketones, Scheme 36.[76]

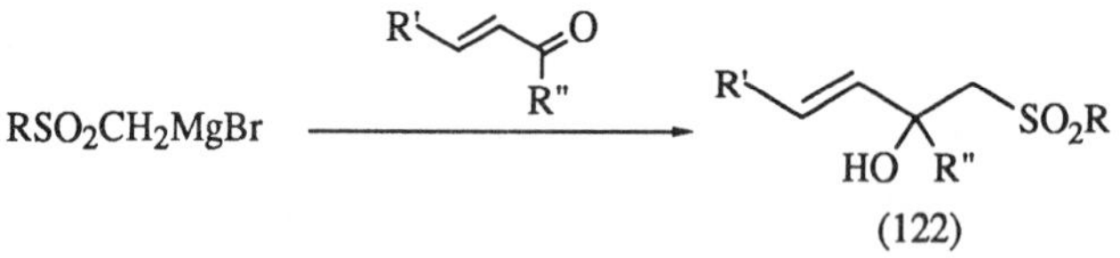

Scheme 36

The behaviour of the magnesiosulphones contrasts with that of ordinary Grignard reagents which give substantial 1,4-addition in some cases. This general reluctance of sulphonyl carbanions to undergo conjugate addition to α,β-unsaturated carbonyls is now well established, although a few examples are collected in the next section. The hydroxy sulphone products (**122**) are rather unstable when $R'' =$ alkyl. This is attributed to dehydration and polymerisation; however, another problem with such products is the great ease of reversion to the ketone and sulphone starting materials, as mentioned for (**119**). Kocienski has described this phenomenon, which is particularly noticeable when attempting derivatisation (i.e. for subsequent Julia elimination) of β-hydroxy sulphones derived from ketones, Scheme 37.[77]

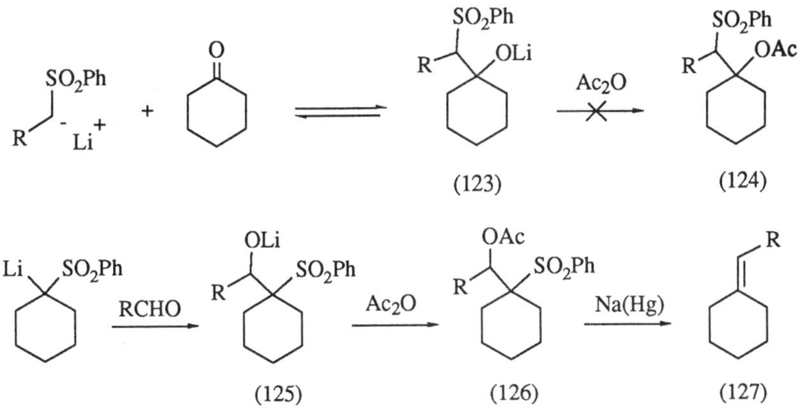

Scheme 37

Thus (**124**), a potential precursor to (**127**), is not available, due to the instability of the tertiary alkoxide (**123**). The problem can be overcome in some cases by reversing the functionality on the two reaction components, in which case (**125**) can be acetylated to give (**126**) and thence (**127**).

The problems inherent in using magnesiosulphone derivatives, particularly the very low solubility of those reagents in the usual solvents, and the extended reaction times often required, stimulated a search for more suitable conditions for the generation and electrophilic quench of sulphonyl carbanions. Although the use of metal amides in liquid ammonia has been examined in detail,[78] by far the most effective, convenient and popular conditions involve the use of ^{n}BuLi as base in THF solvent. Truce and Klingler used such conditions to study the formation of pairs of diastereoisomeric hydroxy sulphones, Scheme 38.[79]

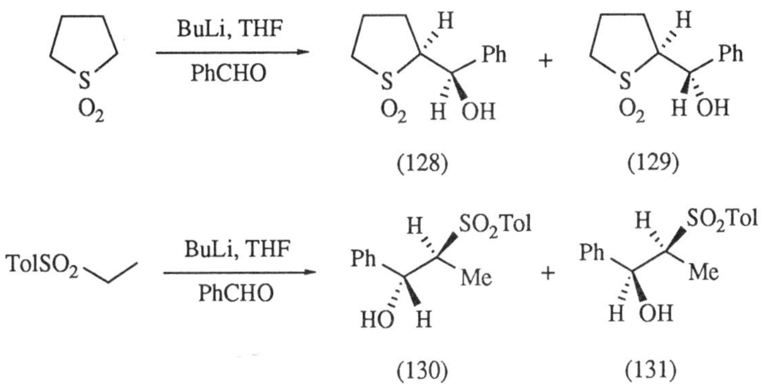

Scheme 38

The first reaction shown gives a 1:1 mixture of (128) and (129) whilst the second favours the formation of the *erythro*-isomer (130) over the *threo*-hydroxy sulphone (131), (ratio 62:38). The formation of mixtures of stereoisomers in such reactions appears quite general, although for applications involving the Julia elimination this is of no consequence.   Truce was able to make unambiguous assignments of configuration of the β-hydroxy sulphone products as well as using NMR in order to establish their preferred conformations. Significantly, the paper also describes the highly stereoselective access to the *threo*-β-hydroxy sulphone isomers by reduction of the corresponding β-ketosulphones with $NaBH_4$ in methanol.

More recent examples of β-hydroxy sulphone preparations include an intramolecular cyclisation of (132)[80] and the addition of oxygenated sulphone (133) to various aldehydes and ketones, Scheme 39.[81]

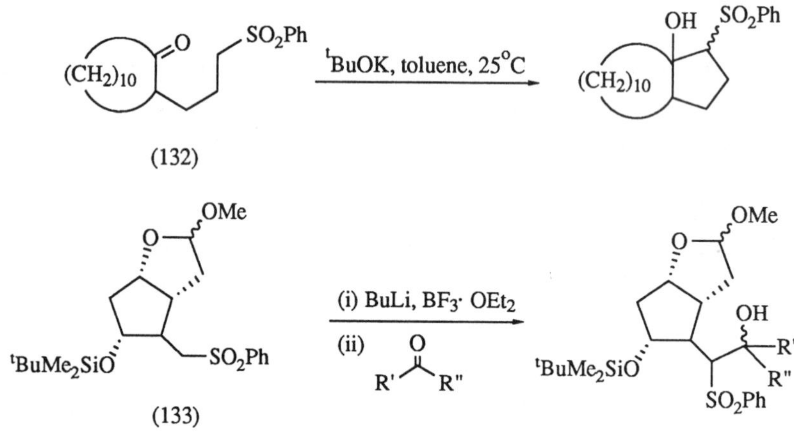

(132)

(133)

Scheme 39

In the reactions of (133) little or no adduct is formed in many cases unless $BF_3 \cdot OEt_2$ is included. The problem here may be one of competing proton abstraction from easily enolisable carbonyl partners by the sulphonyl carbanion. The addition of $BF_3 \cdot OEt_2$ may either form a less basic 'ate'-complex of the sulphonyl carbanion, or simply activate the carbonyl towards nucleophilic attack rather than deprotonation. Kocienski *et al.* solved the same problem by reverting to the use of magnesio- rather than lithiosulphones.[82] Julia and co-workers have shown that β-hydroxy sulphones can be obtained in excellent yield by the reaction of a sulphonyl carbanion with a suitable metal derivative of δ-valerolactol, Scheme 40.[83]

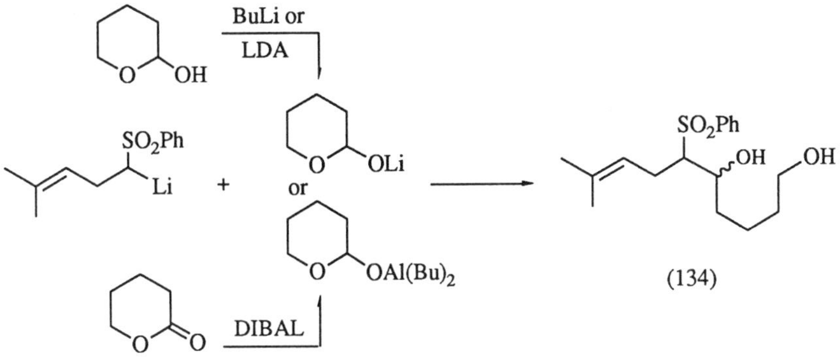

Scheme 40

The required lactol derivatives are prepared either by deprotonation of the parent lactol or by reduction of γ-valerolactone with DIBAL. In both cases a 1:1 mixture of diastereoisomeric products (**134**) is obtained in 86–87% yield.

3.2.4 Michael Additions

The Michael reaction of sulphonyl carbanions with α,β-unsaturated esters has proved to be a useful method of preparing cyclopropane ester products, Scheme 41.[84]

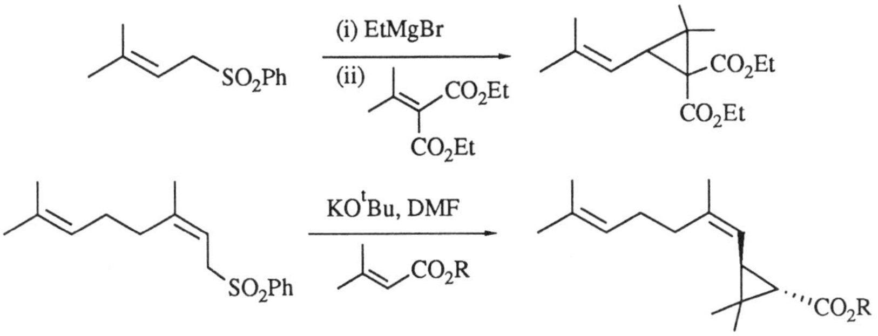

Scheme 41

In such cases the initial conjugate addition is followed by intramolecular displacement of sulphinate. The reactions are *trans*-selective, forming products such as (**135**), in which the original double-bond stereochemistry of the allylic sulphone is retained in the product (*cf.* Scheme 25). Similar reactions occur using benzyl triflones under even milder conditions.

Martel *et al.* showed that when allyl methyl sulphones are employed in this sequence the initial conjugate addition to the unsaturated ester is followed by proton transfer and intramolecular sulphone acylation to give (136), Scheme 42.[85]

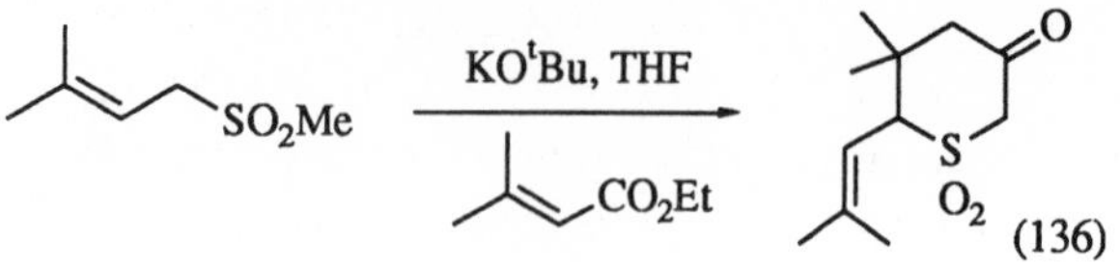

Scheme 42

An alternative mode of cyclopropanation, also involving Michael addition of a sulphonyl carbanion, has been observed by Ghera and Ben-David in reactions of bromocrotonate (137) with certain benzylic sulphones, Scheme 43.[86]

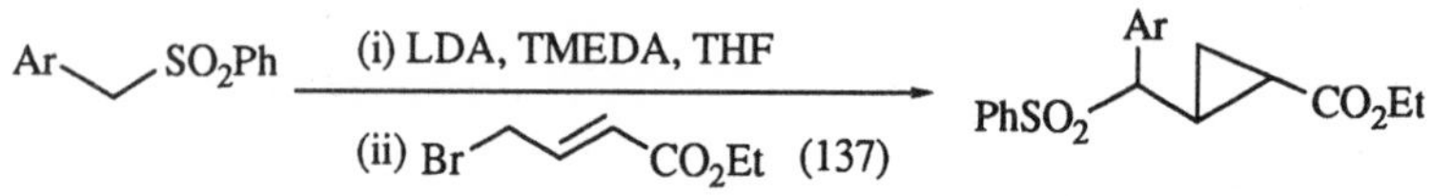

Scheme 43

As will be seen from the examples below the requirement for a relatively stabilised sulphonyl carbanion, particularly allylic and benzylic systems, appears quite general if Michael addition rather than direct 1,2-addition is to be observed. A number of exceptions have been described, however, for example the Michael additions of sulphone (138) to cycloalkenones, described by De Lambaert *et al.*, Scheme 44.[87]

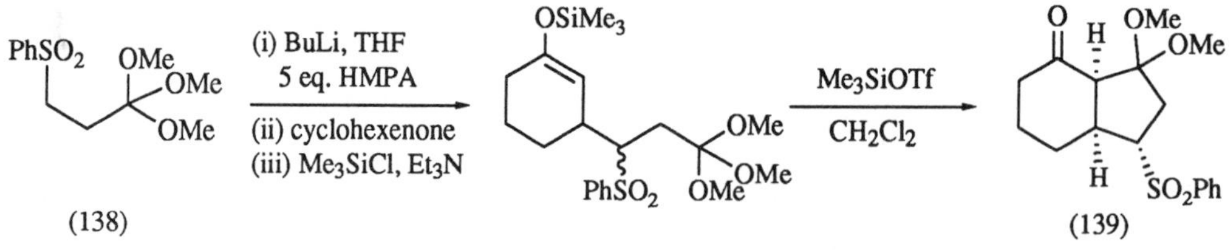

Scheme 44

The addition of 5 equivalents of HMPA to the reaction medium is crucial to the success of the Michael addition, in THF alone mixtures of 1,2- and 1,4-adducts are obtained. As shown, the enolate resulting from the Michael addition can be trapped as an enol silane, and subsequently cyclised to give 'cyclopentannulated' products such as (**139**).

Studies of Michael additions of allylic sulphonyl carbanions to enones have been carried out by several groups.[88–90] These reactions are complicated by the ability of the enone partner to react in either a 1,2- or a 1,4-sense, and by the fact that the sulphone can react α or γ to the sulphur atom. The most detailed and informative account by far is that of Haynes' group,[90] which includes a detailed study of the product distribution and stereochemistry from a good range of sulphone reactions, Scheme 45.

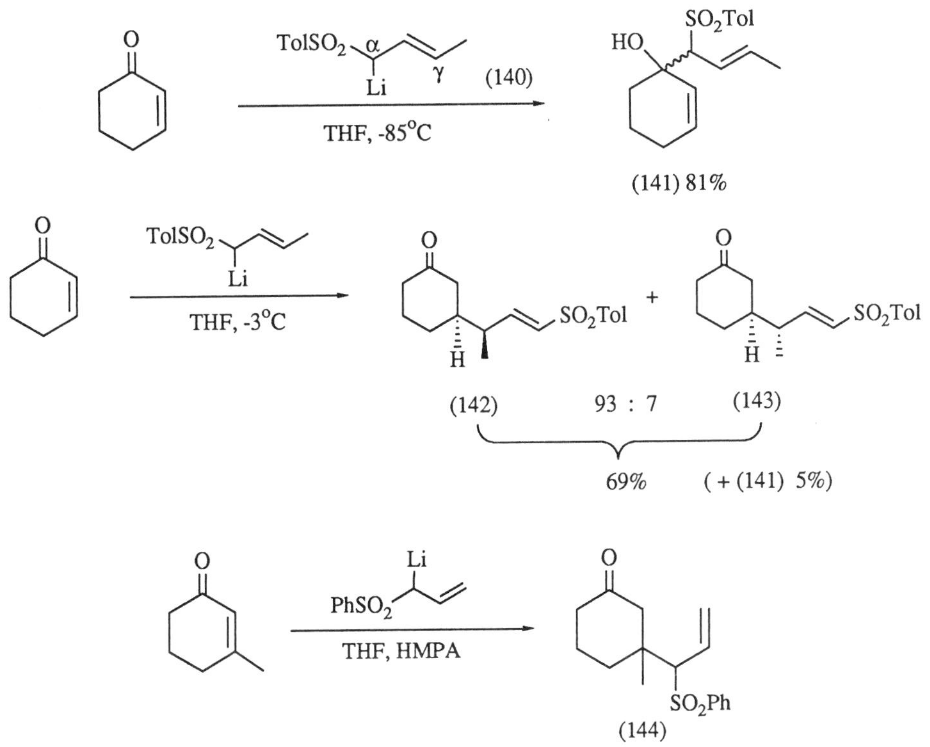

Scheme 45

Using THF as solvent 1,2-adducts such as (**141**) are formed from (**140**) only at low temperature, whereas at around 0°C kinetically controlled conjugate addition of the allyl sulphone occurs to give γ-products such as (**142**) and (**143**). The diastereo-isomeric ratio (**142**):(**143**) reflects the (E):(Z) ratio of the starting allylic sulphone. These products can also be prepared by 'rearrangement' of the initial 1,2-adducts, which was shown to occur by a dissociation (retro-aldol)–recombination route, on warming to 0°C. That (142) and (143) are formed at 0°C by kinetically controlled addition, and not 1,2-addition–rearrangement, is indicated by the much slower rate of the latter process (examined by treatment of (141) with LDA). Contrary to the above, reactions carried out with HMPA as cosolvent occur at -78°C to give Michael adducts by sulphonyl anion reaction at the α-position, e.g. (144).[89] Clearly the ability to tune this type of reaction, simply by changing the solvent, thus making available either type of product at will, is a highly desirable feature.

Two examples of the use of Michael additions of anions derived from benzylic and allylic sulphones in synthesis are shown in Scheme 46.[91,92]

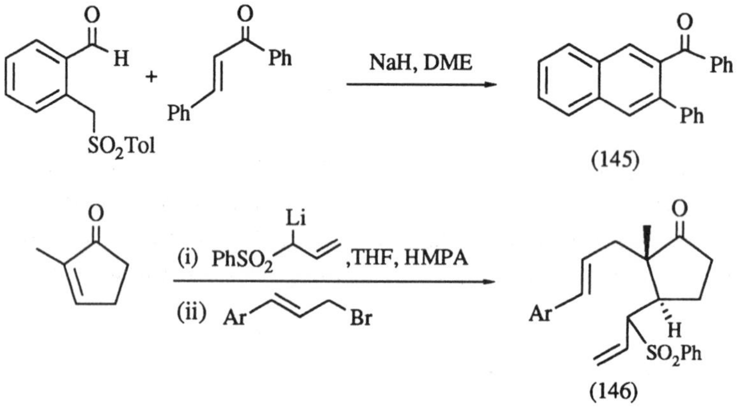

Scheme 46

Both sequences illustrate the potential of a conjugate addition followed by electrophilic trap to introduce both α- and β-substituents onto a suitable Michael acceptor. Many variously substituted naphthalenes such as (145) can be prepared as shown, simply by changing the substitution pattern or oxidation state of the two reaction partners.[91] The preparation of (146) illustrates a three-component coupling which provides stereoselective access to a steroid precursor.[92]

Sulphones having α-nitrogen substitution can also participate in Michael additions, Scheme 47.[93,94] Ylide (147) undergoes addition to maleimides in the presence of alcohols to give products of formula (148).[93] This corresponds to a formal addition of a formyl anion in a Michael sense. The second sequence proceeds by Michael addition of the anion of unsaturated tosyl isocyanide (149) (an alkenyl TosMIC reagent) to an α,β-unsaturated carbonyl compound, followed by intramolecular enolate attack on the isocyanide terminus.[94]

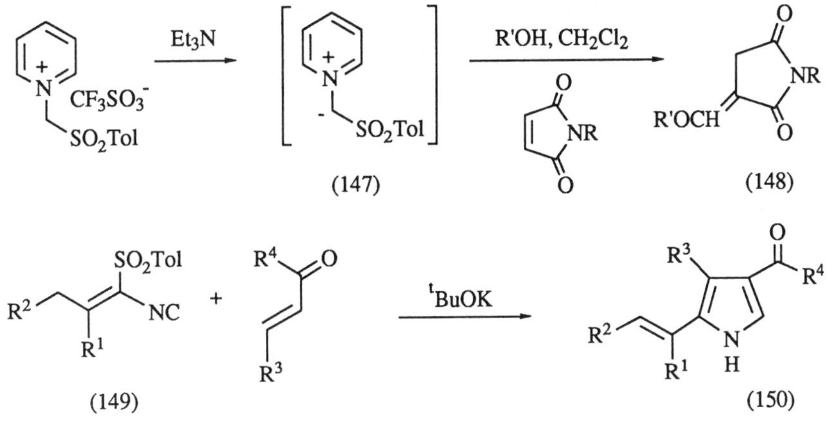

Scheme 47

Elimination of sulphinate then gives the pyrrole product. The method can be extended to allow elaboration of these products to indoles, simply by using a dienone in the initial Michael addition. This gives a product (**150**) having R^3 = CH=CHR, which is set up for an electrocyclic ring-closure–dehydrogenation sequence as outlined in Scheme 48.

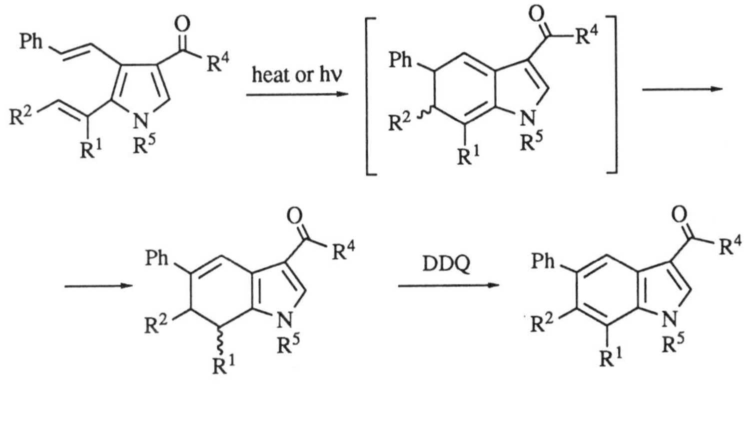

Scheme 48

Hauser and Rhee described a very elegant method for the annulation of aromatic rings which utilises the conjugate addition of a benzylic sulphonyl carbanion, such as that derived from (**151**), Scheme 49.[95]

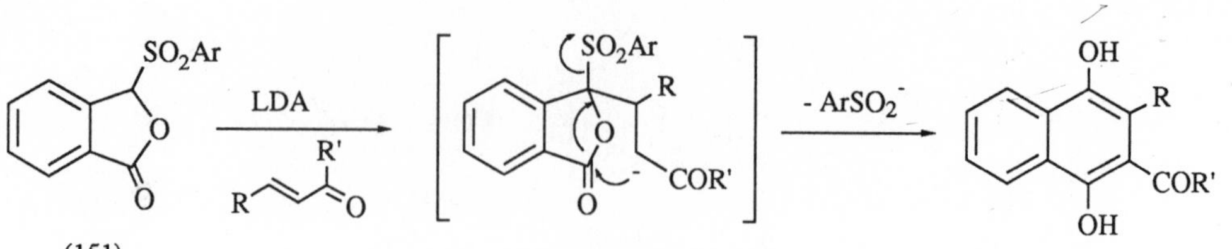

(151)

Scheme 49

This rather neat, regiospecific synthesis has been applied to a number of targets including quinoid pigments having a very unusual benz[*a*]naphthacene ring system[96] and intermediates for anthracyclines.[97]

Finally, acrylonitrile gives the double Michael product (**152**) when heated with an allyl sulphone under the phase-transfer conditions described earlier, Scheme 50.[54]

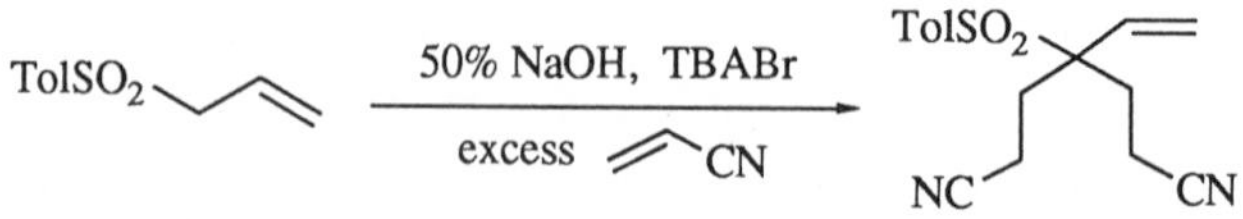

(152) 61%

Scheme 50

3.2.5 Acylations

This topic has been briefly covered in Chapter 2 in the context of simple β-ketosulphone preparation. The first significant studies in this area, conducted by Truce and Knospe, revealed that both inter- and intramolecular acylation of sulphones provides access to β-ketosulphone products, Scheme 51.[98]

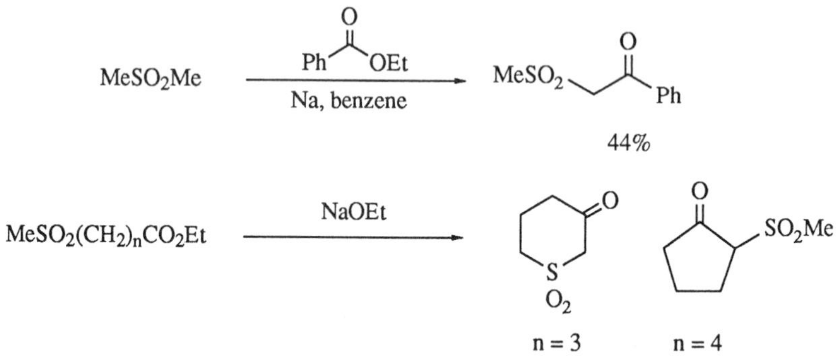

Scheme 51

Russell *et al.* have examined the reactions of dimethyl sulphone with a range of diesters such as diethyl phthalate and obtained mixtures of 2:1 condensation products such as (**153**) and (**154**), Scheme 52.[99]

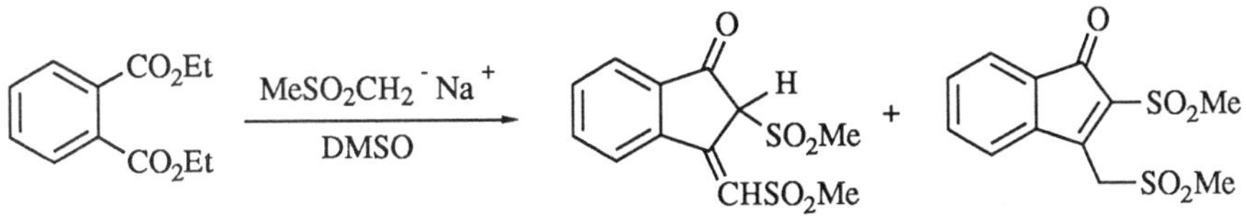

(153) 10%   (154) 70%

Scheme 52

These products most likely arise through initial attack on each of the ester groups by a sulphonyl carbanion to form a bis(ketosulphone), which then cyclises. More recently Trost *et al.* have employed a sulphone acylation reaction in the construction of a precursor to (+)-brefeldin A, Scheme 53.[100]

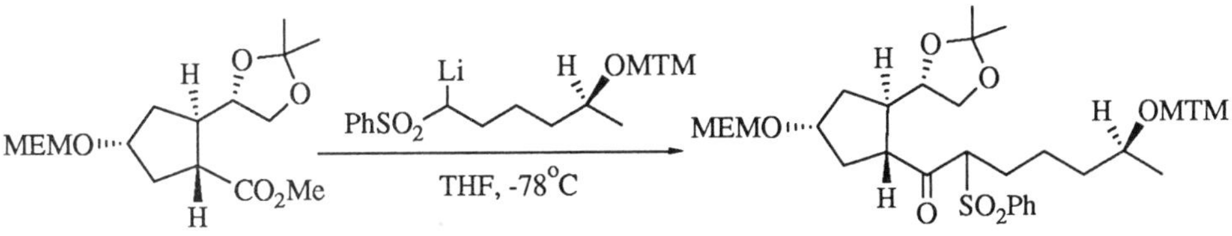

Scheme 53

As is to be expected, either an excess of sulphonyl carbanion or of base (LDA) is required to achieve good yields of the β-ketosulphone product.

Lactones can also be employed in reactions with sulphonyl carbanions, as demonstrated by the synthesis of spiroketals described by Brimble and co-workers, Scheme 54.[101]

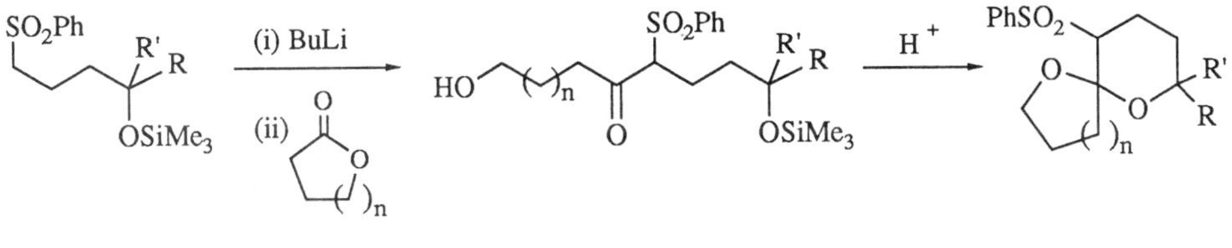

Scheme 54

In this case spiroketal products are obtained in only moderate yield (*ca.* 50%) despite the use of two equivalents of BuLi in the condensation step. Intramolecular acylation of sulphonyl carbanions is a good way to access highly functionalised tricyclic and spirocyclic β-ketosulphones, Scheme 55.[102,103]

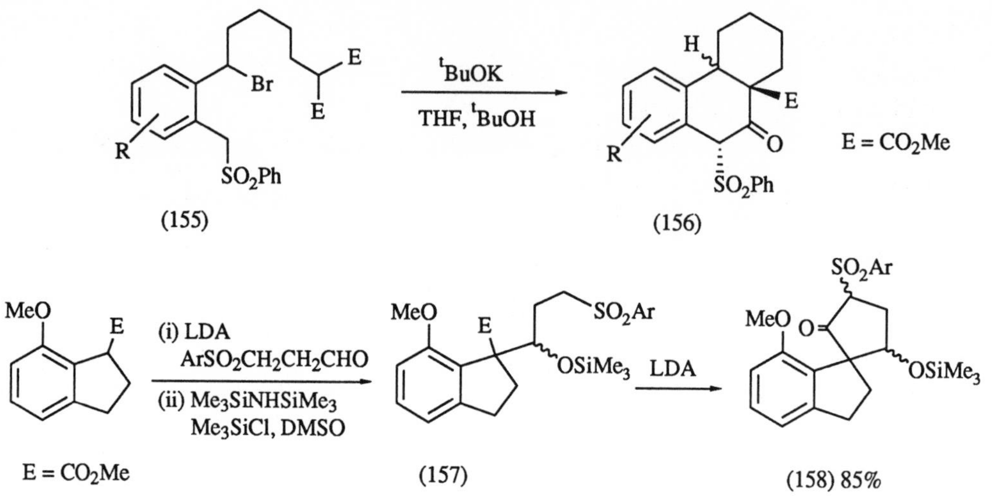

Scheme 55

Treatment of (155) with an excess of base results in successive intramolecular malonate alkylation and sulphone acylation. The tricyclic products such as (156) are obtained stereoselectively, the major products having *trans*-ring fusion.[102] Cyclisation of (157) occurs in similar fashion, using LDA at -78°C with warming to 0°C, to give spirocyclic product (158) in good yield.[103]

Analogous cyclisation chemistry can be carried out using amides and carbonates as the electrophilic sites, Scheme 56.[104–106]

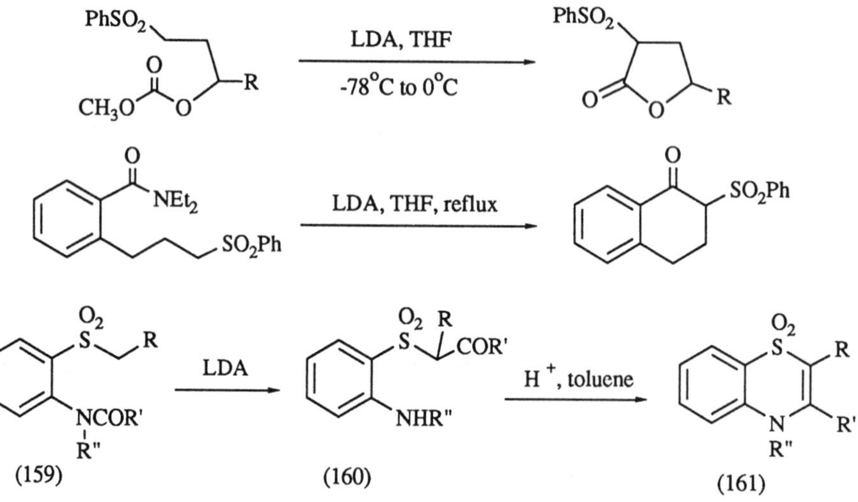

Scheme 56

The formation of (160) from (159) occurs by intramolecular acyl transfer from nitrogen to carbon.[106] The desired cyclic products (161) can be obtained by resorting to acid-catalysed condensation.

Finally, reaction of sulphonyl carbanions with nitriles gives isolable iminosulphone products of structure (162), Scheme 57.[107]

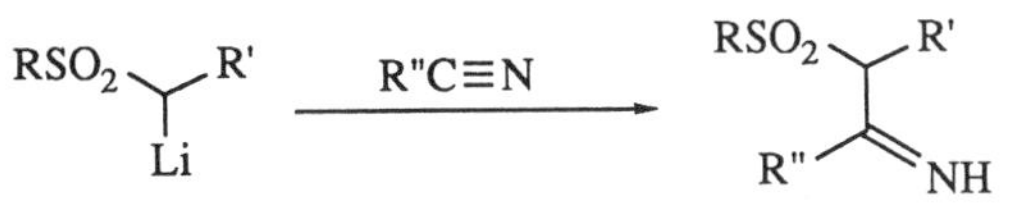

Scheme 57

These products were further transformed into unusual types of heterocyclic compound, such as 1,3-thiazine-2,6-dithiones, 1,2-dithiole-3-thiones and 1,3-thiazole-2-thiones, by reaction with carbon disulphide.

3.2.6 Dimerisation, Coupling and Reactions with Organometallics

Julia has firmly established the coupling of sulphonyl carbanions using cupric salts as a means of obtaining disulphones, and derived products, Scheme 58.[108]

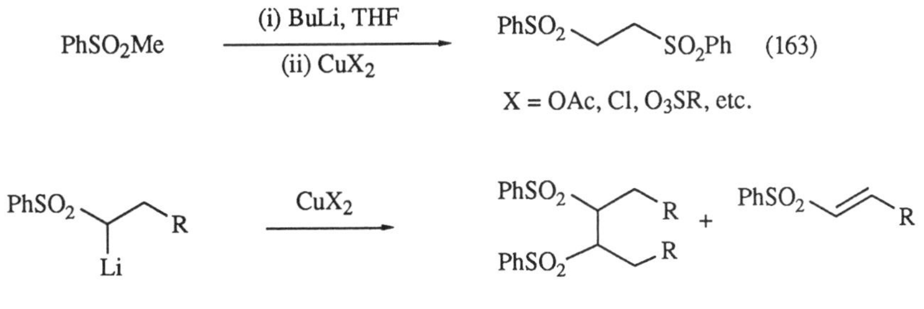

Scheme 58

In the coupling of methyl phenyl sulphone the yield of dimer (163) was found to be highly dependent on the nature of the cupric salt used, with $CuCl_2$ giving the best results (70%). In reactions of higher homologous sulphones dehydrogenation to give vinyl sulphones competes with the coupling. If desired, clean conversion to the vinyl sulphone can be achieved by reaction of the lithiated sulphone with an excess of cupric acetate. Buchi and Freidinger have conducted related studies of the coupling of allylic sulphones using either iodine or ferric chloride, Scheme 59.[109]

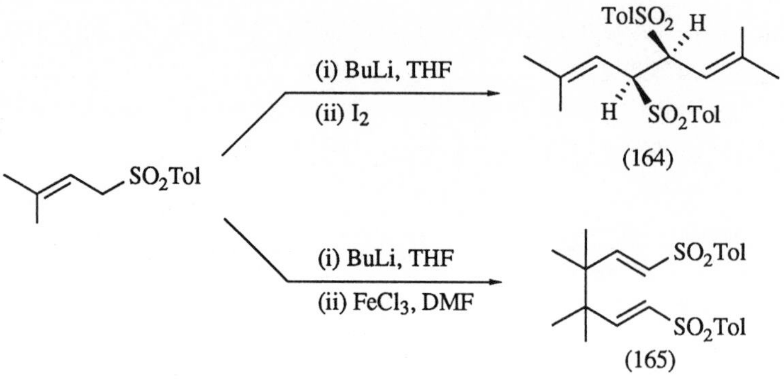

Scheme 59

The two oxidants give complementary results, with iodine giving the α,α-coupled product (164), and with $FeCl_3$ giving bis(vinyl sulphone) (165). The first product arises by sulphonyl carbanion iodination and reaction of the so-formed α-iodo-sulphone with a further equivalent of sulphonyl carbanion. Compound (165) presumably results from a regioselective radical coupling process.

Another sulphone coupling reaction uses the magnesio-derivatives of allylic sulphones and results in the formation of desulphonylated conjugated trienes, Scheme 60.[110]

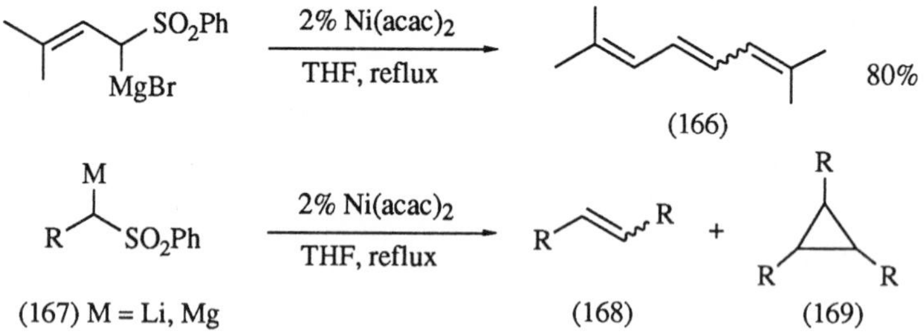

Scheme 60

A range of trienes such as (166) are obtained in good yield as mixtures of stereoisomers about the newly formed double bond. The reaction is less clean with other types of sulphonyl carbanions, although simple alkenes (168) can be formed in reasonable yield. Remarkably, (167), where R = Ph and M = Li, gives cyclopropane (169) in 64% yield, accompanied by only minor amounts of stilbene.

Julia *et al*. have examined the reaction of sulphonyl carbanions with a range of halogenocarbenoids.[111,112] Their first report describes reactions using $HCBr_2Li$ and $RCBr_2Li$ which enable novel syntheses of mono- or *gem*-dihalogenoalkenes respectively, Scheme 61.[111]

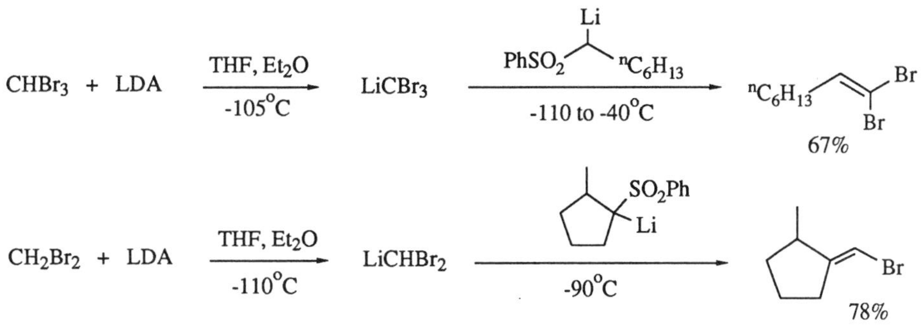

Scheme 61

Evidence presented by Julia clearly points to the involvement of a metal carbenoid, rather than a derived carbene, in the coupling process. The more thermally stable bromocarbenoids are preferred over the chloro analogues, which do not give good results. The reaction is stereoselective, giving predominantly (*E*)-bromoalkenes from reactions of LiCRBr$_2$ with a variety of sulphones. Interestingly, the use of sodium carbenoids, which have greater thermal stability than the lithium derivatives, allows one reaction to be carried out at -45°C, although with somewhat poorer stereoselectivity than at -90°C. A later study by the same group describes the extension of this carbenoid chemistry to iodo analogues, thus providing stereoselective access to iodoalkenes.[112]

Three other examples of reactions of sulphonyl carbanions with organometallic partners are grouped together in Scheme 62.

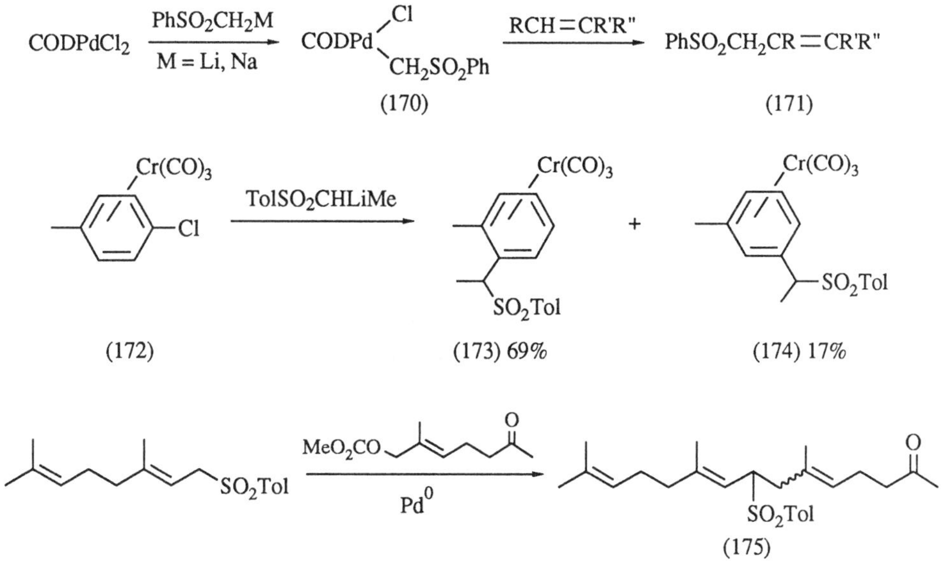

Scheme 62

The remarkable addition of a $PhSO_2CH_2$ unit to a simple unactivated alkene is yet another contribution from Julia's group.[113] The reaction presumably proceeds by addition of the alkene to the COD palladium complex (170), followed by a β-elimination process to give the substituted alkene product (171).

As indicated by the conversion of (172) to (173) and (174), sulphonyl carbanions can react with chlorotoluenetricarbonyl complexes to give products corresponding to various modes of S_NAr reaction.[114] The synthesis of (175) represents one of the very few examples of the reaction of a simple sulphonyl carbanion with a (π-allyl)palladium complex.[115] Usually in such reactions more stabilised anions must be used (e.g. bis-sulphones, see next section) but here allylic or benzylic sulphones (but *not* alkyl sulphones) participate.

3.2.7 Reactions of sp²-Hybridised Sulphonyl Carbanions

The metallation of sulphones described in this chapter largely involves the removal of a hydrogen α to the sp^3-hybridised carbon of a sulphone of general formula $ArSO_2CHRR'$ (R, R' = H, alkyl, etc.). If no such sp^3-carbon centre is available then removal of a hydrogen from an sp^2-carbon centre (i.e. from alkenyl or aryl groups) activated by a sulphone group is possible. Such deprotonations to give sp^2-sulphonyl carbanions are described below.

The original report of Eisch and Galle described the deprotonation of simple aryl alkenyl sulphones using MeLi at -90°C, Scheme 63.[116]

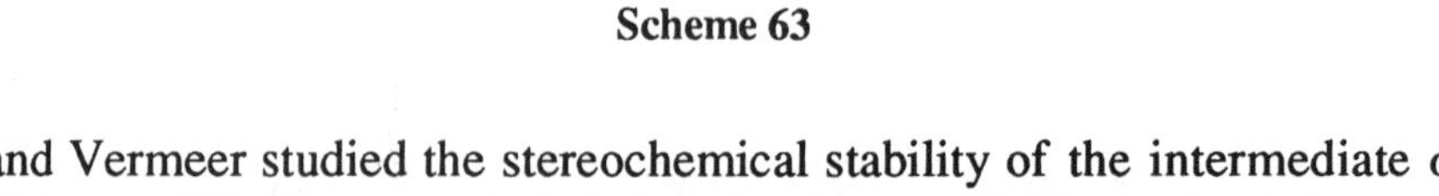

(176)

Scheme 63

Kleijn and Vermeer studied the stereochemical stability of the intermediate α-lithio-vinyl sulphones.[117] They concluded that simple *trans*-lithiovinyl sulphones such as (176) can be formed cleanly, but the corresponding *cis*-isomers suffer isomerisation at -60°C to give an equilibrium mixture of anions in which the *trans*-compound predominates. As shown in Scheme 64, trisubstituted vinyl sulphones also undergo isomerisation under these conditions. Thus, either isomer of the sulphone (177) gives the same mixture of stereoisomers following deprotonation and reprotonation.

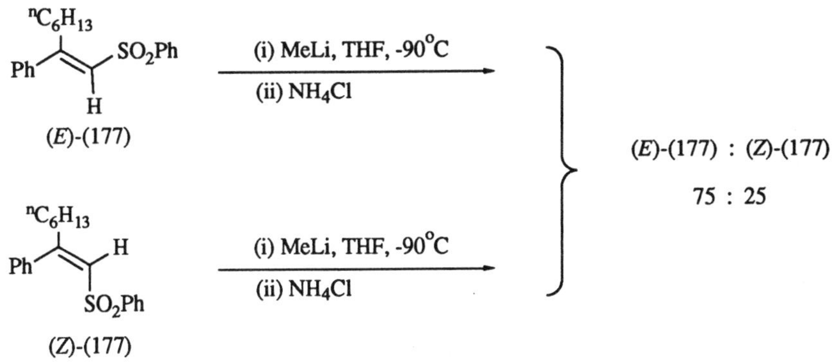

Scheme 64

An intriguing aspect of this report is that the less stable (Z)-isomers were prepared via intermediate α-cupriovinylic sulphones, e.g. Scheme 65.[118]

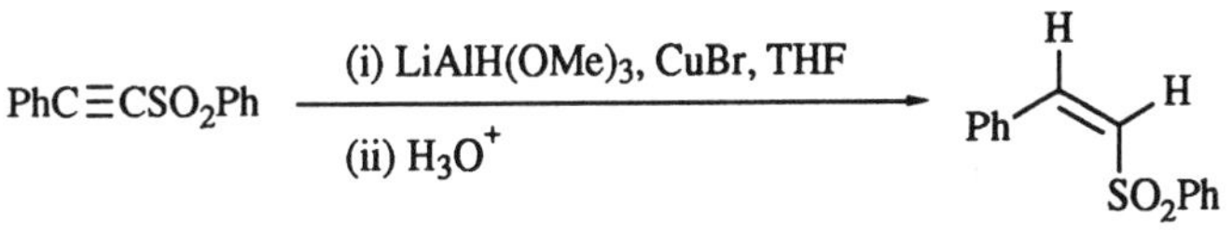

Scheme 65

Thus it would appear that the copper derivatives are more configurationally stable than the lithio-anions.

Variously heterosubstituted vinyl sulphones have been deprotonated as above and found to react with electrophiles to give substituted derivatives, Scheme 66.[119–121]

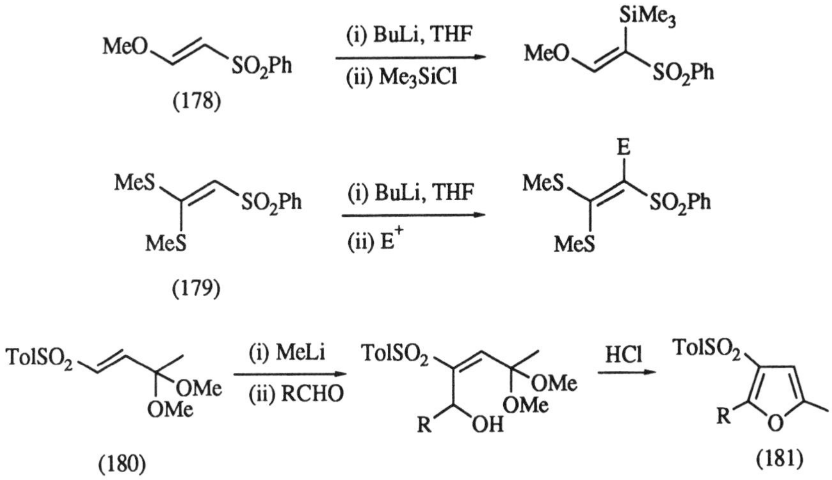

Scheme 66

Of special interest is the great range of substitutions possible using sulphone (**180**), for example with aldehydes, leading to useful products such as furan (**181**).[121] In each of the systems (**178**)-(**180**) (Z)- to (E)-anion isomerisation does not arise since only the more stable (E)-isomers of (**178**) and (**180**) are used.

Cyclisations of α-lithiovinyl sulphones have been carried out by McCombie *et al.*, using either epoxides or ketones as the intramolecular electrophilic trap, Scheme 67.[122,123]

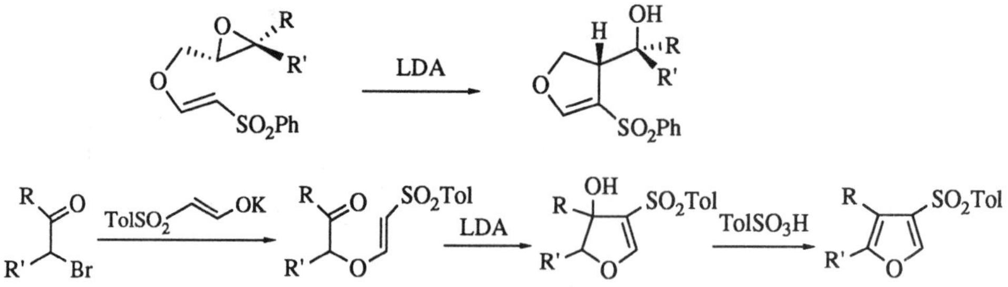

(182)   (183)

Scheme 67

Remarkably, intermediates of structure (**182**) cyclise to give dihydrofurans by abstraction of the α-vinylic sulphone hydrogen, even where R is an alkyl group. Thus there is a strong kinetic preference for formation of the vinyl carbanion over alternative ketone enolisation. This appears to be a useful new approach to 2,3-disubstituted furans of general structure (**183**). The same research group has determined that vinylsulphonyl carbanions having a β-oxygen grouping are even less configurationally stable than simple alkyl derivatives.[124] Thus treatment of (Z)-vinyl sulphones such as (**184**) with BuLi results in rapid and complete isomerisation to the (E)-vinylic carbanion, even at -90°C, as indicated by the stereochemistry of derived products, i.e. (**185**), Scheme 68.

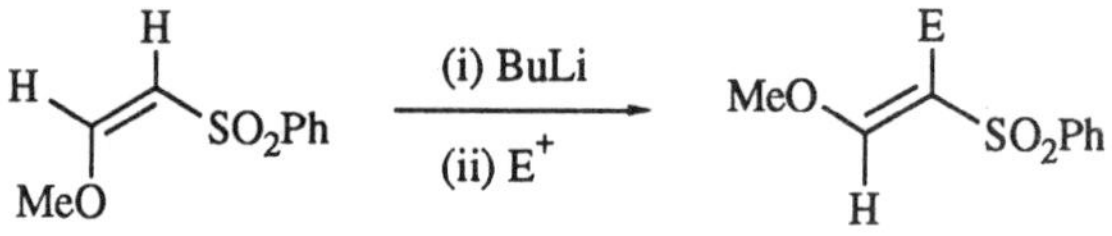

(Z)-(**184**)   (E)-(**185**)

Scheme 68

The type of epoxide cyclisation shown in Scheme 67 has recently been adopted by Pelter *et al.*, using homochiral epoxides, as an approach to lignan natural products.[125]

The metallation of aromatic sulphones on the aromatic nucleus has been known for many years and was studied in detail by Truce and Mandaj.[126] Deprotonation can occur at an *ortho*-benzylic position (e.g. to form **186**), or in simple diaryl sulphones an sp^2-aryl carbanion, e.g. (**187**), can be formed, Scheme 69.

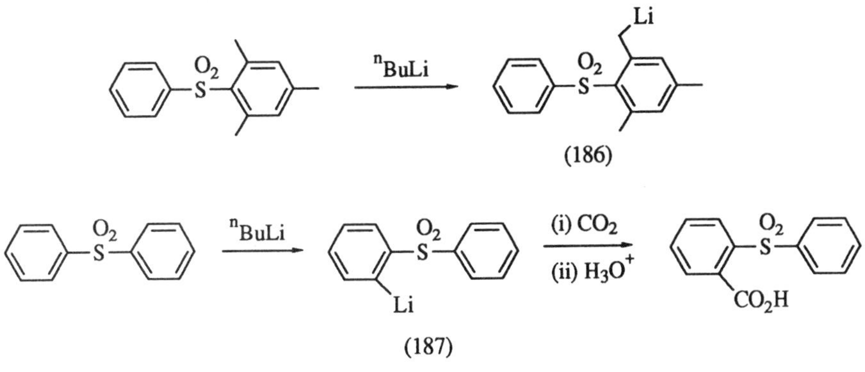

Scheme 69

Lithiosulphones such as (**186**) undergo the Truce–Smiles rearrangement on heating, as described in Chapter 5. Early studies of this type of metallation revealed that systems such as (**188**) can form dianions,[127] whilst deprotonation of phenyl-2-thienyl sulphone occurs preferentially on the thiophene nucleus, to give (**189**), Scheme 70.[128]

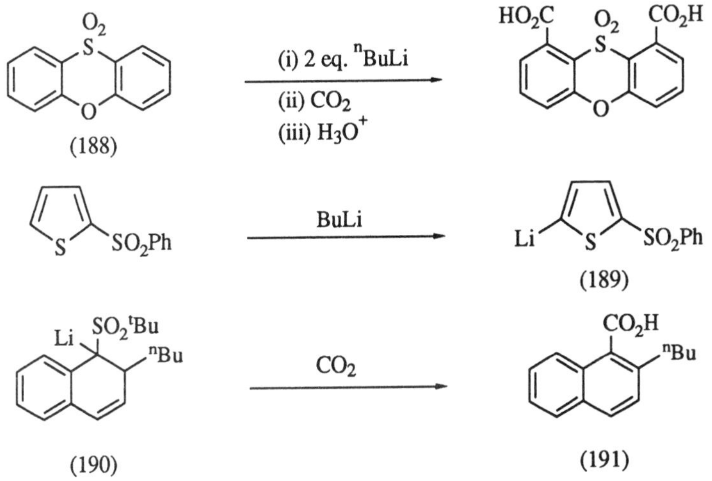

Scheme 70

In addition, naphthyl systems can undergo competing modes of reaction involving addition of BuLi to give lithiosulphones such as (**190**).[129] In this case (**191**) results from addition to CO_2 followed by elimination of sulphinic acid.

The use of two equivalents of base in the presence of an *in situ* Me_2SiCl_2 electrophilic quench has been used to convert diphenyl sulphone to product (192), Scheme 71.[130]

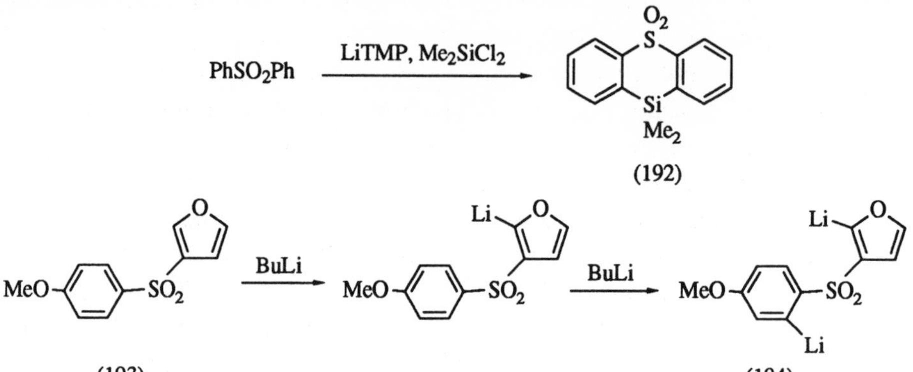

Scheme 71

Deprotonation of furan sulphone (193) occurs initially at C-2 of the furan, and an additional equivalent of BuLi then produces (194).[131]

Snieckus and co-workers have examined the use of the $^tBuSO_2$ group as an activating group for the *ortho*-metallation of aromatic systems.[132] In inter- and intramolecular competition experiments the sulphone proves a powerful directing group, compared to several other known *ortho*-directors such as OMOM and $CONR_2$. A wide variety of electrophiles has been reacted with metallated phenyl *tert*-butyl sulphone. Two sequential metallations can also be conducted, and subsequent desulphonylation, using Raney nickel, provides access to *meta*-substituted aromatics, Scheme 72.

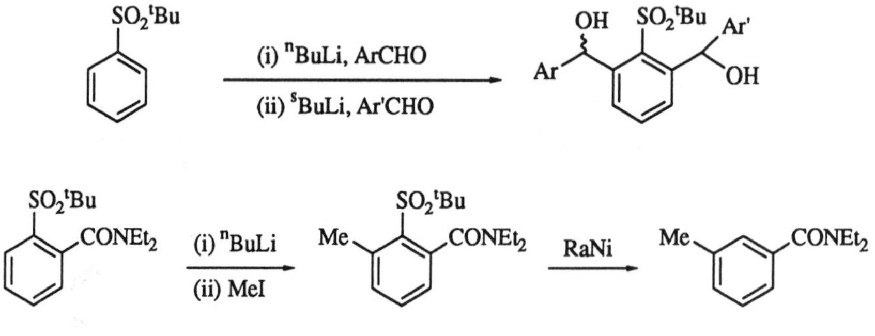

Scheme 72

This methodology offers attractive opportunities for high-yielding and regioselective assembly of polysubstituted aromatics.

3.3 Reactions of Additionally Stabilised Sulphonyl Carbanions

Sulphonyl carbanions having additional functionality directly attached to the anionic centre undergo many reactions analogous to those described earlier in this chapter. However, additional anion stability such as that imparted by a carbonyl, cyano or nitro group (or an additional sulphone group) may significantly modify the reactivity. For example, whereas simple sulphonyl carbanions undergo 1,2-addition to α,β-unsaturated carbonyl compounds, sulphones having additional stabilisation are softer and less basic, and will often undergo 1,4-additions.

The presence of an α-functional group in the sulphonyl carbanion can also lead to unique chemistry, such as the *in situ* desulphonylations possible using α-oxygenated derivatives, or cyclisations observed using TosMIC. Similarly the anion chemistry of β-carbonyl sulphones incorporates the usual complications of enolate chemistry, namely *C*- vs *O*-reactivity, polyalkylation, etc. For these reasons the chemistry of such special sulphonyl carbanions has been given a separate subsection.

3.3.1 Ketosulphones and Related Systems

Kohler and Potter conducted the first detailed study of the alkylation and acylation reactions of β-ketosulphones. As might be expected the use of methyl iodide results in predominant *C*-alkylation, whereas the use of dimethyl sulphate gives the product of *O*-alkylation, Scheme 73.[133]

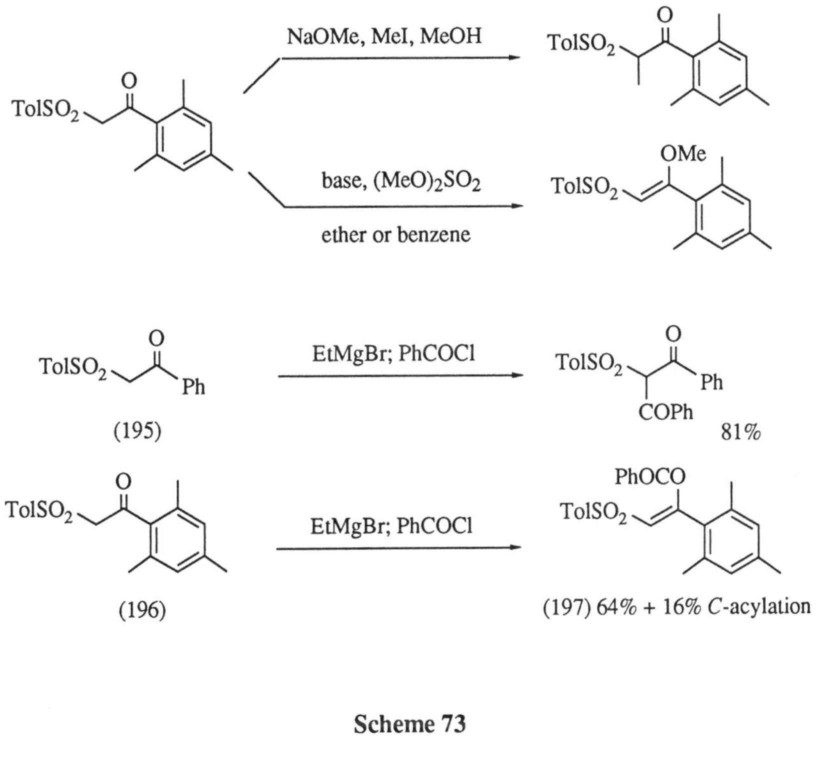

Scheme 73

Sulphones in Organic Synthesis

Interestingly, the magnesio-derivative of phenyl ketone (**195**) gives only the product of *C*-acylation, whereas under the same conditions, the corresponding mesitylene ketosulphone (**196**) gives mainly *O*-acyl product (**197**).

Clean *C*-alkylation of β-ketosulphones has been reported using allyl bromide in the presence of potassium carbonate in DMF or acetonitrile,[134] or by the use of the tetraethylammonium enolate generated using (**198**), Scheme 74.[135]

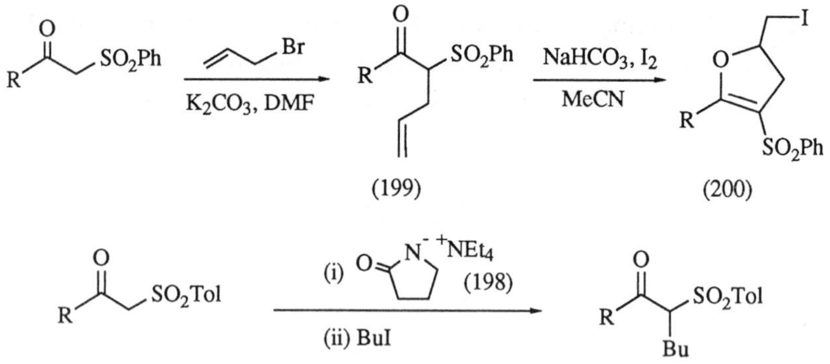

Scheme 74

As shown, the *C*-alkylated product (**199**) is subsequently converted to an *O*-alkylated product (**200**) by an iodoetherification procedure. Elimination of HI using DBU then gives a substituted furan. The formation of (**198**) is carried out electrolytically from 2-pyrrolidone, this species being used predominantly for the *C*-alkylation of β-diketones. Some rather unexpected *C*- vs *O*-alkylation behaviour is observed in cyclisations of a β-ketosulphone on reaction with 1,3-dibromopropane, Scheme 75.[136]

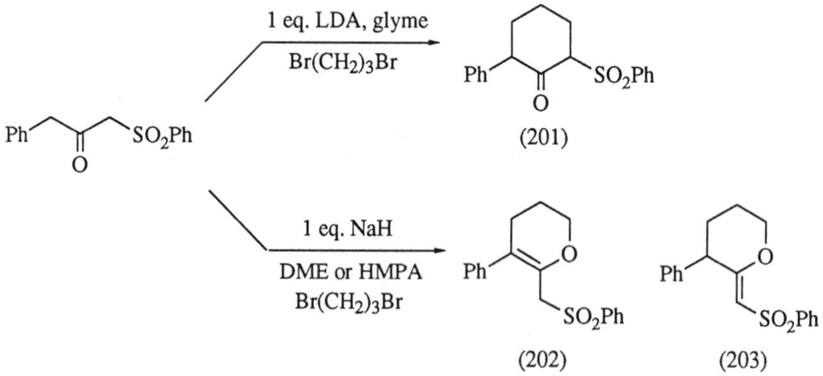

Scheme 75

The use of LDA results in formation of *C*-alkylated product (**201**), whereas NaH in DME or glyme gives (**202**), presumably via (**203**). Although one equivalent of NaH in HMPA gives (**202**), the substitution of DMSO as solvent unexpectedly gives predominantly the *C*-alkylated product (**201**). The anomalous behaviour of this system is attributed to complex conformational possibilities which have not been fully elucidated.

Trost and Urabe have used the alkylation of β-ketosulphones as a means of constructing systems suitable for testing silicon-mediated annulation reactions.[137] Thus alkylation of sulphone (**204**) with the bifunctional reagent (**205**) leads to an intermediate allenylsilane (**206**) which is then cyclised using either Lewis acids, or TBAF in THF, to give dienyl hydroxy sulphone (**207**), Scheme 76.

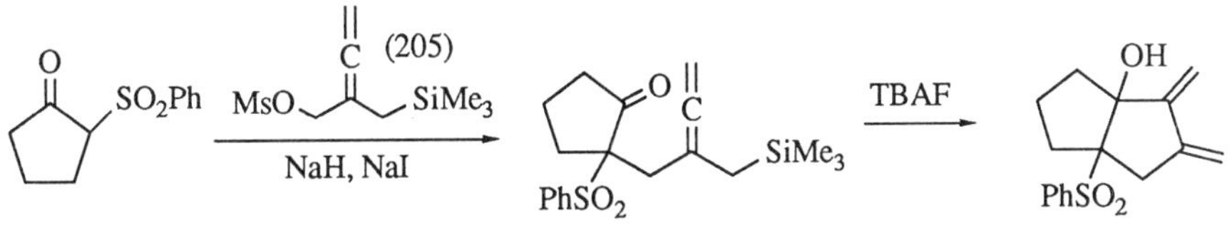

Scheme 76

The β-ketosulphone alkylation presumably occurs by reaction with an *in situ* formed allenyl iodide. The sulphone group in compounds such as (**206**) appears not to interfere with the Lewis acid mediated cyclisation.

One-pot double alkylation of β-keto- or β-ester sulphones is possible, using appropriate dihalogenoalkanes, to give cyclic derivatives, Scheme 77.[138,139]

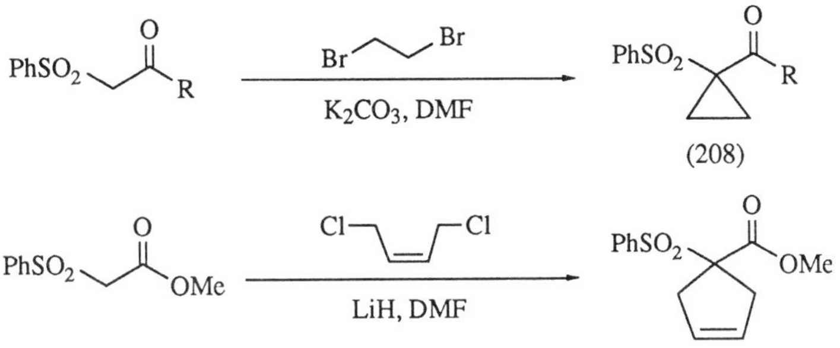

Scheme 77

The rather mild conditions make such alkylations convenient for scale-up. In each of the cases shown subsequent desulphonylations were also investigated. Thus treatment of (**208**) with Raney nickel gives a simple cyclopropylketone in moderate yield.[138] Elimination of sulphinate from (**209**) is possible by treatment with ¹BuOK in THF, to give cyclopentadienylcarboxylic esters.[139] Nantz and Fuchs have also demonstrated the usefulness of sulphones such as (**209**) as precursors to cyclopentadienones, Scheme 78.[140]

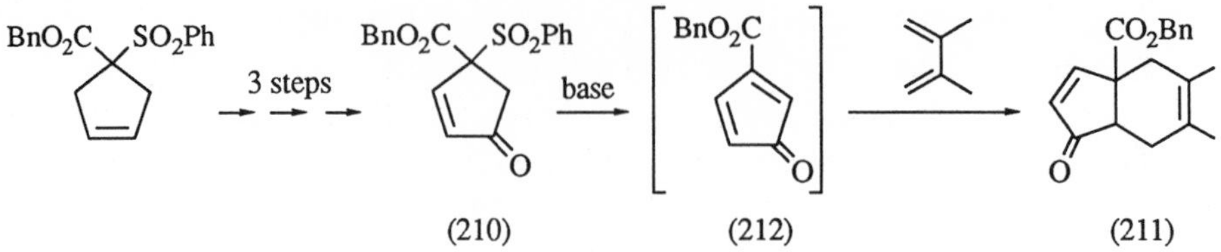

(210)   (212)   (211)

Scheme 78

Thus, treatment of cyclopentenone (**210**) with base in the presence of a suitable diene results in the formation of cycloadducts such as (**211**), derived from the unstable cyclopentadienone intermediate (**212**). Unfortunately plans to extend this process to intramolecular variants were unsuccessful.

α,β-Unsaturated ketones having a γ-sulphone group can be regarded as vinylogous β-ketosulphones. Anions derived from such systems can give either α-alkylation (the normal outcome for simple enones) or γ-alkylation. Since alkylation adjacent to sulphur usually occurs in reactions of unsaturated sulphonyl carbanions the abnormal γ-alkylation was anticipated in such systems. These expectations were partially realised, for example using ketosulphone (**213**), Scheme 79.[141]

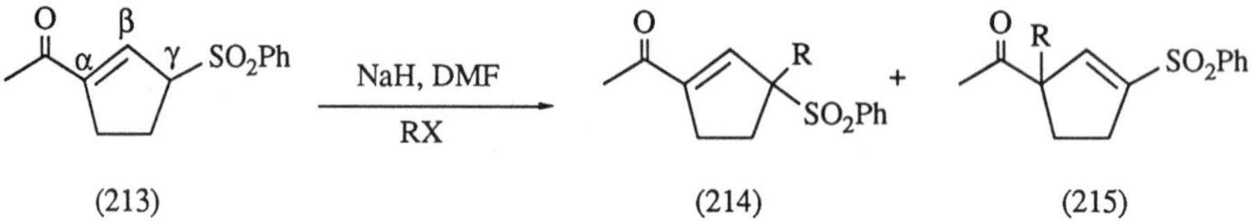

(213)   (214)   (215)

Scheme 79

The γ-alkylation product (**214**) predominates over (**215**) when sterically unencumbered alkylating agents such as methyl iodide or allyl bromide are employed; selectivity, however, is poor with ethyl bromide. This method therefore offers only a partial solution to controlling γ-alkylation in extended enolate systems.

Aldol chemistry of β-ketosulphones is much less widely utilised than alkylation reactions. Nevertheless, this approach can provide a useful entry to aldol products following desulphonylation, and unsaturated ketones following dehydration, Scheme 80.[142,143]

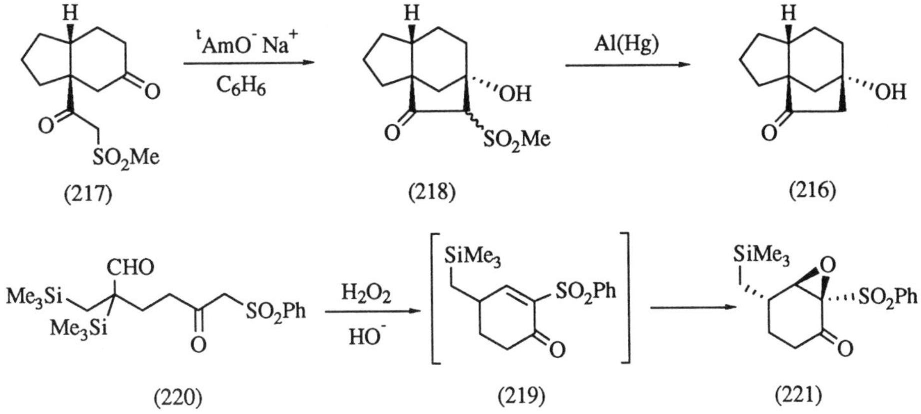

Scheme 80

Cyclisations to form the bridgehead hydroxy ketone (**216**) have been conducted as part of a synthetic study directed towards the gibberellins.[142] Ketosulphone (**217**) is found more satisfactory than either the corresponding ketosulphoxide or simple methyl ketone as a precursor to (**216**). Even so, the yield is rather poor, due to the reversibility of the aldol closure to give (**218**). This problem does not arise in the formation of (**219**), since a Peterson elimination renders the cyclisation of (**220**) irreversible.[143] Under the conditions used the powerful Michael acceptor (**219**) is converted to the epoxide (**221**).

Finally, α-cyanosulphones participate in chemistry analogous to that of ketosulphones, e.g. alkylations using DBU in benzene.[144]

3.3.2 Nitrogen-substituted Sulphones

The alkylation of α-nitrosulphones can be carried out simply by treatment of a derived sodium or lithium salt with an appropriate alkyl halide. Alternatively, allylic acetates can be used as reaction partners if $(PPh_3)_4Pd$ is employed as catalyst, Scheme 81.[145]

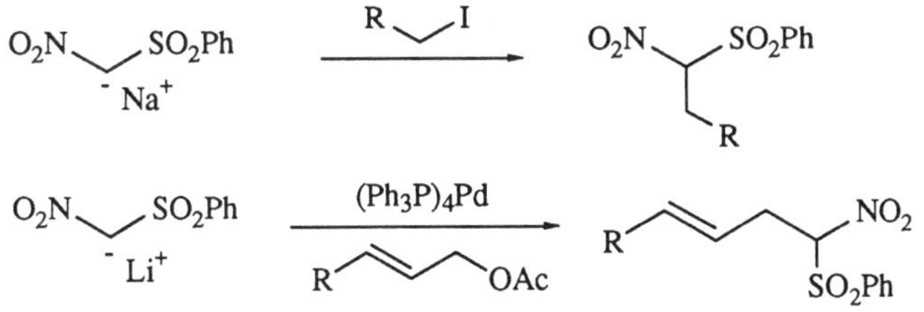

Scheme 81

The metal salts are easy to handle and problems due to *O*-alkylation or polyalkylation are rarely encountered.

Tosylmethylisocyanide (TosMIC) is an easily prepared sulphone with a versatile repertoire of reactions leading to carbocycles and heterocycles.[146] TosMIC carbanion alkylation with alkyl halides allows the introduction of one or two substituents, and by using dihalogenoalkanes as alkylating agents rings of various sizes can be prepared. The isonitrile sulphone functionality can subsequently be converted to a carbonyl group, Scheme 82.[147–150]

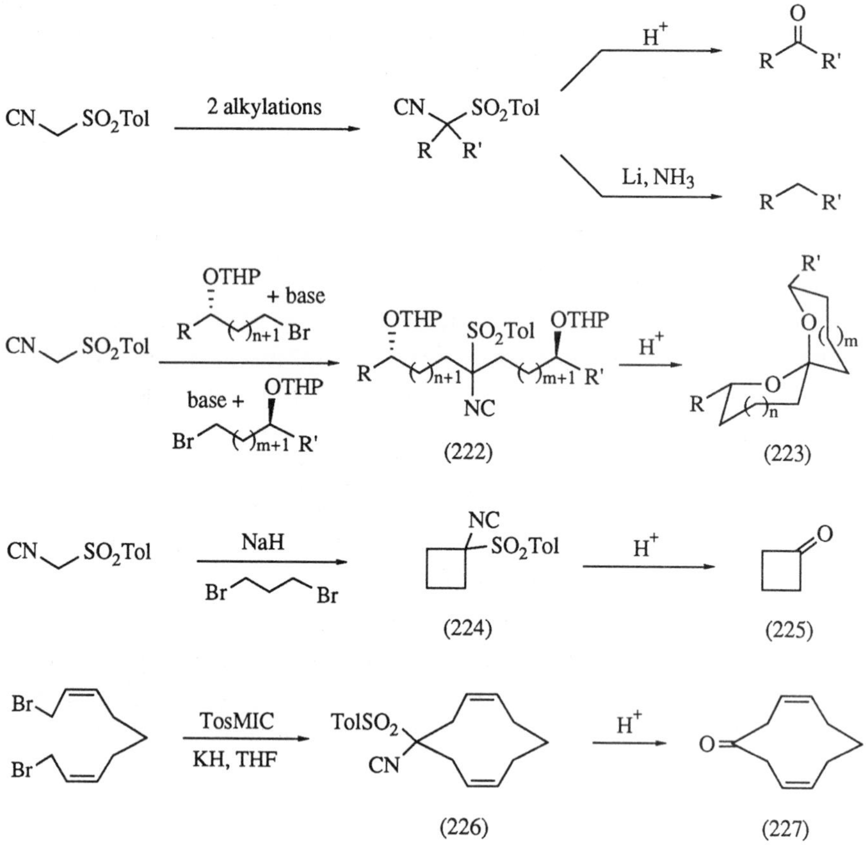

Scheme 82

Both symmetrical and unsymmetrical derivatives (**222**) can be prepared and then converted to spiroketal products (**223**). The ability to prepare small strained rings such as (**224**) which on acid treatment gives cyclobutanone (**225**), and medium rings such as (**226**) which can be transformed into ketone (**227**), further underlines the utility of this approach. In this chemistry the TosMIC unit reacts as an acyl anion or dianion equivalent, which is evident following unmasking of the latent carbonyl functionality. An alternative reductive removal of the two TosMIC functional groups using lithium in ammonia can also be useful.

The central mode of reactivity of TosMIC with aldehydes and ketones involves a stepwise cycloaddition pathway to give a key intermediate anion (**228**). Depending on the precise reaction conditions this can then lead to tosyloxazoline (**229**),[151] *N*-(1-tosyl-1-alkenyl)formamide (**230**),[152] or nitrile (**231**),[153] Scheme 83.

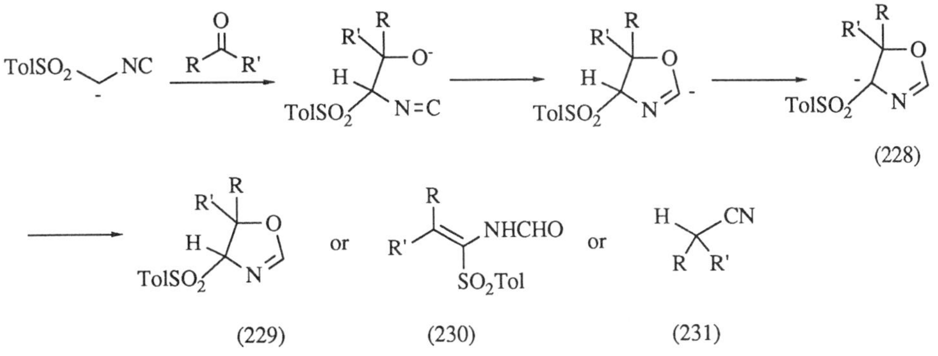

Scheme 83

The use of a weak base, such as K_2CO_3 in MeOH, enables isolation of (**229**), whereas under aprotic conditions (e.g. ^tBuOK in THF) the reaction yields (**230**) and eventually (**231**). If aldehydes are used as the reaction partners then the intermediate tosyloxazolines can also undergo elimination of $TolSO_2H$ to give oxazoles (**232**), Scheme 84.[151]

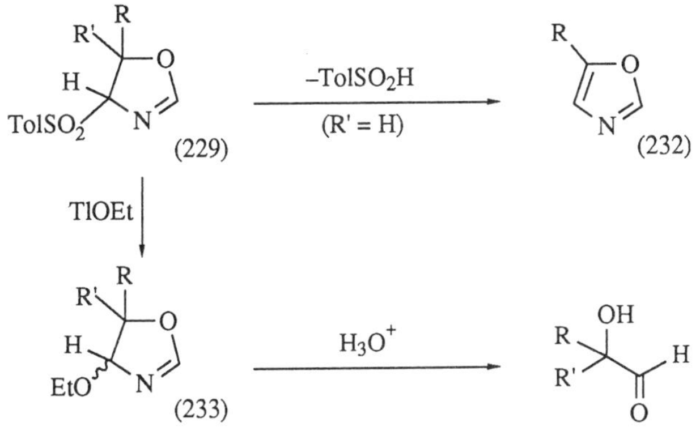

Scheme 84

As shown, 4-ethoxy-2-oxazolines (**233**) may also be formed from (**229**) (TlOEt is simply used as the base in the initial condensation reaction), and their hydrolysis using dilute acid furnishes α-hydroxy aldehydes.[154]

A host of related heterocycle syntheses, including those of imidazoles[155] and thiazoles,[156] follow a similar course to the preparation of oxazoles. Other useful homologations using this reagent have also been developed, including for example, a method for building hydroxyacetyl side chains in 17-oxosteroids, Scheme 85.[157]

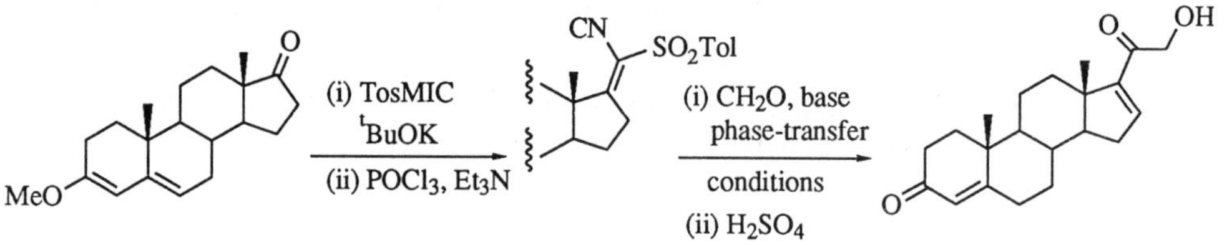

Scheme 85

3.3.3 Oxygen-substituted Sulphones

Studies by Ley and co-workers have established that 2-arenesulphonyl-tetrahydropyrans readily form the corresponding carbanions, which can then undergo useful C–C bond forming reactions, Scheme 86.[158]

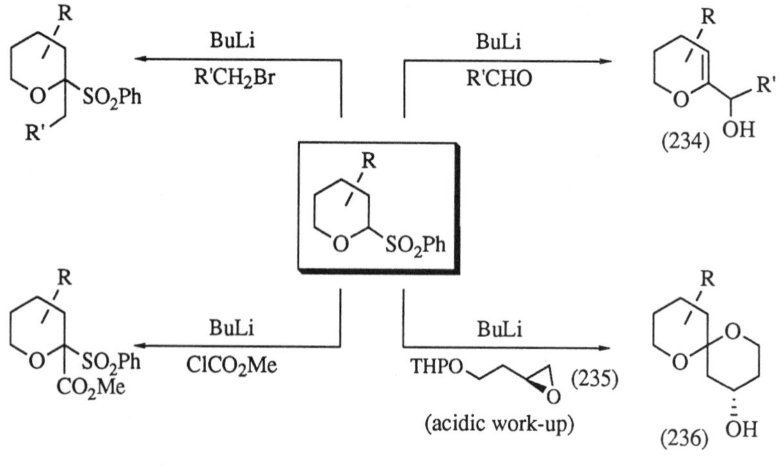

Scheme 86

Reactions with alkylating and acylating agents can be carried out to give derivatives of the parent sulphone. However, with carbonyl compounds as the electrophiles, spontaneous *in situ* elimination of benzenesulphinic acid occurs, to give enol ethers such as (**234**). This behaviour can be exploited, in that a properly positioned pendant hydroxyl group present in R' will cyclise onto the enol ether, under acid catalysis, to give spiroacetal products. A minor variation on this theme uses an epoxide as the electrophile. Thus reaction with (235) gives (236) after acid treatment. This strategy has been further exploited, in the synthesis of the spiroketal portion of the milbemycin natural products.

Studies by Fuji *et al.* on the carbanion chemistry of 1,3-oxathiane 3,3-dioxides such as (28) were highlighted previously in Scheme 4.[27] These indicate that, where possible, deprotonation of a dialkyl sulphone occurs primarily away from an α-oxygen substituent. Thus it appears that an α-oxygen functionality destabilises the sulphonyl carbanion.

An interesting family of α-oxygenated sulphones is formed on treatment of lithiated allylic carbamates with TolSO$_2$F, Scheme 87.[159]

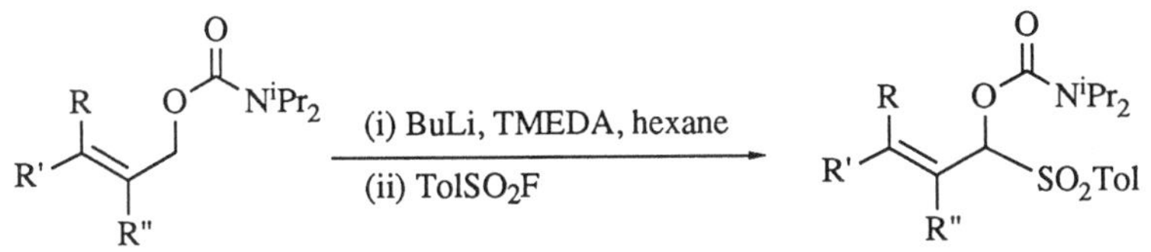

Scheme 87

The product sulphones (237) can be deprotonated and reacted with carbonyl compounds to give α- or γ-adducts, Scheme 88.

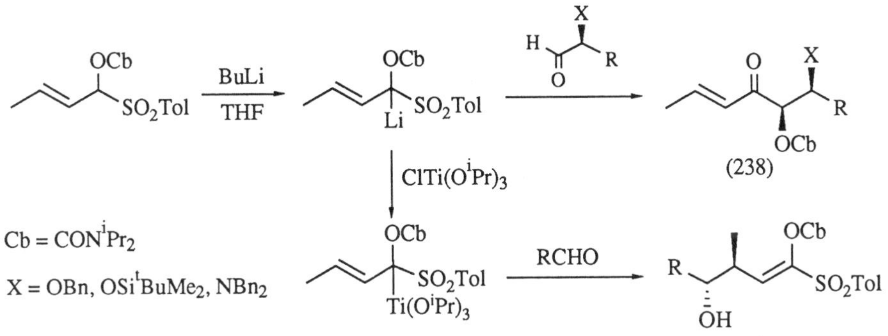

Scheme 88

Notably the derived lithio-anions undergo addition to certain chiral aldehydes to give *syn*-α-adducts (238), in which the carbamate group has undergone a migration. By contrast, titanation of the lithio-carbanion by addition of ClTi(O^{i}Pr)$_3$ allows efficient γ-addition, e.g. to give (239), again in stereoselective fashion.

3.3.4 Sulphur-substituted Sulphones

Sulphonyl carbanions having additional sulphur substituents (at various oxidation states) have been used extensively in synthesis. Carbanions derived from methylthiomethyl *p*-tolyl sulphone (**240**), its derivatives and homologues, provide access to a range of useful products, Scheme 89.[160–162]

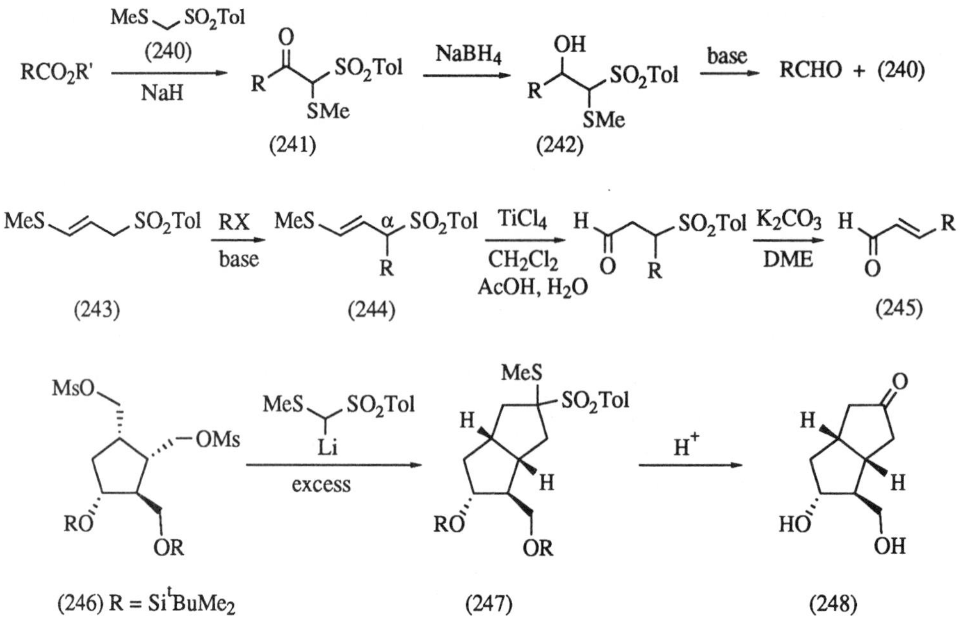

Scheme 89

In the first sequence acylation of (**240**) gives ketosulphone (**241**) which can be reduced to give (**242**), which then undergoes a base-induced retro-aldol process.[160] The overall result is clean conversion of an ester to an aldehyde with regeneration of (**240**). The unsaturated sulphone (**243**) has been used by the same research group to access α,β-unsaturated aldehydes. Thus alkylation, hydrolysis of vinyl sulphide (**244**) and finally sulphinate elimination leads to enal (**245**) in good overall yield.[161] Compounds having an additional substituent at the α-position to the sulphone undergo the same process.

Double alkylation of the anion of (**240**) using the bismesylate (**246**), gives (**247**) which can then be hydrolysed to the known carbacyclin precursor (**248**).[162] In this reaction excess anion presumably acts as base to effect the cycloalkylation.

Geminal bis-sulphones form carbanions very easily and their monoalkylation, e.g. to give (**249**), is facile, Scheme 90.

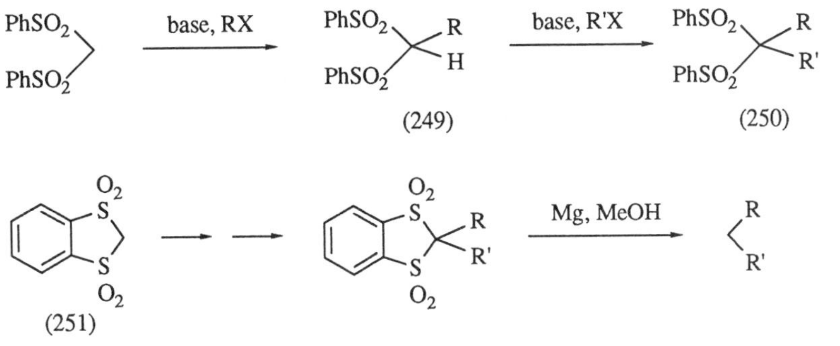

Scheme 90

The formation of the dialkylated compound (**250**) can be problematic, especially with bulky R and R' groups, due to unfavourable interaction with the sterically demanding PhSO$_2$ groups. Kundig and Cunningham have recommended the use of 1,3-benzodithiole tetraoxide (**251**) in such circumstances.[163] This sulphone system undergoes improved alkylations compared with its bis(arenesulphonyl) counterparts and the sulphone groups are easily removed using magnesium in methanol.

The high stability of the bis-sulphonyl carbanion is again demonstrated by the facile opening of cyclopropane (**252**) using a wide range of nucleophiles, Scheme 91.[164]

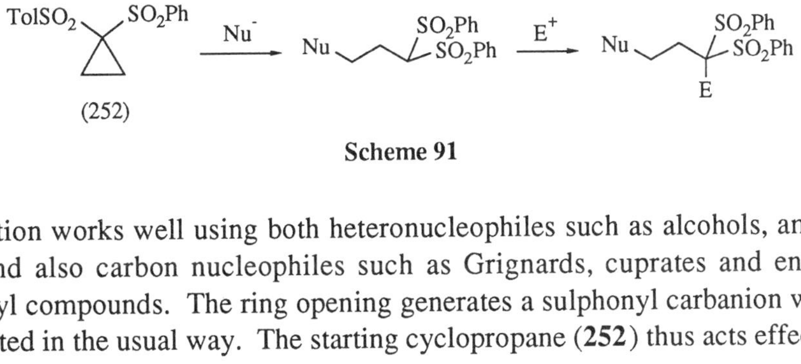

Scheme 91

The reaction works well using both heteronucleophiles such as alcohols, amines and thiols, and also carbon nucleophiles such as Grignards, cuprates and enolates of dicarbonyl compounds. The ring opening generates a sulphonyl carbanion which can be alkylated in the usual way. The starting cyclopropane (**252**) thus acts effectively as a propylene 1,3-dipole.

Carbanions derived from bis-sulphones have great utility in reactions involving nucleophilic attack on π-allyl transition metal complexes.[165] In such reactions these easily generated soft anions have advantages over malonate anions and related species, since the anion-activating groups are so easily removed in subsequent steps. Oppolzer and Gaudin have utilised a palladium-mediated coupling between an alkenyl bis-sulphone such as (**253**) and an allylic chloride or carbonate to assemble substrates for ene-reactions, Scheme 92.[166]

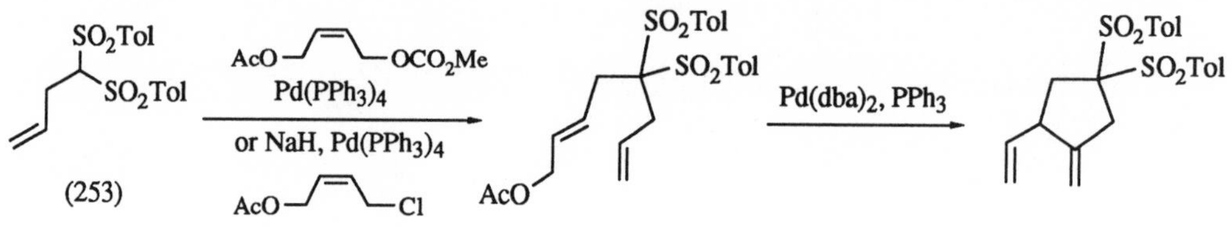

Scheme 92

Sodium hydride is used as the base for coupling with the allylic chloride, whereas if the carbonate is used the MeO⁻ generated by the formation of the π-allyl complex acts as the base. Variations on this theme allow formation of six-membered rings and fused bicyclic products. Both steps can be conveniently combined in a one-pot operation if desired.

Trost's group has made many valuable contributions in the area of transition metal catalysed cyclisation processes, with sulphones featuring prominently amongst the most useful reaction partners. One such report probed the regiochemistry of the reaction of bis-sulphonyl carbanions with pentadienyl metal complexes, e.g. Scheme 93.[167]

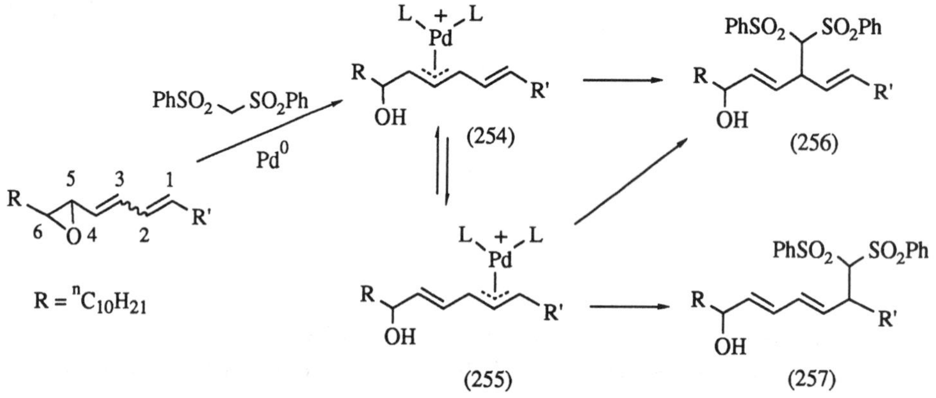

Scheme 93

Complexes (254) and (255) are thought to equilibrate, allowing nucleophilic attack to occur at C-1 or C-3. The electronic effect of the OH group in (254) and (255) is to exclude attack at C-5 and promote attack at the most distal position. Thus when R' = H the sole product is (257). The effect can be countered by changing R' to CO_2Et, in which case (256) is the only product. The regiochemical outcome of this last reaction can again be overturned if the central carbon (C-3) is substituted, in which case attack occurs at C-5. Thus steric effects can intervene and override the electronic bias of the system.

Cyclisations using bis-sulphones provide a good route to medium and large ring carbocycles and macrolides, Scheme 94.[168,169]

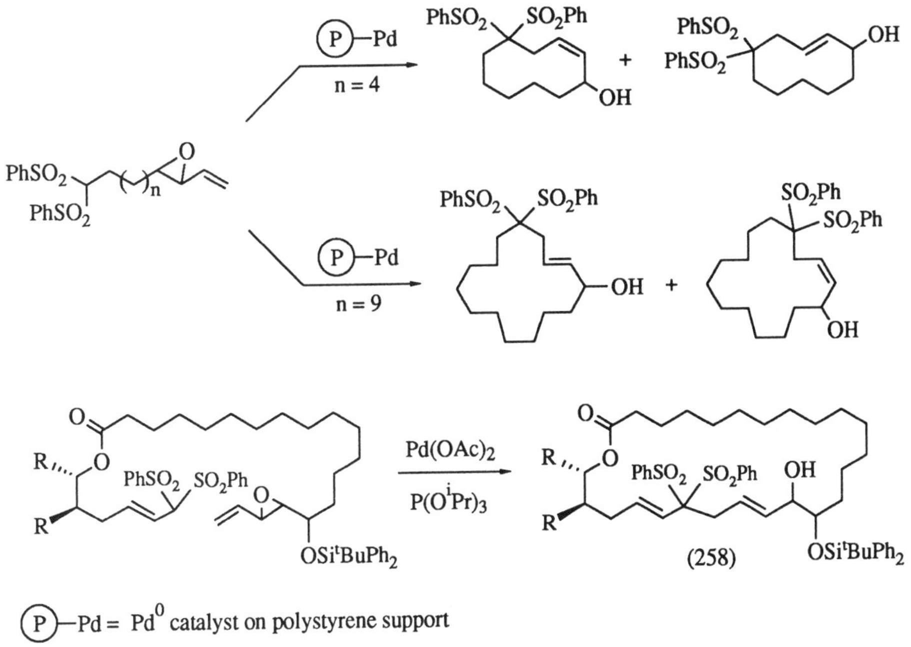

Scheme 94

The use of a polymer-bound catalyst system allows the formation of 10- and 15-membered ring systems at 0.1–0.5M concentration. In related studies aimed at the synthesis of tetrin A, a macrocyclisation to give (258) is achieved in 92% isolated yield. Subsequent manipulation of the diene bis-sulphone system into a tetraene is also possible by means of a desulphonylation followed by palladium-mediated elimination.

The facility of these palladium-mediated reactions contrasts with analogous hexacarbonylmolybdenum(0) processes, in which bis-sulphones are found to be ineffective, presumably for steric reasons.[170] Cyclisation of cationic η^4-diene complexes of molybdenum using bis-sulphonyl carbanions is, however, highly effective, Scheme 95.[171]

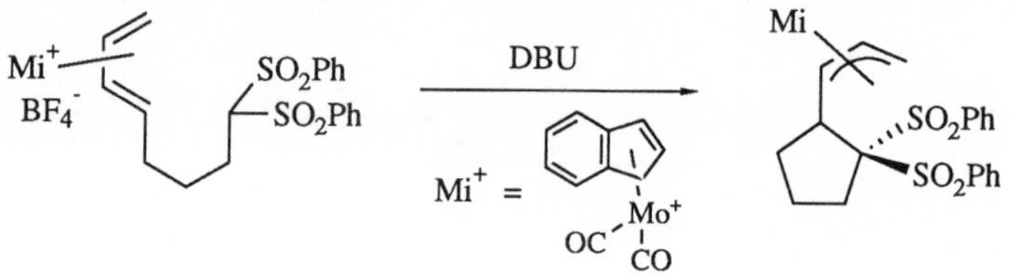

Scheme 95

Here the use of a geminal bis-sulphone gives vastly improved results over the corresponding malonate derivative, presumably owing to the greater acidity of the former system.

3.3.5 Halogen-substituted Sulphones

The reaction of α-halogenosulphonyl carbanions with carbonyl compounds was highlighted in Chapter 2 as a direct route to α,β-epoxy sulphones. This method can be used to homologate ketones to α-hydroxy aldehydes, Scheme 96.[172]

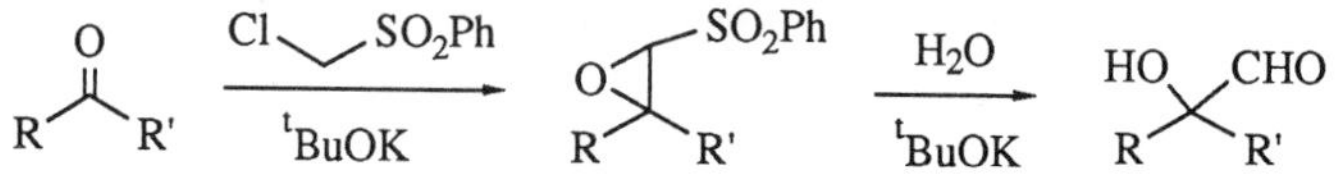

Scheme 96

Thus treatment of the epoxy sulphones, obtained in the usual fashion, with ^{t}BuOK and water gives the hydroxy aldehyde products, sometimes in dimeric form. The sequence proceeds stereoselectively with some steroidal substrates, the second step involving attack of hydroxide at the α-carbon.

Analogous attack of a chlorosulphonyl carbanion on an imine results in the formation of a 2-sulphonylaziridine product, Scheme 97.[173]

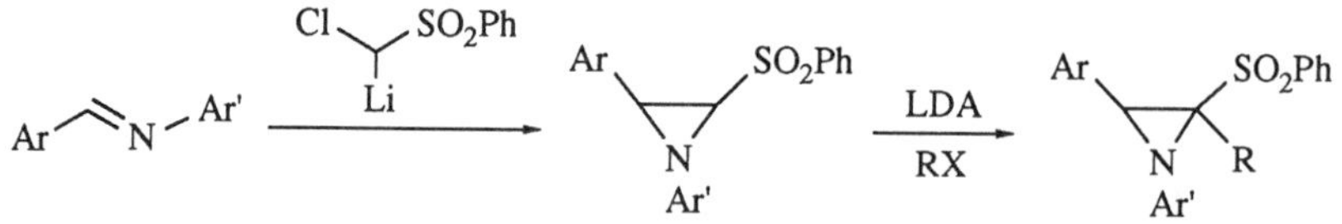

Scheme 97

The initially formed aziridines can be further substituted at the α-position via derived carbanions.

Makosza and co-workers have been largely responsible for the development of the carbanion chemistry of chloroalkyl sulphones. Scheme 98 illustrates some aspects of their work involving vicarious nucleophilic substitution (VNS) reactions of aromatic systems.[174]

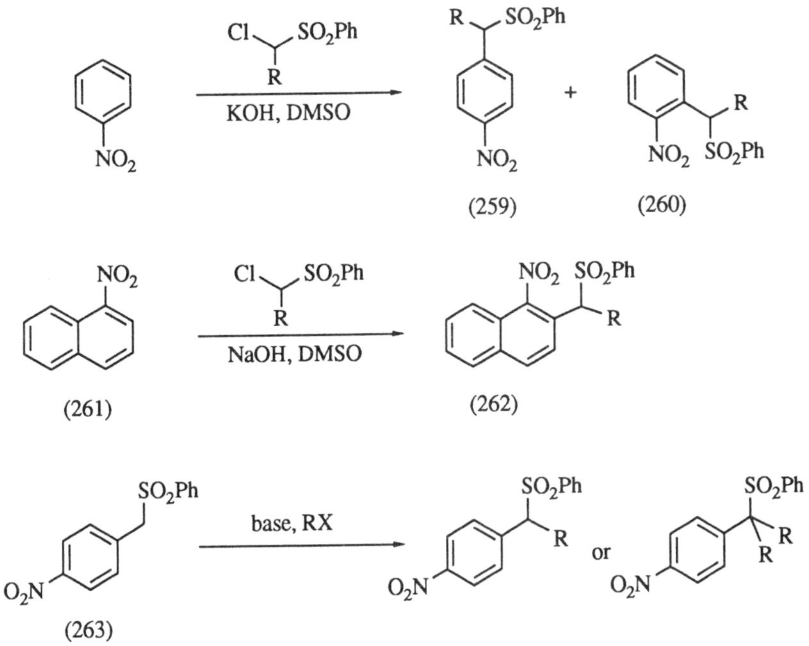

Scheme 98

This reaction, for example to give (259) and (260), involves nucleophilic addition to the nitroarene, followed by loss of HCl and finally protonation. Attack at a hydrogen-substituted ring position is highly favoured and *para*-substitution usually predominates, especially when utilising substituted chlorosulphones, i.e. (259) R ≠ H. Substitution in nitronaphthalenes occurs mainly at the 2-position for the 1-nitro compounds, e.g. (261) gives (262), and at the 1-position for 2-nitro compounds, unless these positions are blocked. Finally, the products of these reactions, e.g. (263), can be further substituted using conventional alkylation conditions.

3.4 Sulphonyl Polyanions

The sulphone group is unique in its ability to promote the formation of a wide range of polyanionic species by multiple proton abstraction. This section describes the chemistry of such polycarbanionic species, as well as polyanion chemistry involving species having one sulphonyl carbanion and an additional non-carbon-centred anion. The latter category encompasses the chemistry of doubly deprotonated hydroxy sulphones and sulphonyl carboxylic acids. Since in these cases the polyanion formation is due to an additional acidic functionality remote from the sulphone this chemistry has been grouped under the heading of remote polyanions.

3.4.1 Polyanions from Simple Sulphones

The preparation of a range of *gem*-dianions has been investigated by Kaiser *et al.*[175] For example, treatment of benzyl phenyl sulphone with two equivalents of BuLi in THF results in the formation of the α,α-dilithiosulphone (264), Scheme 99.[176]

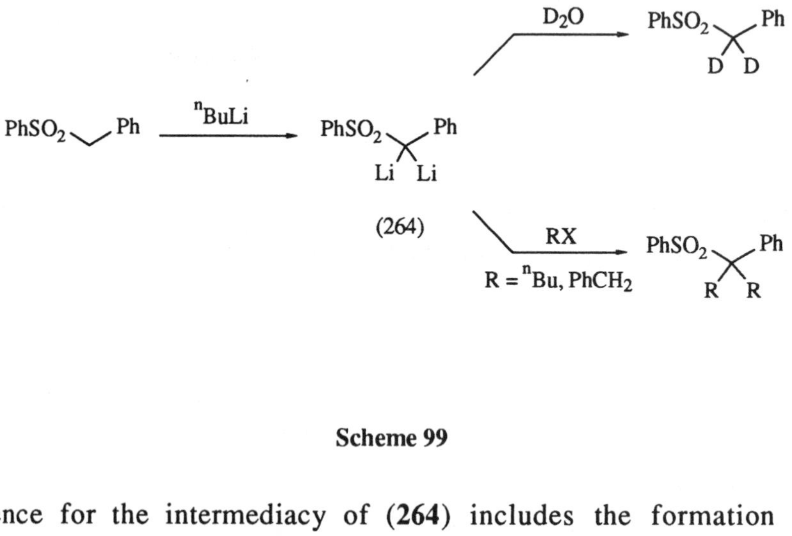

Scheme 99

Evidence for the intermediacy of (264) includes the formation of *gem*-dideuterated and *gem*-dialkylated products. Also the IR spectrum of (264) displays no S–O stretching frequencies at 1325 and 1150 cm^{-1}, characteristic of the starting sulphone, and still evident in the IR spectrum of the sulphonyl monoanion.

One synthetic use of α,α-dianions is simply as 'super-nucleophilic' sulphonyl carbanions, for example in reactions with relatively poor electrophiles. Thus in attempts to couple sulphone (265) with iodide (266) difficulties encountered include oxidative anion coupling and very sluggish alkylation.[177] Use of the sulphonyl dianion in THF/HMPA enables clean formation of the desired product, Scheme 100.

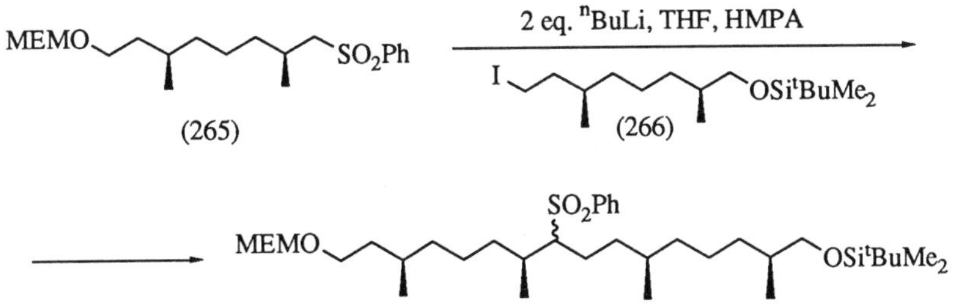

Scheme 100

An alternative application of α,α-dianions involves geminal dialkylation, nicely illustrated by a synthesis of *trans*-2,6-disubstituted piperidines, Scheme 101.[178]

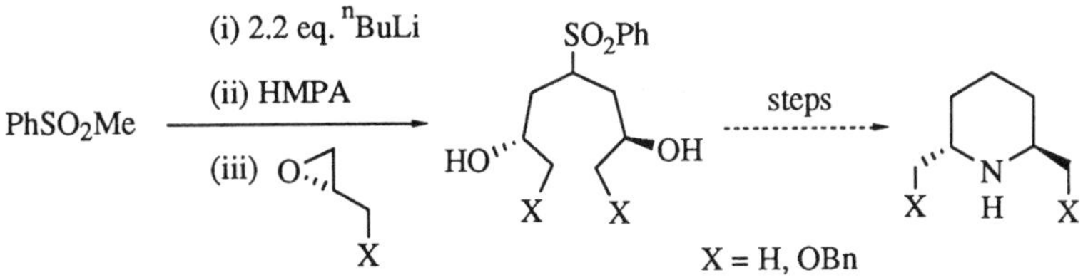

Scheme 101

A variety of interesting and useful applications of sulphonyl α,α-dianion chemistry has been demonstrated by Eisch's group, Scheme 102.[179]

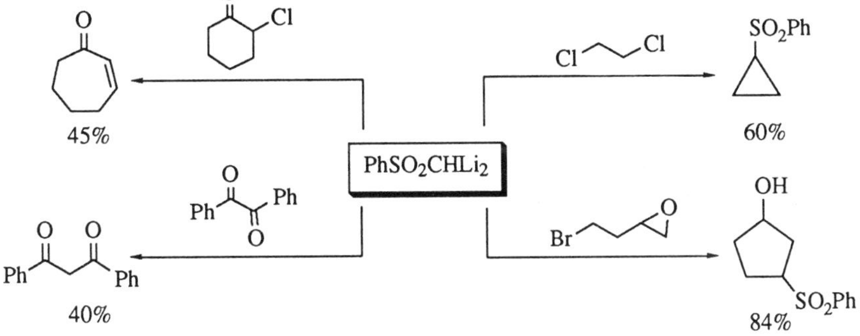

Scheme 102

Reactions with dihalogenoalkanes and halogenoepoxides give the anticipated cyclic products, with the latter process proceeding by initial attack on the epoxide terminus to give the larger of the two possible cyclic products. Reaction with benzil or 2-chlorocyclohexanone involves attack on both electrophilic sites followed by ring-opening of the so-formed cyclopropanolate involving the expulsion of sulphinate anion. Remarkably, conjugate addition is observed on reaction with chalcone.

Addition of carbonyl compounds to α,α-dilithiosulphones can furnish either bis-aldol products or vinyl sulphones, Scheme 103.[180]

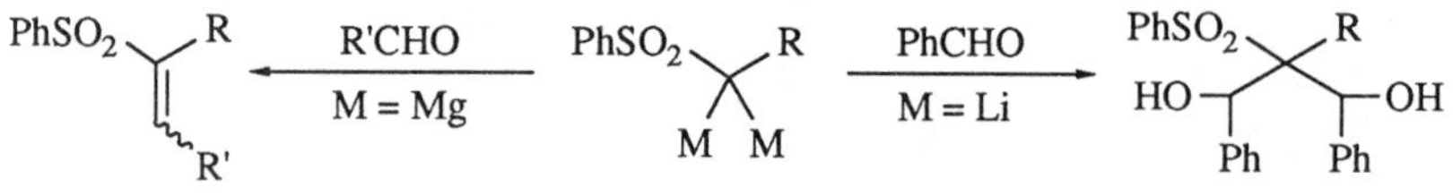

Scheme 103

Although no direct evidence for a *g e m*-dimagnesio-derivative is presented, the efficient formation of vinyl sulphones is found to require two equivalents of magnesium iodide.

In acylation reactions the use of geminal sulphonyl dianions offers significant advantages over the simple monoanions. With simple sulphonyl anions the acidic β-ketosulphone product consumes up to half of the starting sulphonyl carbanion, requiring its use in excess. By contrast, acylation of a *gem*-dilithiosulphone leads directly to the anion of the β-ketosulphone product.[181] This reaction was utilised in a recent synthesis of a fragment required for the total synthesis of boromycin.[182] An *in situ* acylation–alkylation sequence was found to be possible by reacting α,α-dilithiosulphones with a bromoester such as ethyl 4-bromobutyrate, Scheme 104.[183]

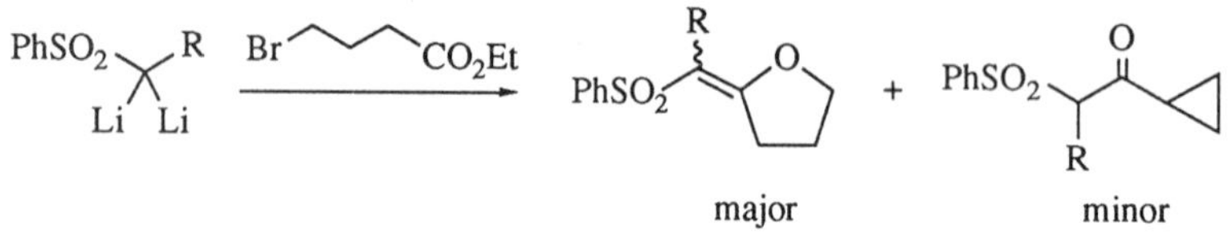

Scheme 104

Initial *C*-acylation is followed by *O*-alkylation of the so-formed β-ketosulphonyl anion. The minor cyclopropane product presumably arises by equilibration between the β-ketosulphonyl anion and the much less stable (and therefore more reactive) simple enolate anion. The same group of workers have also generalised the reaction of lactones with α,α-dilithiosulphones, furnishing various hydroxy ketone products.[184]

Bosworth and Magnus have explored the deprotonation chemistry of sulphone (267) consistent with the formation of an α,α-dianion, Scheme 105.[185]

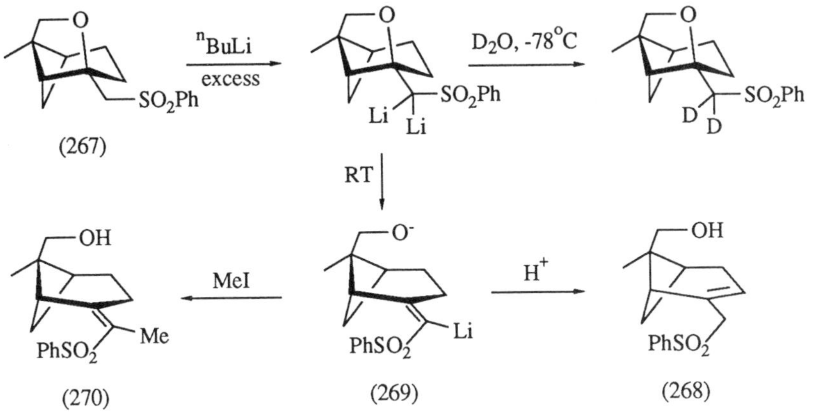

Scheme 105

Thus at low temperature the β-alkoxy sulphonyl dianion could be doubly deuterated, whereas on warming, ring opening occurred to give allylic sulphone (**268**). In the presence of methyl iodide the intermediate vinylsulphonyl anion (**269**) could be trapped to give (**270**).

Deuteration studies have shown that the carbanion of allyl phenyl sulphone undergoes a second kinetically controlled deprotonation at the *ortho*-position to give mainly 1,*ortho*-dilithio-species (**271**), Scheme 106.[186]

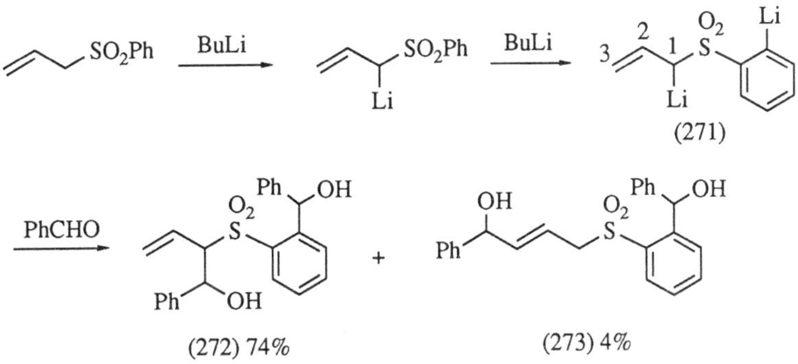

Scheme 106

Reaction with benzaldehyde gives mainly the 1,*ortho*-adduct (**272**) along with minor amounts of (**273**). However, on warming to 50°C (**271**) undergoes complete isomerisation to the 1,1-dilithio-species (**274**). Reaction with benzaldehyde in this case gives the (*E*)-1,3-diadduct (**275**), whereas dialkylation and cycloalkylation processes give 1,1-substituted products, Scheme 107.

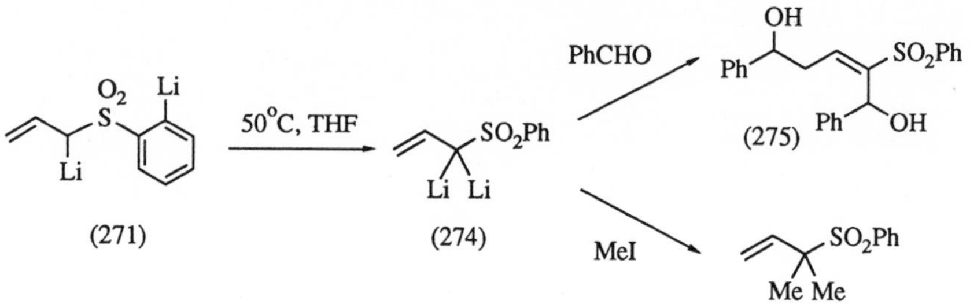

Scheme 107

Surprisingly, the results of the alkylation of 1,*ortho*-dianion (**271**) are very similar. Presumably initial reaction of (**271**) at the α-position is followed by transmetallation to the substituted 1-lithioallyl sulphone, Scheme 108.

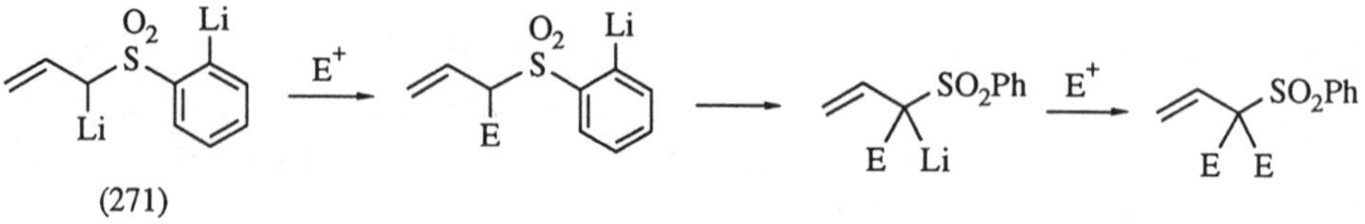

Scheme 108

A remarkable change in the reactivity of (**274**) towards carbonyl compounds occurs on titanation with (ᶦPrO)₂TiCl₂. Instead of diadduct (**275**), either a γ-adduct, e.g. (**276**), or a dienyl sulphone, e.g. (**277**), or a mixture of the two, is obtained, Scheme 109.

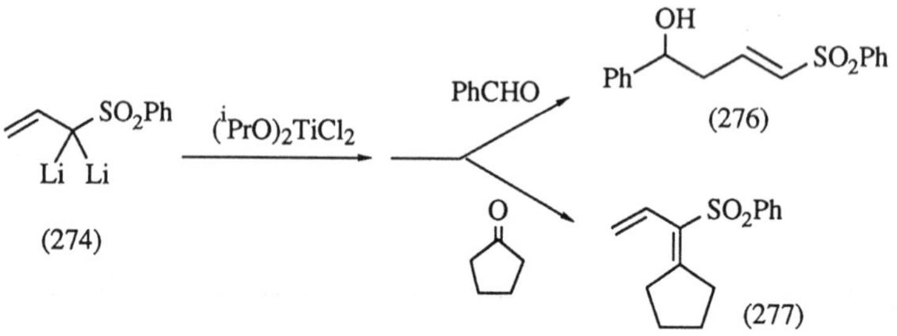

Scheme 109

Other reports from the same research group describe further structural investigation of these dianions[187] and their application to synthetic problems.[188]

A highly diastereoselective coupling of dianions derived from certain branched allylic sulphones with terminal epoxides has been described by Trost and Merlic, Scheme 110.[189]

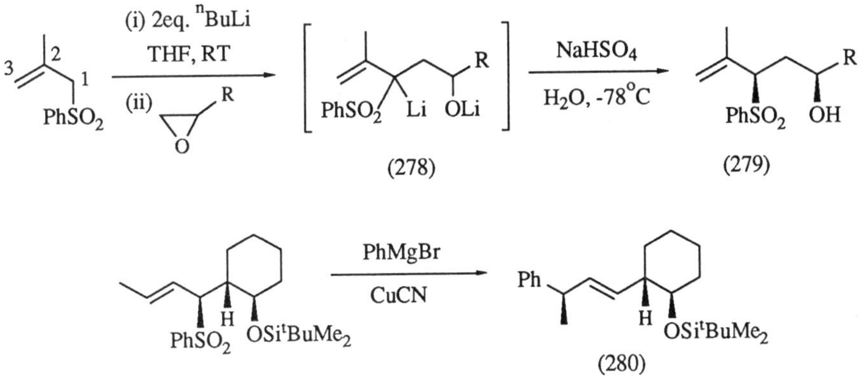

Scheme 110

The product outcome is the result of kinetically controlled protonation of the intermediate alkoxide carbanion (**278**), the *syn*-isomer (**279**) being formed highly selectively. Analogous reactions involving non-terminal epoxides, or allylic sulphones not having a substituent at C-2 are kinetically less selective, although in such cases base equilibration often gives good isomer ratios. The significance of these results is underlined by the stereoselective conversion of the allylic sulphone products into alkenes such as (**280**).

A range of dialkyl sulphones have also been converted to the corresponding 1,3-dianions, Scheme 111.[190–192]

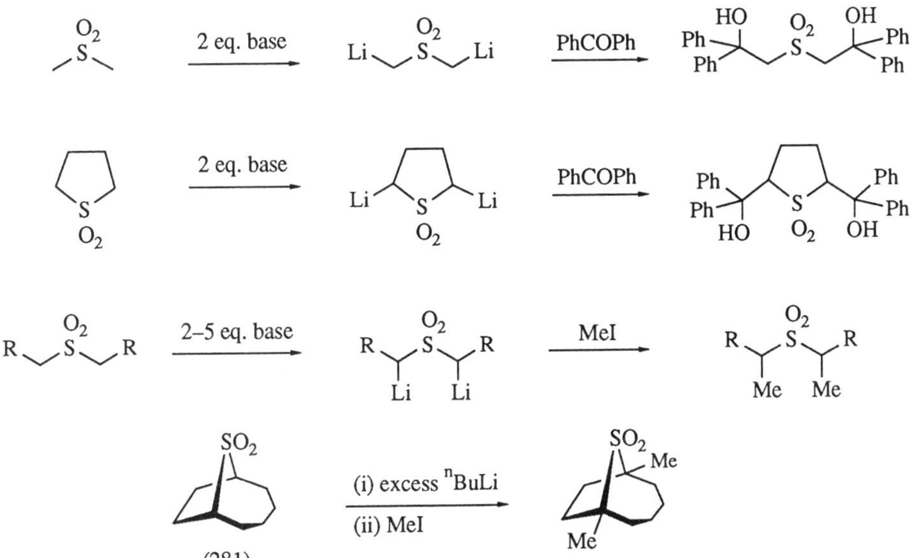

Scheme 111

Such 1,3-dianions, generated using either metal amides or [n]BuLi, react with a few simple electrophiles such as benzophenone or methyl iodide. Remarkably, the dianion of (281) and unsaturated analogues are readily formed and alkylated in high yield. These results indicate that the formation of 1,3-dianions is preferred over 1,1-dianions. An additional interesting reaction of these 1,3-dianions is their direct conversion to alkenes by reaction with copper salts, Scheme 112.[191]

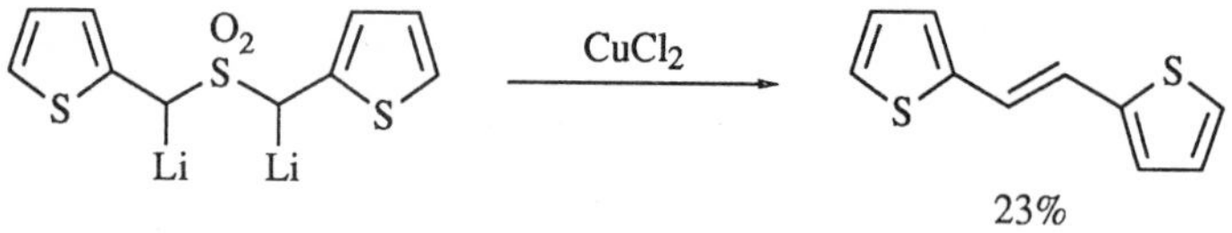

23%

Scheme 112

This rather low yielding method is presumably closely related to the Ramberg–Bäcklund rearrangement.

Finally, results have been obtained which are consistent with the formation of aromatic dianions such as (282), Scheme 113.[193]

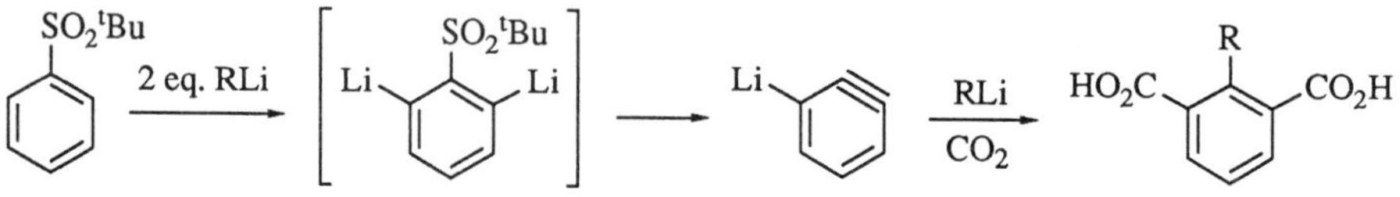

(282)

Scheme 113

Elimination of sulphinate is proposed to give a lithiated benzyne intermediate, which on reaction with a further equivalent of alkyllithium and quenching with CO_2 gives the observed product. Depending on the reaction conditions other products can arise by coupling of (282) with a benzyne.

3.4.2 Polyanions from Functionalised Sulphones

The formation of the geminal dimagnesium salt of bis(phenylsulphonyl)methane was reported by Kohler and Tishler as early as 1935.[194] Useful synthetic applications of polyanions derived from disulphones have, however, only been explored very recently. Most notably, Hendrickson and collaborators have examined mesyl triflone (283) as a reagent for alkene construction using polyanion chemistry, Scheme 114.[195]

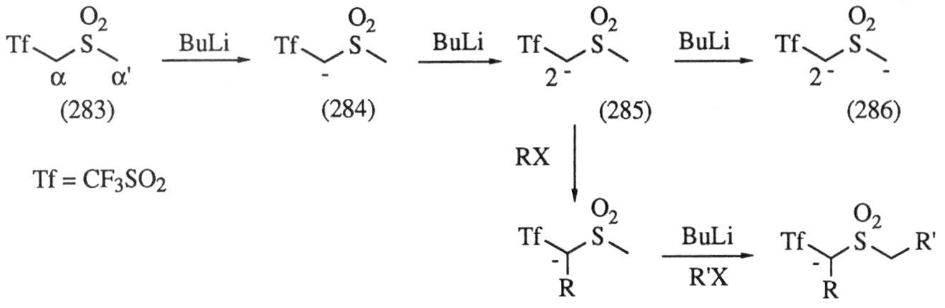

(283) (284) (285) (286) (287) (288)

Scheme 114

The triflone (**283**) is readily deprotonated to form a mono-, di- or trianion. The monoanion (**284**) formed at the α-position is very unreactive and alkylation at this position is best carried out via the α-dianion (**285**). The reaction with three equivalents of base gives trianion (**286**); however, the selectivity in the reaction of this species at the α- and α'-positions is poor. Selective alkylation at the α'-position can be achieved once one α-substituent has been introduced, i.e. (**287**) → (**288**). Reactions of anions and dianions derived from (**283**) are effective with a range of electrophiles, including alkyl halides, carbonyl compounds and acylating agents. Providing the α-position is fully substituted the products can then undergo a Ramberg–Bäcklund-type elimination to give alkene products, Scheme 115.

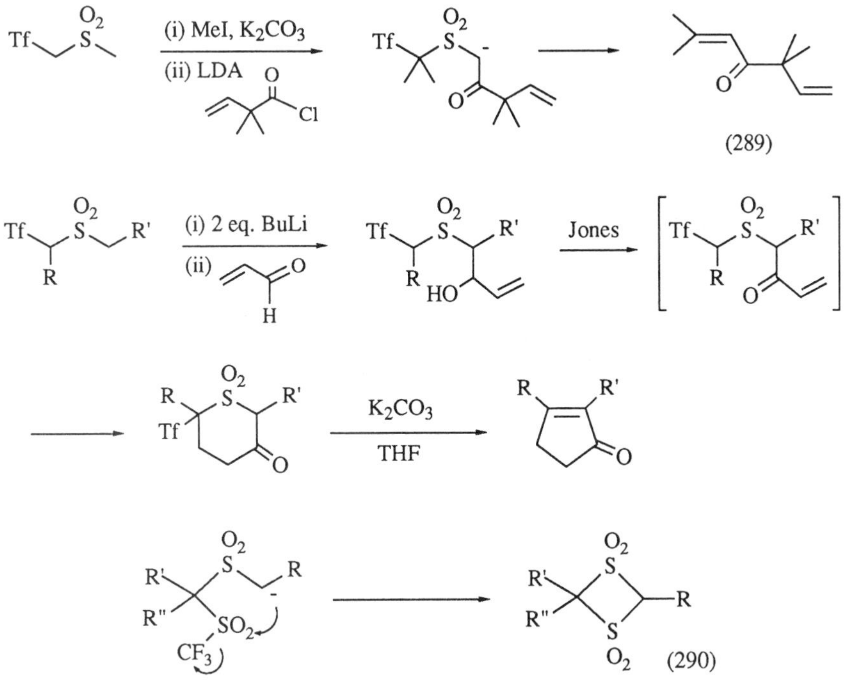

(289) (290)

Scheme 115

The method was applied to the synthesis of artemesia ketone (**289**) and dihydrojasmone. Interestingly, in some examples the elimination produces cyclic disulphone products (**290**), presumably by intramolecular displacement of CF_3^-.

Doubly lithiated TosMIC undergoes reactions which the monoanion itself does not, e.g. Scheme 116.[196]

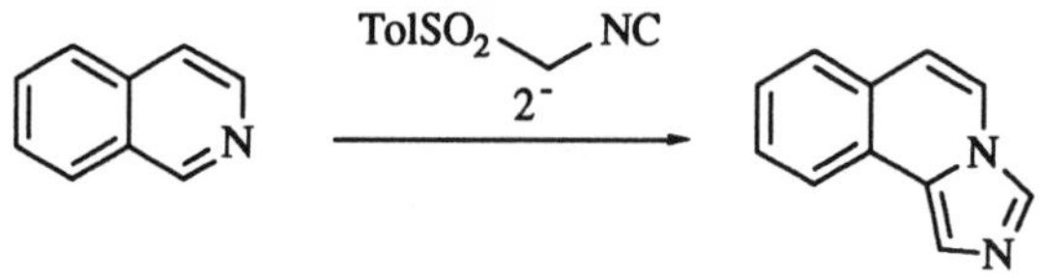

Scheme 116

The dianion also contrasts with the monoanion in that it undergoes reaction with unsaturated esters at the carbonyl group rather than undergoing Michael addition.

A range of chemistry involving polyanions derived from β-ketosulphones has been explored. Whilst treatment of β-ketosulphones such as (**291**) with NaH allows α-alkylation, the generation of a dianion provides a means of selective substitution at the γ-position, Scheme 117.[197]

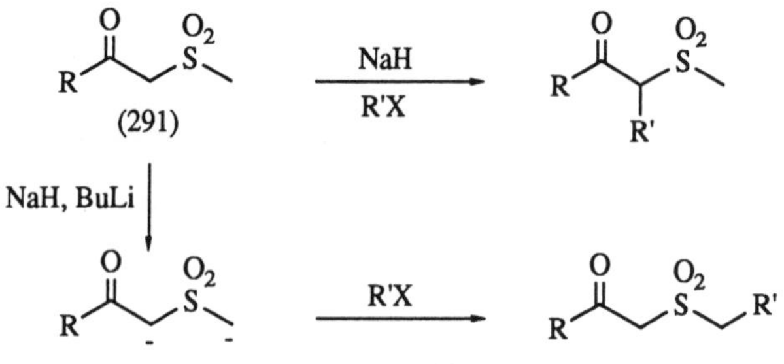

Scheme 117

This approach can also be used to prepare cyclic derivatives such as (**292**). Subsequent halogenative deacylation can then be used to access halogenosulphones for Ramberg–Bäcklund rearrangement, Scheme 118.[198]

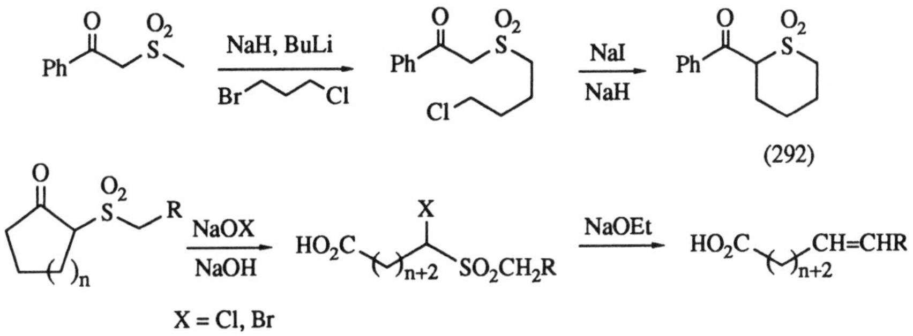

Scheme 118

Lygo and O'Connor found that β-ketosulphonyl dianions react with epoxides in moderate yield (30–49%), Scheme 119.[199]

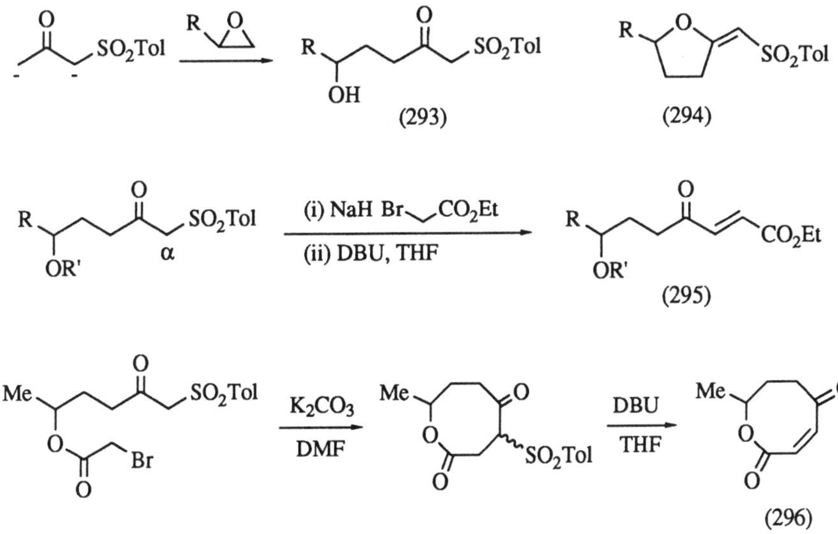

Scheme 119

The yield of product is reduced due to the instability of these systems on purification and particularly their tendency to cyclise to tetrahydrofurans of type (**294**). The hydroxy β-ketosulphones (**293**) can be further transformed into useful products by α-alkylation with ethyl bromoacetate and elimination of sulphinic acid to give ketoester (**295**). An intramolecular version of this sequence also allows the preparation of medium-ring lactones such as (**296**). In some systems it is possible to carry out double deprotonation of a β-ketosulphone or β-ester sulphone at the α-position, Scheme 120.[200,201]

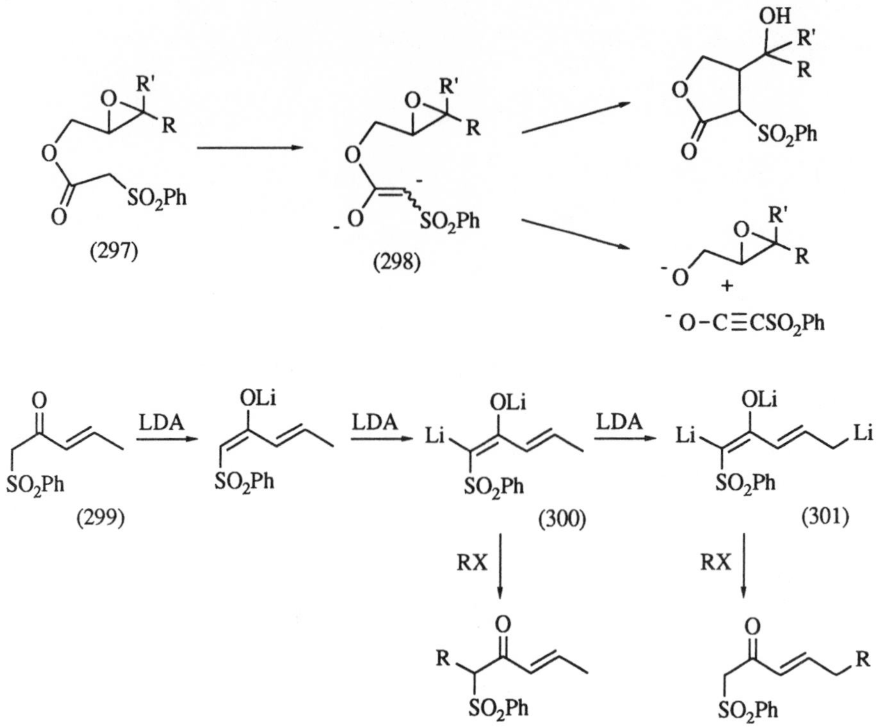

Scheme 120

Cyclisation of (**297**) requires double deprotonation to form (**298**). Whereas the desired sulphonyl lactone products are obtained in good yield starting with *cis*-epoxides, the reaction is less satisfactory with *trans*-epoxides, as competitive fragmentation of (**298**) occurs.[200] Enone sulphone (**299**) forms either the dilithiosulphone (**300**) or trilithio species (**301**) on treatment with the appropriate amount of LDA, as reflected by the formation of α- or γ-alkylated products.[201] A later report shows that the polyanion chemistry of such sulphones is dependent on the substitution pattern.[202]

3.4.3 Remote Polyanions

If a sulphone possessing a relatively acidic group such as an alcohol is treated with base, then clearly the alcohol will be deprotonated first. If a further equivalent of base is added the sulphone can then be deprotonated, thus forming a dianion. Such dianions can usually be made to react α to sulphur in C–C bond forming processes, without concomitant ether formation at the alkoxide anionic site. This approach is attractive in that the need for hydroxyl protection is avoided. Furthermore, sulphonyl anions having a β-oxygen functionality can be alkylated in this way, whereas facile β-elimination would normally be expected to cause problems. This section describes the chemistry of such doubly deprotonated hydroxy sulphones, along with analogous chemistry of remote dianions generated from sulphonyl carboxylic acids.

Julia *et al.* showed that a range of hydroxy sulphones can be *C*-alkylated via the corresponding dianions, formed using two equivalents of BuLi, Scheme 121.[203]

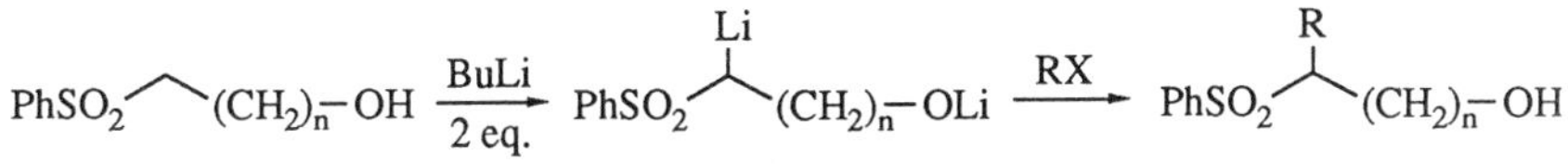

Scheme 121

This method allows easy access to a range of substituted alcohols, following removal of the sulphone. This report sets the stage for examination of the stereochemical aspects of this type of dianion chemistry and also for synthetic applications. The dianions of simple hydroxy sulphones, e.g. (**302**), undergo alkylation with rather poor selectivity, Scheme 122.[204]

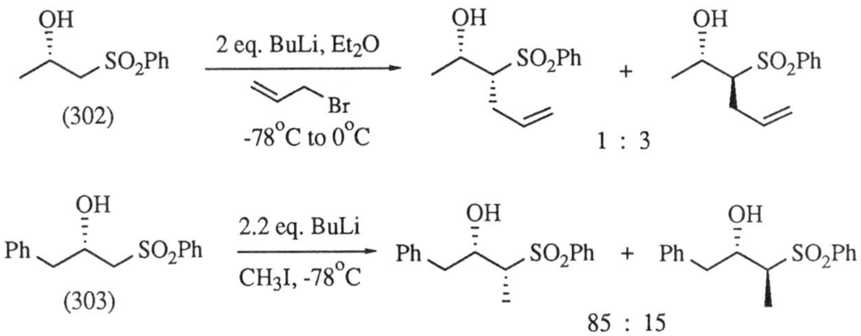

Scheme 122

The analogous methylation reaction of the bulkier sulphone (**303**) gives a slightly better selectivity, this time favouring the *syn*-isomer.[205] Comparison of these two reports is complicated by the possibility of equilibration in the reaction of (**302**) due to the relatively high reaction temperature. This was ruled out in the second case and the stereochemical result was rationalised in terms of kinetically controlled attack on a non-chelated dianion. Analogous studies on the aldol reaction of such dianions were also carried out, Scheme 123.[206]

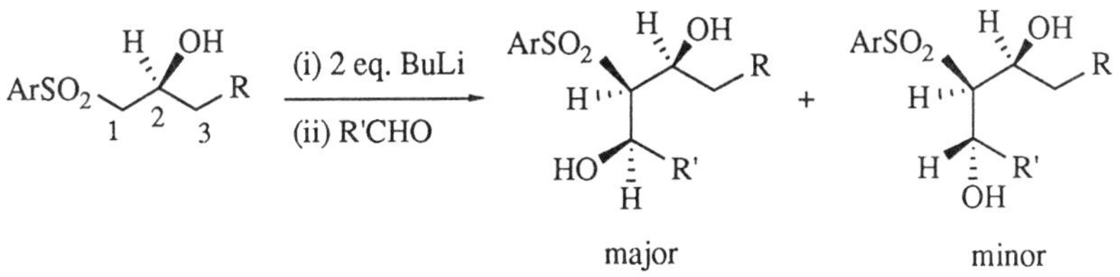

Scheme 123

Good stereocontrol at C-1 can be achieved, with one diastereoisomer predominating overall. The results appear to conform to a model in which the sulphone group and the deprotonated hydroxyl function prefer an *anti*-orientation. Further evidence for such a preference is found in reactions of dianions derived from cyclic β-hydroxy sulphones, Scheme 124.[207]

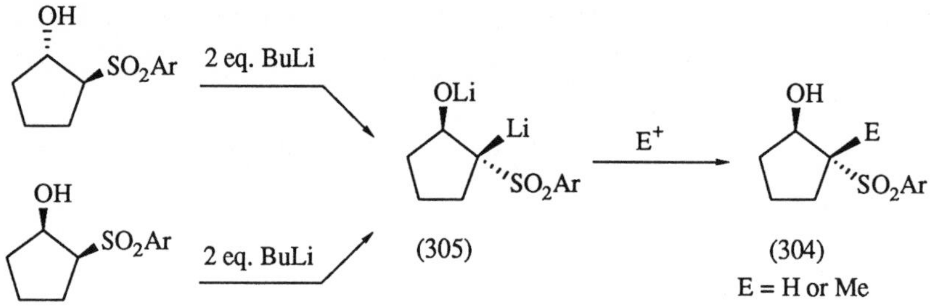

Scheme 124

Thus, only *trans*-products (**304**) are obtained, starting from either isomeric hydroxy sulphone. Simplistically this result can be rationalised as reflecting a preference for configuration (**305**) of the intermediate dianion. However, this explanation takes little account of the complexities of sulphonyl carbanion configuration and alkylation behaviour outlined in Section 3.1.

Alkylation and aldol reactions have also been carried out on more complex optically active sulphone systems, Scheme 125.[208,209]

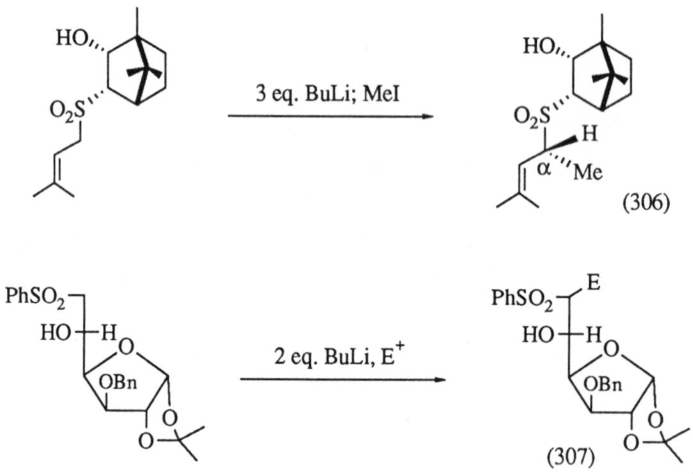

Scheme 125

The camphor derivative (306) is obtained as a 96:4 mixture with the epimer at the sulphone α-position.[208] Depending on the amount of methyl iodide used the product can be isolated as either the free secondary alcohol or a derived methyl ether. Considering that three equivalents of base are employed in this case the possibility of trianion involvement has to be considered. The diastereoselectivities observed in the reactions of a sugar-derived hydroxy sulphonyl dianion, leading to (307), are rather poor.[209] Here, effects such as the overall stereochemical bias of the chiral sugar and the possibility of chelation involving the nearby benzyl ether, complicate any rationalisation of the outcome.

A series of reports has shown how dianions of hydroxy sulphones can be used to prepare lactones, Scheme 126.[210–211]

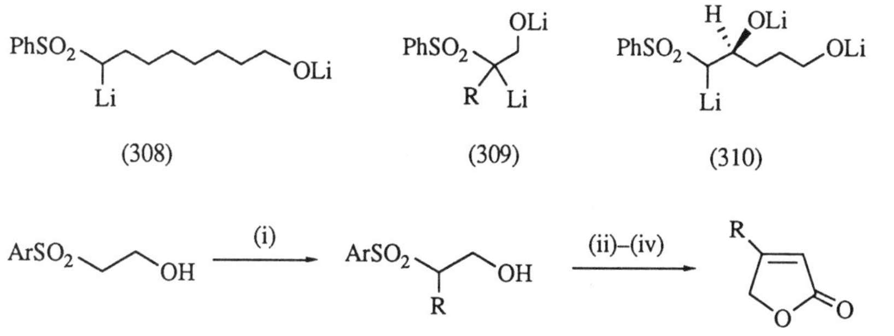

(308) (309) (310)

reagents:
(i) 2.2 eq. BuLi, THF, TMEDA; RX   (ii) 2.2 eq. BuLi, THF, TMEDA; ICH₂CO₂Na
(iii) pTSA, benzene, reflux   (iv) Et₃N

Scheme 126

In this type of sequence a functionalised electrophile, here ICH_2CO_2Na, reacts with the dianion. Lactonisation and elimination of sulphinic acid gives the observed product, in this case a butenolide.[210] In reactions with ICH_2CO_2Na, the dianion (308) with very remote anionic centres behaves poorly compared to (309). Other reports describe similar chemistry, including in one case the reactions of trianion (310).[211]

The reaction of remote dianions generated from 4-(phenylsulphonyl)butanoic acid with either carbonyl compounds or imines furnishes lactone and lactam products respectively, Scheme 127.[212]

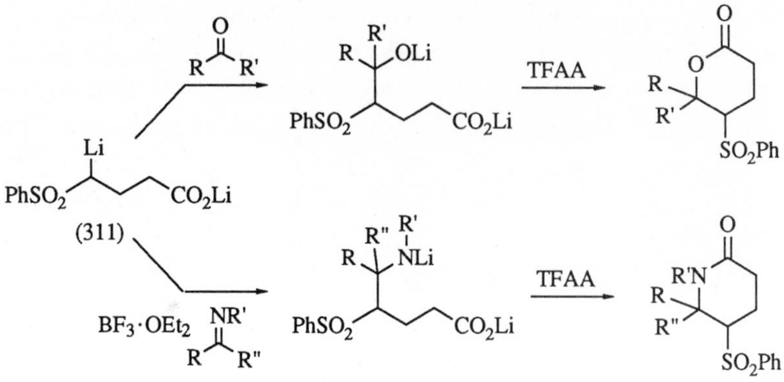

Scheme 127

The soluble dianion (**311**) is again generated using two equivalents of BuLi. Reaction with either a carbonyl or imine partner (Lewis-acid activation is required in the latter case) is then followed by cyclisation using trifluoroacetic anhydride (TFAA). Further chemistry of the products was also investigated, including a synthesis of sulcatole.[213]

The reactions of remote dianions generated from a bis(phenyl sulphone) with dihalogenoalkanes such as (**311**) provide a novel entry to medium-ring products such as (**312**), Scheme 128.[214]

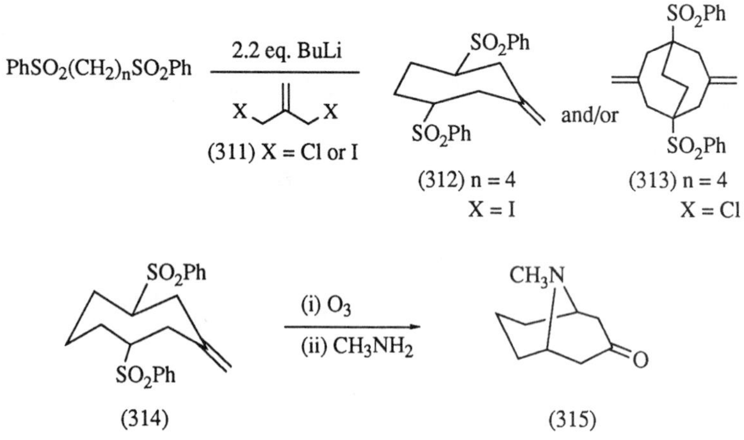

Scheme 128

In one case (n = 4), a propensity for formation of the bicyclic product (**313**) is observed, when the dichloride (**311**) (X = Cl) is used. This problem can be overcome simply by using the corresponding diiodide as alkylating agent. By appropriate choice of alkylating agent and reaction conditions the monocyclic products such as (**312**) can be obtained in good yield for n = 3, 4 and 5. Further useful transformations of the products were also examined including the conversion of (**314**) into the alkaloid pseudopelletierine (**315**).

Chapter 3 References

1. See, for example, the general sulphone references given in Chapter 1, particularly S. Oae and Y. Uchida, in *The Chemistry of Sulphones and Sulphoxides,* Ed. S. Patai, Z. Rappoport, and C. J. M. Stirling, John Wiley and Sons, **1988**, p.583; K. Ogura, in *Comprehensive Organic Synthesis*, Ed. B. M. Trost and I. Fleming, Pergamon Press, Oxford, **1991**, 1, 505; A. Krief, in *Comprehensive Organic Synthesis*, Ed. B. M. Trost and I. Fleming, Pergamon Press, Oxford, **1991**, 3, 85; see also: B. S. Thyagarajan, *Mech. React. Sulphur Comp.,* **1969**, *4*, 115; M. Gresser, *Mech. React. Sulphur Comp.,* **1969**, *4*, 29; A. Krief, *Tetrahedron*, **1980**, *36*, 2531; J-F. Biellmann and J-B. Ducep, *Org. React.,* **1982**, *27*, 1; S. Wolfe, *Stud. Org. Chem.,* **1985**, *19*, 133.

2. For an excellent review of sulphone carbanions with emphasis on X-ray crystal structures see: G. Boche, *Angew. Chem., Int. Ed. Engl.,* **1989**, *28*, 277.

3. E. J. Corey and E. T. Kaiser, *J. Am. Chem. Soc.,* **1961**, *83*, 490; D.J. Cram, W. D. Nielson, and B. Rickborn, *J. Am. Chem. Soc.,* **1960**, *82*, 6415; D.J. Cram, D. A. Scott and W. D. Nielson, *J. Am. Chem. Soc.,* **1961**, *83*, 3696; H. L. Goering, D. L. Towns, and B. Dittmar, *J. Org. Chem.,* **1962**, *27*, 736.

4. E. J. Corey and T. H. Lowry, *Tetrahedron Lett.,* **1965**, 803; D. J. Cram and A. S. Wingrove, *J. Am. Chem. Soc.,* **1962**, *84*, 1496.

5. D. J. Cram and A. S. Wingrove, *J. Am. Chem. Soc.,* **1963**, *85*, 1100.

6. E. J. Corey, H. Konig, and T. H. Lowry, *Tetrahedron Lett.,* **1962**, 515.

7. L. A. Paquette, J. P. Freeman, and M. J. Wyvratt, *J. Am. Chem. Soc.,* **1971**, *93*, 3216.

8. W. T. van Wijnen, H. Steinberg, and T. J. De Boer, *Tetrahedron,* **1972**, *28*, 5423.

9. H. E. Zimmerman and B. S. Thyagarajan, *J. Am. Chem. Soc.,* **1960**, *82*, 2505.

10. A. Ratajczak, F. A. L. Anet, and D. J. Cram, *J. Am. Chem. Soc.,* **1967**, *89*, 2072.

11. W. Kirmse and U. Mrotzeck, *J. Chem. Soc., Chem. Commun.,* **1987**, 709.

12. W. von E. Doering and L. K. Levy, *J. Am. Chem. Soc.,* **1955**, *77*, 509.

13. F. G. Bordwell, N. R. Vanier, W. S. Matthews, J. B. Hendrickson, and P. L. Skipper, *J. Am. Chem. Soc.,* **1975**, *97*, 7160; F. G. Bordwell, J. C. Branca, C. R. Johnson, and N. R. Vanier, *J. Org. Chem.,* **1980**, *45*, 3884.

14. G. Chassaing, A. Marquet, J. Corset, and F. Froment, *J. Organomet. Chem.,* **1982**, *232*, 293; R. Lett, G. Chassaing, and A. Marquet, *J. Organomet. Chem.,* **1976**, *111*, C17; G. Chassaing and A. Marquet, *Tetrahedron*, **1978**, *34*, 1399; A. Berlin, S. Bradamante, R. Ferraccioli, and G. A. Pagani, *J. Chem. Soc., Chem. Commun.,* **1986**, 1191.

15. W. Hollstein, K. Harms, M. Marsch, and G. Boche, *Angew. Chem., Int. Ed. Engl.,* **1988**, *27*, 846.

16. H-J. Gais, G. Hellmann, H. Gunther, F. Lopez, H. J. Lindner, and S. Braun, *Angew. Chem., Int. Ed. Engl.*, **1989**, *28*, 1025; H-J. Gais, G. Hellmann, and H. J. Lindner, *Angew. Chem., Int. Ed. Engl.*, **1990**, *29*, 100.

17. H-J. Gais, J. Vollhardt, H. J. Lindner, and H. Paulus, *Angew. Chem., Int. Ed. Engl.*, **1988**, *27*, 1540.

18. H-J. Gais, J. Vollhardt, and C. Kruger, *Angew. Chem., Int. Ed. Engl.*, **1988**, *27*, 1092.

19. J. S. Grossert, J. Hoyle, T. S. Cameron, S. P. Roe, and B. R. Vincent, *Can. J. Chem.*, **1987**, *65*, 1407.

20. S. Wolfe, L. A. LaJohn, and D. F. Weaver, *Tetrahedron Lett.*, **1984**, *25*, 2863; S. Wolfe, A. Stolow, and L. A. LaJohn, *Tetrahedron Lett.*, **1983**, *24*, 4071.

21. D. A. Bors and A. Streitwieser Jr., *J. Am. Chem. Soc.*, **1986**, *108*, 1397.

22. N. S. Simpkins, unpublished results.

23. E. J. Corey and A. W. Gross, *Tetrahedron Lett.*, **1984**, *25*, 495.

24. F. G. Bordwell, E. Doomes, and P. W. R. Corfield, *J. Am. Chem. Soc.*, **1970**, *92*, 2581.

25. E. J. Corey and T. H. Lowry, *Tetrahedron Lett.*, **1965**, 793.

26. M. D. Brown, M. J. Cook, B. J. Hutchinson, and A. R. Katritzky, *Tetrahedron*, **1971**, *27*, 593; R. R. Fraser and F. J. Schuber, *J. Chem. Soc., Chem. Commun.*, **1969**, 1474; see also N. H. Werstiuk and S. Banerjee, *Can. J. Chem.*, **1985**, *63*, 2100.

27. K. Fuji, Y. Usami, K. Sumi, M. Ueda, and K. Kajiwara, *Chem. Lett.*, **1986**, 1655.

28. J. F. King and R. Rathore, *J. Am. Chem. Soc.*, **1990**, *112*, 2001.

29. T. Durst, *Tetrahedron Lett.*, **1971**, 4171.

30. H. E. Zimmerman and B. J. Thyagarajan, *J. Am. Chem. Soc.*, **1958**, *80*, 3060.

31. B. M. Trost and N. R. Schmuff, *J. Am. Chem. Soc.*, **1985**, *107*, 396; see also K. Hayakawa, H. Nishiyama, and K. Kanematsu, *J. Org. Chem.*, **1985**, *50*, 512.

32. R. V. Williams, G. W. Kelley, J. Loebel, D. van der Helm, and P. C. Bulman Page, *J. Org. Chem.*, **1990**, *55*, 3840.

33. A. Padwa, M. W. Wannamaker, and A. D. Dyszlewski, *J. Org. Chem.*, **1987**, *52*, 4760.

34. R. Tanikaga, K. Hamamura, K. Hosoya, and A. Kaji, *J. Chem. Soc., Chem. Commun.*, **1988**, 817.

35. J. Kattenberg, E. R. De Waard, and H. O. Huisman, *Tetrahedron*, **1974**, *30*, 463.

36. S. H. Pine, G. Shen, J. Bautista, C. Sutton Jr., W. Yamada, and L. Apodaca, *J. Org. Chem.*, **1990**, *55*, 2234.

37. F. G. Bordwell, M. Van Der Puy, and N. R. Vanier, *J. Org. Chem.*, **1976**, *41*, 1883 and 1885; F. G. Bordwell, J. E. Bares, J. E. Bartmess, G. E. Drucker, J. Gerhold, G. J. McCollum, M. Van Der Puy, N. R. Vanier, and W. S. Matthews, *J. Org. Chem.*, **1977**, *42*, 326; W. S. Matthews, J. E. Bares, J. E. Bartmess, F. G. Bordwell, F. J. Cornforth, G. E. Drucker, Z. Margolin, R. J. Callum, G. J.

McCollum, and N. R. Vanier, *J. Am. Chem. Soc.*, **1975**, *97*, 7006; see also R. G. Pearson and R. L. Dillon, *J. Am. Chem. Soc.*, **1953**, *75*, 2439; K. P. Ang and T. W. S. Lee, *Aust. J. Chem.*, **1977**, *30*, 521; J. Hine, J. C. Philips, and J. I. Maxwell, *J. Org. Chem.*, **1970**, *35*, 3943.

38. This type of interaction has been likened to the anomeric effect, see reference 20.
39. R. L. Shriner and S. O. Greenlee, *J. Org. Chem.*, **1939**, *4*, 242.
40. M. Julia and C. Blasioli, *Bull. Soc. Chim. Fr.*, **1976**, 1941.
41. M. Larcheveque, C. Sanner, R. Azerad, and D. Buisson, *Tetrahedron*, **1988**, *44*, 6407.
42. E. Schrotter, B. Schonecker, U. Hauschild, P. Droescher, and H. Schick, *Synthesis*, **1990**, 193.
43. S. Wershofen and H-D. Scharf, *Synthesis*, **1988**, 854.
44. R. M. Kennedy, A. Abiko, T. Takemasa, H. Okumoto, and S. Masamune, *Tetrahedron Lett.*, **1988**, *29*, 451.
45. M. Pohmakotr and S. Pisutjaroenpong, *Tetrahedron Lett.*, **1985**, *26*, 3613.
46. K. Kondo and D. Tunemoto, *Tetrahedron Lett.*, **1975**, 1007.
47. K. Kondo and D. Tunemoto, *Tetrahedron Lett.*, **1975**, 1397.
48. Y-H. Chang and H. W. Pinnick, *J. Org. Chem.*, **1978**, *43*, 373.
49. R. Bird, G. Griffiths, G. F. Griffiths, and C. J. M. Stirling, *J. Chem. Soc., Perkin Trans. 2*, **1982**, 579.
50. F. Benedetti and C. J. M. Stirling, *J. Chem. Soc., Chem. Commun.*, **1983**, 1374.
51. H. Takayanagi, T. Uyehara, and T. Kato, *J. Chem. Soc., Chem. Commun.*, **1978**, 359.
52. J. J. Burger, T. B. R. A. Chen, E. R. De Waard, and H. O. Huisman, *Tetrahedron*, **1980**, 1847.
53. D. Savoia, C. Trombini, and A. Umani-Ronchi, *J. Chem. Soc., Perkin Trans. 1*, **1977**, 123.
54. A. Jonczyk and T. Radwan-Pytlewski, *J. Org. Chem.*, **1983**, *48*, 910.
55. J. Golinski, A. Jonczyk, and M. Makosza, *Synthesis*, **1979**, 461.
56. K. Sato, O. Miyamoto, S. Inoue, T. Yamamoto, and Y. Hirasawa, *J. Chem. Soc., Chem. Commun.*, **1982**, 153.
57. T. Mandai, Y. Iuchi, K. Suzuki, M. Kawada, and J. Otera, *Tetrahedron Lett.*, **1982**, *23*, 4721.
58. G. M. P. Giblin, S. H. Ramcharitar, and N. S. Simpkins, *Tetrahedron Lett.*, **1988**, *29*, 4197; see also A. J. Bridges and J. W. Fischer, *J. Chem. Soc., Perkin Trans. 1*, **1983**, 2359; K. Hartke and H-U. Gleim, *Liebigs Ann. Chem.*, **1976**, 716; K. Hartke and M-H. Lee, *Arch. Pharm. (Weinheim)*, **1991**, *324*, 373.
59. J. B. Hendrickson, A. Giga, and J. Wareing, *J. Am. Chem. Soc.*, **1974**, *96*, 2275.
60. M. Julia and D. Uguen, *Bull. Soc. Chim. Fr.*, **1976**, 513.
61. S. O. Nwaukwa, S. Lee, and P. M. Keehn, *Synth. Commun.*, **1986**, *16*, 309.
62. J. C. Carretero and L. Ghosez, *Tetrahedron Lett.*, **1988**, *29*, 2059.

63. J. A. Marshall and R. C. Andrews, *J. Org. Chem.*, **1985**, *50*, 1602.

64. S. Hanessian, N. G. Cooke, B. DeHoff, and Y. Sakito, *J. Am. Chem. Soc.*, **1990**, *112*, 5276.

65. T. Nakata, K. Saito, and T. Oishi, *Tetrahedron Lett.*, **1986**, *27*, 6345.

66. B. Achmatowicz, S. Marczak, and J. Wicha, *J. Chem. Soc., Chem. Commun.*, **1987**, 1226.

67. J. Scherkenbeck, M. Barth, U. Thiel, K-H. Metten, F. Heinemann, and P. Welzel, *Tetrahedron*, **1988**, *44*, 6325.

68. C. Greck, P. Grice, S. V. Ley, and A. Wonnacott, *Tetrahedron Lett.*, **1986**, *27*, 5277.

69. D. Craig and A. M. Smith, *Tetrahedron Lett.*, **1990**, *31*, 2631.

70. B. Crobel, J. M. Decesare, and T. Durst, *Can. J. Chem.*, **1978**, *56*, 505; see also V. Cere, C. Paolucci, S. Pollicino, E. Sandri, and A. Fava, *J. Org. Chem.*, **1991**, *56*, 4513.

71. J. M. Decesare, B. Corbel, T. Durst, and J. F. Blount, *Can. J. Chem.*, **1981**, *59*, 1415.

72. F. Benedetti, S. Fabrissin, T. Gianferrara, and A. Risaliti, *J. Chem. Soc., Chem. Commun.*, **1987**, 406.

73. Y. Gaoni, *J. Org. Chem.*, **1982**, *47*, 2564, and references therein.

74. T. Satoh, Y. Kawase, and K. Yamakawa, *J. Org. Chem.*, **1990**, *55*, 3962. For a related example which forms a piperidine ring, see D. Tanner, H. H. Ming, and M. Bergdahl, *Tetrahedron Lett.*, **1988**, *29*, 6493.

75. L. Field, *J. Am. Chem. Soc.*, **1952**, *74*, 3919.

76. J. W. McFarland and D. N. Buchanan, *J. Org. Chem.*, **1965**, *30*, 2003.

77. P. J. Kocienski, *Chem. Ind.*, **1981**, 548.

78. D. F. Tavares and P. F. Vogt, *Can. J. Chem.*, **1967**, *45*, 1519.

79. W. E. Truce and T. C. Klingler, *J. Org. Chem.*, **1970**, *35*, 1834; see also C. A. Kingsbury, *J. Org. Chem.*, **1972**, *37*, 102.

80. C. Fehr, *Helv. Chim. Acta*, **1983**, *66*, 2519.

81. B. Achmatowicz, E. Baranowska, A. R. Daniewski, J. Pankowski, and J. Wicha, *Tetrahedron*, **1988**, *44*, 4989.

82. P. J. Kocienski, B. Lythgoe, and S. Ruston, *J. Chem. Soc., Perkin Trans. 1*, **1978**, 829.

83. E. Alvarez, T. Cuvigny, C. Herve du Penhoat, and M. Julia, *Tetrahedron*, **1988**, *44*, 111.

84. M. Julia and A. Guy-Rouault, *Bull. Soc. Chim. Fr.*, **1967**, 1410; R. V. M. Campbell, L. Crombie, D. A. R. Findley, R. W. King, G. Pattenden, and D. A. Whiting, *J. Chem. Soc., Perkin Trans. 1*, **1975**, 897; J. Martel and C. Huynh, *Bull. Soc. Chim. Fr.*, **1967**, 985.

85. J. Martel, C. Huynh, E. Toromanoff, and G. Nomine, *Bull. Soc. Chim. Fr.*, **1967**, 982.

86. E. Ghera and Y. Ben-David, *Tetrahedron Lett.*, **1979**, 4603.

87. S. De Lombaert, I. Nemery, B. Roekens, J. C. Carretero, T. Kimmel, and L. Ghosez, *Tetrahedron Lett.,* **1986**, *27,* 5099. For other examples of Michael reactions of sulphone carbanions, see P. A. Bartlett, F. R. Green III, and E. H. Rose, *J. Am. Chem. Soc.,* **1978**, *100,* 4852; K. Fuji, M. Node, and Y. Usami, *Chem. Lett.,* **1986**, 961.

88. G. A. Kraus and K. Frazier, *Synth. Commun.,* **1978**, *8,* 483.

89. M. Hirama, *Tetrahedron Lett.,* **1981**, *22,* 1905.

90. M. R. Binns, R. K. Haynes, A. G. Katsifis, P. A. Schober, and S. C. Vonwiller, *J. Org. Chem.,* **1989**, *54,* 1960.

91. J. Wildeman, P. C. Borgen, and H. Pluim, *Tetrahedron Lett.,* **1978**, 2213.

92. D. N. Jones, in *Perspectives in the Organic Chemistry of Sulphur,* Ed. B. Zwanenburg and A. J. H. Klunder, Elsevier, Amsterdam, **1987**, p. 189; J. P. Adams, J. Bowler, M. A. Collins, D. N. Jones, and S. Swallow, *Tetrahedron Lett.,* **1990**, *31,* 4355.

93. R. A. Abramovitch, S. S. Mathur, D. W. Saunders, and D. P. Vanderpool, *Tetrahedron Lett.,* **1980**, *21,* 705.

94. J. Moskal and A. M. van Leusen, *J. Org. Chem.,* **1986**, *51,* 4131.

95. F. M. Hauser and R. P. Rhee, *J. Org. Chem.,* **1978**, *43,* 178.

96. F. M. Hauser and Y. Caringal, *J. Org. Chem.,* **1990**, *55,* 555.

97. M. Azadi-Ardakani, R. Hayes, and T. W. Wallace, *Tetrahedron,* **1990**, *46,* 6851.

98. W. E. Truce and R. H. Knospe, *J. Am. Chem. Soc.,* **1955**, *77,* 5063.

99. G. A. Russell, E. T. Sabourin, and G. Hamprecht, *J. Org. Chem.,* **1969**, *34,* 2339.

100. B. M. Trost, J. Lynch, P. Renaut, and D. H. Steinman, *J. Am. Chem. Soc.,* **1986**, *108,* 284.

101. M. A. Brimble, C. J. Rush, G. M. Williams, and E. N. Baker, *J. Chem. Soc., Perkin Trans. 1,* **1990**, 414; M. A. Brimble, D. L. Officer, and G. M. Williams, *Tetrahedron Lett.,* **1988**, *29,* 3609; see also A. Alzerreca, M. Aviles, L. Collazo, and A. Prieto, *J. Heterocyclic Chem.,* **1990**, *27,* 1729.

102. E. Ghera, R. Maurya, and Y. Ben-David, *J. Org. Chem.,* **1988**, *53,* 1912.

103. S. M. Bennet and D. L. J. Clive, *J. Chem. Soc., Chem. Commun.,* **1986**, 878.

104. V. Reutrakul, P. Tuchinda, and K. Kusamran, *Chem. Lett.,* **1979**, 1055; see also Y. Gaoni, *Tetrahedron Lett.,* **1976**, 503.

105. M. Date, M. Watanabe, and S. Furukawa, *Chem. Pharm. Bull.,* **1990**, *38,* 902.

106. F. Babudri, S. Florio, A. M. Vitrani, and L. Di Nunno, *J. Chem. Soc., Perkin Trans. 1,* **1984**, 1899.

107. M. Muraoka, T. Yamamoto, T. Ebisawa, W. Kobayashi, and T. Takeshima, *J. Chem. Soc., Perkin Trans. 1,* **1978**, 1017. For a much earlier report, see L. Field, J. E. Lawson, and J. W. McFarland, *J. Am. Chem. Soc.,* **1956**, *78,* 4389.

108. J-B. Baudin, M. Julia, C. Rolando, and J-N. Verpeaux, *Bull. Soc. Chim. Fr.,* **1987,** 493; M. Julia, G. Le Thuillier, C. Rolando, and L. Saussine, *Tetrahedron Lett.,* **1982,** *23,* 2453; T. Kauffmann and R. Jousen, *Chem. Ber.,* **1977,** *110,* 3930.

109. G. Buchi and R. M. Freidinger, *Tetrahedron Lett.,* **1985,** *26,* 5923.

110. M. Julia and J-N. Verpeaux, *Tetrahedron Lett.,* **1982,** *23,* 2457; see also L. Engman, *J. Org. Chem.,* **1984,** *49,* 3559.

111. P. Charreau, M. Julia, and J-N. Verpeaux, *J. Organomet. Chem.,* **1989,** *379,* 201.

112. P. Charreau, M. Julia, and J-N. Verpeaux, *Bull. Soc. Chim. Fr.,* **1990,** *127,* 275.

113. M. Julia and L. Saussine, *Tetrahedron Lett.,* **1974,** 3443.

114. R. Khourzom, F. Rose-Munch, and E. Rose, *Tetrahedron Lett.,* **1990,** *31,* 2011.

115. J. Tsuji, I. Shimizu, I. Minami, Y. Ohashi, T. Sugiura, and K. Takahashi, *J. Org. Chem.,* **1985,** *50,* 1523.

116. J. J. Eisch and J. E. Galle, *J. Org. Chem.,* **1979,** *44,* 3279; M. Yamamoto, K. Suzuki, S. Tanaka, and K. Yamada, *Bull. Chem. Soc. Jpn.,* **1987,** *60,* 1523.

117. H. Kleijn and P. Vermeer, *J. Organomet. Chem.,* **1986,** *302,* 1.

118. J. Meijer and P. Vermeer, *Recl. Trav. Chim. Pays-Bas,* **1975,** *94,* 14.

119. N. S. Simpkins, *Tetrahedron Lett.,* **1987,** *28,* 989.

120. M. Yamamoto, T. Takemori, S. Iwasa, S. Kohmoto, and K. Yamada, *J. Org. Chem.,* **1989,** *54,* 1757.

121. C. Najera and M. Yus, *J. Org. Chem.,* **1988,** *53,* 4708; C. Najera and M. Yus, *Tetrahedron Lett.,* **1987,** *28,* 6709.

122. S. W. McCombie, B. B. Shankar, and A. K. Ganguly, *Tetrahedron Lett.,* **1985,** *26,* 6301.

123. S. W. McCombie, B. B. Shankar, and A. K. Ganguly, *Tetrahedron Lett.,* **1987,** *28,* 4123.

124. S. W. McCombie, B. B. Shankar, A. K. Ganguly, A. Padwa, W. H. Bullock, and A. D. Dyszlewski, *Tetrahedron Lett.,* **1987,** *28,* 4127.

125. A. Pelter, R. S. Ward, and G. M. Little, *J. Chem. Soc., Perkin Trans. 1,* **1990,** 2775.

126. W. E. Truce and E. J. Mandaj, Jr., *Sulphur Reports,* **1983,** *3,* 259.

127. H. Gilman and S. H. Eidt, *J. Am. Chem. Soc.,* **1956,** *78,* 2633.

128. W. E. Truce and M. F. Amos, *J. Am. Chem. Soc.,* **1951,** *73,* 3013.

129. F. M. Stoyanovich, R. G. Karpenko, and Y. L. Gol'dfarb, *Tetrahedron,* **1971,** *27,* 433.

130. T. D. Krizan and J. C. Martin, *J. Am. Chem. Soc.,* **1983,** *105,* 6155.

131. G. D. Hartman and W. Halczenko, *Tetrahedron Lett.,* **1987,** *28,* 3241.

132. M. Iwao, T. Iihama, K. K. Mahalanabis, H. Perrier, and V. Snieckus, *J. Org. Chem.,* **1989,** *54,* 24.

133. E. P. Kohler and H. A. Potter, *J. Am. Chem. Soc.,* **1936,** *58,* 2166.

134. J. W. Lee and D. Y. Oh, *Heterocycles,* **1990,** *31,* 1417.

135. T. Shono, S. Kashimura, M. Sawamura, and T. Soejima, *J. Org. Chem.*, **1988**, *53*, 907.
136. F. Cooke and P. Magnus, *J. Chem. Soc., Chem. Commun.*, **1976**, 519.
137. B. M. Trost and H. Urabe, *J. Am. Chem. Soc.*, **1990**, *112*, 4982.
138. M. W. Thomsen, B. M. Handwerker, S. A. Katz, and S. A. Fisher, *Synth. Commun.*, **1988**, *18*, 1433.
139. M. H. Nantz, X. Radisson, and P. L. Fuchs, *Synth. Commun.*, **1987**, *17*, 55.
140. M. H. Nantz and P. L. Fuchs, *J. Org. Chem.*, **1987**, *52*, 5298.
141. P. T. Lansbury, R. W. Erwin, and D. A. Jeffrey, *J. Am. Chem. Soc.*, **1980**, *102*, 1602.
142. H. O. House and J. K. Larson, *J. Org. Chem.*, **1968**, *33*, 61.
143. M. Ochiai, K. Sumi, E. Fujita, and M. Shiro, *Tetrahedron Lett.*, **1982**, *23*, 5419.
144. N. Ono, R. Tamura, R. Tanikaga, and A. Kaji, *Synthesis*, **1977**, 690.
145. P. A. Wade, H. R. Hinney, N. V. Amin, P. D. Vail, S. D. Morrow, S. A. Hardinger, and M. S. Saft, *J. Org. Chem.*, **1981**, *46*, 765.
146. A. M. van Leusen, in *Perspectives in the Organic Chemistry of Sulphur*, Ed. B. Zwanenburg and A. J. H. Klunder, Elsevier, Amsterdam, **1987**, p.119; A. M. van Leusen, G. J. M. Boerma, R. B. Helmholdt, H. Siderius, and J. Strating, *Tetrahedron Lett.*, **1972**, 2367.
147. O. Possel and A. M. van Leusen, *Tetrahedron Lett.*, **1977**, 4229; J. S. Yadav and P. S. Reddy, *Tetrahedron Lett.*, **1984**, *25*, 4025.
148. J. S. Yadav and V. R. Gadgil, *Tetrahedron Lett.*, **1990**, *31*, 6217.
149. D. van Leusen and A. M. van Leusen, *Synthesis*, **1980**, 325.
150. M. S. Frazza and B. W. Roberts, *Tetrahedron Lett.*, **1981**, *22*, 4193.
151. A. M. van Leusen, B. E. Hoogenboom, and H. Siderius, *Tetrahedron Lett.*, **1972**, 2369. For an asymmetric variant of the TosMIC aldol reaction, see M. Sawamura, H. Hamashima, and Y. Ito, *J. Org. Chem.*, **1990**, *55*, 5935.
152. U. Schollkopf, R. Schroder, and E. Blume, *Liebigs Ann. Chem.*, **1972**, *766*, 130.
153. O. H. Oldenziel, D. van Leusen, and A. M. van Leusen, *J. Org. Chem.*, **1977**, *42*, 3114; A. M. van Leusen and P. G. Oomkes, *Synth. Commun.*, **1980**, *10*, 399.
154. O. H. Oldenziel and A. M. van Leusen, *Tetrahedron Lett.*, **1974**, 163, O. H. Oldenziel and A. M. van Leusen, *Tetrahedron Lett.*, **1974**, 167.
155. A. M. van Leusen and O. H. Oldenziel, *Tetrahedron Lett.*, **1972**, 2373.
156. A. M. van Leusen and J. Wildeman, *Synthesis*, **1977**, 501.
157. D. van Leusen and A. M. van Leusen, *Tetrahedron Lett.*, **1984**, *25*, 2581; see also D. van Leusen and A. M. van Leusen, *Synthesis*, **1991**, 531.
158. S. V. Ley, B. Lygo, F. Sternfeld, and A. Wonnacott, *Tetrahedron*, **1986**, *42*, 4333; C. Greck, P. Grice, S. V. Ley, and A. Wonnacott, *Tetrahedron Lett.*, **1986**, *27*, 5277. For an earlier report describing the anion and dianion chemistry

of acyclic alkoxy sulphones, see K. Tanaka, S. Matsui, and A. Kaji, *Bull. Chem. Soc. Jpn.*, **1980**, *53*, 3619. See also the series of three consecutive communications from Julia's group, starting with F. Chemla, M. Julia, D. Uguen, and D. Zhang, *Synlett*, **1991**, 501.

159. M. Reggelin, P. Tebben, and D. Hoppe, *Tetrahedron Lett.*, **1989**, *30*, 2915; P. Tebben, M. Reggelin, and D. Hoppe, *Tetrahedron Lett.*, **1989**, *30*, 2919.

160. K. Ogura, N. Yahata, K. Takahashi, and H. Iida, *Tetrahedron Lett.*, **1983**, *24*, 5761.

161. K. Ogura, T. Iihama, K. Takahashi, and H. Iida, *Tetrahedron Lett.*, **1984**, *25*, 2671.

162. H-J. Gais, G. Schmiedl, W. A. Ball, J. Bund, G. Schmiedl, and I. Erdelmeier, *Tetrahedron Lett.*, **1988**, *29*, 1773.

163. E. P. Kundig and A. F. Cunningham, Jr., *Tetrahedron*, **1988**, *44*, 6855.

164. B. M. Trost, J. Cossy, and J. Burks, *J. Am. Chem. Soc.*, **1983**, *105*, 1052.

165. B. M. Trost, *Angew. Chem., Int. Ed. Engl.*, **1989**, *28*, 1173.

166. W. Oppolzer and J-M. Gaudin, *Helv. Chim. Acta*, **1987**, *70*, 1477.

167. B. M. Trost, C. J. Urch, and M-H. Hung, *Tetrahedron Lett.*, **1986**, *27*, 4949.

168. B. M. Trost and R. W. Warner, *J. Am. Chem. Soc.*, **1982**, *104*, 6112.

169. B. M. Trost, J. T. Hane, and P. Metz, *Tetrahedron Lett.*, **1986**, *27*, 5695; B. M. Trost, P. Metz, and J. T. Hane, *Tetrahedron Lett.*, **1986**, *27*, 5691.

170. Y. Masuyama, H. Hirai, Y. Kurusu, and K. Segawa, *Bull. Chem. Soc. Jpn.*, **1987**, *60*, 1525.

171. J. S. Baxter, M. Green, and T. V. Lee, *J. Chem. Soc., Chem. Commun.*, **1989**, 1595.

172. M. Adamczyk, E. K. Dolence, D. S. Watt, M. R. Christy, J. H. Reibenspies, and O. P. Anderson, *J. Org. Chem.*, **1984**, *49*, 1378.

173. V. Reutrakul, V. Prapansiri, and C. Panyachotipun, *Tetrahedron Lett.*, **1984**, *25*, 1949.

174. J. Golinski and M. Makosza, *Tetrahedron Lett.*, **1978**, 3495; M. Makosza, J. Golinski, and J. Baran, *J. Org. Chem.*, **1984**, *49*, 1488; M. Makosza, W. Danikiewicz, and K. Wojciechowski, *Liebigs Ann. Chem.*, **1987**, 711; M. Makosza and A. Tyrala, *Synth. Commun.*, **1986**, *16*, 419.

175. E. M. Kaiser, J. D. Petty, and P. L. A. Knutson, *Synthesis*, **1977**, 509. For another review of dianion chemistry, see C. M. Thompson and D. L. C. Green, *Tetrahedron*, **1991**, *47*, 4223.

176. E. M. Kaiser, L. E. Solter, R. A. Schwarz, R. D. Beard, and C. R. Hauser, *J. Am. Chem. Soc.*, **1971**, *93*, 4237.

177. C. H. Heathcock, B. L. Finkelstein, E. T. Jarvi, P. A. Radel, and C. R. Hadley, *J. Org. Chem.*, **1988**, *53*, 1922.

178. S. Najdi and M. J. Kurth, *Tetrahedron Lett.*, **1990**, *31*, 3279.

179. J. J. Eisch, S. K. Dua, and M. Behrooz, *J. Org. Chem.*, **1985**, *50*, 3674.

180. V. Pascali, N. Tangari, and A. Umani-Ronchi, *J. Chem. Soc., Perkin Trans. 1*, **1973**, 1166; A. Bongini, D. Savoia, and A. Umani-Ronchi, *J. Organomet. Chem.*, **1976**, *112*, 1.

181. M. W. Thomsen, B. M. Handwerker, S. A. Katz, and R. B. Belser, *J. Org. Chem.*, **1988**, *53*, 906.

182. J. D. White, M. A. Avery, S. C. Choudhry, O. P. Dhingra, M. Kang, and A. J. Whittle, *J. Am. Chem. Soc.*, **1983**, *105*, 6517.

183. M. C. Mussatto, D. Savoia, C. Trombini, and A. Umani-Ronchi, *J. Chem. Soc., Perkin Trans. 1*, **1980**, 260.

184. S. Cavicchioli, D. Savoia, C. Trombini, and A. Umani-Ronchi, *J. Org. Chem.*, **1984**, *49*, 1246.

185. N. Bosworth and P. Magnus, *J. Chem. Soc., Perkin Trans. 1*, **1973**, 2319.

186. J. Vollhardt, H-J. Gais, and K. L. Lukas, *Angew. Chem., Int. Ed. Engl.*, **1985**, *24*, 610.

187. H-J. Gais, J. Vollhardt, H. Gunther, D. Moskau, H. J. Lindner, and S. Braun, *J. Am. Chem. Soc.*, **1988**, *110*, 978; H-J. Gais and J. Vollhardt, *Tetrahedron Lett.*, **1988**, *29*, 1529.

188. H-J. Gais, W. A. Ball, and J. Bund, *Tetrahedron Lett.*, **1988**, *29*, 781.

189. B. M. Trost and C. A. Merlic, *J. Am. Chem. Soc.*, **1988**, *110*, 5216.

190. E. M. Kaiser, R. D. Beard, and C. R. Hauser, *J. Organomet. Chem.*, **1973**, *59*, 53; E. M. Kaiser and C. R. Hauser, *Tetrahedron Lett.*, **1967**, 3341; see also W. E. Truce and T. C. Klingler, *J. Org. Chem.*, **1970**, *35*, 1834.

191. J. S. Gossert, J. Buter, E. W. H. Asveld, and R. M. Kellogg, *Tetrahedron Lett.*, **1974**, 2805.

192. L. A. Paquette, R. H. Meisinger, and R. Gleiter, *J. Am. Chem. Soc.*, **1973**, *95*, 5414.

193. F. M. Stoyanovich, R. G. Karpenko, G. I. Gorushkina, and Y. L. Gol'dfarb, *Tetrahedron*, **1972**, *28*, 5017; F. M. Stoyanovich and B. P. Fedorov, *Angew. Chem., Int. Ed. Engl.*, **1966**, *5*, 127.

194. E. P. Kohler and M. Tishler, *J. Am. Chem. Soc.*, **1935**, *57*, 217.

195. J. B. Hendrickson, G. J. Boudreaux, and P. S. Palumbo, *J. Am. Chem. Soc.*, **1986**, *108*, 2358; J. B. Hendrickson and P. S. Palumbo, *J. Org. Chem.*, **1985**, *50*, 2110.

196. S. P. J. M. van Nispen, C. Mensink, and A. M. van Leusen, *Tetrahedron Lett.*, **1980**, *21*, 3723.

197. D. Scholz, *Liebigs Ann. Chem.*, **1983**, 98; see also D. Laduree, D. Paquer, and P. Rioult, *Recl. Trav. Chim. Pays-Bas*, **1977**, *96*, 254.

198. J. S. Grossert, J. Hoyle, and D. L. Hooper, *Tetrahedron*, **1984**, *40*, 1135; D. Scholz, *Liebigs Ann. Chem.*, **1984**, 264.

199. B. Lygo and N. O'Connor, *Synlett*, **1990**, *1*, 282.

200. S. W. McCombie, B. B. Shankar, and A. K. Ganguly, *Tetrahedron Lett.*, **1989**, *30*, 7029.

201. P. T. Lansbury, G. E. Bebernitz, S. C. Maynard, and C. J. Spagnuolo, *Tetrahedron Lett.*, **1985**, *26*, 169.
202. P. T. Lansbury, C. J. Spagnuolo, and E. L. Grimm, *Tetrahedron Lett.*, **1986**, *27*, 2725.
203. M. Julia, D. Uguen, and A. Callipolitis, *Bull. Soc. Chim. Fr.*, **1976**, 519.
204. A. P. Kozikowski, B. B. Mugrage, C. S. Li, and L. Felder, *Tetrahedron Lett.*, **1986**, *27*, 4817.
205. R. Tanikaga, K. Hosoya, K. Hamamura, and A. Kaji, *Tetrahedron Lett.*, **1987**, *28*, 3705.
206. R. Tanikaga, K. Hosoya, and A. Kaji, *Chem. Lett.*, **1987**, 829.
207. I. Rothberg, B. Sundoro, G. Balanikas, and S. Kirsch, *J. Org. Chem.*, **1983**, *48*, 4345.
208. N. P. Singh, B. Metz, and J. F. Biellmann, *Bull. Soc. Chim. Fr.*, **1990**, *127*, 98.
209. K. S. Kim, J-K. Sohng, S. B. Ha, C. S. Cheong, D. I. Jung, and C. S. Hahn, *Tetrahedron Lett.*, **1988**, *29*, 2847.
210. K. Tanaka, K. Ootake, K. Imai, N. Tanaka, and A. Kaji, *Chem. Lett.*, **1983**, 633; R. Tanikaga, K. Hosoya, and A. Kaji, *Synthesis*, **1987**, 389.
211. T. Sato, Y. Okumura, J. Itai, and T. Fujisawa, *Chem. Lett.*, **1988**, 1537.
212. C. M. Thompson, D. L. C. Green, and R. Kubas, *J. Org. Chem.*, **1988**, *53*, 5389; C. M. Thompson, *Tetrahedron Lett.*, **1987**, *28*, 4243.
213. C. M. Thompson, J. A. Frick, and C. E. Woytowicz, *Synth. Commun.*, **1988**, 889.
214. P. T. Lansbury, C. J. Spagnuolo, B. Zhi, and E. L. Grimm, *Tetrahedron Lett.*, **1990**, *31*, 3965.

CHAPTER 4

Additions to Unsaturated Sulphones

α,β-Unsaturated sulphones are extremely useful as Michael acceptors with a host of nucleophilic partners. Reaction with suitable heteroatom nucleophiles such as alcohols, thiols and amines yields β-heterosubstituted sulphones, some of which are useful in further synthesis. More importantly, reactions with organometallics and more stabilised carbon nucleophiles, such as enolates, are also possible.

Although Michael additions to a vinyl sulphone were first described in 1935 the full synthetic value of this reaction has only recently been exploited.[1] Vinyl sulphones can be particularly valuable in Michael chemistry since, in contrast with unsaturated carbonyl compounds, competing addition to the sulphone functional group is not possible.

4.1 Heteroatom Nucleophiles

This section describes the additions of nitrogen, oxygen and sulphur nucleophiles to alkenyl, allenyl and alkynyl sulphones. Brief mention of conjugate hydride additions has also been included here.

4.1.1 Nitrogen Nucleophiles

McDowell and Stirling have examined the addition of amines to α,β-unsaturated sulphones in some detail.[2] Additions to phenyl vinyl sulphone take place readily, but substituents on the sulphone at either the α- or β-position greatly reduce the rate of this reaction. Similar reactions of propargyl phenyl sulphone and allenyl phenyl sulphone give the same product. This was shown to be due to isomerisation of the propargyl sulphone to the allenyl isomer which then shows typical vinyl sulphone behaviour.

A rather interesting double addition process occurs when either propargyl or allenyl sulphones are reacted with an amino alcohol such as (-)-ephedrine, Scheme 1.[3]

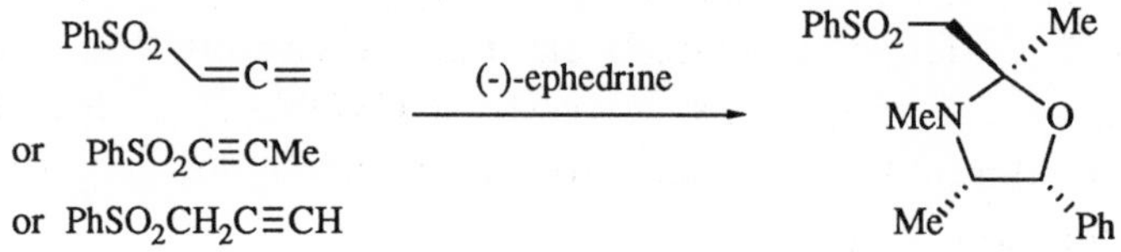

Scheme 1

The 1,3-oxazolidine product is formed as a single diastereoisomer in each case. An intermediate enamine sulphone can be isolated in one case, indicating that the initial addition occurs through the nitrogen centre.

The conjugate addition of amines to cyclic vinyl sulphones has been applied to the synthesis of some unusual nucleosides, Scheme 2.[4]

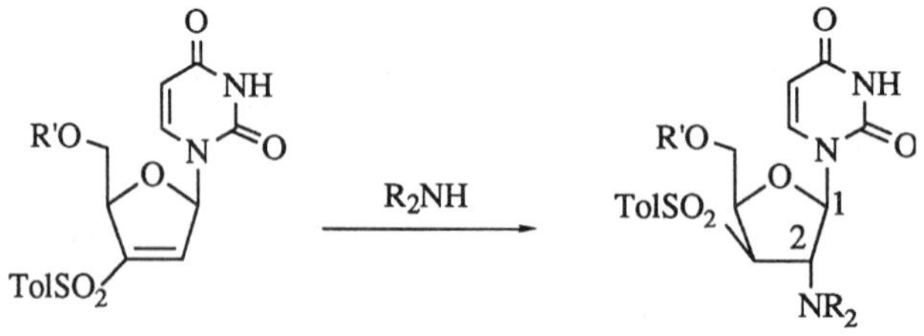

Scheme 2

Both purine and pyrimidine derivatives react to give *trans*-addition products in which the nitrogen substituents at C-1 and C-2 are also *trans*. Intramolecular addition of nucleophilic nitrogen groups to vinyl sulphones is also possible, Scheme 3.[5,6]

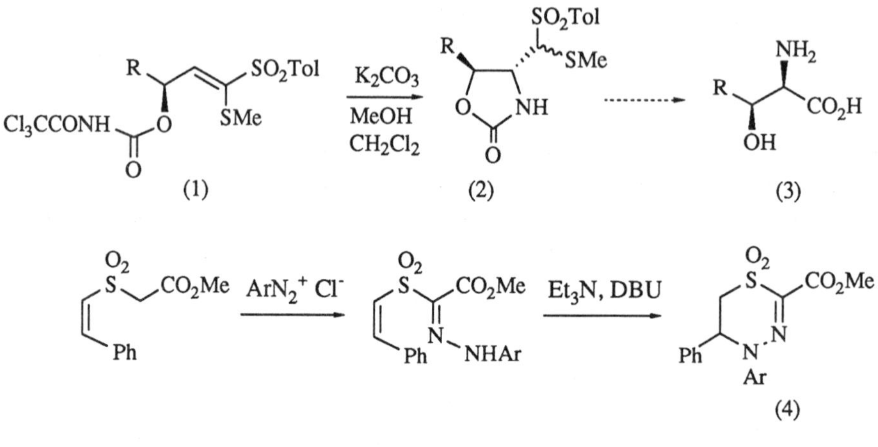

Scheme 3

Highly stereoselective cyclisation of vinyl sulphones (**1**) occurs to give *trans*-oxazolidinones (**2**).[5] These compounds can then be converted to a range of *syn*-β-hydroxy-α-amino acids (**3**), where R = Me, CH_2OH, etc. Novel heterocyclic sulphones of structure (**4**) are readily formed in moderate to good yield through 6-*endo-trig* cyclisation, as shown.[6]

Padwa and Norman have examined a range of cyclisation reactions utilising 2,3-bis(phenylsulphonyl)-1,3-butadiene (**5**), Scheme 4.[7]

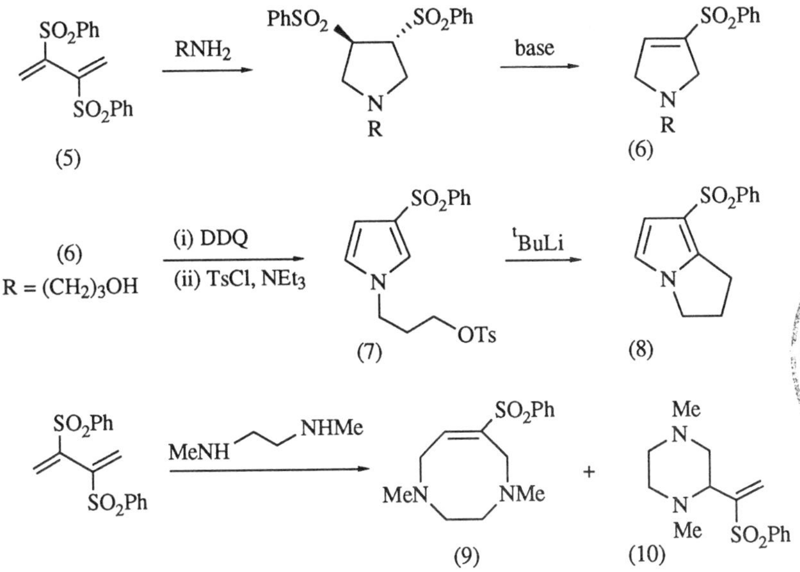

Scheme 4

Reaction of (**5**) with a simple amine gives a bis(phenylsulphonyl)pyrrolidine by two sequential Michael processes, the second of which is a disfavoured 5-*endo-trig* cyclisation. Facile base elimination of benzenesulphinate then gives the 3-pyrroline (**6**) which on heating with DDQ forms the corresponding pyrrole, leading to (**7**). In turn, one such compound could be further elaborated to form the pyrrolizidine-type product (**8**). Michael reactions of (**5**) with diamines were also studied. Thus treatment with *N,N′*-dimethylethylenediamine gives (**10**) which appears to arise, at least partially, by formation of the larger-ring isomer (**9**), followed by rearrangement.

4.1.2 Oxygen Nucleophiles

Addition of alkoxide groups to vinyl and allenyl sulphones usually occurs at the β-position of the sulphone in analogous fashion to the amine additions.[8] This reaction has been used to construct 2-alkoxyallylic sulphones, required in a study of intramolecular [4+3] furan cycloadditions, Scheme 5.[9]

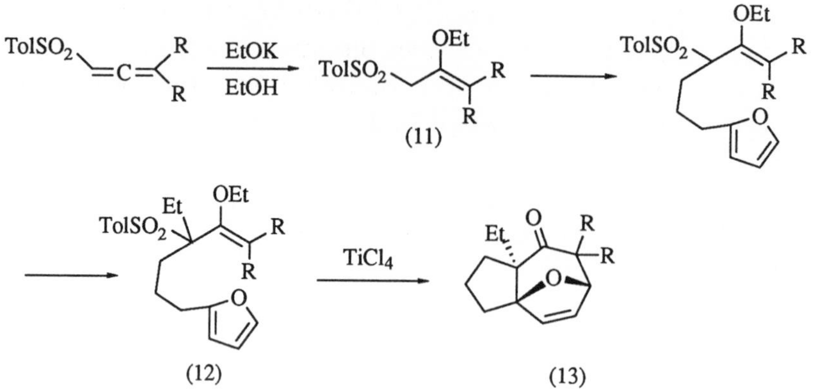

(11)

(12)   (13)

Scheme 5

The allyl sulphone (**11**) undergoes two consecutive alkylations to give (**12**), which on treatment with TiCl$_4$ forms an allylic cation leading to (**13**). Cyclisations of vinyl sulphones described by Knochel and co-workers follow similar lines to the chemistry in Scheme 4, for example resulting in the formation of spirocyclic ether (**14**), Scheme 6.[10]

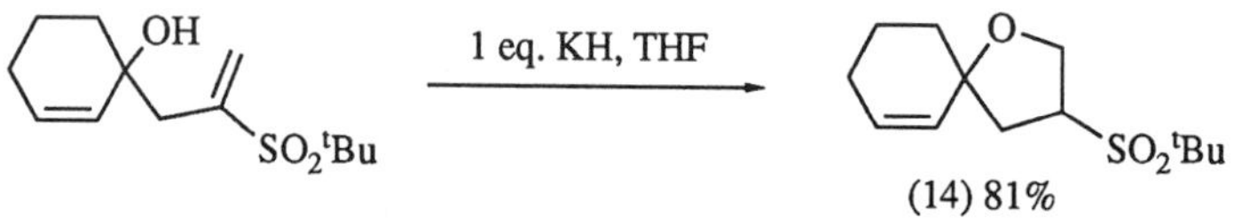

(14) 81%

Scheme 6

Here the substrate undergoes rapid cyclisation on treatment with KH, despite the opportunity for an alternative oxy-Cope rearrangement. Attempts to cyclise analogous sulphoxides were unsuccessful.

Certain styryl sulphones can undergo overall substitution of the sulphone group rather than furnishing Michael adducts, Scheme 7.[11]

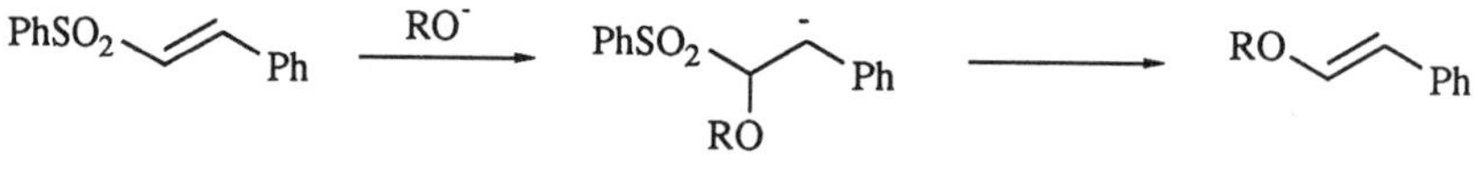

Scheme 7

Presumably, under the conditions used, the normal Michael addition is reversible. Thus the abnormal mode of addition to give a benzylic anion, even if only a minor pathway, can eventually lead to the observed vinyl ether products.

The addition of peroxide species to vinyl sulphones, which normally results in the formation of α,β-epoxy sulphones, has been discussed in Section 2.4.2. By careful control of the reaction conditions a variety of β-hydroperoxy sulphones can be isolated from this type of reaction.[12]

4.1.3 Sulphur Nucleophiles

The addition of thiols to vinyl sulphones has been examined as a means of protecting the thiol group, Scheme 8.[13]

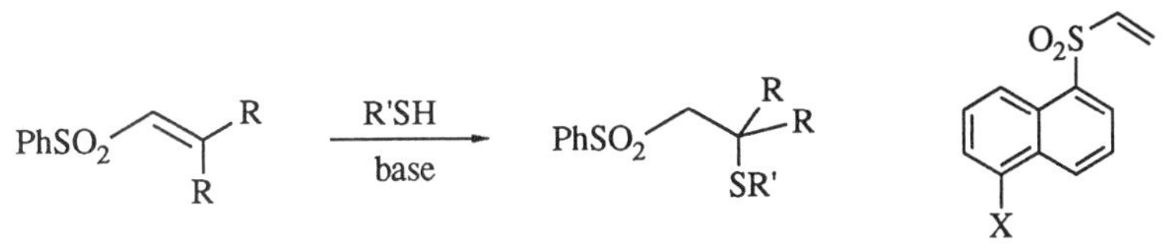

PhSO₂ — R R'SH/base → PhSO₂ — R SR'

R = H, Me   (15)   (16) X = OMe, NMe₂

Scheme 8

The thiol adducts (15) are stable to strongly acidic conditions and to reductions using borohydride, borane, etc. Deprotection to regenerate the thiol is readily achieved using dilute base, for example ᵗBuOK in ᵗBuOH. The fluorescent vinyl sulphones (16) can be used for the quantitative determination of thiol groupings in polypeptides.

The reaction of thiols with alkynyl sulphones follows a similar course to the additions involving oxygen and nitrogen nucleophiles, Scheme 9.[14,15]

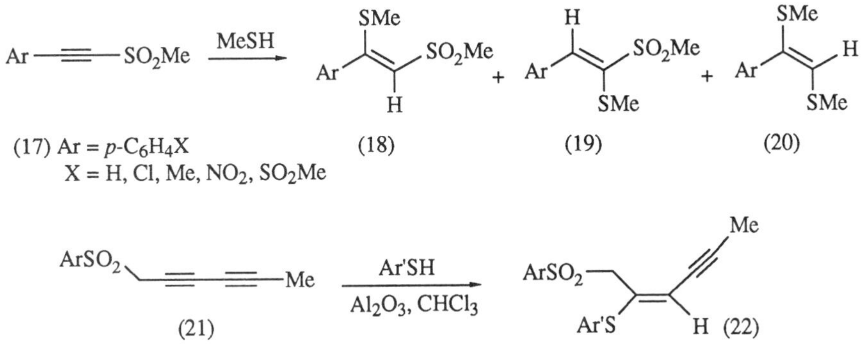

Ar≡–SO₂Me MeSH → (18) + (19) + (20)

(17) Ar = p-C₆H₄X
X = H, Cl, Me, NO₂, SO₂Me   (18)   (19)   (20)

ArSO₂—≡≡—Me Ar'SH / Al₂O₃, CHCl₃ → ArSO₂ / Ar'S / H (22)

(21)

Scheme 9

Three products are obtained from the reaction of alkynyl sulphone (17) with methanethiol.[14] The 'normal' addition product (18) is accompanied by the regioisomer (19), especially significant when X = NO₂ and SO₂Me, and the double addition product (20). Thus, the inductive effect of the aromatic substituent can be tuned to favour either α- or β-addition, *trans*-addition being favoured in all cases. The conversion of (21) to (22) presumably represents another example of addition to an allenyl sulphone formed *in situ*.

The addition of a nucleophile to an allenyl sulphone generates a sulphonyl carbanion, which can be used in a subsequent carbon–carbon bond-forming process. This type of chemistry has been developed by Padwa and Yeske using a variety of nucleophiles, including benzenesulphinate, e.g. Scheme 10.[16]

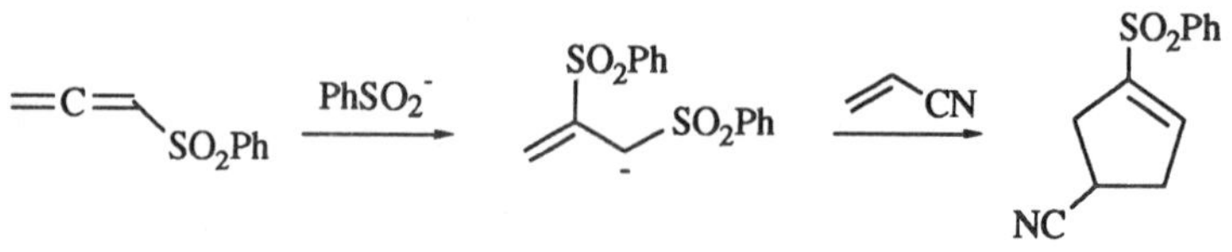

Scheme 10

The initial addition process is followed by attack on another Michael acceptor, such as acrylonitrile, and then cyclisation, which results in expulsion of benzenesulphinate. Along with the cyclisations shown in Schemes 4 and 6, this process highlights the ability of vinyl sulphones to facilitate 5-*endo-trig* ring closures.

4.1.4 Hydride Addition

A few scattered reports describe the use of hydride reducing agents for the conversion of vinyl sulphones to their saturated analogues, e.g. Scheme 11.[17,18]

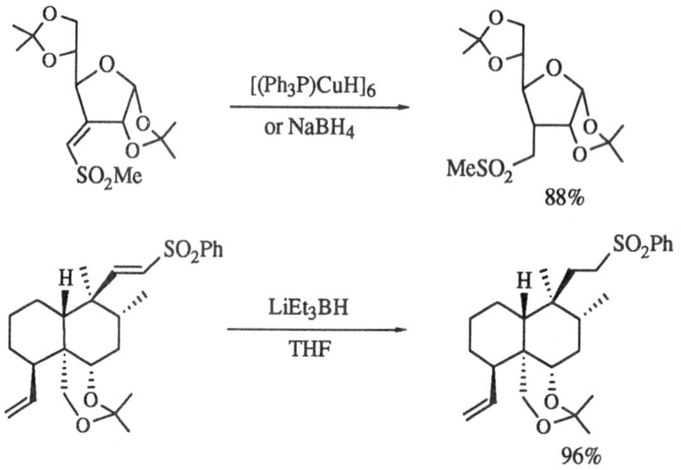

Scheme 11

These reductions presumably occur by Michael addition of hydride, although no detailed study of this type of reaction of vinyl sulphones has appeared.

4.2 Non-stabilised Organometallics

Since the seminal report by Posner and Brunelle concerning the reaction of organocuprates with simple vinyl sulphones,[19] other groups have reported analogous chemistry using alkynyl sulphones[20] and vinyl cuprates,[21] Scheme 12.

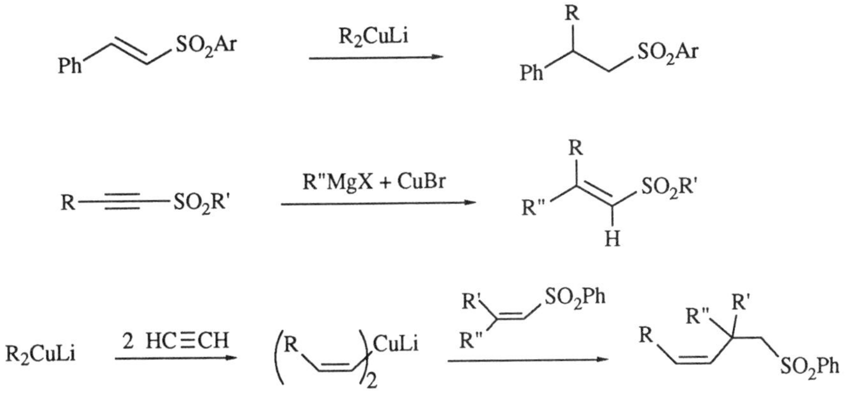

Scheme 12

General trends in such reactions include the predominant *syn*-addition to alkynyl sulphones and the reduced rate of addition upon increasing the degree of substitution on the vinyl sulphone. Whilst synthetic applications of these reactions using simple vinyl sulphones have been few, the corresponding chemistry of cyclic vinyl sulphones has been explored in detail by Fuchs and co-workers.[22] One such study probed the use of a variety of organometallic reagents in reactions with cyclooctenyl phenyl sulphones, Scheme 13.[23]

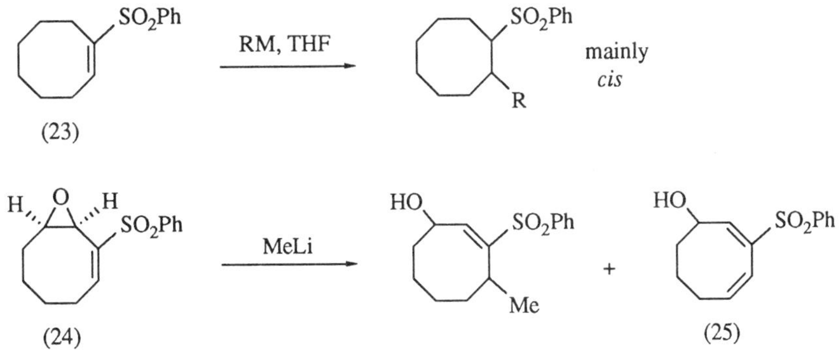

Scheme 13

In reactions with simple sulphone (23), cuprate species such as Me₂CuLi and Ph₂CuLi prove reasonably effective, whilst Grignard reagents, alkylcerium and alkyltitanium species are unreactive. Perhaps surprisingly alkyllithiums are highly efficient, even ᵗBuLi giving the addition product in 75% yield, despite the possibility of competing sulphone deprotonation. However, using the epoxyvinyl sulphone (24) deprotonation is a serious side reaction, resulting in the formation of (25). Here direct addition to the epoxide is also observed using reagents such as MeMgBr, MeCu and Me₃ZnLi. In general, Fuchs has used alkyllithiums to good effect in such additions, cuprates often offering no advantages. With simple acyclic vinyl sulphones Julia and co-workers have shown that Grignard reagents can undergo effective Michael addition, depending on the substitution pattern of the sulphone.[24]

A series of papers has described additions and SN2' reactions of a number of cyclopentenyl sulphones having additional amine or epoxide functionality, e.g. Scheme 14.[25]

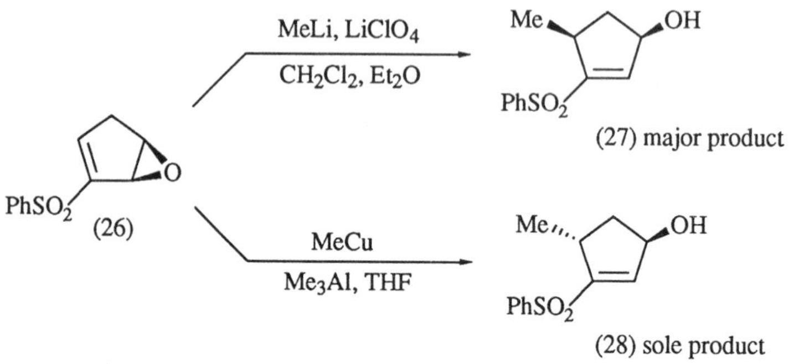

Scheme 14

In initial experiments involving reaction of (26) with MeLi, competing sulphone deprotonation was found to be a serious problem. Firstly, the use of MeLi in the presence of LiClO₄ gives predominantly *cis*-alcohol (27), presumably through some chelation-controlled addition. In contrast, the use of a MeCu/Me₃Al combination in THF results in the exclusive formation of (28), within the limits of detection.

Later studies have shown that for addition of groups other than methyl to such sulphones, alternative solutions to the deprotonation problem are needed.[26] To this end, both the aminosulphone (29) and the corresponding quaternary salt (30) have been found to be readily accessible and to undergo highly stereoselective reactions with appropriate organometallics, Scheme 15.[27] Again, complementary sequences can be devised in order to access either diastereoisomeric product. Interestingly, extension of these reactions to the analogous cyclohexenyl sulphones has also enabled stereoselective additions to be conducted, but with the opposite stereochemical trends to those shown for five-membered systems.[28]

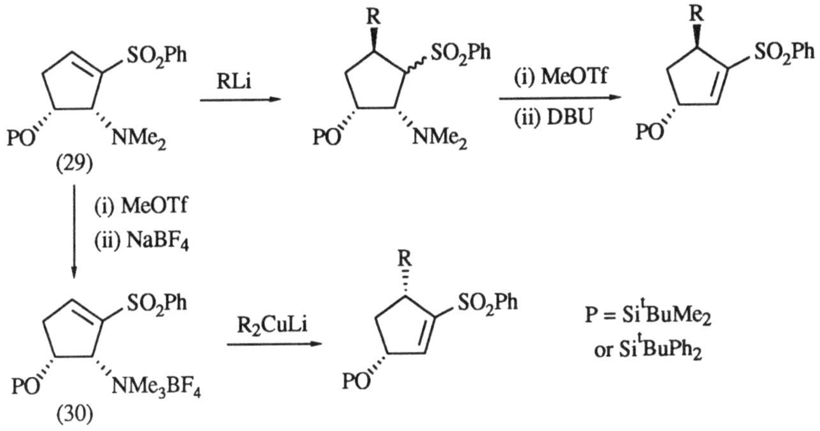

Scheme 15

The total synthesis of (+)-carbacyclin, reported by Hutchinson and Fuchs, demonstrates the effectiveness of this approach for the regio- and stereoselective introduction of carbon substituents onto a cyclopentane ring, Scheme 16.[29]

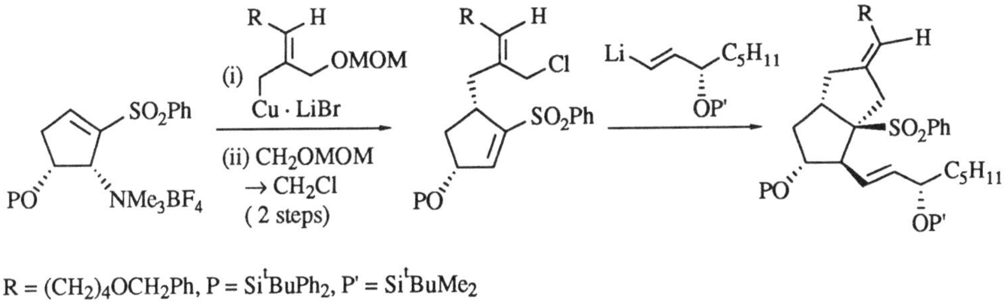

$R = (CH_2)_4OCH_2Ph, P = Si^tBuPh_2, P' = Si^tBuMe_2$

Scheme 16

The first addition generates a transposed vinyl sulphone intermediate which can then undergo another conjugate addition, followed by intramolecular alkylation of the resulting sulphonyl carbanion. Fuchs has developed a number of variants on the theme of vinyl sulphone conjugate addition followed by intramolecular trapping. Perhaps the most spectacular involves intermediates such as (31), which, on treatment with BuLi, furnishes tetracyclic sulphone (32), Scheme 17.[30]

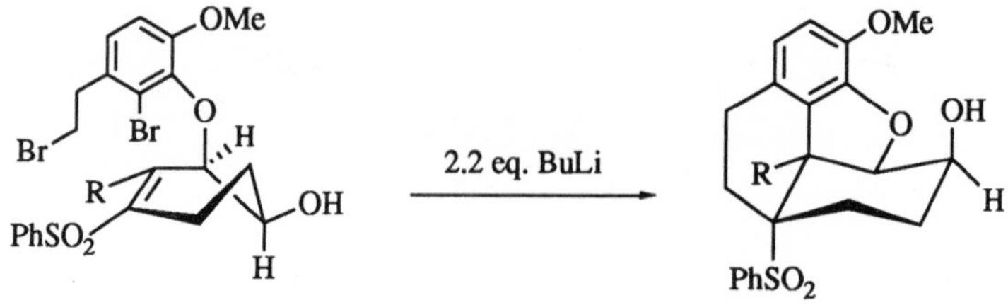

(31)   (32)

Scheme 17

Some competing sulphone deprotonation is observed but the reaction is efficient enough to provide the cornerstone of a total synthesis of morphine.

Another elegant contribution from the same group involves the *in situ* generation of a vinyl sulphone which then undergoes Michael addition and electrophilic trapping, Scheme 18.[31]

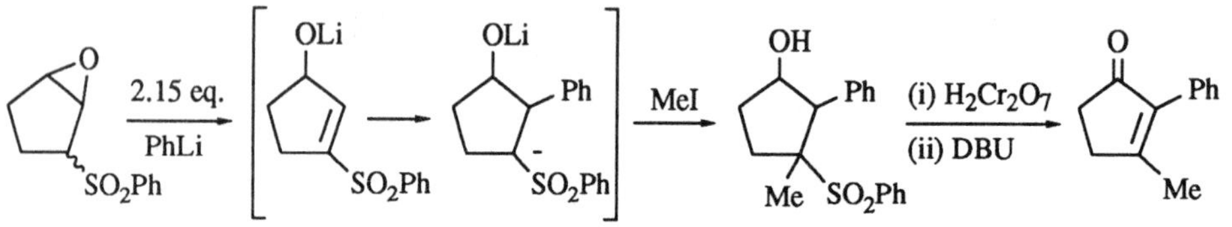

Scheme 18

The overall process allows the construction of a cycloalkenone in which the α- and β-substituents have been introduced as nucleophilic and electrophilic groups respectively – the inverse polarity to that normally obtained. Further studies have shown that vinyl sulphones can also act as acceptors in reactions with rather hindered dithiane anions.[32]

These studies have convincingly established the vinyl sulphone grouping as a significant and versatile functionality for the convergent construction of highly complex carbocycles.

Much of the significant Michael addition chemistry of acyclic vinyl sulphones has focused on compounds having a silicon group at the α-centre.[33] In particular, Isobe *et al.* have examined the diastereocontrol possible in additions to chiral α-silylvinyl sulphones, e.g. Scheme 19.[34]

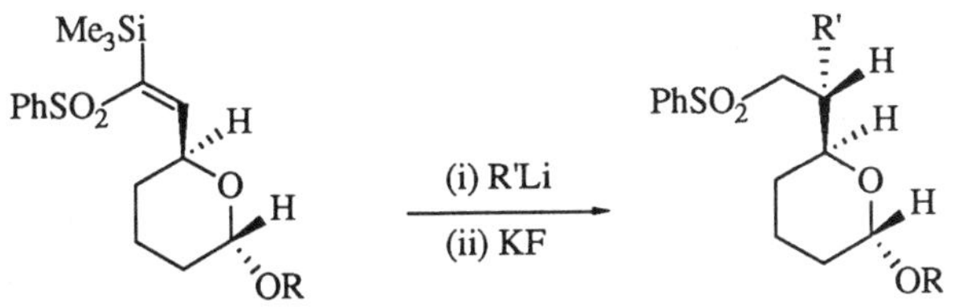

Scheme 19

The addition of organolithiums to sulphones such as (**33**) is highly diastereocontrolled and has enabled the construction of densely functionalised chiral fragments for the synthesis of maytansine[35] and okadaic acid.[36] The presence of the silicon grouping in these vinyl sulphones presumably facilitates the Michael addition, whilst also eliminating any possibility of vinylic hydrogen removal. A recent development of this chemistry has involved vinyl sulphone substrates incorporating a homochiral auxiliary. Both camphor-derived[37] and L-valinol-derived[38] systems have been shown to be effective, Scheme 20.

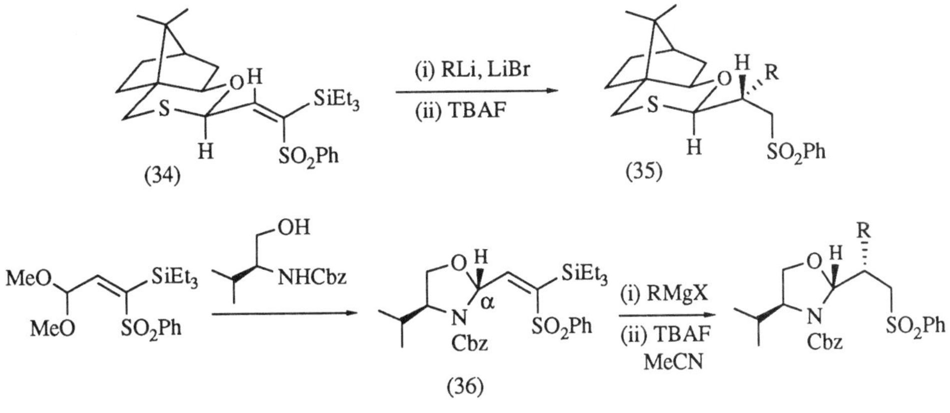

Scheme 20

Addition to either type of chiral vinyl sulphone is highly diastereoselective under optimal conditions. The stereochemical outcome of the addition to (**34**), resulting in the selective formation of (**35**), was rationalised in terms of the directing effect of the allylic ether oxygen atom. A similar effect due to the NCbz group was invoked to explain the result with (**36**). The versatility of the valinol systems is increased by the finding that the sulphone epimeric at the α-carbon can also be prepared, although this compound is somewhat more reluctant to undergo Michael addition. In both cases the chiral auxiliary can be removed from the alkylated products, yielding aldehydes and related derivatives in high optical purity.

The Michael addition to an α-silylated vinyl sulphone has also found application in a route to benzocyclobutanes, Scheme 21.[39]

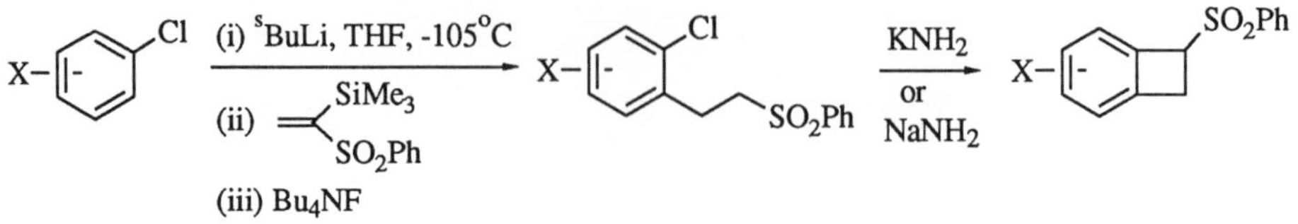

Scheme 21

The first stage involves *ortho*-metallation of a chlorobenzene (X = various substituents) at very low temperature followed by addition to the sulphone. Remarkably, the initial metallation is regioselective, and free of side reactions due to benzyne formation. Only in the second step is a benzyne intermediate formed and then trapped intramolecularly by the neighbouring sulphonyl carbanion.

4.3 Enolates, Related Anions and Miscellaneous Carbon Nucleophiles

Hamann and Fuchs have studied the reactions of vinyl sulphone (37) with a range of nucleophiles including metallated acetonitrile and enolates derived from ketones, esters and amides, Scheme 22.[40]

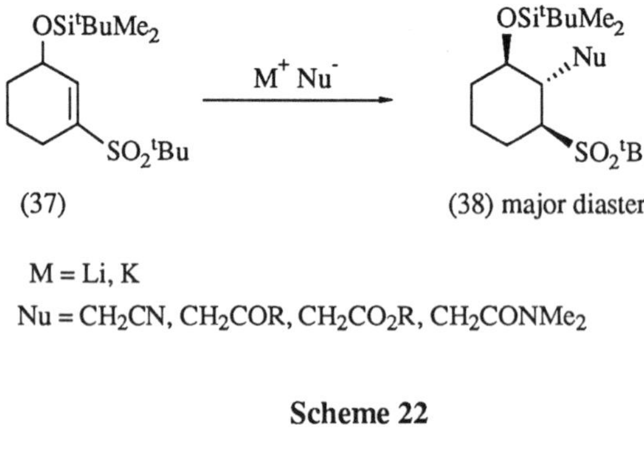

$M = Li, K$
$Nu = CH_2CN, CH_2COR, CH_2CO_2R, CH_2CONMe_2$

Scheme 22

The major isomer formed on protic work-up is the all-equatorial sulphone (38). In some cases a dramatic improvement in the yield of product is observed on changing the counterion from lithium to potassium. However, the precise nature of this effect is uncertain, since the potassium species are not generated free of lithium salts.

A very attractive feature of the Michael addition of enolates to vinyl sulphones is the possibility of devising tandem reaction sequences which allow cyclisation or annulation in one pot. Two examples which nicely illustrate this point are shown in Scheme 23.[41,42]

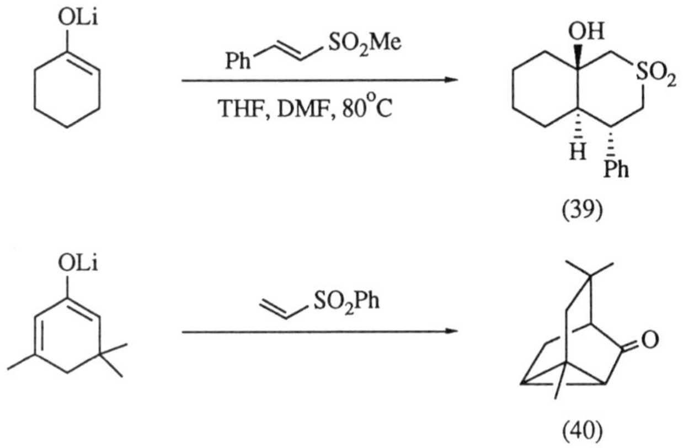

Scheme 23

The reaction of methyl styryl sulphone with lithium enolates derived from a selection of ketones gives cyclic β-hydroxy sulphones such as (39), usually in good yield.[41] Subsequent attempts to effect Ramberg–Bäcklund reactions using these products were unsuccessful; however, other types of desulphonylation were described. Compound (40) is the result of a double Michael reaction followed by intramolecular displacement of the sulphone group.[42] The reaction allows a very rapid construction of this particular tricyclic skeleton, although yields are moderate at best. Enolates derived from β-dicarbonyl compounds may react either on oxygen or on carbon with certain vinyl sulphone partners.[43] The use of NaOMe as base and methanol as solvent allows *C*-alkylation of such enolates, for example to furnish (41), Scheme 24.

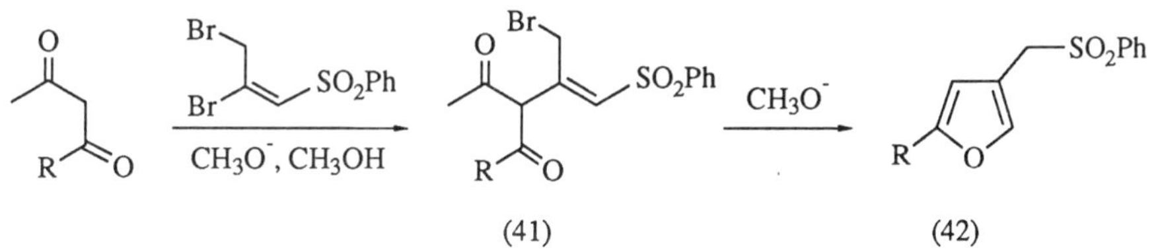

Scheme 24

This product arises by an initial conjugate addition–elimination process. Subsequent deacetylation then occurs, followed by intramolecular enolate alkylation of oxygen, leading eventually to furan sulphone (42).

A rather elegant intramolecular enolate addition to a vinyl sulphone allows the stereoselective preparation of an intermediate for dihydrocompactin synthesis, Scheme 25.[44]

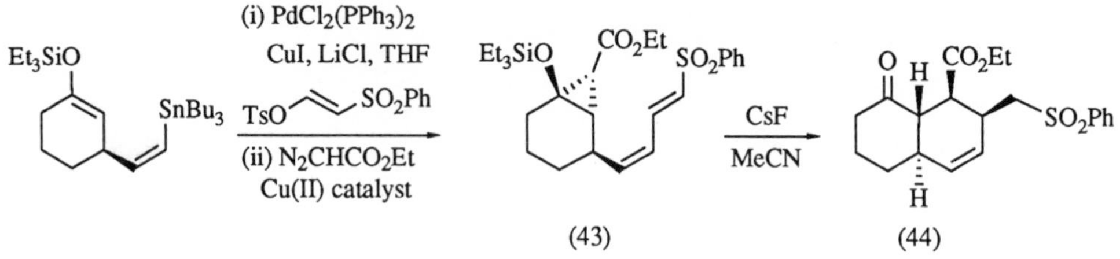

(43)   (44)

Scheme 25

The key intermediate (43) required for this study was constructed by a Stille-type vinyl stannane coupling to form the required dienyl sulphone, followed by regioselective cyclopropanation of the silyl enol ether group. Treatment of this compound with fluoride then generates the required enolate, thus leading to (44).

Michael addition to vinyl sulphones is made even more facile by the incorporation of additional carbanion-stabilising groups at the α-vinylic position. An example of this is the condensation of ketones with 1,1-bis(phenylsulphonyl)ethylene, which can be conducted under neutral conditions, Scheme 26.[45]

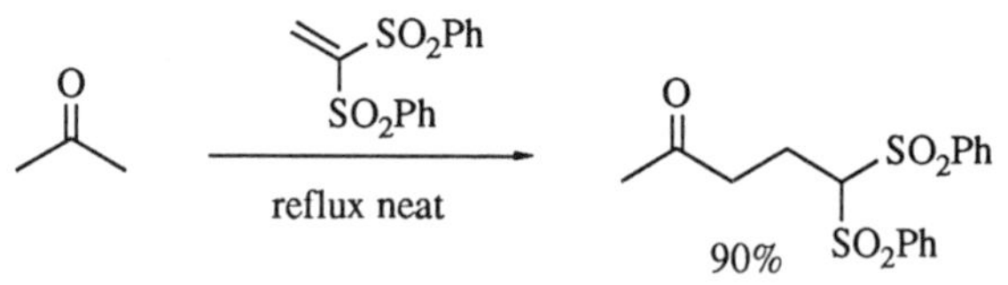

Scheme 26

This reaction can be carried out by using the ketone as solvent or by combining the starting materials in equimolar amounts in acetonitrile as the solvent.

An indirect route to this type of Michael addition product employs the corresponding enamines in place of ketones.[46] A focus of recent interest in this area has been the unexpected regiochemistry of the attack of cyclohexanone-derived enamines on some β-phenyl-substituted vinyl sulphones.[47] Thus, whilst most vinyl sulphones give the expected Michael products, e.g. (45), the phenyl-substituted derivatives can give products arising from attack at the α-position of the vinyl sulphone, e.g. (46), Scheme 27.

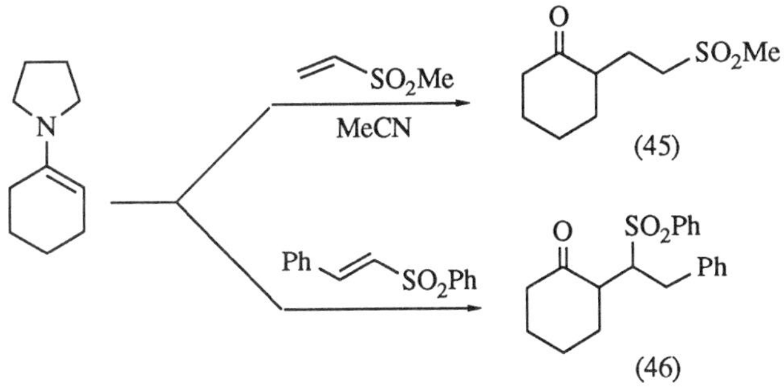

Scheme 27

Another report describes similar results, the regioselectivity of attack depending on the substituents on the enamine and on the geometry of the starting sulphone.[48] Analogous reactions using allenyl phenyl sulphone give 'normal' adducts arising by addition to the central carbon of the allene unit, Scheme 28.[49]

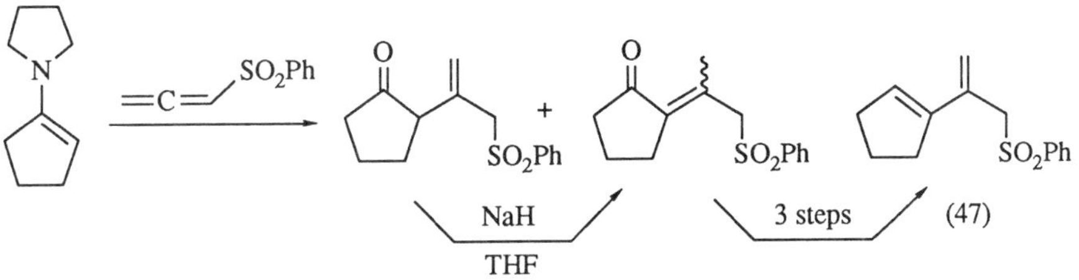

Scheme 28

The unsaturated ketosulphone products can then be further transformed into dienyl sulphones such as (**47**). The chemistry of these dienes was also probed, including their alkylation, desulphonylation and Diels–Alder cycloadditions.

The Michael addition of nitronate anions to vinyl sulphones has been achieved by means of a solid–liquid phase-transfer method.[50] Although the reaction of metallated nitriles with vinyl sulphones can give simple addition products (e.g. Scheme 22), other modes of reaction can be observed, depending on the structure of the two reaction partners, Scheme 29.[51]

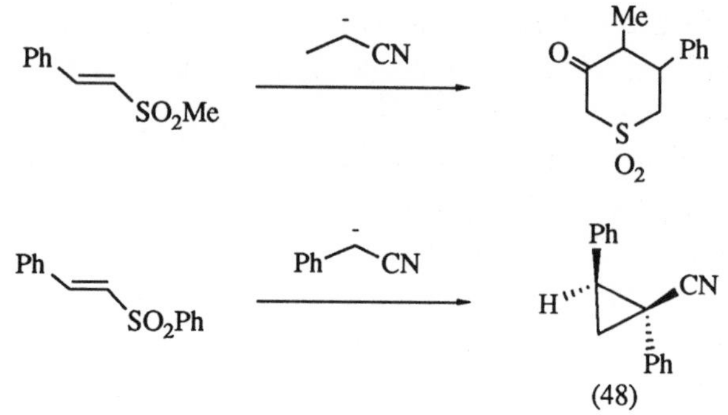

Scheme 29

Thus with methyl styryl sulphone the initial Michael addition is followed by proton transfer to give a new sulphonyl carbanion which undergoes intramolecular attack on the nitrile group. Cyclopropane nitrile (**48**) arises by intramolecular sulphinate displacement, this being favoured by the phenyl substituent on the starting nitrile.

Finally, addition of cyanide to vinyl sulphones followed by *in situ* elimination of benzenesulphinate results in overall conversion to an α,β-unsaturated nitrile, Scheme 30.[52]

Scheme 30

In this conversion of one Michael acceptor into another the electron demand at the two olefinic centres is reversed.

4.4 Cyclopropanations

One method for the synthesis of cyclopropyl sulphones is to react a nucleophile with a vinyl sulphone which incorporates an additional halogen leaving group. Two different sequences which adopt this type of approach are shown in Scheme 31.[53,54]

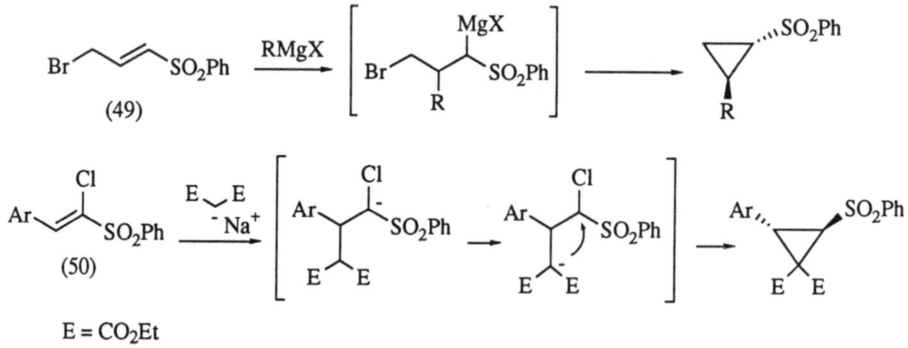

E = CO₂Et

Scheme 31

Both methods afford the *trans*-products in stereoselective fashion, but both have discernable limitations. Thus cyclopropane products are obtained from (**49**) only when certain unsaturated Grignard reagents are used (i.e. vinyl, allyl, phenyl, benzyl and propargyl).[53] With saturated Grignard reagents cyclopropanes are not obtained, alternative coupling products being formed if cuprous salts are included. In reactions of (**50**) with ethyl acetoacetate again no cyclopropane is observed, and alternative ring closure through the acetyl oxygen leads to a dihydrofuran.[54]

Vinyl sulphones can be converted to cyclopropyl sulphones when treated with the usual reagents for cyclopropanation of electron-deficient double bonds such as enones. Sulphonium ylides carry out this conversion efficiently, for example using a ^{t}BuOK/DMSO system, Scheme 32.[55]

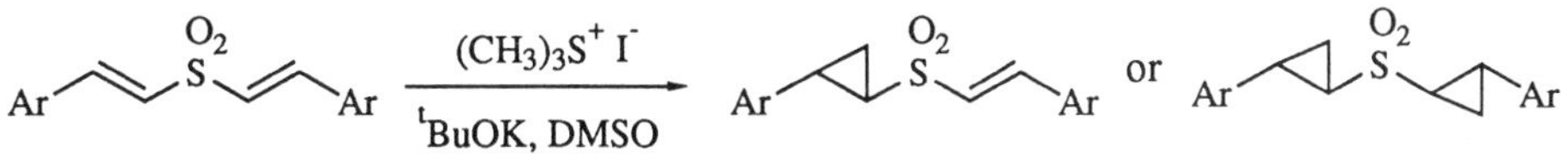

Scheme 32

In reactions of bis(styryl) sulphones either the *trans*-mono- or bis-cyclopropanated products can be obtained. Phase-transfer conditions have provided a useful means of preparing a variety of cyclopropyl sulphones, as illustrated in Scheme 33.[56–58]

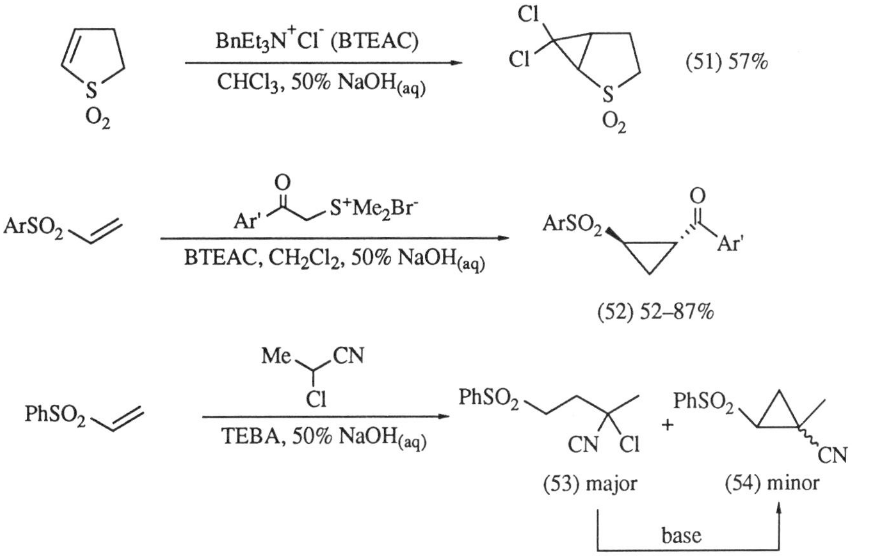

Scheme 33

In the report describing the successful preparation of (51) and other dihalogenocyclopropyl sulphones, alternative cyclopropanation conditions such as the Simmons–Smith procedure were found not to yield cyclopropanated products.[56] Very similar reaction conditions have also been employed in the preparation of ketosulphone products (52), starting with a range of phenacyl sulphonium salts.[57]

Finally, a similar process aimed at the synthesis of cyanosulphones such as (54) gives mainly the simple addition product (53).[58] This compound can be cyclised to give (54) on re-exposure to base under more drastic reaction conditions, but only in moderate yield.

4.5 Radical Additions

Several groups have studied the intermolecular addition of ether- and alcohol-derived carbon-centred radicals to vinyl sulphones, e.g. to form products such as (55)-(59) Scheme 34.[59-62]

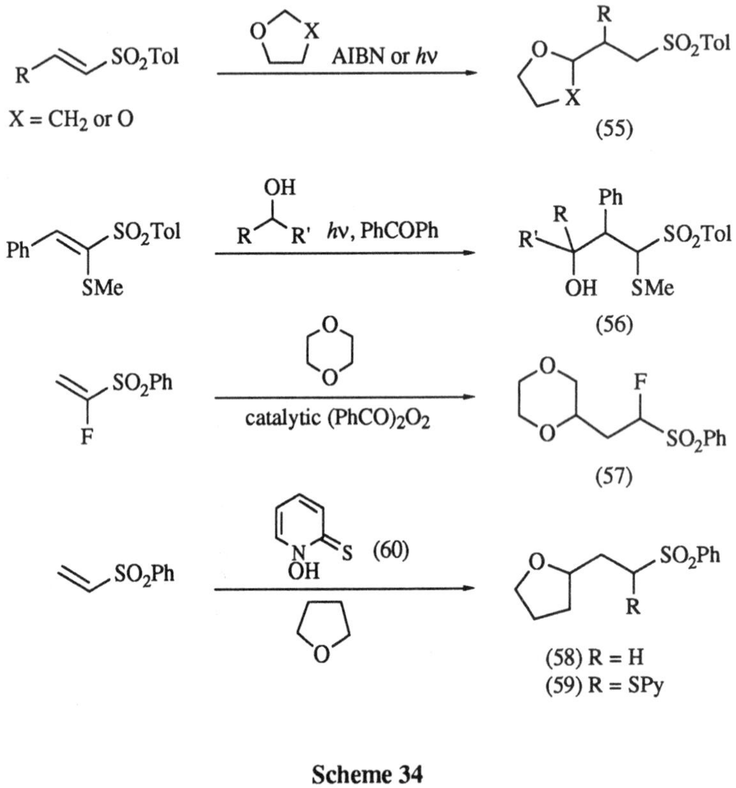

Scheme 34

Each of these examples involves generating the initial carbon-centred radical by hydrogen atom abstraction. This usually requires at least one equivalent of some radical-initiating species such as AIBN or dibenzoyl peroxide, since the radical chain-transfer step is usually inefficient. The method is really only applicable to the addition of simple radicals which can be generated from readily available alcohols or ethers, since these are required in excess, and indeed are often used as the solvent. The use of *N*-hydroxy-2-pyridinethione (60) has some advantages in this chemistry, since the hydroxyl radical needed for hydrogen abstraction is generated in a chain mechanism. Thus, the predominant product is (59), resulting from attack of the α-sulphonyl radical on the thiocarbonyl group of (60). Usually, however, lesser amounts of (58) are also formed.

A number of other notable contributions to this area have come from Barton's group, e.g. Scheme 35.[63,64]

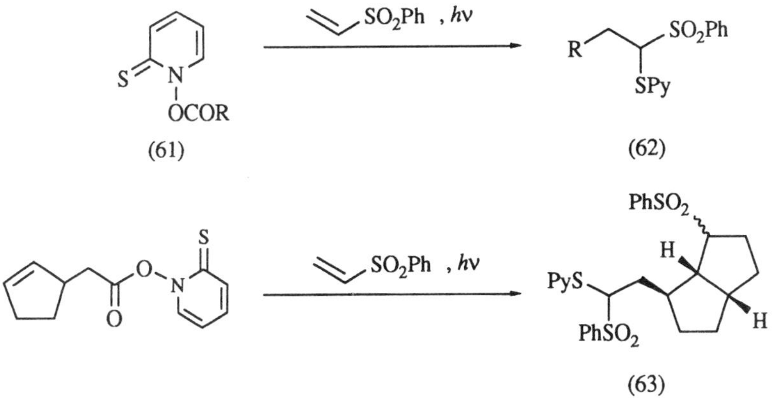

Scheme 35

In this chemistry thiohydroxamate esters (61), derived from (60), act as a source of radicals which add to electrophilic acceptors including phenyl vinyl sulphone. As in the sequence leading to (59) the product radical is further functionalised by reaction with a molecule of the starting ester. The thiopyridyl group in (62) can subsequently be oxidatively eliminated, thus providing a synthesis of substituted vinyl sulphones.[63] Two molecules of vinyl sulphone are incorporated in the generation of (63), which is formed stereoselectively with respect to three of the four asymmetric centres on the bicyclic nucleus.[64]

A more common method of forming carbon-centred radicals involves treating alkyl halides or selenides with a tin hydride, usually Bu_3SnH, in the presence of a catalytic amount of initiator such as AIBN. This is a particularly effective method for the formation of rings. Scheme 36 gives two such examples in which vinyl sulphones are used as the radical acceptors.[65,66]

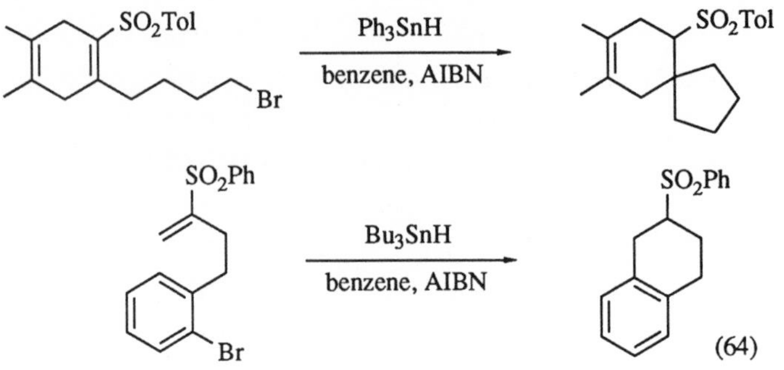

Scheme 36

Most commonly this method is used to synthesise cyclopentanes, since the extremely rapid radical 5-*exo-trig* ring closure usually overwhelms the alternative direct radical reduction. The synthesis of (**64**) in 88% yield represents one of a smaller number of successful examples of 6-*endo-trig* closures.

Both intermolecular and intramolecular additions to vinyl sulphones have been utilised in studies aimed at the total synthesis of naturally occurring targets, e.g. Scheme 37.[67,68]

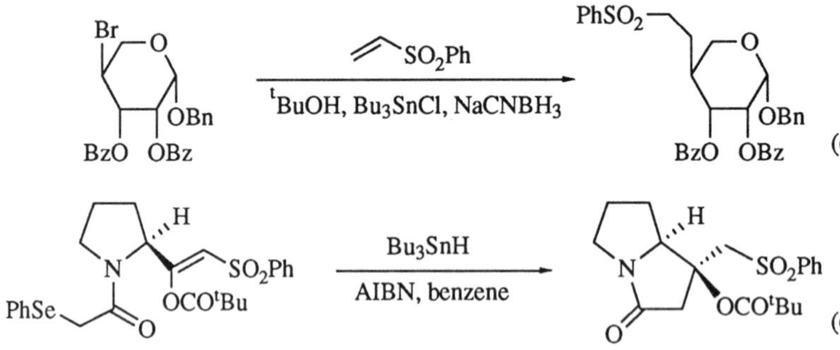

Scheme 37

The preparation of (**65**) requires the use of a large excess of phenyl vinyl sulphone and the maintenance of a very low concentration of tin hydride by using Stork's Bu$_3$SnCl/NaCNBH$_3$ method.[67] Compound (**66**) has been prepared as an intermediate for pyrrolizidine alkaloid synthesis.[68] In both examples the products are formed stereoselectively due to the bias exerted by the chiral starting material.

Curran has included vinyl sulphones amongst the acceptor alkenes which participate in a radical-mediated synthesis of methylenecyclopentenes, Scheme 38.[69]

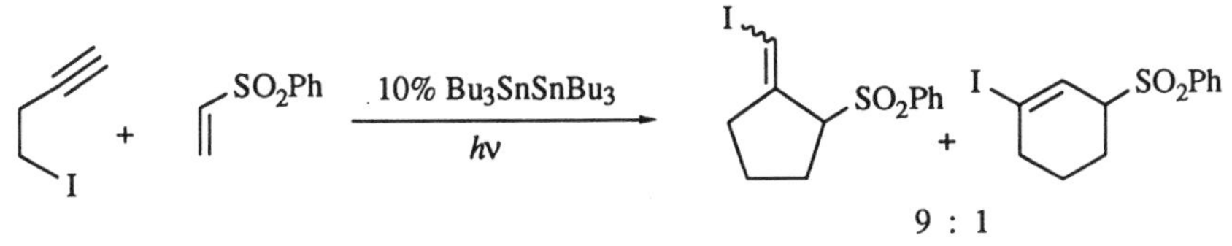

Scheme 38

The key difference between this reaction and those described previously is that the chain-transfer step involves iodine, and not hydrogen, atom transfer. As in the Barton chemistry this has synthetically appealing consequences, since no great loss of functionality occurs during the reaction.

Chapter 4 References

1. E. P. Kohler and H. Potter, *J. Am. Chem. Soc.*, **1935**, *57*, 1316; G. A. Russell, H-D. Becker, and J. Schoeb, *J. Org. Chem.*, **1963**, *28*, 3584.
2. S. T. McDowell and C. J. M. Stirling, *J. Chem. Soc. (B)*, **1967**, 351.
3. M. Cinquini, F. Cozzi, and M. Pelosi, *J. Chem. Soc., Perkin Trans. 1*, **1979**, 1430; see also M. Cinquini, S. Colonna, and F. Cozzi, *J. Chem. Soc., Perkin Trans. 1*, **1978**, 247.
4. J-C. Wu, T. Pathak, W. Tong, J. M. Vial, G. Remaud, and J. Chattopadhyaya, *Tetrahedron*, **1988**, *44*, 6705.
5. M. Hirama, H. Hioki, and S. Ito, *Tetrahedron Lett.*, **1988**, *29*, 3125.
6. M. Takahashi and I. Yamashita, *Heterocycles*, **1990**, *31*, 1537.
7. A. Padwa and B. H. Norman, *J. Org. Chem.*, **1990**, *55*, 4801.
8. C. J. M. Stirling, *J. Chem. Soc.*, **1964**, 5856.
9. M. Harmata and C. B. Gamlath, *J. Org. Chem.*, **1988**, *53*, 6154; M. Harmata, C. B. Gamlath and C. L. Barnes, *Tetrahedron Lett.*, **1990**, *31*, 5981.
10. P. Auvray, P. Knochel, and J. F. Normant, *Tetrahedron Lett.*, **1985**, *26*, 4455.
11. M. Julia, A. Righini, and D. Uguen, *J. Chem. Soc., Perkin Trans. 1*, **1978**, 1646.
12. H. Kropf and F. Wohrle, *J. Chem. Res. (S)*, **1987**, 387. For an alternative route to hydroperoxy sulphones, see E. L. Clennan and X. Chen, *J. Am. Chem. Soc.*, **1989**, *111*, 5787.
13. L. Horner and H. Lindel, *Liebigs Ann. Chem.*, **1985**, 22; Y. Kuroki and R. Lett, *Tetrahedron Lett.*, **1984**, *25*, 197; L. Horner and H. Lindel, *Phosphorus and Sulphur*, **1983**, *15*, 1.
14. H. A. Selling, *Tetrahedron*, **1975**, *31*, 2387.
15. B. S. Thyagarajan and R. A. Chandler, *Synth. Commun.*, **1990**, *20*, 53.
16. A. Padwa and P. E. Yeske, *J. Am. Chem. Soc.*, **1988**, *110*, 1617.
17. B. Musicki and T. S. Widlanski, *Tetrahedron Lett.*, **1991**, *32*, 1267.
18. S. V. Ley, N. S. Simpkins, and A. J. Whittle, *J. Chem. Soc., Chem. Commun.*, **1983**, 503; P. S. Jones, S. V. Ley, N. S. Simpkins, and A. J. Whittle, *Tetrahedron*, **1986**, *42*, 6519. For reductions of sugar-derived vinyl sulphones, see T. Sakakibara, I. Takai, A. Yamamoto, H. Iizuka, K. Hirasawa, and Y. Ishido, *Tetrahedron Lett.*, **1990**, *31*, 3749.
19. G. H. Posner and D. J. Brunelle, *J. Org. Chem.*, **1973**, *38*, 2747.
20. J. Meijer and P. Vermeer, *Recl. Trav. Chim. Pays-Bas*, **1975**, *94*, 14.
21. G. De Chirico, V. Fiandanese, G. Marchese, F. Naso, and O. Sciacovelli, *J. Chem. Soc., Chem. Commun.*, **1981**, 523.
22. P. L. Fuchs and T. F. Braish, *Chem. Rev.*, **1986**, *86*, 903.
23. S. A. Hardinger and P. L. Fuchs, *J. Org. Chem.*, **1987**, *52*, 2739.
24. J-L. Fabre, M. Julia, and J-N. Verpeaux, *Bull. Soc. Chim. Fr.*, **1985**, 762.
25. J. C. Saddler and P. L. Fuchs, *J. Am. Chem. Soc.*, **1981**, *103*, 2112.
26. D. K. Hutchinson and P. L. Fuchs, *J. Am. Chem. Soc.*, **1985**, *107*, 6137.

27. D. K. Hutchinson, S. A. Hardinger, and P. L. Fuchs, *Tetrahedron Lett.,* **1986,** *27,* 1425; Y. Pan, D. K. Hutchinson, M. H. Nantz, and P. L. Fuchs, *Tetrahedron,* **1989,** *45,* 467.

28. Y. Pan, S. A. Hardinger, and P. L. Fuchs, *Synth. Commun.,* **1989,** *19,* 403.

29. D. K. Hutchinson and P. L. Fuchs, *J. Am. Chem. Soc.,* **1987,** *109,* 4755.

30. J. E. Toth and P. L. Fuchs, *J. Org. Chem.,* **1987,** *52,* 473; see also: C. R. Nevill, Jr. and P. L. Fuchs, *Synth. Commun.,* **1990,** *20,* 761; P. R. Hamann, J. E. Toth, and P. L. Fuchs, *J. Org. Chem.,* **1984,** *49,* 3865.

31. P. C. Conrad and P. L. Fuchs, *J. Am. Chem. Soc.,* **1978,** *100,* 346.

32. T. F. Braish, J. C. Saddler, and P. L. Fuchs, *J. Org. Chem.,* **1988,** *53,* 3647.

33. M. Isobe, in *Perspectives in the Organic Chemistry of Sulphur,* Ed. B. Zwanenburg and A. J. Klunder, Elsevier, Amsterdam, **1987,** p.209. For examples involving addition to allenyl sulphones see S. Braverman, P. F. T. M. van Asten, J. B. van der Linden, and B. Zwanenburg, *Tetrahedron Lett.,* **1991,** *32,* 3867.

34. M. Isobe, Y. Funabashi, Y. Ichikawa, S. Mio, and T. Goto, *Tetrahedron Lett.,* **1984,** *25,* 2021; see also C. Alcaraz, J. C. Carreterol, and E. Dominguez, *Tetrahedron Lett.,* **1991,** *32,* 1385.

35. M. Kitamura, M. Isobe, Y. Ichikawa, and T. Goto, *J. Org. Chem.,* **1984,** *49,* 3517.

36. See the series of four consecutive papers, starting with Y. Ichikawa, M. Isobe, D-L. Bai, and T. Goto, *Tetrahedron,* **1987,** *43,* 4737.

37. M. Isobe, J. Obeyama, Y. Funabashi, and T. Goto, *Tetrahedron Lett.,* **1988,** *29,* 4773.

38. M. Isobe, Y. Hirose, K. Shimokawa, T. Nishikawa, and T. Goto, *Tetrahedron Lett.,* **1990,** *31,* 5499.

39. M. Iwao, *J. Org. Chem.,* **1990,** *55,* 3622.

40. P. R. Hamann and P. L. Fuchs, *J. Org. Chem.,* **1983,** *48,* 914.

41. K. Takaki, K. Nakagawa, and K. Negoro, *J. Org. Chem.,* **1980,** *45,* 4789.

42. R. M. Cory and R. M. Renneboog, *J. Org. Chem.,* **1984,** *49,* 3898.

43. A. Padwa, S. S. Murphree, and P. E. Yeske, *J. Org. Chem.,* **1990,** *55,* 4241. For a related addition–elimination reaction see B. W. Metcalf and E. Bonilavri, *J. Chem. Soc., Chem. Commun.,* **1978,** 914.

44. J. P. Marino and J. K. Long, *J. Am. Chem. Soc.,* **1988,** *110,* 7916.

45. O. De Lucchi, L. Pasquato, and G. Modena, *Tetrahedron Lett.,* **1984,** *25,* 3647.

46. A. Risaliti, S. Fatutta, and M. Forchiassin, *Tetrahedron,* **1967,** *23,* 1451.

47. S. Fatutta and A. Risaliti, *J. Chem. Soc., Perkin Trans. 1,* **1974,** 2387.

48. F. Benedetti, S. Fabrissin, and A. Risaliti, *Tetrahedron,* **1984,** *40,* 977.

49. K. Hayakawa, M. Takewaki, I. Fujimoto, and K. Kanematsu, *J. Org. Chem.,* **1986,** *51,* 5100.

50. H. Galons, S. Labidalle, M. Miocque, B. Ligniere, and G. Bram, *Phosphorus and Sulphur,* **1988,** *39,* 73.

51. T. Agawa, Y. Yoshida, M. Komatsu, and Y. Ohshiro, *J. Chem. Soc., Perkin Trans. 1*, **1981**, 751.
52. D. F. Taber and S. A. Saleh, *J. Org. Chem.*, **1981**, *46*, 4817.
53. J. J. Eisch and J. E. Galle, *J. Org. Chem.*, **1979**, *44*, 3277.
54. I. Yamamoto, T. Sakai, K. Ohta, K. Matsuzaki, and K. Fukuyama, *J. Chem. Soc., Perkin Trans. 1*, **1985**, 2785.
55. M. S. R. Naidu and R. M. Rani, *Phosphorus and Sulphur*, **1984**, *19*, 259; see also D. B. Reddy, B. Sankaraiah, and T. Balaji, *Ind. J. Chem.*, **1980**, *19B*, 563.
56. S. M. Liebowitz and H. J. Johnson, *Synth. Commun.*, **1986**, *16*, 1255.
57. D. B. Reddy, P. S. Reddy, B. V. Reddy, and P. A. Reddy, *Synthesis*, **1987**, 74.
58. A. Jonczyk and M. Makosza, *Synthesis*, **1976**, 387.
59. K. Inomata, H. Suhara, H. Kinoshita, and H. Kotake, *Chem. Lett.*, **1988**, 813; see also A. Weichert and H. M. R. Hoffmann, *J. Org. Chem.*, **1991**, *56*, 4098.
60. K. Ogura, A. Yanagisawa, T. Fujino, and K. Takahashi, *Tetrahedron Lett.*, **1988**, *29*, 5387.
61. D. P. Matthews and J. R. McCarthy, *J. Org. Chem.*, **1990**, *55*, 2973.
62. J. Boivin, E. Crepon, and S. Z. Zard, *Tetrahedron Lett.*, **1990**, *31*, 6869.
63. D. H. R. Barton, H. Togo, and S. Z. Zard, *Tetrahedron Lett.*, **1985**, *26*, 6349; D. H. R. Barton, N. Ozbalik, and J. C. Sarma, *Tetrahedron Lett.*, **1988**, *29*, 6581; D. H. R. Barton, C-Y. Chern, and J. C. Jaszberenyi, *Tetrahedron Lett.*, **1991**, *32*, 3309; D. H. R. Barton, J. Boivin, E. Crepon, J. Sarma, H. Togo, and S. Z. Zard, *Tetrahedron*, **1991**, *47*, 7091.
64. D. H. R. Barton, E. da Silva, and S. Z. Zard, *J. Chem. Soc., Chem. Commun.*, **1988**, 285; see also E. Lee and D. S. Lee, *Tetrahedron Lett.*, **1990**, *31*, 4341.
65. D. L. J. Clive and R. J. Bergstra, *J. Org. Chem.*, **1990**, *55*, 1786; see also J. K. Crandall and T. A. Ayers, *Tetrahedron Lett.*, **1991**, *32*, 3659.
66. A. Padwa, S. S. Murphree, and P. E. Yeske, *Tetrahedron Lett.*, **1990**, *31*, 2983.
67. M. V. Rao and M. Nagarajan, *J. Org. Chem.*, **1988**, *53*, 1432.
68. N. S. Simpkins, *Chem. Soc. Rev.*, **1990**, *19*, 335.
69. D. P. Curran and M-H. Chen, *J. Am. Chem. Soc.*, **1987**, *109*, 6558.

CHAPTER 5

Rearrangements of Sulphones

Sulphones have provided a rich source of rearrangement reactions. In particular, the ability of the sulphone to stabilise an adjacent carbanion facilitates a range of rearrangements in which an α-sulphonyl carbanion acts as either the initiator or terminator. In addition, rearrangement termination due to the sulphone acting as a leaving group has become a useful process.

5.1 Ring Expansions and Fragmentations

A range of ring-expansion processes has been reported in which the sulphone group is intimately involved. A simple example is a variant of the diazomethane ring expansion of cycloalkanones employing benzylsulphonyldiazomethane, e.g. Scheme 1.[1]

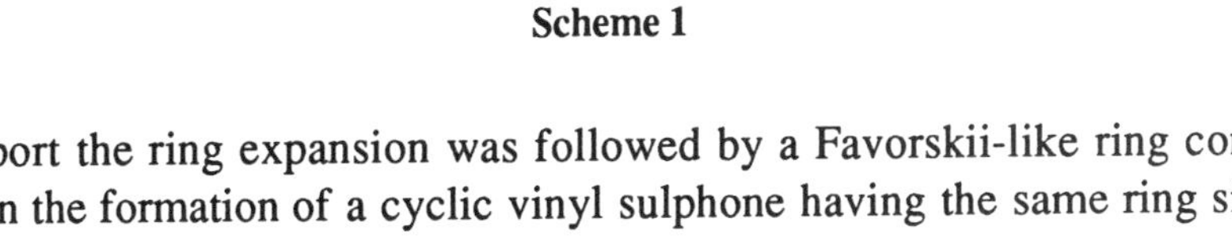

Scheme 1

In this report the ring expansion was followed by a Favorskii-like ring contraction, resulting in the formation of a cyclic vinyl sulphone having the same ring size as the starting ketone.

Two further examples of ring expansion processes are illustrated in Scheme 2.[2,3] In the conversion of cyclic β-ketosulphones to ring-expanded systems such as (2) the sulphone serves simply as an acidifying grouping; the reaction works equally well starting with β-ketoesters. The sequence is initiated by Michael addition to an α,β-unsaturated selenone (1), leading to cyclopropane (2) which then undergoes ring opening to give the product ketosulphone (3). Depending on the stereochemistry of the starting selenones the cyclopropane intermediates can be isolated in some cases.

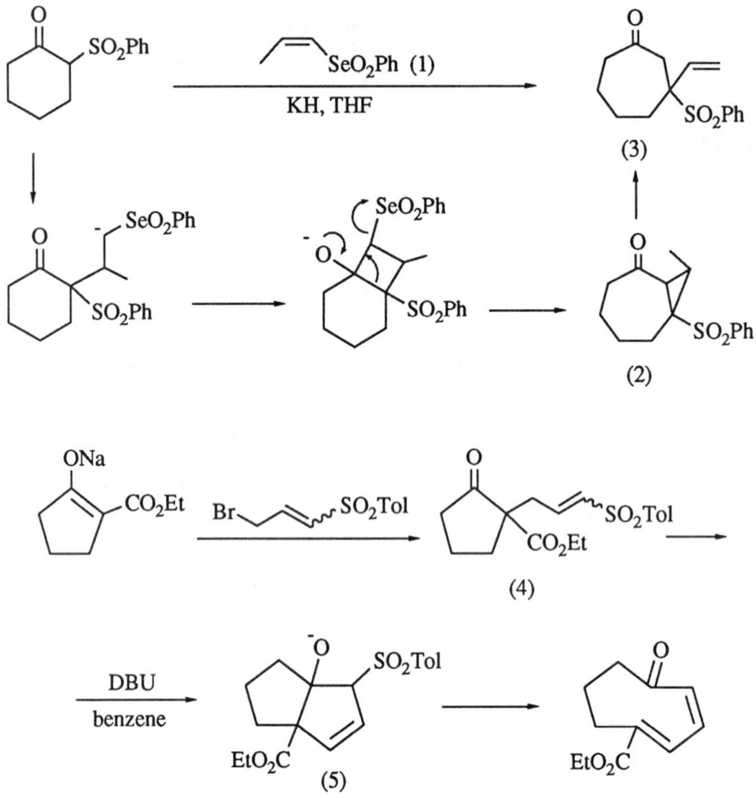

Scheme 2

In the second process the vinyl sulphone group in (**4**) is deprotonated and subsequently undergoes ring closure to give (**5**). This intermediate can then fragment as shown, expulsion of tolylsulphinate resulting in the formation of a ring-expanded dienone system. Depending upon the ring size of the starting material and the precise conditions used, other reaction pathways, leading to acyclic products, are also observed.

Trost and Mikhail have reported a one-carbon ring-expansion process which relies on the ability of the sulphone to act as a leaving group under the influence of a Lewis acid.[4] The scope and limitations of this reaction have been probed by other workers, using ring-fused cyclobutanones as the substrates, Scheme 3.[5]

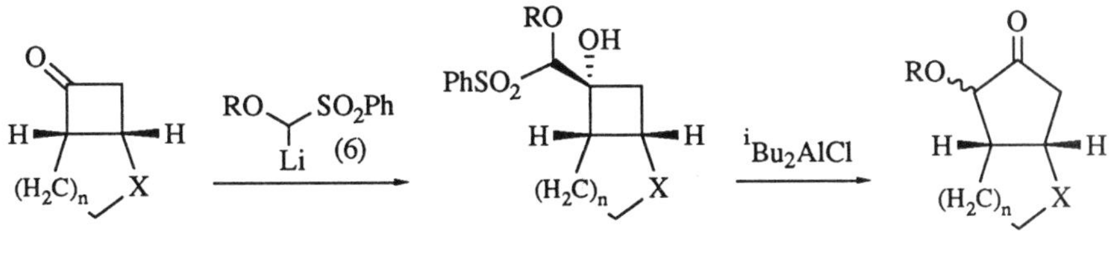

Scheme 3

Several sulphone reagents (**6**) having different R groups were employed, and both the overall efficiency of the ring expansion and the stereocontrol attainable were examined in some detail. The final alkoxy ketone products were found to be formed in much higher yield using a two-stage procedure, rather than combining the anion addition and rearrangement in one pot. Some correlation was found between the stereochemistry of the products (**8**) and that of the intermediates (**7**). However, relatively easy epimerisation of the α-alkoxy epimer of (**8**) to the corresponding β-isomer complicates the situation somewhat.

Trost's group has also examined the chemistry of ketosulphones such as (**9**) and (**10**) and have found that highly complementary reactions can be initiated, depending on the reagents employed, Scheme 4.[6,7]

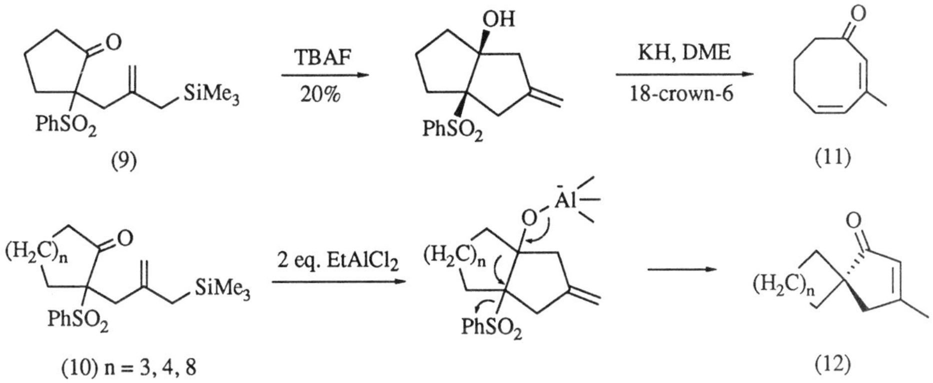

Scheme 4

Upon reaction with TBAF (**9**) undergoes ring closure to form a bicyclic hydroxy sulphone. Subsequent base treatment results in ring opening, double-bond isomerisation and elimination of sulphinate to give the eight-membered dienone (**11**).[6] In some cases the fluoride treatment is sufficient to effect both the ring closure and ring opening steps, thus yielding products retaining the sulphone group. Under Lewis acid conditions similar substrates gave spiroannulated products such as (**12**).[7]

This process proceeds well only for compounds of type (**10**) having a flexible ring, due to the stereoelectronic requirement of the pinacol-type rearrangement involved. What amounts to a heteroatom version of this process has been described by Bhat and Cookson, Scheme 5.[8]

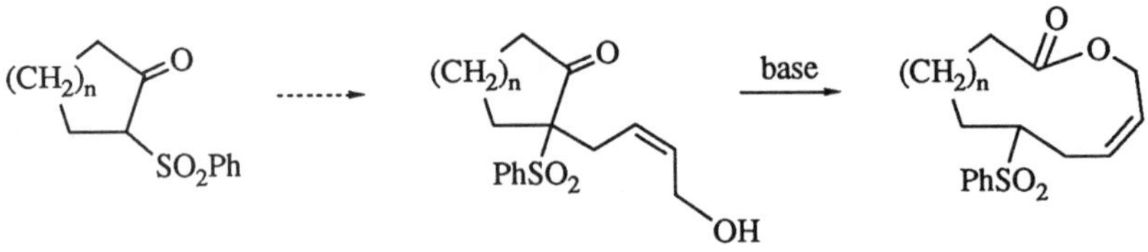

Scheme 5

Again, larger ring compounds react smoothly, whilst a cyclopentanone substrate resisted attempts at rearrangement.

A more complex sequence of transformations can be initiated by treatment of sulphones such as (**13**) with MeOK or ^{t}BuOK, Scheme 6.[9]

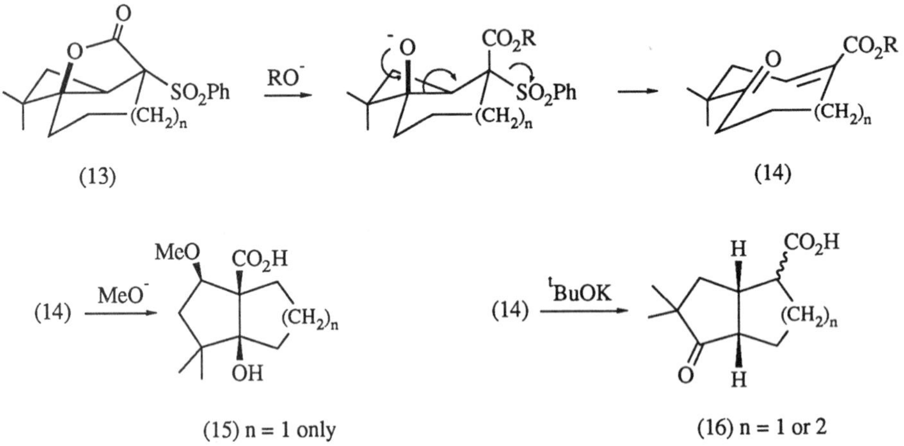

(13)   (14)

(15) n = 1 only   (16) n = 1 or 2

Scheme 6

Initial lactone opening is followed by a Grob-type fragmentation to give a medium ring α,β-unsaturated ester (**14**), which can be isolated in some cases. However, in the presence of MeO⁻, a Michael addition and transannular aldol sequence ensues, giving (**15**). Alternatively, in the presence of ^{t}BuOK, ketone deprotonation and intramolecular Michael addition to the unsaturated ester furnishes (**16**).

Two related sequences, both involving a sulphonyl carbanion reaction prior to a carbon–carbon bond cleavage and an elimination, are shown in Scheme 7.[10,11]

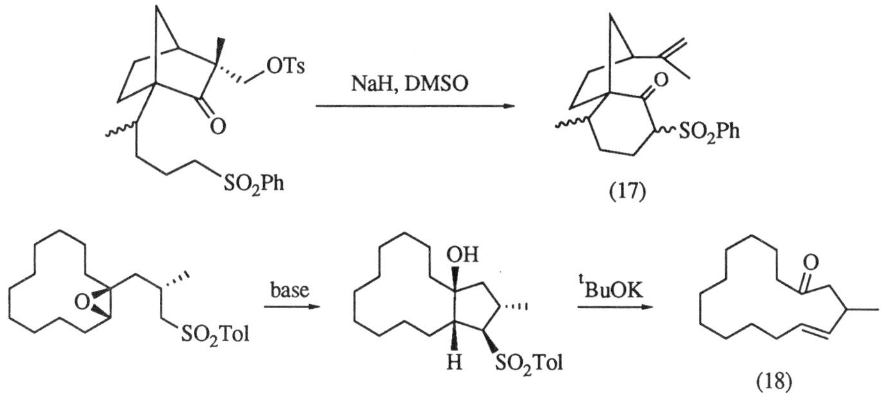

Scheme 7

In the first case the fragmentation of an intermediate β-hydroxy sulphone is directed towards (**17**) by the appropriate positioning of a tosylate leaving group. The Grob fragmentation leading to (**18**) bears obvious similarities to the previously described sequence which gives (**14**).

A most spectacular deep-seated rearrangement occurs on treatment of (**19**) with ᵗBuOK, Scheme 8.[12]

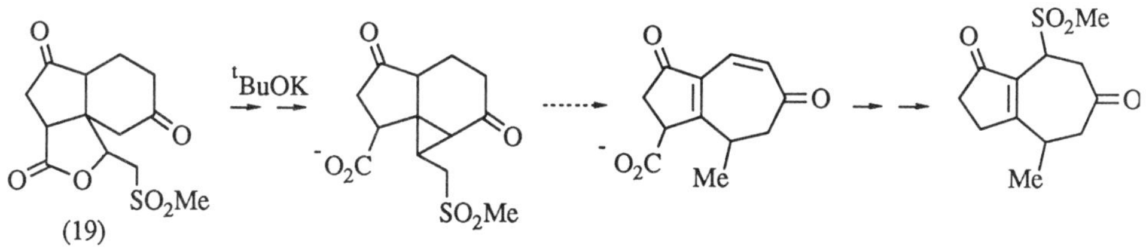

Scheme 8

The authors' rationalisation involves ring enlargement brought about through an intramolecular cyclopropanation of a vinyl sulphone formed *in situ*. Conversion to the observed product requires sulphinate elimination and regioselective readdition to one of a number of possible dienone intermediates.

Radical-initiated cyclopropane to cyclopentane rearrangements have been conducted on a range of suitably substituted vinyl cyclopropanes, including cyclopropane sulphones, Scheme 9.[13]

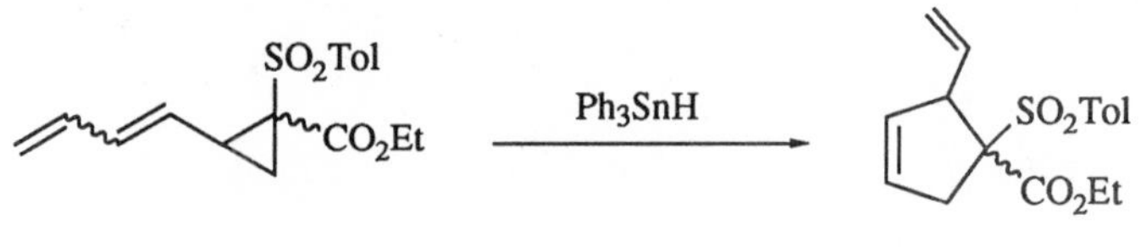

Scheme 9

Attack of a radical on the dienyl system results in cyclopropane cleavage to give a stabilised radical which can then recyclise in 5-*exo-trig* fashion. A number of radical sources have been investigated, including Bu_3SnH, Ph_3GeH and PhSH.

5.2 Sulphinate to Sulphone Rearrangements

A number of researchers, most notably Braverman, have made important contributions to the understanding of sulphinate ester rearrangements.[14] A significant finding is that allylic sulphinates, derived from simple acyclic allylic alcohols, rearrange with clean 1,3-transposition of functionality, i.e. (20) gives only (21), and (22) gives only (23), Scheme 10.

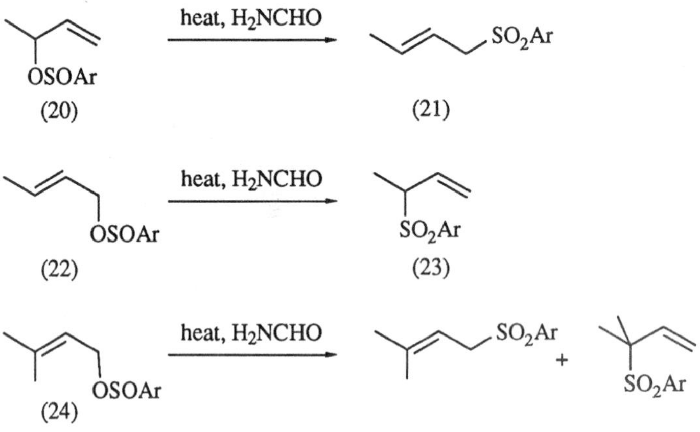

Scheme 10

In such cases a [2,3] sigmatropic rearrangement is operative. In contrast, the rearrangements of more heavily substituted systems such as (24), and sulphinates derived from cyclic allylic alcohols, occur (at least partly) by an ionisation–ion-pair recombination mechanism.

Whitham and co-workers have examined this aspect in some detail by using labelled compounds and by conducting crossover experiments with externally added sulphinate.[15] When the rearrangement of (24) (Ar = *p*-tolyl) is carried out in the presence of benzenesulphinate, the products are formed as mixtures of *p*-tolyl and phenyl sulphones. In order to access some cyclic allylic sulphones Whitham has used the alternative stereospecific rearrangement of the analogous sulphenates. Oxidation of the so-formed sulphoxide products then gives the desired sulphones.[16]

Recent studies have concentrated on finding the best conditions for effecting clean rearrangements in simple systems, and, most importantly, examining the chirality transfer from sulphur to carbon which accompanies the process. Hiroi *et al.* have been instrumental in clarifying this area.[17] Dipolar aprotic solvents such as formamide or dimethylformamide (DMF) appear to give the best results. The use of alcoholic solvents has also been described but transesterification problems can intrude under the fairly vigorous reaction conditions required.

Optically active sulphinates can be prepared and rearranged with very good overall stereospecificity (i.e. a high level of fidelity in the transfer of chirality from sulphur to carbon), Scheme 11.

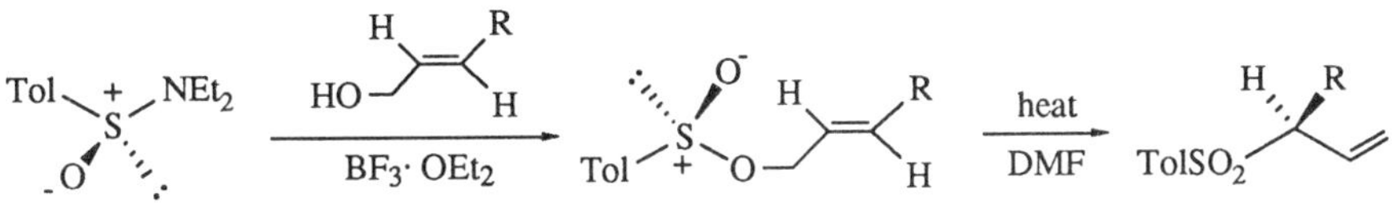

Scheme 11

Treatment of allylic alcohols with optically active sulphinamide **(25)** in the presence of $BF_3{\cdot}OEt_2$ is apparently the best way to prepare the sulphinate esters with clean inversion at sulphur. Thermolysis in DMF then gives the allylic sulphone, the process occurring with a high level of stereospecificity (*ca.* 90%). Analogous rearrangements to give allenyl sulphones have also been studied.[18] As in the simple allylic systems much evidence points to such rearrangements proceeding by a [2,3] sigmatropic mechanism, Scheme 12.

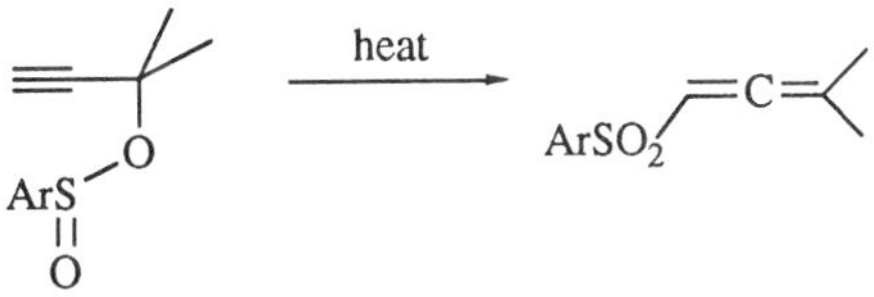

Scheme 12

Whilst the above studies have concentrated on systems having one 'dummy' aryl group attached to sulphur, this chemistry is also amenable to variations in which two groups attached to sulphur can be rearranged. In this way dimethylallyl alcohol can be converted to diprenyl sulphone, Scheme 13.[19]

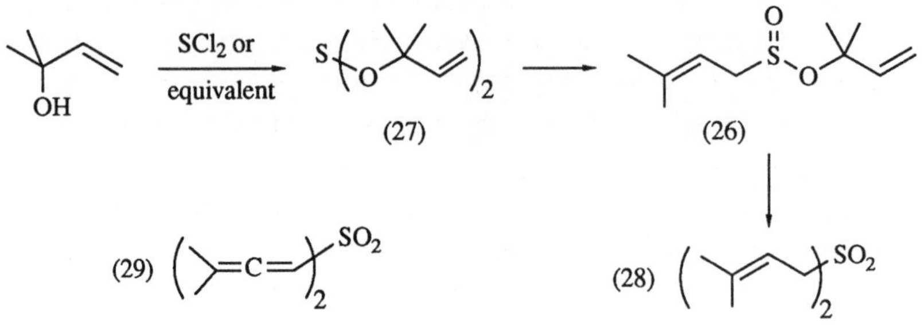

Scheme 13

The reaction with sulphur dichloride first gives the intermediate diallyl sulphoxylate (**27**) which very rapidly rearranges to sulphinate (**26**). The second rearrangement to give (**28**) requires either thermolysis or treatment with silica gel. In the latter case acid-catalysed ion-pair formation presumably occurs. A range of symmetrical sulphones has been prepared in this way, including a diallenyl sulphone (**29**). Similar rearrangements of benzylic sulphinates, including furfuryl systems, can be carried out, these also occurring by the ionisation–ion-pair recombination mechanism.[20]

When allylic sulphinates are heated with zero-valent palladium catalysts such as (Ph₃P)₄Pd, sulphone products are obtained.[21] This rearrangement is very different to the sigmatropic process described previously, and involves the formation of an intermediate π-allyl palladium species, Scheme 14.

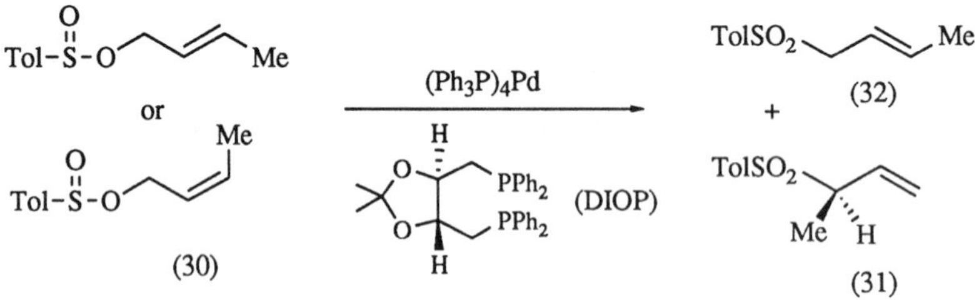

Scheme 14

In this process either stereoisomer of allylic sulphinate (**30**) provides mainly (*R*)-(**31**) in good enantiomeric excess (up to 87%). This is because each isomer is transformed via the same π-allyl intermediate, asymmetric induction being due to the homochiral ligand on palladium, rather than to chirality in the starting sulphinate. This result is in striking contrast to the corresponding sigmatropic process in which (*Z*)- and (*E*)-isomers give stereocomplementary results. A further difference from the thermal rearrangement is the tendency of this palladium-mediated reaction to give unwanted non-transposed sulphone products (**32**). This is a particularly acute problem when larger groups are attached to the allylic chain, i.e. if the methyl group in (30)-(32) is replaced by some larger group. Nevertheless, this reaction is an attractive one for the rapid preparation of certain optically active allylic sulphones. A further advantage is that allylic acetates can be used in place of the starting allylic sulphinates, provided, of course, that sodium benzenesulphinate is included in the mixture.

The potential of the allylic sulphinate to sulphone conversion appears not to have been much exploited in synthesis. One interesting development in this area, however, involves utilising this rearrangement to construct systems then capable of undergoing a Lewis-acid mediated ene reaction, Scheme 15.[22]

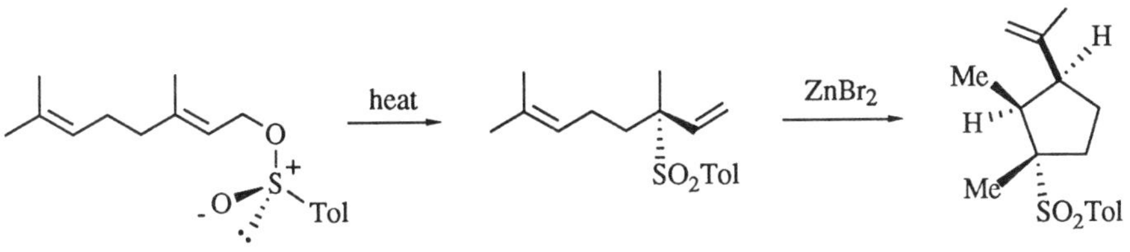

Scheme 15

Rearrangement of (S)-(33) under either thermal or palladium-catalysed conditions gives (S)-(34), which then undergoes intramolecular ene reaction to give (35). Thus a functionalised cyclopentane having three contiguous asymmetric centres is available in only three steps starting from geraniol. In terms of maintaining the configurational integrity of these intermediates the thermal process is superior (*ca.* 90% stereospecific) to the palladium-mediated process (60–65% stereospecific).

That increased functionality is tolerated around the reactive allylic system involved in the rearrangement is shown by two final examples, Scheme 16.[23,24]

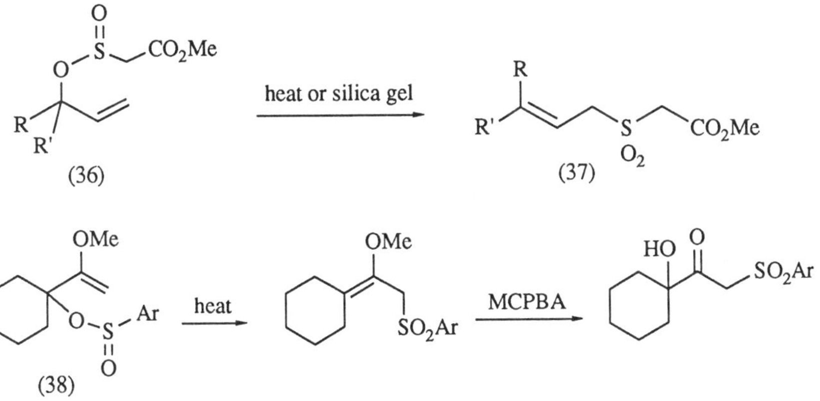

Scheme 16

The use of ester sulphinate (**36**) to give the rearranged sulphone (**37**) is then followed by Ramberg–Bäcklund rearrangement to give dienoic acids as the ultimate products.[23] In the case of (**38**) a method for the introduction of the α-hydroxy ketone unit was developed by oxidising the initially formed rearrangement product, in this case a vinyl ether, with MCPBA.[24]

5.3 1,3-Rearrangements of Allylic Sulphones

A range of simple allylic sulphones has been observed to undergo rearrangement in the presence of radical initiators, e.g. Scheme 17.[25] Reactions under either set of conditions are thought to involve radical chain mechanisms.

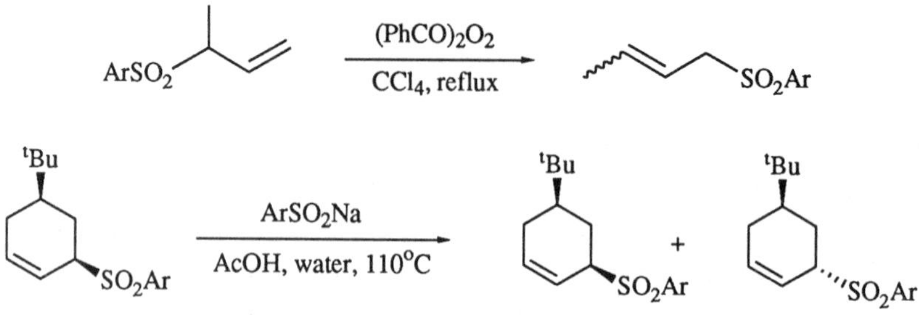

Scheme 17

Rearrangement is initiated by addition of a radical (Ph·, PhCO₂·, or ArSO₂·) to the allylic sulphone. Subsequent expulsion of ArSO₂· then establishes the chain process, Scheme 18.

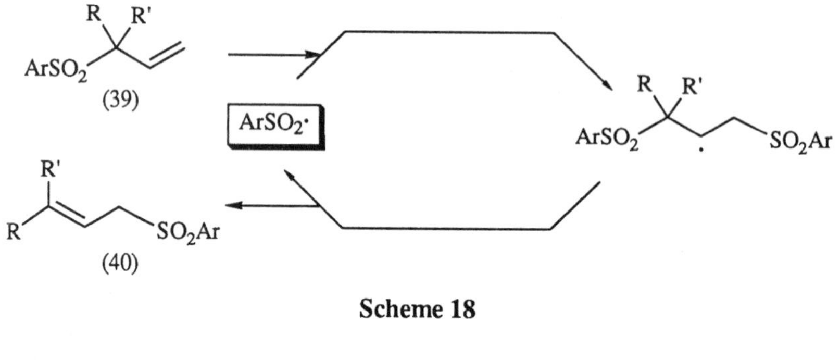

Scheme 18

The net result is conversion of the less-substituted double-bond isomer (**39**) into the more stable compound (**40**). Not surprisingly this process usually furnishes mixtures of double-bond isomers.

Further evidence for a radical chain mechanism is the effective inhibition of these rearrangements by hydroquinone. Also, some hindered substrates, in which radical addition would be expected to be suppressed for steric reasons, give correspondingly sluggish reactions.

Analogous rearrangements of allylic sulphones have been observed in epoxidation reactions, using MCPBA in the presence of sodium hydrogen carbonate, Scheme 19.[26]

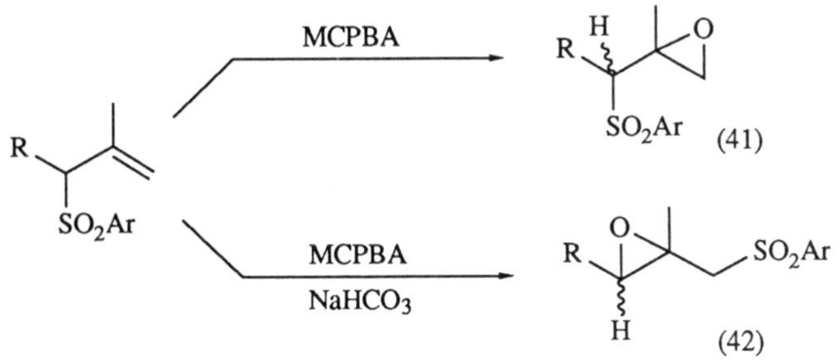

Scheme 19

Using MCPBA alone the expected β,γ-epoxy sulphone (**41**) is obtained. However, inclusion of sodium hydrogen carbonate results in the formation of the isomeric epoxide (**42**). This was shown to arise by 1,3-rearrangement of the starting allylic sulphone prior to epoxidation, in a manner similar to the reactions in Scheme 17.

The realisation that allylic sulphones are susceptible to radical attack and that ArSO$_2$ radicals can act as effective radical chain carriers has led Whitham and co-workers to construct substrates in which these processes result in useful cyclisation, Scheme 20.[27]

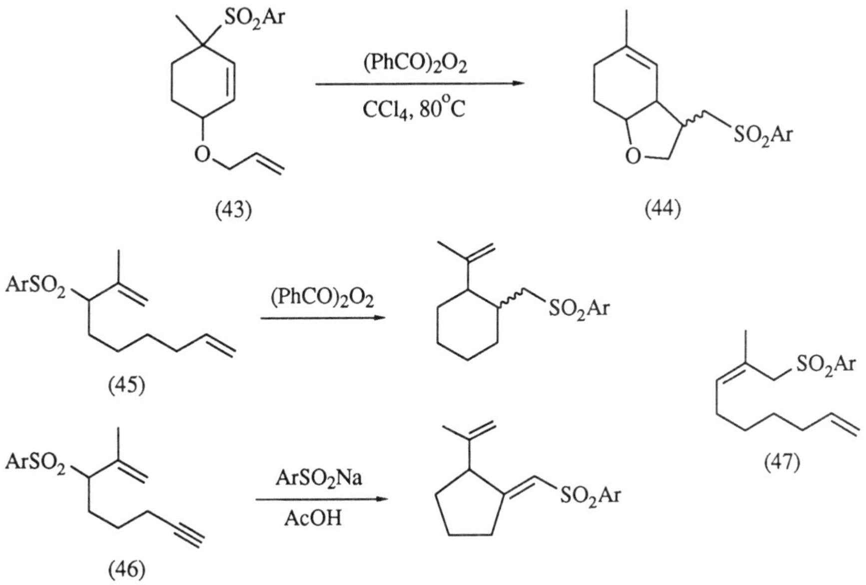

Scheme 20

The conversion of (43) to (44) is a simple example of this type of radical cyclisation. Attack of the ArSO$_2$ radical at the terminus of the allylic ether gives a carbon-centred radical. This undergoes 5-*exo-trig* cyclisation onto the cyclohexene, thus regenerating the ArSO$_2$ radical. In the reactions of (45) and (46) the cyclisation is cleverly combined with an initial 1,3-rearrangement, e.g. to (47). As indicated the reaction is capable of furnishing six-membered rings and vinyl sulphones, although stereocontrol is generally poor.

Extension of this type of SH2' reaction to examples in which intermolecular carbon–carbon bond formation occurs has also proved possible, Scheme 21.[28]

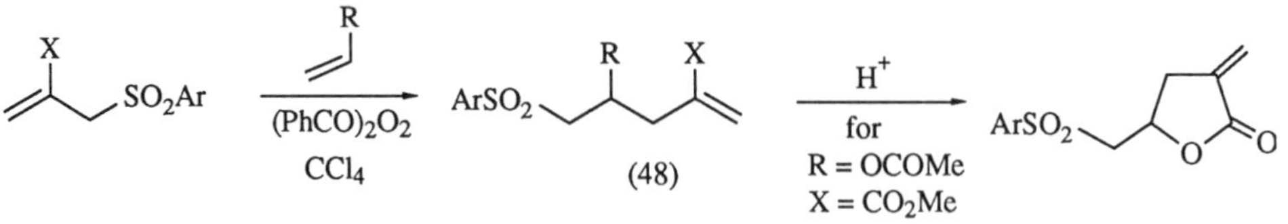

Scheme 21

In this sequence the addition of an ArSO$_2$ radical to the simple alkene gives an intermediate carbon radical which then undergoes intermolecular addition to the allylic sulphone partner to give product (48). The reaction is limited to sulphones having an electron withdrawing group X, and the alkenyl partner may be only monosubstituted (although R = alkyl, OR' and CO$_2$Me are all tolerated). One of the products can be very simply transformed into an α-methylene-γ-lactone by treatment with acid.

Addition–fragmentation reactions of allylic sulphones can also be used to form carbon–carbon bonds by employing radicals generated from alkyl iodides, e.g. Scheme 22.[29]

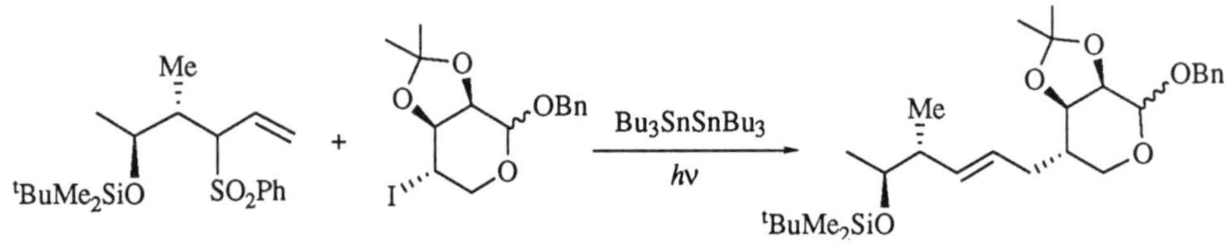

Scheme 22

The coupling shown proved rather difficult to achieve, since the system is not properly designed to undergo a normal chain reaction, radical generation from the starting alkyl iodide being maintained by slow (syringe pump) addition of Bu$_3$SnSnBu$_3$ to the irradiated mixture.

Rearrangements analogous to those shown above in Scheme 17 and 18 for simple sulphones have also been observed with allylic sulphones having an additional sulphide grouping, Scheme 23.[30,31]

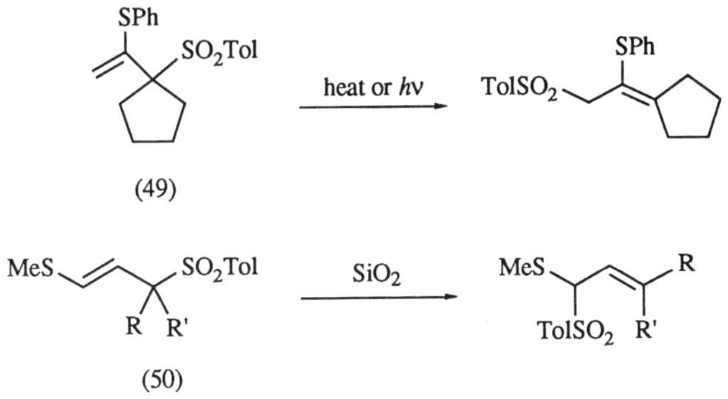

(49)

(50)

Scheme 23

The rearrangement of systems such as (**49**) is even more facile than that of compounds not containing the extra sulphur group.[30] This is to be expected if a radical pathway is followed, since a more electron-rich π-bond favours the addition of the somewhat electrophilic $PhSO_2$ radical. As in the previous work by Whitham's group, tandem rearrangement–cyclisation procedures were also explored. The rearrangement of (**50**) mediated by silica gel is another example of an ion-pair formation–recombination reaction.[31]

Finally, the rearrangement of allylic sulphones via ion pairs can be promoted by certain palladium catalysts, Scheme 24.[32]

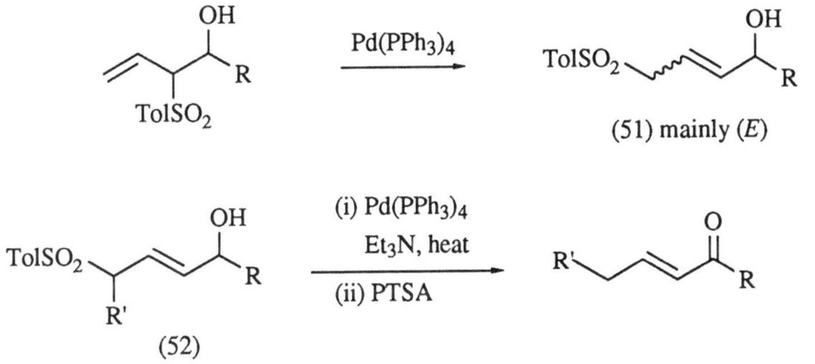

(51) mainly (*E*)

(52)

Scheme 24

As before, conversion to a more stable isomer is observed, here facilitated by the reversible formation of an intermediate π-allyl palladium species. Interestingly, the products of these conversions, e.g. (**51**), can be further alkylated via the corresponding dianion to give (**52**). Treatment of these alkylated products with a palladium catalyst again forms a π-allyl complex, but in this case elimination occurs to give mainly α,β-unsaturated ketone products.

5.4 The Truce–Smiles Rearrangement

On treatment with butyllithium in ether diaryl sulphones having an *ortho*-methyl group undergo benzylic deprotonation to give anions such as (**53**), Scheme 25.

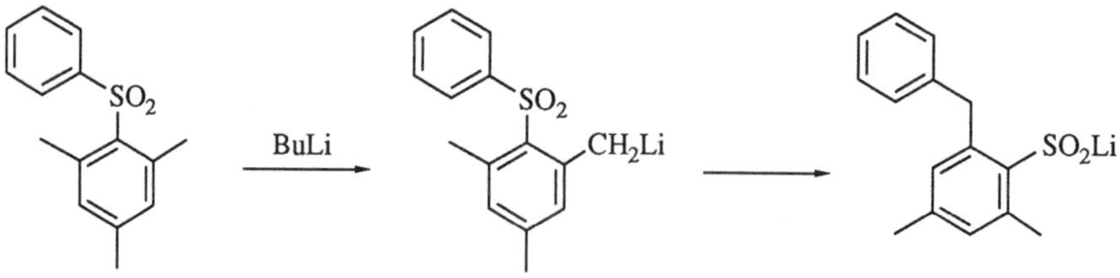

BuLi

(53)

(54)

Scheme 25

On refluxing in ether this anion undergoes rearrangement to form the sulphinate (**54**). The reaction is similar to the well-known Smiles rearrangement and has become known as the Truce–Smiles rearrangement.[33] Significantly, in this modification the migrating aromatic group is attacked by a carbanionic group, rather than a heteroatom nucleophile, as is the case in the simple Smiles reaction. Another notable difference from the usual Smiles reaction is the lack of any need for electronic activation of the migrating aryl group. The reaction is quite general, a second example being the conversion of naphthyl sulphone (**55**) to (**56**), Scheme 26.

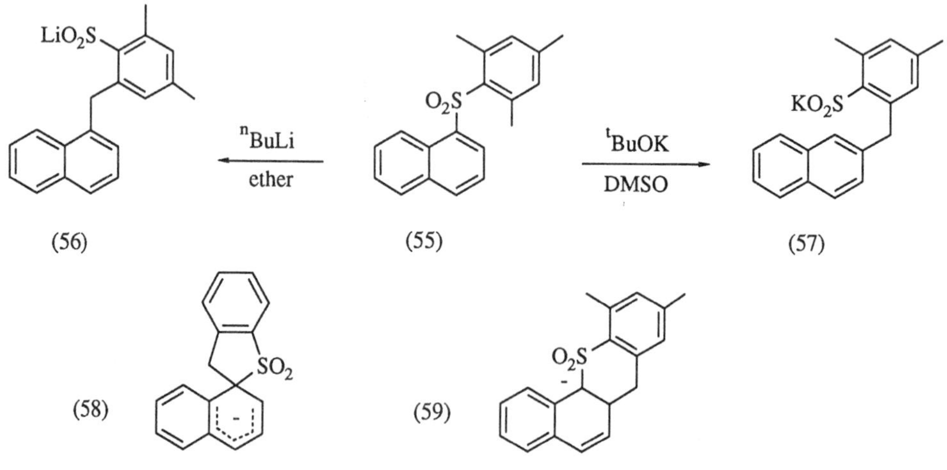

^{n}BuLi
ether

^{t}BuOK
DMSO

(56)

(55)

(57)

(58)

(59)

Scheme 26

In some systems, however, other modes of rearrangement can occur. This alternative type of reaction, leading to (57), is observed when (55) is treated with ^tBuOK as base in DMSO, in place of the usual $^nBuLi/Et_2O$ system. The two kinds of rearrangement are thought to involve intramolecular attack of the benzylic carbanion on the neighbouring aromatic nucleus. In the case of the 'normal' Truce–Smiles rearrangement, *ipso*-attack leading to the Meisenheimer-type intermediate (58) explains the outcome, while the alternative pathway occurs by a type of internal Michael addition to give (59).

The appropriate positioning of a leaving group on the aromatic nucleus which would normally migrate in this reaction can divert the process to a simple internal nucleophilic aromatic substitution, e.g. to give (60), Scheme 27.[33]

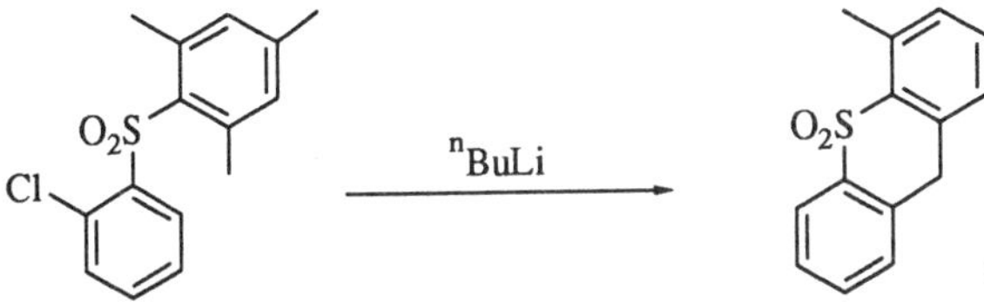

Scheme 27

Further studies on the rearrangements of such metallated sulphones have shown that migration of an alkyl rather than an aryl group can also occur, Scheme 28.[33,34]

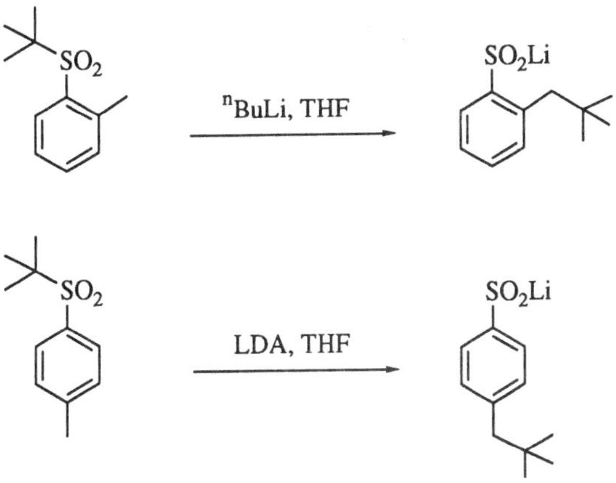

Scheme 28

The rearrangement of *para*-substituted sulphones is particularly striking and demands a different mechanistic interpretation from that for the aryl group migrations. A variety of studies, including crossover experiments, points to the intermediacy of radicals in these reactions. The most plausible reaction pathway is a radical–radical anion chain mechanism.[34]

5.5 Miscellaneous Rearrangements

The facile formation of carbanions from sulphones can be very useful in initiating or accelerating well-known rearrangements such as the [2,3] Wittig rearrangement or the Claisen rearrangement. Bruckner and Peiseler described an elegant modification of the first of these two processes in which formation of a carbanion derived from an α-oxygenated sulphone initiates rearrangement, Scheme 29.[35]

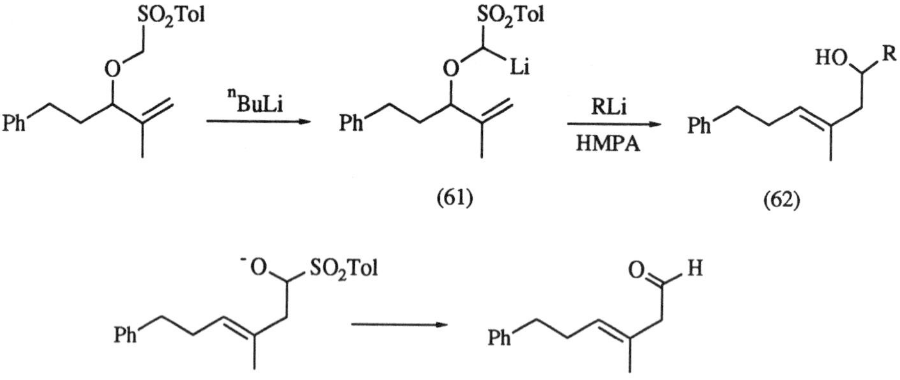

(61)   (62)

(63)   (64)

Scheme 29

In THF the sulphonyl carbanion (61) is sufficiently stable to undergo alkylation. However, addition to another alkyllithium in the presence of HMPA gives (62). The reaction occurs by Wittig rearrangement to give (63), which collapses to give an aldehyde (64) which is then intercepted by the alkyllithium present. This novel and stereoselective process was applied to the synthesis of artemesia alcohol.

The introduction of a sulphonyl carbanionic centre into an allyl vinyl ether has a marked accelerating effect on the Claisen rearrangements of such systems, Scheme 30.[36]

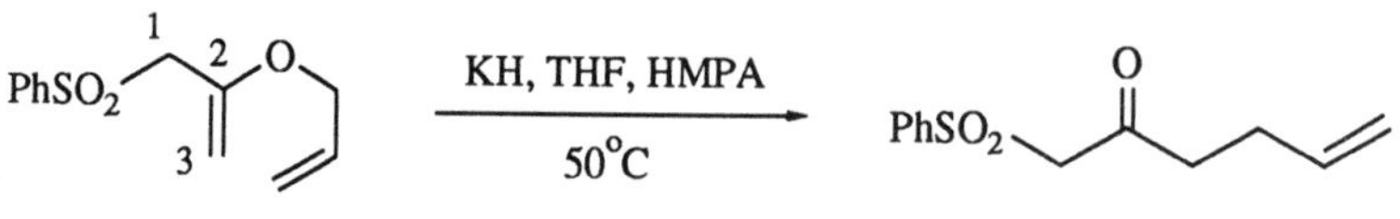

Scheme 30

The rearrangements occur in good yield at temperatures between 20°C and 50°C depending on the substituent pattern. The reaction is highly regioselective, bonding to the vinyl unit occurring at C-3 (not C-1) to give only the ketosulphone product bearing the sulphone group at the end of the chain. As in the Wittig chemistry described above, the use of the dissociating solvent HMPA is important for optimum results.

Even more dramatic acceleration can be brought about by incorporating a sulphonyl carbanion into substrates for vinylcyclopropane rearrangement.[37] While this process usually requires pyrolysis, the analogous carbanion-accelerated rearrangement occurs at -30°C, Scheme 31.

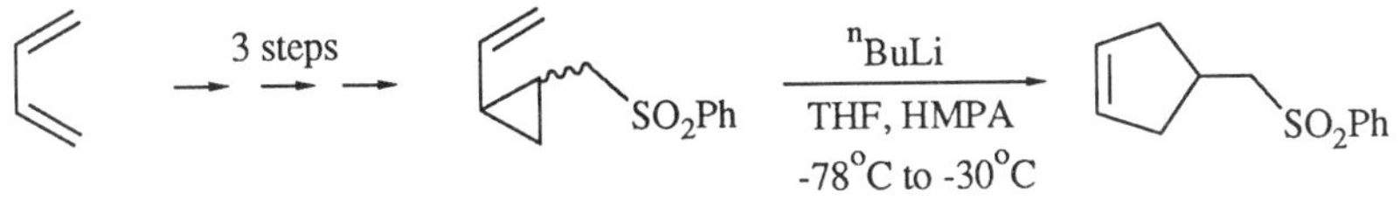

Scheme 31

The process can be extended to incorporate *in situ* electrophilic trapping and subsequent desulphonylation of the product cyclopentenylmethyl sulphonyl carbanions.

Certain *N*-alkylamidosulphones undergo rearrangement with loss of SO_2 when treated with base, e.g. Scheme 32.[38]

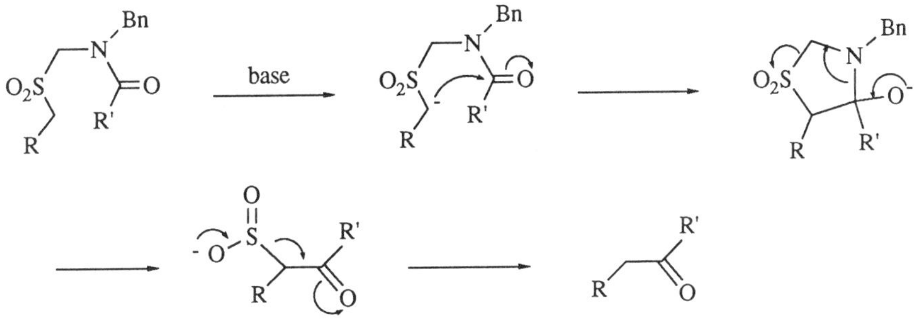

Scheme 32

The reaction is typically conducted at room temperature using LDA as base or at reflux in THF with NaH. The yields of product are rather modest, with better results being obtained using sulphides rather than sulphones, provided that R is an aromatic group.

Two groups of workers have reported on the rearrangements of α,β-epoxy sulphones involving a formal 1,2-migration of the sulphone group, Scheme 33.[39]

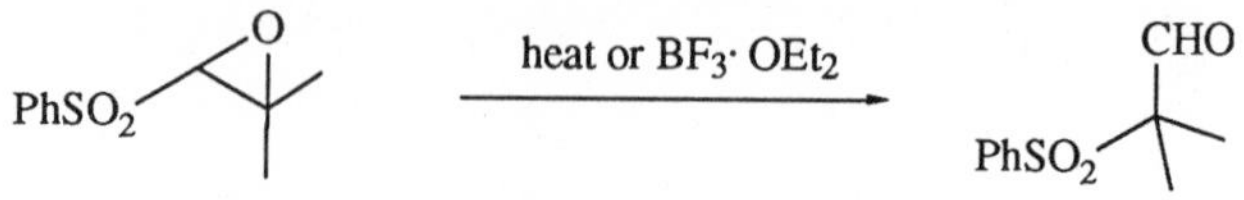

Scheme 33

This rearrangement is very high yielding and is equally applicable to the ring expansion of epoxy sulphones derived from cyclic sulphones.

Finally, another 1,2-sulphone migration has been observed during studies of the condensation of bis(ethylsulphonyl)methane with benzaldehyde, Scheme 34.[40]

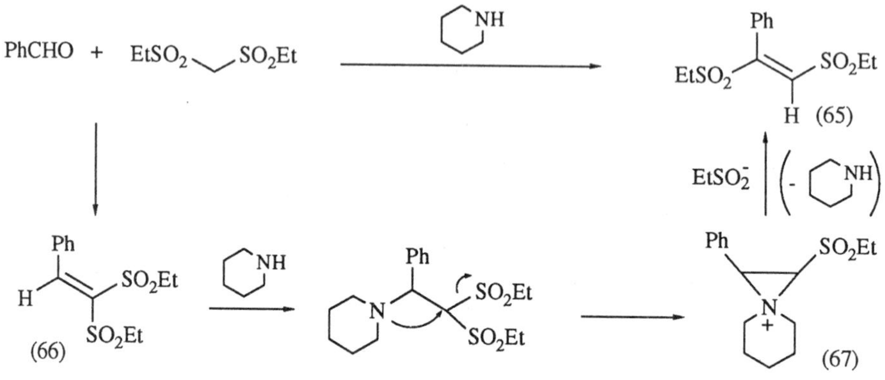

Scheme 34

The product from the condensation is (**65**) and not (**66**). As shown, a rationale has been proposed involving participation by piperidine to produce an aziridinium intermediate (**67**). Additional support for this hypothesis is that independently synthesised (**66**) gives (**65**) on treatment with a catalytic amount of piperidine in refluxing benzene.

Chapter 5 References

1. S. Toyama, T. Aoyama, and T. Shiori, *Chem. Pharm. Bull.*, **1982**, *30*, 3032.
2. T. Sugawara and I. Kuwajima, *Tetrahedron Lett.*, **1985**, *26*, 5571.
3. Y. Masuyama, K. Yamada, H. Tanaka, and Y. Kurusu, *Synth. Commun.*, **1987**, *17*, 1525.
4. B. M. Trost and G. K. Mikhail, *J. Am. Chem. Soc.*, **1987**, *109*, 4124.
5. H. Finch, A. M. M. Mjalli, J. G. Montana, S. M. Roberts, and R. J. K. Taylor, *Tetrahedron*, **1990**, *46*, 4925.
6. B. M. Trost and J. E. Vincent, *J. Am. Chem. Soc.*, **1980**, *102*, 5680; see also B. M. Trost and H. Hiemstra, *J. Am. Chem. Soc.*, **1982**, *104*, 886.
7. B. M. Trost and B. R. Adams, *J. Am. Chem. Soc.*, **1983**, *105*, 4849.
8. V. Bhat and R. C. Cookson, *J. Chem. Soc., Chem. Commun.*, **1981**, 1123.
9. A. S. Kende and I. Kaldor, *Tetrahedron Lett.*, **1989**, *30*, 7329.
10. D. A. Chass, D. Buddhasukh, and P. Magnus, *J. Org. Chem.*, **1978**, *43*, 1750.
11. A. Fischli, Q. Branca, and J. Daly, *Helv. Chim. Acta*, **1976**, *59*, 2443.
12. J. Fayos, J. Clardy, L. J. Dolby, and T. Farnham, *J. Org. Chem.*, **1977**, *42*, 1349.
13. K. Miura, K. Fugami, K. Oshima, and K. Utimoto, *Tetrahedron Lett.*, **1988**, *29*, 1543; see also D. A. Singleton and K. M. Church, *J. Org. Chem.*, **1990**, *55*, 4780.
14. S. Braverman, *Int. J. Sulphur. Chem., Part C.*, **1971**, *6*, 149; S. Braverman, in *The Chemistry of Sulphoxides and Sulphones*, Ed. S. Patai, Z. Rappoport, and C. J. M. Stirling, John Wiley and Sons Ltd, Chichester, **1988**, p.665.
15. D. J. Knight, G. H. Whitham, and J. G. Williams, *J. Chem. Soc., Perkin Trans. 1*, **1987**, 2149 and references therein.
16. D. J. Knight, P. Lin, S. T. Russell, and G. H. Whitham, *J. Chem. Soc., Perkin Trans. 1*, **1987**, 2701.
17. K. Hiroi, R. Kitayama, and S. Sato, *Chem. Pharm. Bull.*, **1984**, *32*, 2628; K. Hiroi, R. Kitayama, and S. Sato, *J. Chem. Soc., Chem. Commun.*, **1983**, 1470.
18. G. Smith and C. J. M. Stirling, *J. Chem. Soc. (C.)*, **1971**, 1530; S. Braverman and H. Mechoulam, *Tetrahedron*, **1974**, *30*, 3883. For some related work on the chemistry of allenyl sulphinates see S. Braverman, *Phosphorus and Sulphur*, **1985**, *23*, 297.
19. G. Buchi and R. M. Freidinger, *J. Am. Chem. Soc.*, **1974**, *96*, 3332.
20. S. Braverman and T. Globerman, *Tetrahedron*, **1974**, *30*, 3873 and references therein.
21. K. Hiroi and K. Makino, *Chem. Pharm. Bull.*, **1988**, *36*, 1744; K. Hiroi, R. Kitayama, and S. Sato, *J. Chem. Soc., Chem. Commun.*, **1984**, 303; K. Hiroi and K. Makino, *Chem. Lett.*, **1986**, 617.

22. K. Hiroi, M. Yamamoto, Y. Kurihara, and H. Yonezawa, *Tetrahedron Lett.,* **1990**, *31*, 2619; K. Hiroi and Y. Kurihara, *J. Chem. Soc., Chem. Commun.,* **1989**, 1778. For related work on intermolecular ene reactions see B. B. Snider, T. C. Kirk, D. M. Roush, and D. Gonzalez, *J. Org. Chem.,* **1980**, *45*, 5015.

23. P. A. Grieco and D. Boxler, *Synth. Commun.,* **1975**, *5*, 315.

24. J. E. Baldwin, O. W. Lever, Jr., and N. R. Tzodikov, *J. Org. Chem.,* **1976**, *41*, 2312.

25. D. J. Knight, P. Lin, and G. H. Whitham, *J Chem. Soc., Perkin Trans. 1,* **1987**, 2707; see also S. O. Myong, L. W. Linder, Jr., S. C. Seike, and R. D. Little, *J. Org. Chem.,* **1985**, *50*, 2244.

26. P. Kocienski, *J. Chem. Soc., Perkin Trans. 1,* **1983**, 945.

27. T. A. K. Smith and G. H. Whitham, *J. Chem. Soc., Perkin Trans. 1,* **1989**, 313; T. A. K. Smith and G. H. Whitham, *J. Chem. Soc., Perkin Trans. 1,* **1989**, 319.

28. I. W. Harvey, E. D. Phillips, and G. H. Whitham, *J. Chem. Soc., Chem. Commun.,* **1990**, 481.

29. G. E. Keck and A. M. Tafesh, *J. Org. Chem.,* **1989**, *54*, 5845.

30. A. Padwa, W. H. Bullock, and A. D. Dyszlewski, *J. Org. Chem.,* **1990**, *55*, 955.

31. K. Ogura, T. Iihama, S. Kiuchi, T. Kajiki, O. Koshikawa, K. Takahashi, and H. Iida, *J. Org. Chem.,* **1986**, *51*, 700; see also K. Ogura, N. Yahata, T. Fujimori, and M. Fujita, *Tetrahedron Lett.,* **1990**, *31*, 4621.

32. K. Inomata, Y. Murata, H. Kato, Y. Tsukahara, H. Kinoshita, and H. Kotake, *Chem. Lett.,* **1985**, 931.

33. W. E. Truce and E. J. Madaj, Jr., *Sulphur Rep.,* **1983**, *3*, 259; W. E. Truce, E. M. Kreider, and W. W. Brand, *Org. React.,* **1970**, *18*, 99; V. N. Drozd, *Int. J. Sulphur. Chem.,* **1973**, *8*, 443; S. W. Schneller, *Int. J. Sulphur Chem.,* **1976**, 579.

34. E. J. Madaj, Jr., D. M. Snyder, and W. E. Truce, *J. Am. Chem. Soc.,* **1986**, *108*, 3466.

35. R. Bruckner and B. Peiseler, *Tetrahedron Lett.,* **1988**, *29*, 5233.

36. S. E. Denmark and M. A. Harmata, *J. Am. Chem. Soc.,* **1982**, *104*, 4972; S. E. Denmark, M. A. Harmata, and K. S. White, *J. Org. Chem.,* **1987**, *52*, 4031 and references therein.

37. R. L. Danheiser, J. J. Bronson, and K. Okano, *J. Am. Chem. Soc.,* **1985**, *107*, 4579.

38. Y. Ishikawa, Y. Kurebayashi, K. Suzuki, Y. Terao, and M. Sekiya, *Chem. Pharm. Bull.,* **1981**, *29*, 2496.

39. D. F. Tavares, R. E. Estep, and M. Blezard, *Tetrahedron Lett.,* **1970**, 2373; T. Durst and K-C. Tin, *Tetrahedron Lett.,* **1970**, 2369.

40. A. R. Friedman and D. R. Graber, *J. Org. Chem.,* **1972**, *37*, 1902.

CHAPTER 6

Cycloaddition Chemistry of Unsaturated Sulphones

Unsaturated sulphones in which the sulphone is directly attached to an alkenyl or alkynyl group undergo a range of cycloadditions, including [2+2], [3+2] and [4+2] processes.[1] This chemistry incorporates both concerted cycloadditions, e.g. Diels–Alder reactions of vinyl sulphones, as well as other stepwise cycloaddition reactions, such as tandem Michael reactions of allenyl sulphones.

6.1 [2+2] Cycloadditions

The cycloadditions of enamines and ynamines with the strained cyclic vinyl sulphone thiete 1,1-dioxide, to give bicyclic products such as (1), were originally reported by Paquette *et al.*[2] Analogous chemistry using simple vinyl and dienyl sulphones has more recently been described by Eisch and *et al.*, Scheme 1.[3]

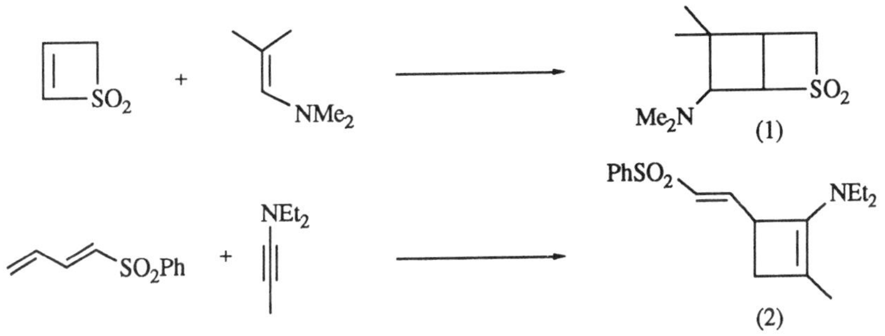

Scheme 1

The reactions are regiospecific, presumably proceeding by stepwise attack of the nucleophilic enamine onto the electron-poor vinyl sulphone partner. The four-membered ring products are obtained in high yield, the enamine products such as (2) being easily hydrolysed with dilute acid, thereby providing a good route to substituted cyclobutanones.

The products of a formal [2+2] cycloaddition are also obtained from the reaction of 1,1-bis(phenylsulphonyl)ethylene with norbornadiene and its benzo-fused analogue (**3**).[4] The isolation of other products such as (**4**) in the latter reaction is good evidence for a zwitterionic intermediate in such processes, Scheme 2.

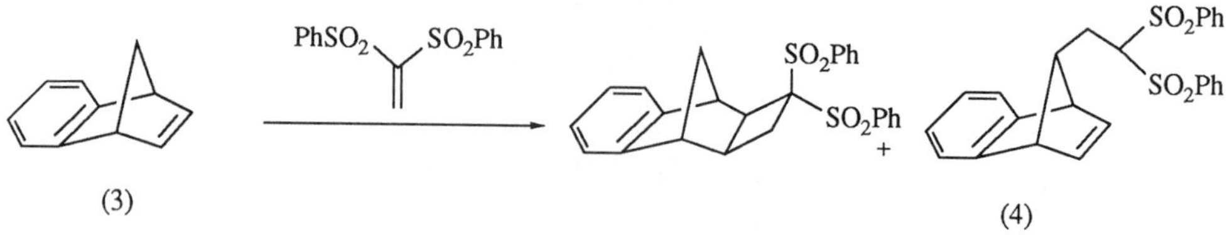

Scheme 2

Photochemical cycloadditions to vinyl sulphones and cyclodimerisations of such systems are relatively rare. On irradiation (*E*)-phenyl styryl sulphones simply undergo photoisomerisation to mixtures of (*Z*)- and (*E*)-isomers.[5] However, under the same conditions 2-phenylsulphonylindene undergoes dimerisation or cycloaddition with simple alkenes, e.g. Scheme 3.

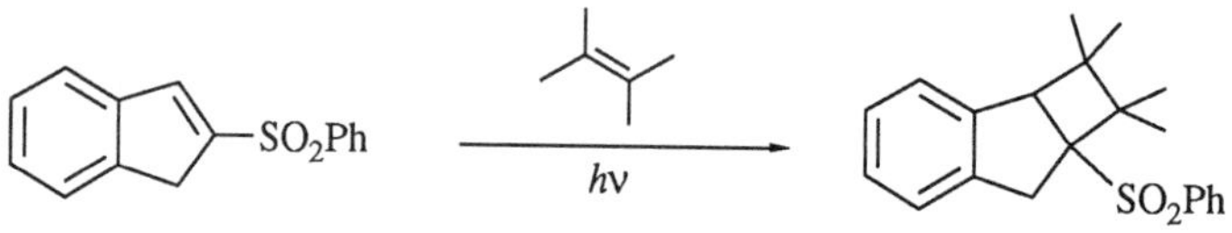

Scheme 3

Similar intermolecular photochemical cycloadditions can be carried out using benzo-fused thiophene 1,1-dioxides, Scheme 4.[6]

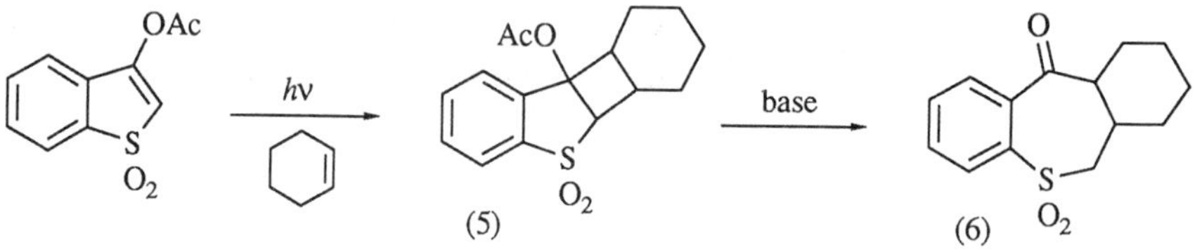

Scheme 4

Treatment of the initial [2+2] cycloadduct (**5**) with sodium hydroxide in aqueous methanol results in acetate hydrolysis and retro-aldol ring-opening to give (**6**). These latter two cycloadditions may well proceed via diradical intermediates.

An electrochemical dimerisation of aryl vinyl sulphones provides moderate to good yields of disulphonyl cyclobutane products, Scheme 5.[7]

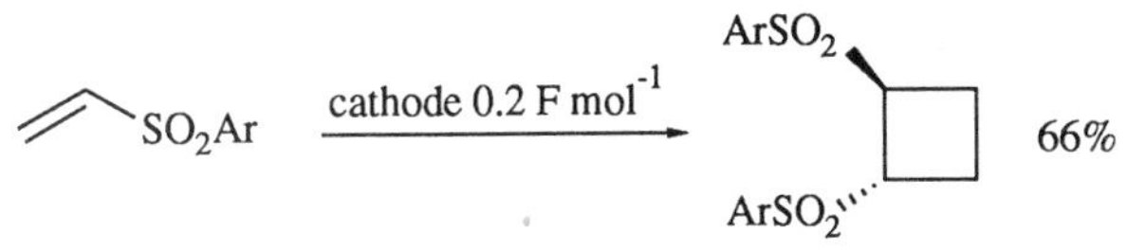

Scheme 5

The 1,2-*trans*-substituted products are produced selectively and the reaction appears to be electrocatalysed, since only 0.1–0.2 Fmol^{-1} gives reasonable chemical yields of dimer. The reaction mechanism remains uncertain but presumably involves coupling of radical anions, followed by chain-carrying electron transfer and cyclisation.

Intramolecular [2+2] cycloaddition can compete with an alternative [4+2] mode of reaction in certain allenyl sulphones, Scheme 6.[8]

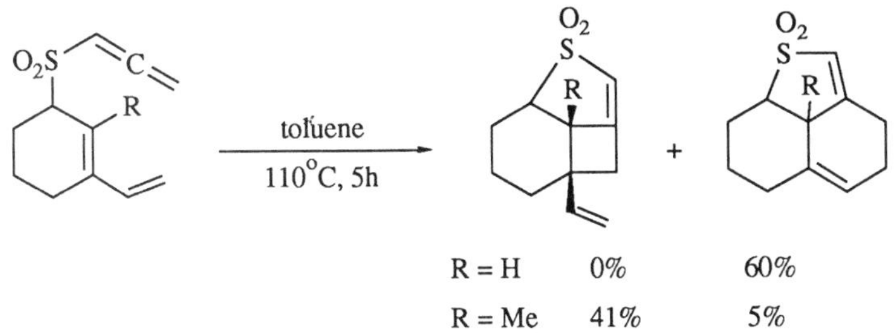

R = H	0%	60%
R = Me	41%	5%

Scheme 6

The periselectivity is dependent on the nature of R which exerts an influence on the conformation of the 1,3-diene. Fuchs' group has studied the intramolecular [2+2] cycloaddition of some highly functionalised complex vinyl sulphones as a route to cytochalasin C intermediates.[9] The photolysis of (7) in acetonitrile at room temperature gives two products, resulting from attack on either face of the vinyl sulphone, Scheme 7.

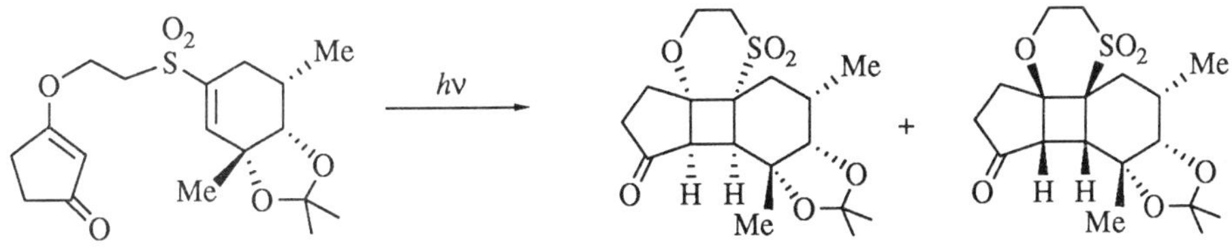

(7)   major   minor

Scheme 7

Unfortunately, subsequent chemistry aimed at opening the cyclobutane ring proved unsuccessful.

Finally, a variety of cycloaddition reactions of thiapyrone dioxides is known, e.g. Scheme 8.[10]

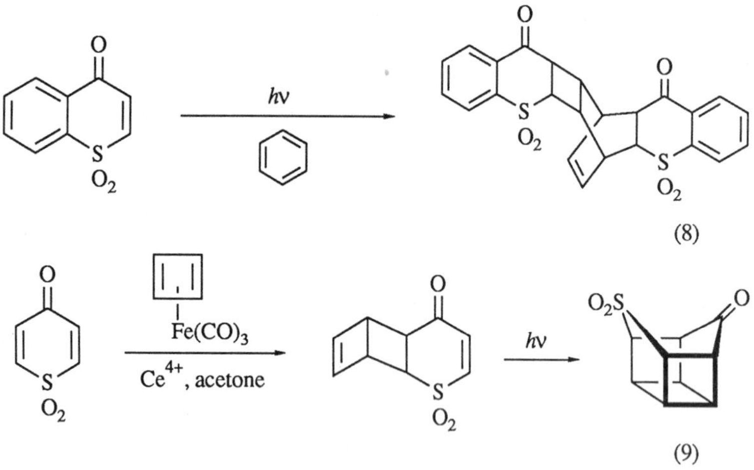

(8)

(9)

Scheme 8

Diadduct (**8**) arises by initial 1,2-photoaddition to benzene, followed by a Diels–Alder cycloaddition of the so-formed cyclohexa-1,3-diene intermediate. Ketosulphone (**9**) is just one of several related caged sulphone products assembled in order to study transannular S–C_{CO} interactions in the corresponding sulphides.

6.2 [3+2] Cycloadditions

In one of a series of significant contributions to the cycloaddition chemistry of vinyl sulphones Padwa's group has examined the reaction of 2-diazopropane with the bicycloheptadiene (**10**).[11] The reaction follows an unexpected course, the cycloaddition occurring across the electronically unactivated alkene, Scheme 9.

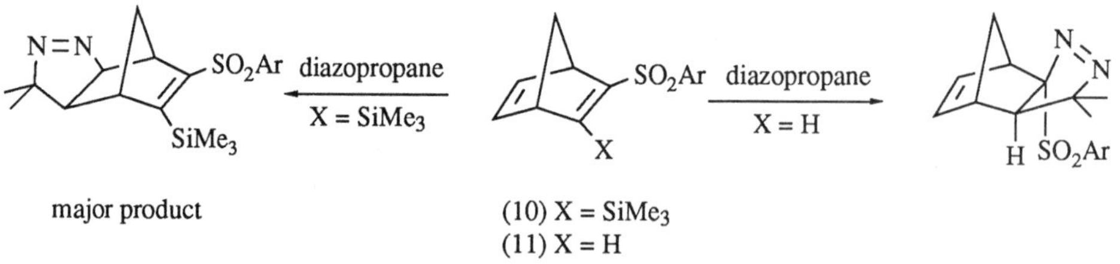

major product

(10) X = SiMe₃
(11) X = H

Scheme 9

This result was rationalised in terms of steric inhibition of the cycloaddition due to the bulky silicon group, the simple vinyl sulphone (**11**) reacting as expected. Similar control of regioselectivity can be exercised in additions of diazoalkanes to alkynyl sulphones.[12]

The reaction of diazomethane with allenyl phenyl sulphone proceeds regioselectively to give mainly pyrazole (**12**), Scheme 10.[13]

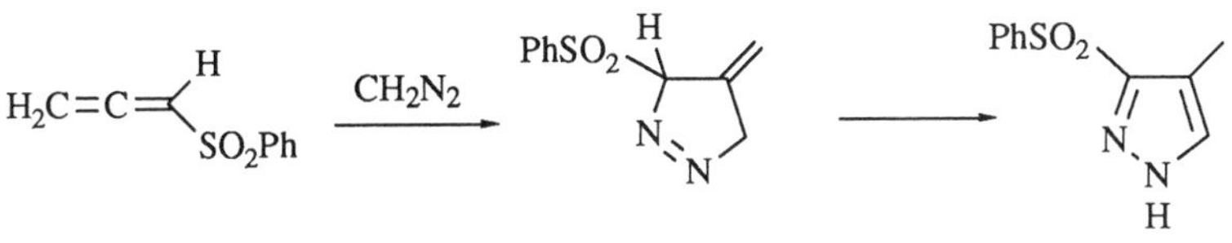

Scheme 10

A minor product, derived either from addition to the non-activated allenyl double bond or by a 1,3-sulphone shift, is also obtained. Allenyl phenyl sulphone also gives the expected cycloadducts when treated with nitrones, Scheme 11.[14]

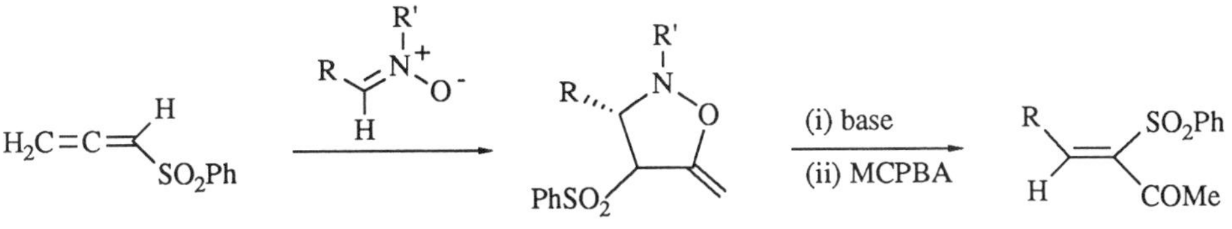

Scheme 11

As shown, further transformation of this type of cycloadduct into acyclic α,β-unsaturated carbonyl compounds is possible using MCPBA. An interesting development of this cycloaddition chemistry is the finding that the use of vinyl sulphone (**13**) allows easy access to cycloadducts regioisomeric with those obtained with allenyl sulphones, Scheme 12.[15]

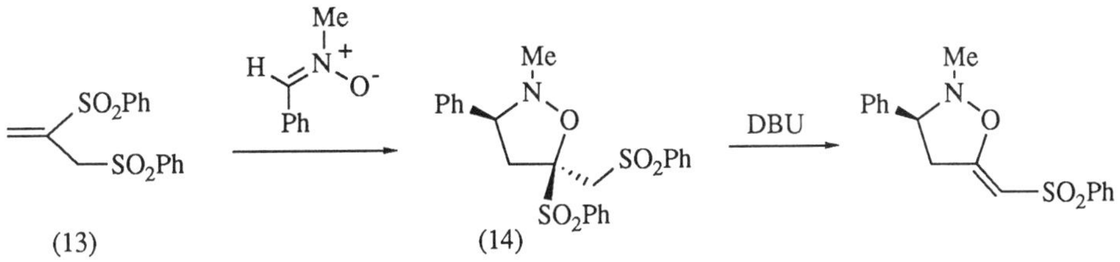

Scheme 12

Even more remarkable is the fact that conducting the cycloaddition under high pressure (3 kbar) at room temperature not only accelerates the reaction but gives the other diastereoisomer of (**14**). Although the 'pressure cycloadduct' was calculated to be the less stable of the two diastereoisomers it was not rearranged to the more stable isomer on heating. These intriguing results await further rationalisation.

The anion derived from (**13**) can be generated by conjugate addition of sodium benzenesulphinate to allenyl phenyl sulphone.[16] This allows for a number of interesting cyclisation–elimination reactions with Michael acceptors to give [3+2] type products, e.g. Scheme 13.

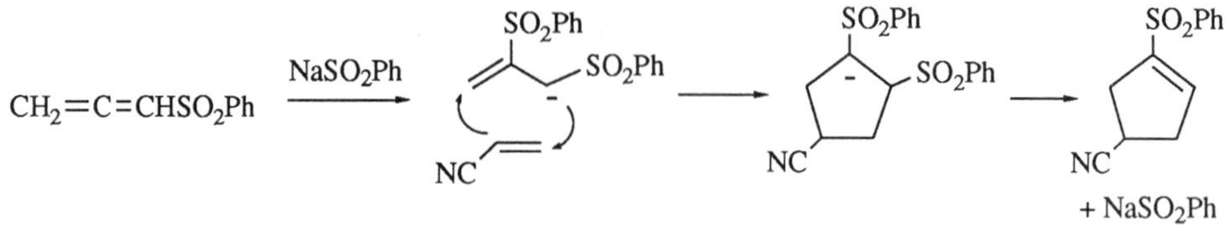

Scheme 13

In this sequence the benzenesulphinate is required in only trace amounts, since it is regenerated in the final elimination step. Analogous processes are possible by using the addition of other anions to the starting allenyl sulphone to trigger cyclisation.

Grigg and co-workers have described the use of nitrones, generated from oximes by Michael addition, for cycloaddition reactions.[17] Vinyl sulphones can be used as reaction partners, either to generate the nitrone, or to participate in the cycloaddition, or in both steps, e.g. Scheme 14.

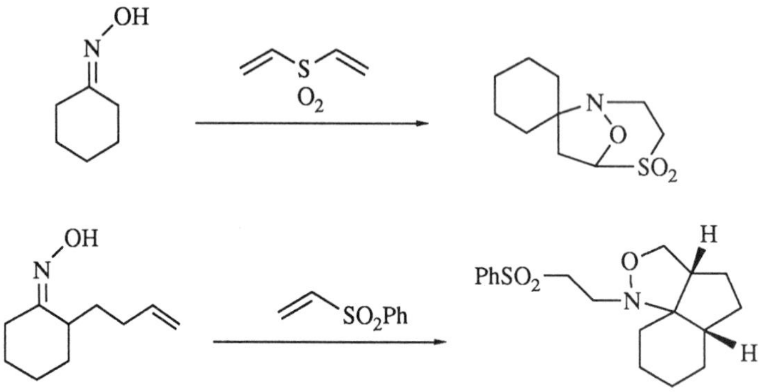

Scheme 14

Both the initial Michael reaction and the cycloaddition can be carried out intra- or intermolecularly, allowing a broad range of interesting functionalised polycyclic products to be prepared.

In general the regiochemistry of cycloadditions to unsaturated sulphones with 1,3-dipoles, including diazoalkanes, nitrones and nitrile oxides, is rather substrate dependent. However, examination of the regioisomer distributions for some cycloadditions reveals some apparent trends, e.g. (15)-(20), Scheme 15.

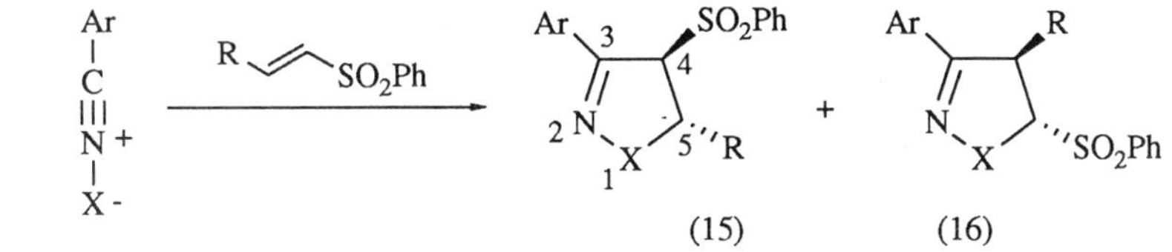

	(15)		(16)
X = NPh	R = H	100 : 0	
	= Me	65 : 35	
	= Ph	90 : 10	
	= COPh	95 : 5	
X = O	R = H	9 : 91	
	= Me	90 : 10	
	= Ph	60 : 40	
	= COPh	75 : 25	

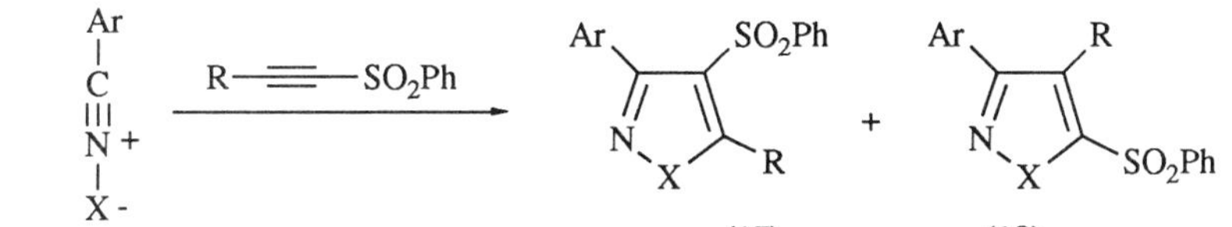

	(17)		(18)
X = NPh	R = H	70 : 30	
	= Me	100 : 0	
	= Ph	100 : 0	
X = O	R = H	80 : 20	
	= Me	100 : 0	
	= Ph	100 : 0	

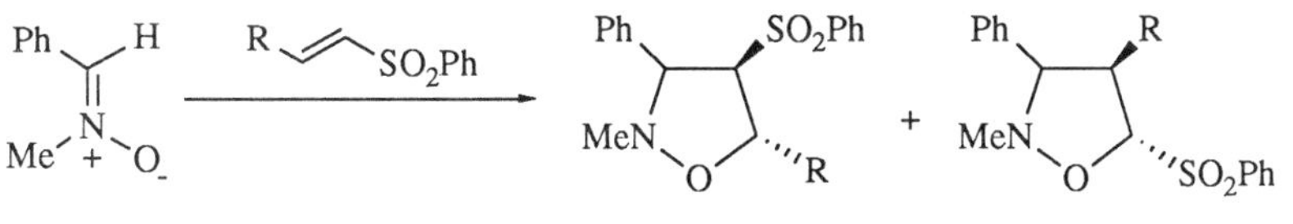

	(19)		(20)
	R = H	68 : 32	
	= Me	100 : 0	

Scheme 15

In reactions of vinyl sulphones with either diphenylnitrilimine or 3,5-dichloromesitylnitrile oxide the 5-alkyl-substituted cycloadduct (15), rather than (16), is usually the major product.[18] This trend is even more pronounced with alkynyl sulphones as the dipolarophiles, isomer (17) often being the sole product.[19] Such product outcomes can be explained by invoking a HOMO(dipole)–LUMO(dipolarophile) interaction as the overriding controlling factor. The exceptional preferential formation of (16), where R = H and X = O, may be attributable to the intervention of steric effects. Cycloadditions with nitrones are less predictable, often producing a mixture of isomers, although, as shown, the 5-alkyl-substituted product (19) again predominates in simple cases.[20]

An interesting Michael addition–dipolar cycloaddition process occurs when divinyl sulphone is treated with imines derived from α-amino esters in the presence of lithium bromide and triethylamine, Scheme 16.[21]

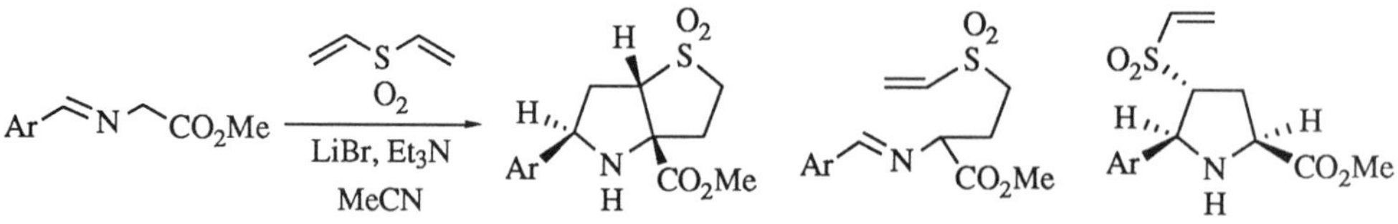

Scheme 16

The products (21) are thought to arise via initial Michael adducts (22). These intermediates were synthesised independently and shown to cyclise to (21) under the usual reaction conditions. Changing the metal present to silver and using DMSO as solvent results in a dramatic change in reactivity, leading to the predominant formation of simple cycloadducts such as (23).

Cycloadditions of vinyl sulphones with pyridinium species (24) occur with a high level of regio- and stereoselectivity, Scheme 17.[22]

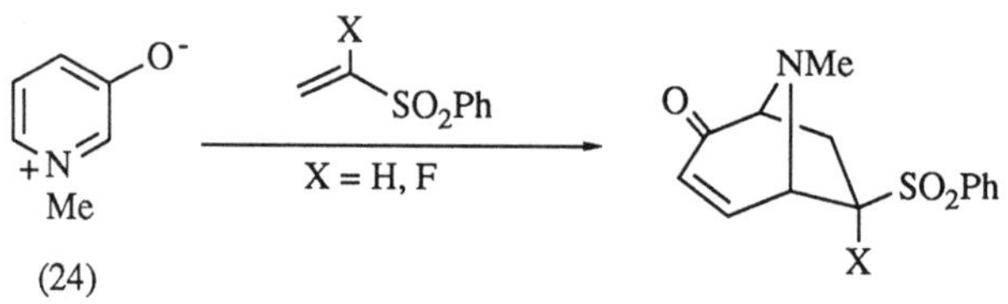

Scheme 17

The cycloadducts were converted to known 2-tropanol derivatives by enone reduction, hydrogenation and desulphonylation.

Trost has included vinyl sulphones amongst the partners that react with trimethylenemethane (TMM) palladium complexes to give [3+2] adducts, Scheme 18.[23]

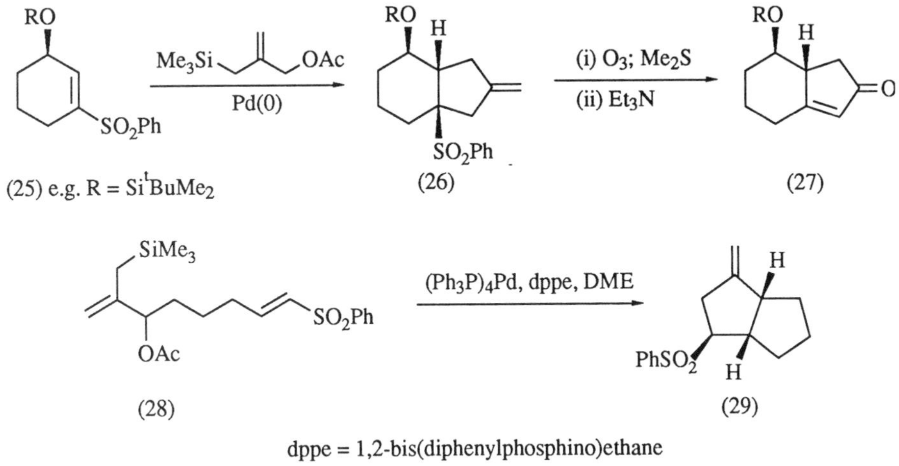

(25) e.g. R = Si^{t}BuMe$_2$ (26)   (27)

(28)   (29)

dppe = 1,2-bis(diphenylphosphino)ethane

Scheme 18

Cycloaddition to vinyl sulphones such as (25) occurs in good yield, the major isomer (26) resulting from attack *trans* to the alkoxy substituent. This result is in contrast to the poor results usually obtained in TMM palladium cycloadditions using cycloalkenones other than cyclopentenone. Subsequent conversion of (26) to (27) completes an effective cyclopentenone annulation. The conversion of (28) to (29) illustrates the power of the intramolecular variant of this reaction. Two carbocyclic rings are constructed in a single step, and with a high degree of stereocontrol.

Finally, Little and Myong have used phenyl vinyl sulphone to trap a 1,3-diyl, generated from a diazo precursor, Scheme 19.[24]

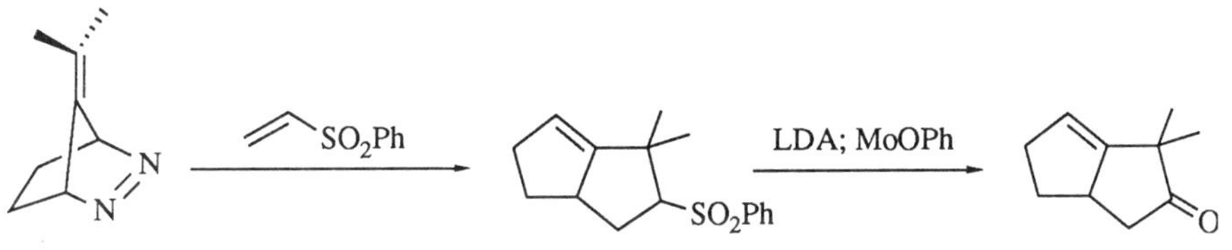

(30)

Scheme 19

The yield in this unusual radical 1,3-cycloaddition, followed by oxidative desulphonylation to give ketone (30), is a respectable 59%.

6.3 [4+2] Cycloadditions

Unsaturated sulphones take part in a host of [4+2] cycloadditions. Most of these reactions involve vinyl sulphones as 2π-components in classical Diels–Alder cycloadditions. Indeed, cycloadditions involving α,β-unsaturated sulphones were carried out as early as 1938 by Alder.[25] However, dienyl sulphones can also act as the 4π-components in this chemistry, for example enabling reverse electron demand Diels–Alder-type reactions.

6.3.1 Sulphones as 2π-Components

The first systematic examination of phenyl vinyl sulphone as an ethylene or terminal alkene equivalent in [4+2] cycloadditions was carried out by Paquette's group.[26] Cycloaddition takes place smoothly with a variety of dienes to give sulphone products, which can be directly desulphonylated, or first alkylated and then desulphonylated, Scheme 20.

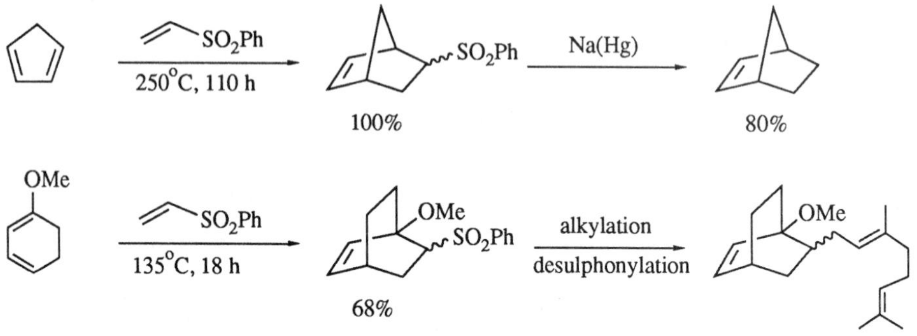

Scheme 20

endo-Selectivity is observed, although this is the result of steric rather than electronic effects. These reactions were found not to be accelerated by Lewis acids.

Regioselective cycloaddition of the unusual thiete 1,1-dioxide (**31**) occurs on treatment with 1,3-diphenylisobenzofuran, Scheme 21.[27]

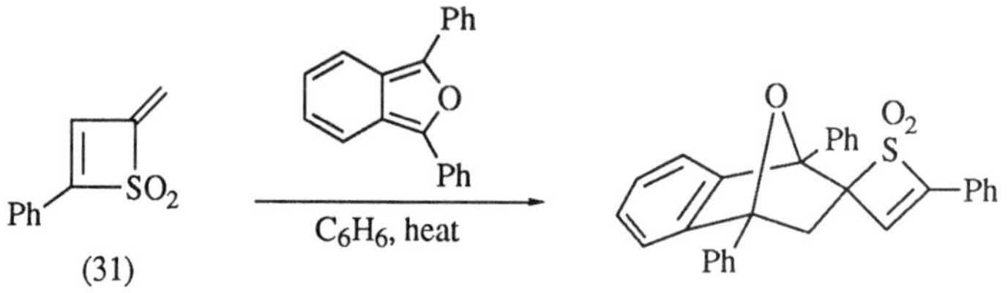

Scheme 21

Preferential reaction occurs on the exocyclic double bond to give a single cycloadduct in high yield. Other reactions, including partial hydrogenation and photodimerisation, also involve selective reaction at the exocyclic double bond.

Heterosubstituted dienes such as Danishefsky's diene are particularly effective in Diels–Alder reactions with vinyl sulphones, e.g. Scheme 22.[28,29]

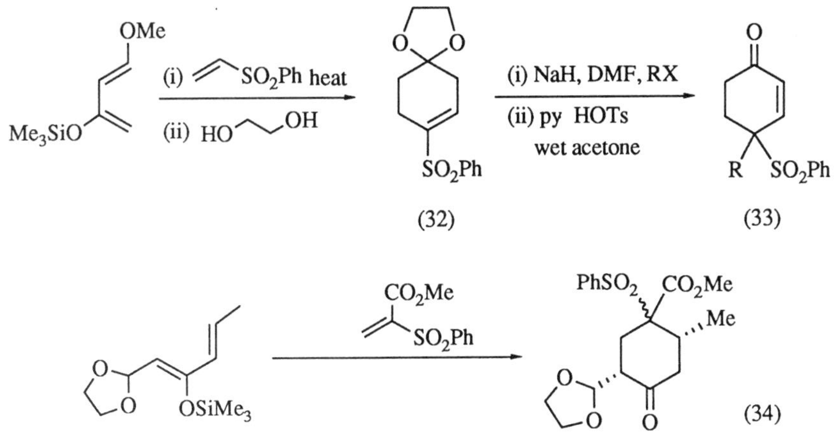

Scheme 22

Cycloadditions such as that leading to (32) have been used to access a range of polysubstituted cyclohexenones.[28] This is effected by regioselective alkylation, followed by either ketal hydrolysis, e.g. to give (33), or desulphonylation followed by hydrolysis. The synthesis of (34) shows how more highly substituted dienes can lead directly into complex functionalised cyclohexanones.[29]

This type of chemistry has been applied to total synthesis, for example to the synthesis of sterpuric acid (35) and 14α-formyl estrone, Scheme 23.[30,31]

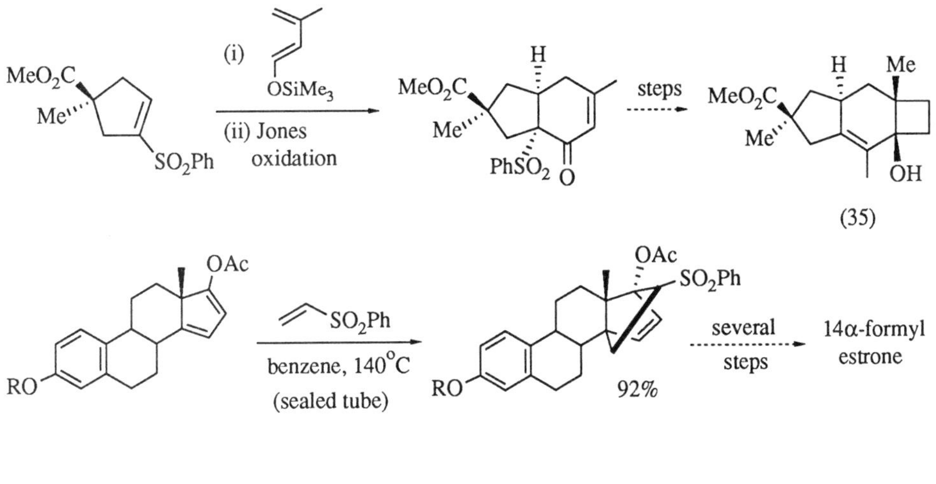

Scheme 23

The reaction of phenyl vinyl sulphone takes places readily with most simple and activated dienes at temperatures ranging from benzene reflux to about 130°C. The use of less reactive or more substituted sulphones usually requires more forcing conditions, although results can still be very good.

Details of the reactivity of various silyl-substituted vinyl sulphones and bis-sulphonyl ethylenes can be found in papers from the groups of Paquette[32] and De Lucchi.[33] The adducts formed using these compounds as dienophiles can undergo subsequent elimination reactions, enabling the starting vinyl sulphones to function as masked acetylene equivalents, e.g. Scheme 24.

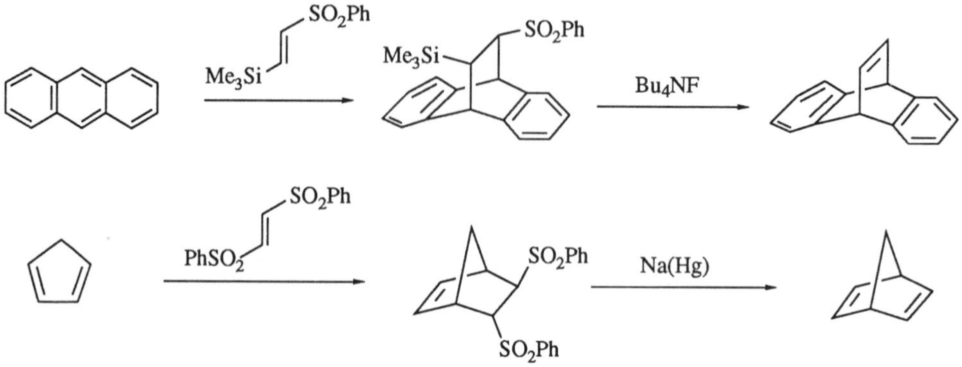

Scheme 24

The cycloadditions of β-nitrovinyl phenyl sulphone are apparently even more facile than those of the bis-sulphonyl derivative and again subsequent elimination can be carried out, this time using Bu_3SnH.[34]

Vinyl sulphones having additional activation in the form of a carbonyl group give good results in cycloadditions. A particularly nice example is the reaction of either (*E*)- or (*Z*)-β-phenylsulphonylacrylates with dienes, which occur with contrasting regiocontrol, Scheme 25.[35]

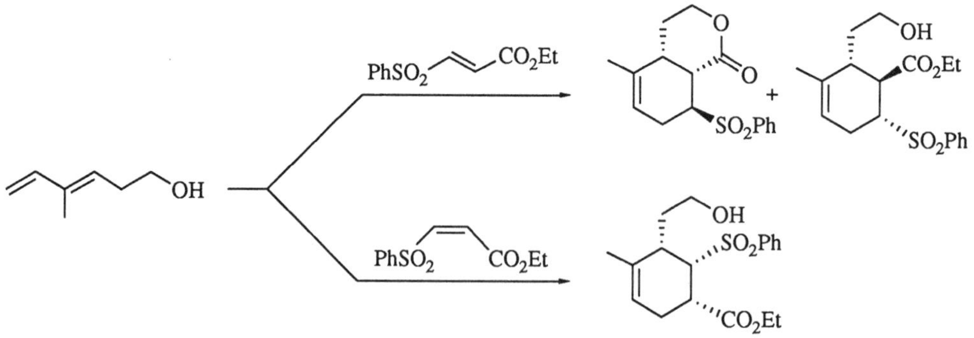

Scheme 25

In each of the examples examined the (*E*)-isomer gives the product arising from the CO_2Et group controlling the regiochemistry, whereas in reactions of the (Z)-isomer the sulphone appears to exert the dominating directing effect.

Intramolecular Diels–Alder (IMDA) reactions involving vinyl sulphones are rather rare. Simple examples were carried out by Craig *et al.*, to give sulphonyl-substituted indane and decalin products such as (**36**), Scheme 26.[36]

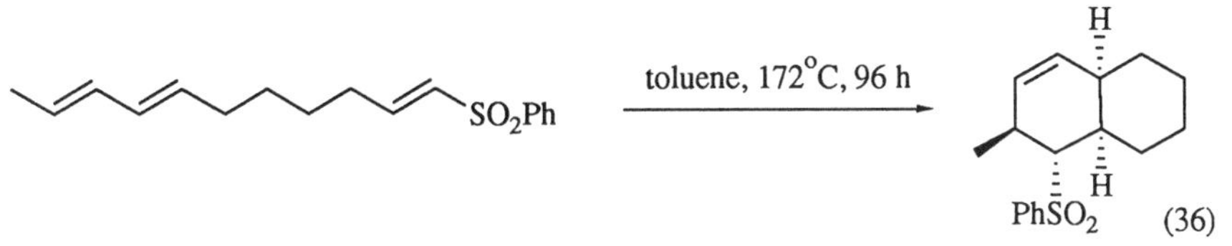

Scheme 26

Other examples of intramolecular cycloadditions using unsaturated sulphones include the use of quinodimethanes and furans as the diene partner, Scheme 27.[37,38]

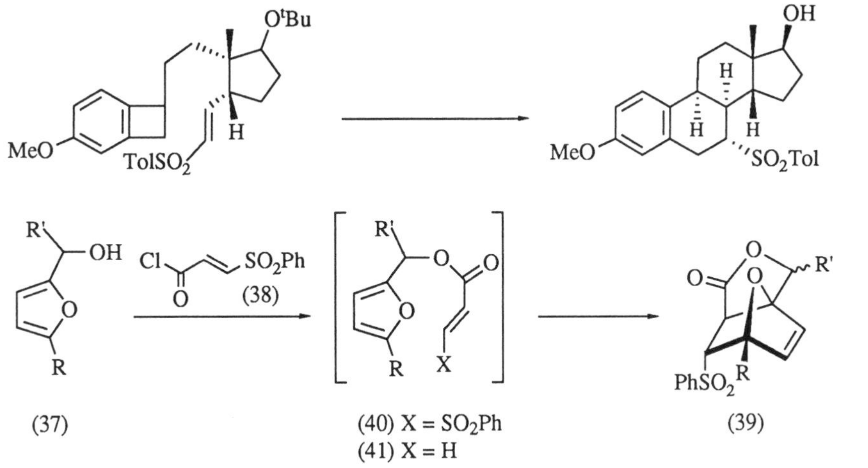

Scheme 27

Remarkably, the reaction of furan (**37**) with (*E*)-2-(phenylsulphonyl)acryloyl chloride (**38**) at room temperature gives the cycloadducts (**39**) directly in good yield. The presumed intermediate acrylate (**40**) X = SO_2Ph was not isolated. By contrast, the acrylate (**41**) X = H could not be cyclised, thus underlining the dramatic acceleration of the cycloaddition afforded by the phenylsulphonyl substituent.

Diastereoselective cycloadditions have been carried out using vinyl sulphones incorporating a homochiral sulphoxide grouping, Scheme 28.[39,40]

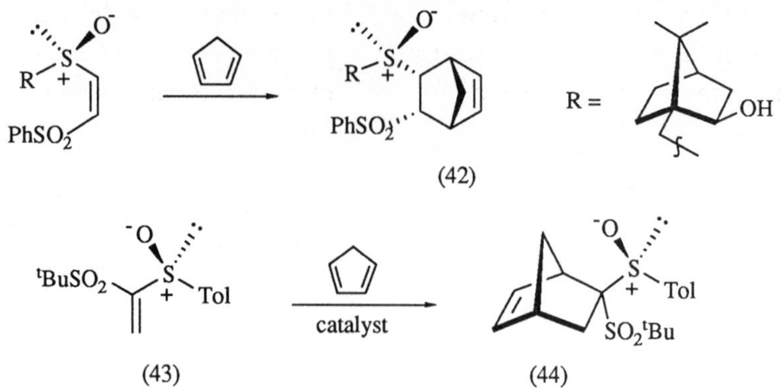

(42)

R =

(43)   (44)

Scheme 28

In both cases one diastereoisomeric product can be isolated in reasonable yield, the bulky sulphone group being *endo* in both (**42**) and (**44**). In the cycloadditions of (**43**) with cyclopentadiene the use of mild Lewis acids such as $ZnBr_2$ or $Eu(fod)_3$ gives best results.[40] Removal of the sulphoxide from both types of cycloadducts is possible to give the optically active norbornadienes.

Diels–Alder reactions of allenyl phenyl sulphone (phenylsulphonylpropadiene) are site selective, regioselective, and often stereoselective. Thus, in cycloadditions with typical dienes reaction occurs on the sulphone-activated double bond to give the anticipated *ortho-* or *para*-type products, e.g. (**45**) and (**46**), Scheme 29.[41]

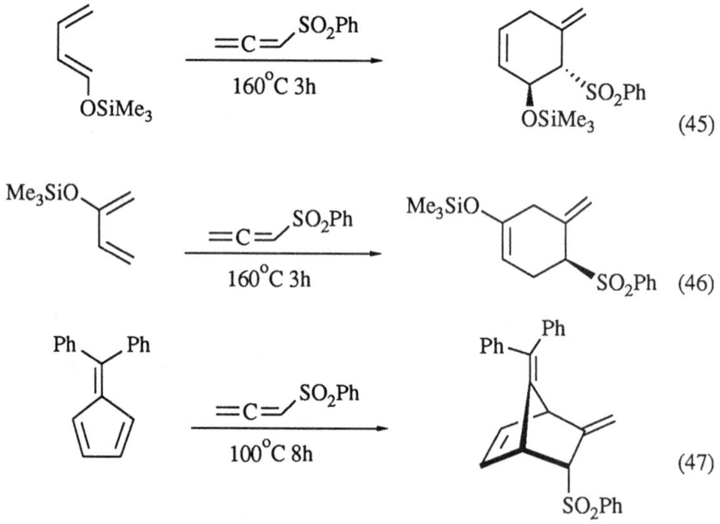

(45)

(46)

(47)

Scheme 29

The regioselectivity is easily understood in terms of the expected LUMO(dienophile)–HOMO(diene) interaction. The frequently observed *endo*-selectivity, e.g. in the formation of (**47**), is apparently a steric effect rather than the result of secondary orbital interactions. Clearly, in some situations the same steric effects can result in apparent *exo*-selectivity, e.g. to give (45).

Cycloadditions with furan have also been explored, as have reactions involving more substituted allenyl sulphones, Scheme 30.[42,43]

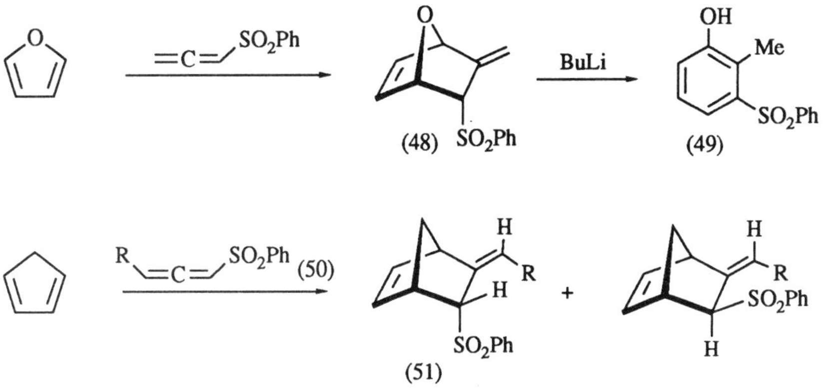

Scheme 30

Treatment of *endo*-adduct (48) with BuLi results in β-elimination to form phenol (49) after protonation.[42] Substituted allenes such as (50) give mainly *endo*-adduct (51) having the (Z)-double-bond geometry.[43]

Alkynyl sulphones readily undergo [4+2] cycloaddition with dienes, e.g. Scheme 31.[44,45]

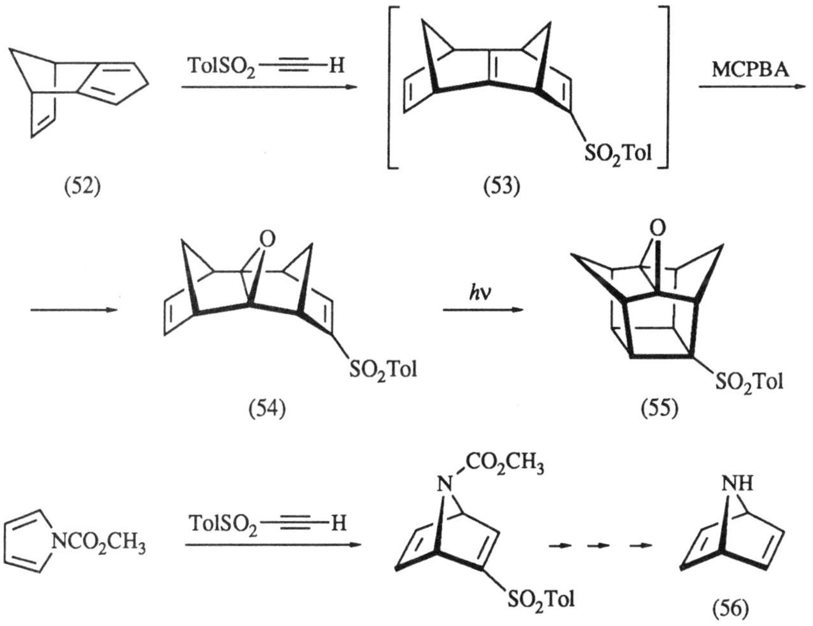

Scheme 31

Reaction of fused cyclopentadiene (**52**) with ethynyl tolyl sulphone, (*p*-tolyl-sulphonyl)acetylene, gives rise to (**53**), which, because of its propensity for air oxidation, was isolated as the derived epoxide (**54**).[44] This provides a very short route to caged compounds such as (**55**) by subsequent intramolecular [2+2] cycloaddition. Similar [4+2] cycloaddition of ethynyl tolyl sulphone to a protected pyrrole provides straightforward access to 7-azanorbornadiene (**56**).

Trifluoromethanesulphonyl alkynes (alkynyl triflones) undergo cycloadditions much faster than ordinary sulphones, Scheme 32.[46]

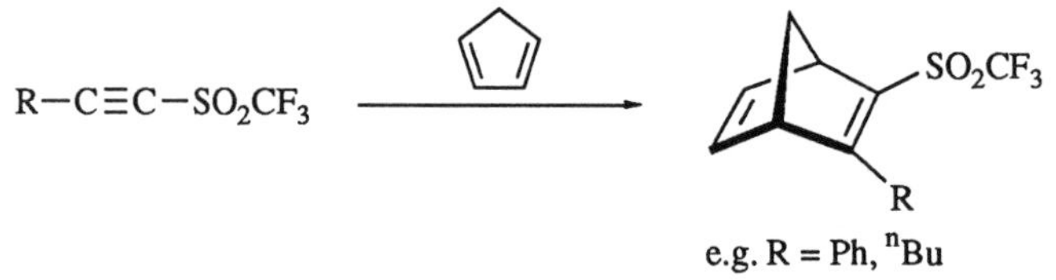

$$R-C{\equiv}C-SO_2CF_3$$

e.g. R = Ph, ^nBu

Scheme 32

In comparative studies using phenylalkynes, the SO_2CF_3 group was found to accelerate the reaction with tetraphenylcyclopentadienone more than a carbonyl or nitrile group. This allows such reactions to be conducted under mild conditions, for example at room temperature. Similar electronic activation is observed with the $SO_2{^nC_4F_9}$ group, but in this case cycloadditions are inhibited by steric effects.[47]

Other doubly activated alkynyl sulphones which undergo smooth [4+2] cycloaddition include ethyl β-phenylsulphonylpropiolate and bis(*tert*-butylsulphonyl)acetylene, Scheme 33.[48,49]

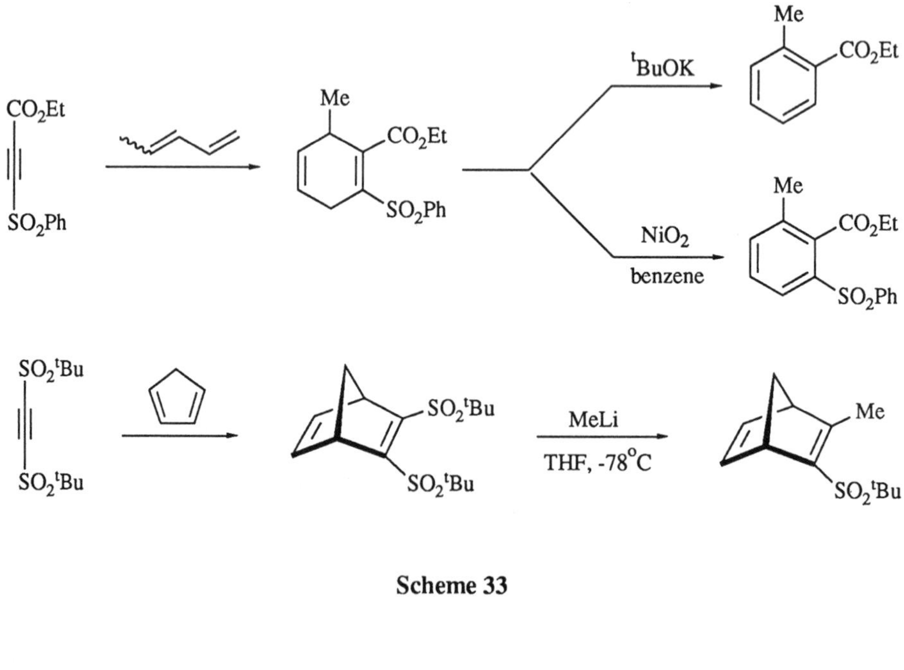

Scheme 33

In each case the authors demonstrated useful further transformations of the initial cycloadducts. Other alkynyl sulphones bearing silicon,[50] tin,[51] or phosphorus[52] substituents have also been examined. These sulphones react in much the same way as the examples already described, although not under such mild conditions.

With aryl phenylethynyl sulphones somewhat anomalous behaviour is observed in reactions with cyclic dienes.[53] Thus, in addition to the expected [4+2] adducts (**57**), these reactions also yield [2+2] cycloadduct (**58**) and fragmentation–addition product (**59**), depending on the particular diene employed, Scheme 34.

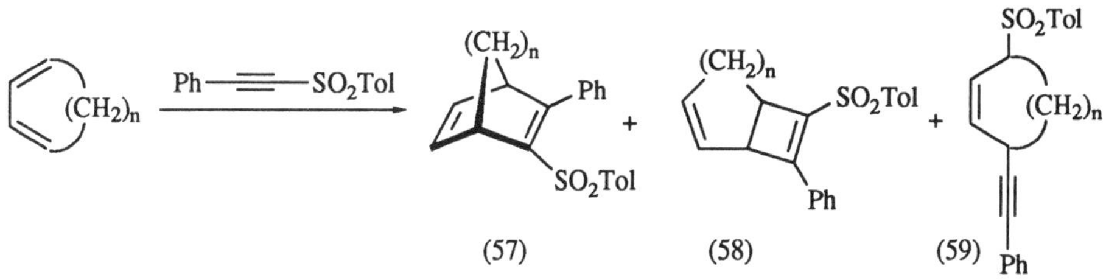

Scheme 34

With cyclopentadiene (n = 1) only the [4+2] adduct (**57**) is obtained. However, with cyclohexadiene and cycloheptadiene (n = 2,3) substantial amounts of [2+2] cycloadduct are also isolated, along with some fragmentation product for n = 3. With cyclooctadiene only (**59**)(n = 4) is obtained. Notice that the [2+2] cycloadducts are not the regioisomers expected, assuming dipolar character in the cycloaddition.

Cycloadditions of a variety of unsaturated sulphones with electron-rich 2-pyridones have been examined by Herdeis and co-workers, Scheme 35.[54]

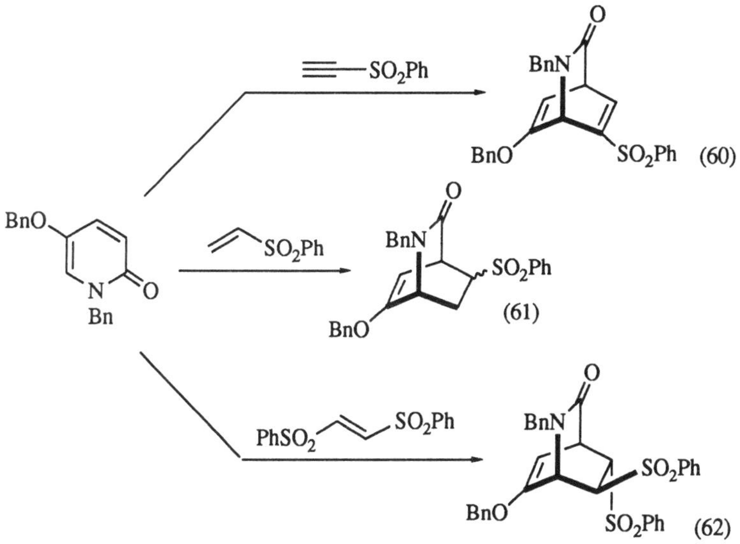

Scheme 35

A switch in regioselectivity is observed on changing the dienophile from the alkynyl sulphone, which gives (60), to phenyl vinyl sulphone, which gives (61). Compounds (60) and (62) can be used to prepare the two ketal vinyl sulphones (63) and (64), which are regioisomeric, Scheme 36.

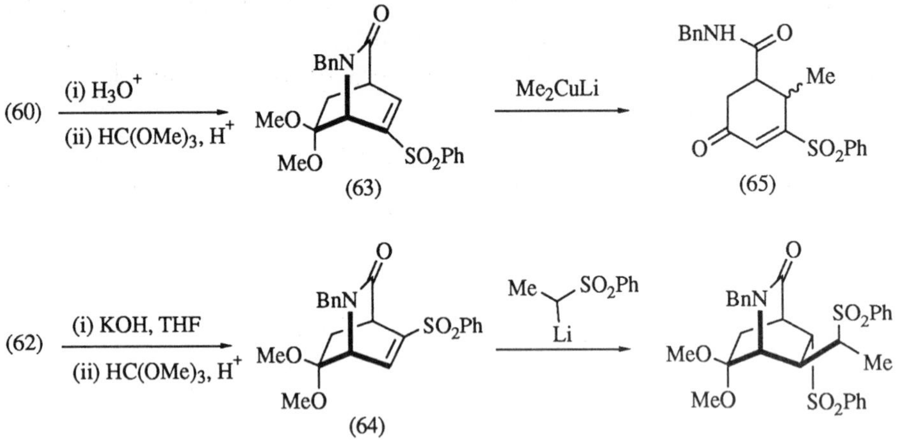

Scheme 36

These compounds react with a range of nucleophiles, including a sulphonyl carbanion, to give addition products. In the case of (63) ring opening occurs with carbon nucleophiles such as Grignard reagents or organocuprates to furnish cyclohexenones such as (65).

The [4+2] cycloadditions of some pyrano[3,4-b]indol-3-ones with doubly activated alkene acceptors, including 1,2-bis(phenylsulphonyl)ethene gives carbazole products, e.g. Scheme 37.[55]

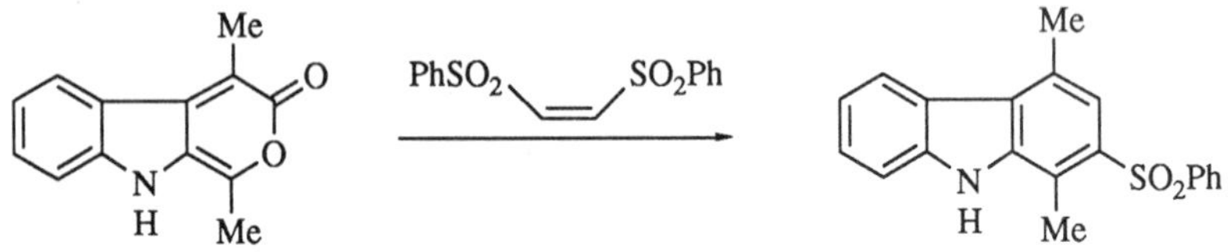

Scheme 37

This 'cycloaddition' is proposed to proceed in a stepwise fashion via zwitterionic or diradical intermediates. In the case of the (Z)-disulphone shown, the first step of the reaction was found to be reversible, resulting in isomerisation of the unsaturated disulphone to the (E)-isomer.

Finally, van Leusen and co-workers have shown that sulphonyl cyanides can participate in [4+2] cycloadditions, Scheme 38.[56]

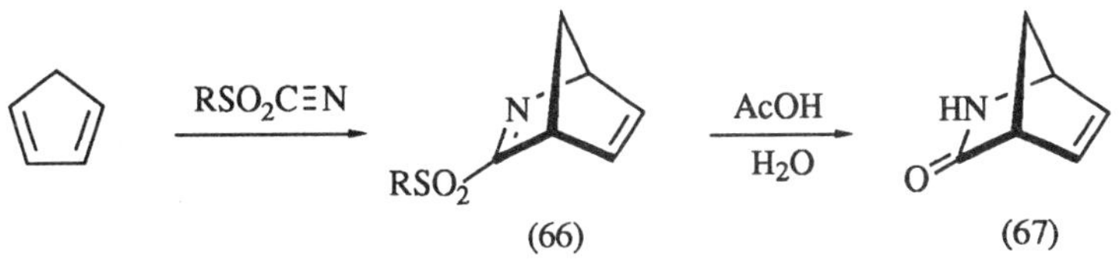

Scheme 38

Reaction occurs at room temperature to furnish the primary cycloadduct (66). This compound is rather sensitive, being easily hydrolysed to give the bicyclic lactam (67).

6.3.2 Sulphones as 4π-Components

Dienyl sulphones are able to serve effectively as the 4π-components in [4+2] cycloadditions with either electron-rich or electron-poor 2π-partners. This reaction was first examined in detail by Inomata et al., using a sulpholene (68) to generate the dienyl sulphone in situ, Scheme 39.[57]

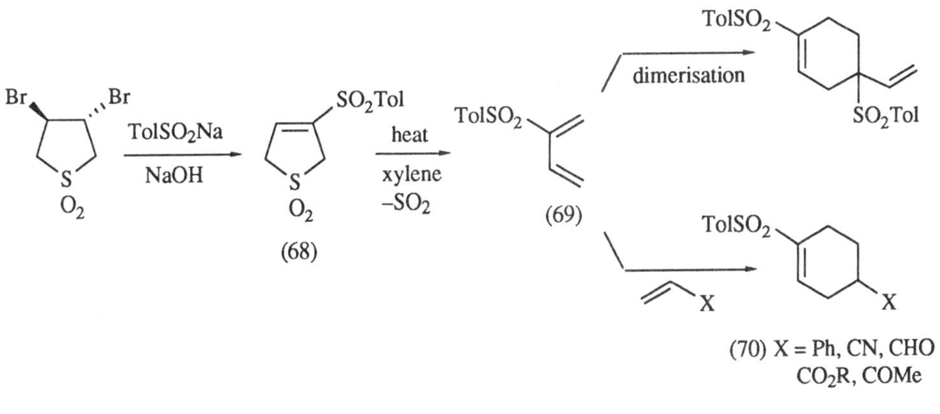

Scheme 39

In the absence of a suitable partner the 2-tolylsulphonylbutadiene (69) simply dimerises. However, with a range of suitable electron-poor 2π-partners, the cycloadducts (70) are formed regioselectively and in high yield. Further study of the dimerisation of substituted dienyl sulphones has shown that in general the reactions are highly regio- and stereoselective, giving para-type endo-products.[58]

Although the simple butadienyl sulphones such as (69) are very prone to dimerisation, it is possible to prepare monomeric diene by brief thermolysis of a dilute solution of an appropriate sulpholene precursor such as (68).[59] This allows the cycloadditions to be studied under conditions in which SO_2 extrusion from a sulpholene would not occur, e.g. Scheme 40.

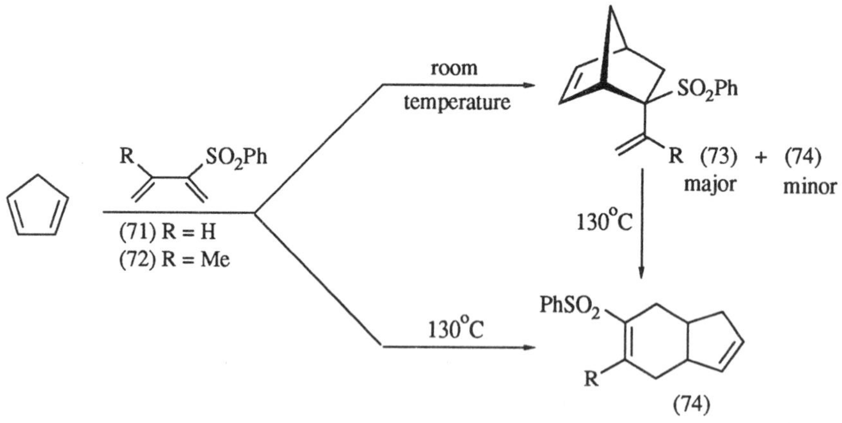

Scheme 40

Reaction of cyclopentadiene with (71), generated *in situ* from the corresponding 3-sulpholene at 130°C, gives (74). In contrast, the use of the free diene (71) at room temperature results in the formation of (73) as the major product, although (74) is still formed to a minor extent. It appears that at room temperature the diene participates as both a 2π-(i.e. a vinyl sulphone) and a 4π-component. The product of vinyl sulphone cycloaddition (73) rearranges to (74) on heating. The formation of (74) at high temperature is therefore probably a result of a mixture of direct diene 4π-cycloaddition and diene 2π-cycloaddition followed by Cope rearrangement. Substituted diene (72) is more stable than the parent compound (71) and participates analogously in cycloaddition reactions. A final point to note is that cycloadducts such as (73) are formed as single stereoisomers, presumably due to favourable secondary orbital interactions.

Studies using dienes bearing both an SPh and an SO_2Ph group show that the sulphenyl group dominates the regioselectivity in cycloadditions with electron-poor dienophiles.[60]

Significant attention has been paid to the potential of dienyl sulphones in reverse electron demand Diels–Alder reactions. Bäckvall and Juntunen reported a number of such cycloadditions, for example using ethyl vinyl ether or a cyclohexanone-derived enamine as the electron-rich 2π-component, Scheme 41.[61]

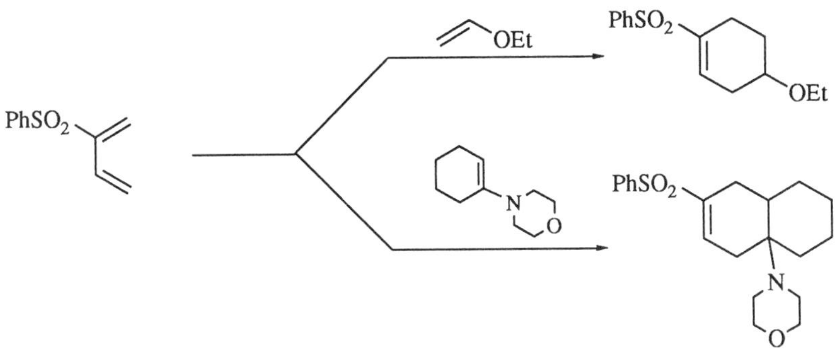

Scheme 41

The usefulness of this type of cycloaddition was further demonstrated by the use of the magnesium salt of indole as 2π-component.[62] This chemistry was then used as a key step in the synthesis of the carbazole alkaloids ellipticine and olivacine.[63] Both 1,3- and 2,3-bis(phenylsulphonyl)butadienes (**75**) and (**76**) can be prepared and are useful 4π-components in imino Diels–Alder reactions, Scheme 42.[64]

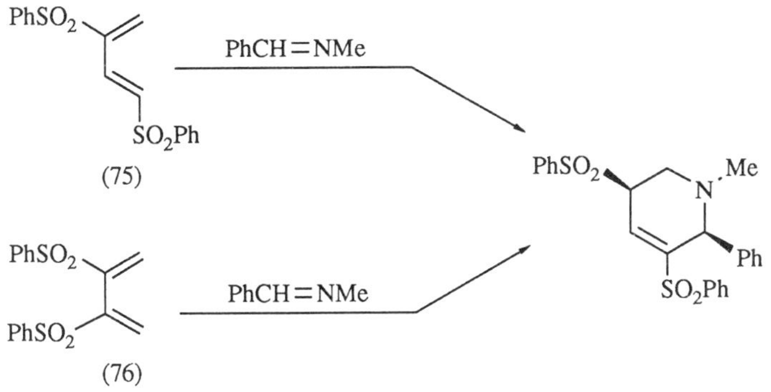

Scheme 42

Remarkably, both types of diene give the same adduct. Evidence points to a rearrangement of 2,3-disubstituted diene (**76**) to the more reactive 1,3-isomer (**75**) prior to cycloaddition. More highly substituted dienes and other dienophiles were also examined.

Three other examples of intermolecular Diels–Alder reactions using dienyl sulphones are grouped in Scheme 43.[65–67]

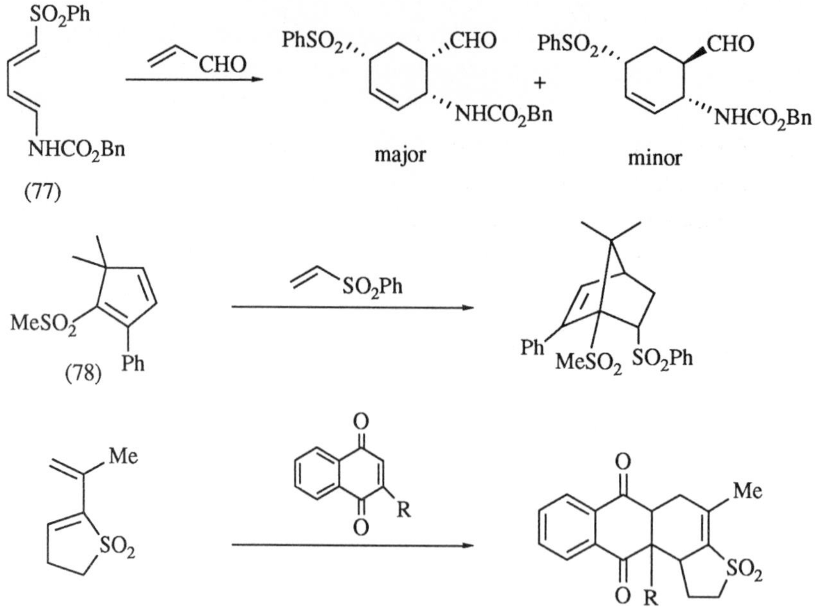

Scheme 43

Again *endo*-products predominate, for example starting from (**77**) and (**78**). In the reactions of (**77**) the nitrogen substituent directs the regiochemistry of the cycloaddition.

Intramolecular variants of this type of cycloaddition have also been examined, e.g. Scheme 44.[68,69]

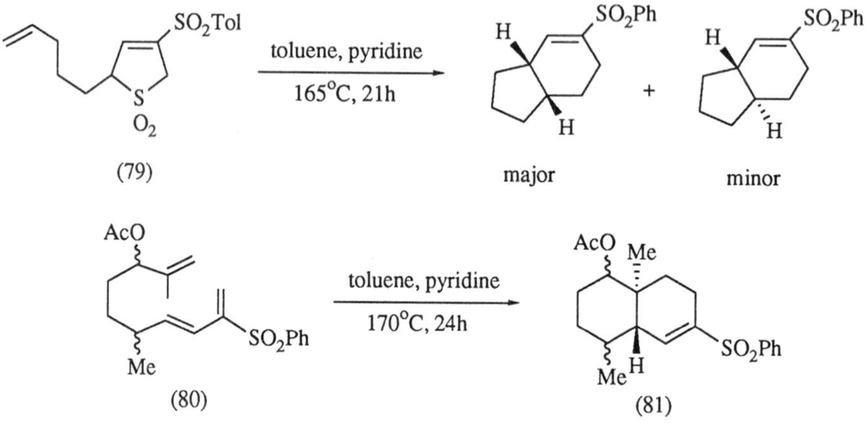

Scheme 44

Construction of (**79**) is very straightforwardly achieved by alkylation of a simple phenylthio-substituted 3-sulpholene followed by oxidation. Extrusion of SO_2 from (**79**), followed by cycloaddition, leads to two cycloadducts as shown. Reactions of homologues and more substituted derivatives can be carried out in a similar fashion. Thermolysis of sulphone (**80**) results in the formation of decalin (**81**) as a mixture of diastereoisomers having the *trans*-fused ring junction.

Dienyl sulphones have been employed in cycloadditions with 2π–partners bearing chiral auxiliaries, e.g. Scheme 45.[70,71]

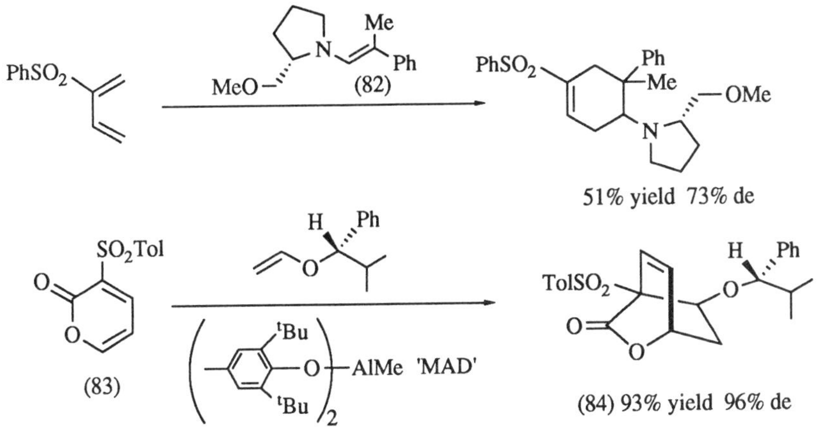

Scheme 45

In reactions involving 2-phenylsulphonylbutadiene, enamines such as (**82**) give better yields and diastereoselectivities than chiral vinyl ethers.[70] Full assignment of the configurations of the newly formed asymmetric centres, as well as development of methods for recovery of the chiral pyrrolidine auxiliary, await further studies. The cycloaddition between chiral vinyl ethers and sulphonyl pyrone (**83**) is high yielding and highly diastereoselective.[71] The degree of asymmetric induction in the cycloaddition is considerably improved when carried out in the presence of Yamamoto's 'MAD' Lewis acid. Cycloadduct (**84**) was converted to a known A-ring precursor for 1α,25-dihydroxyvitamin D3 by a route involving sacrificial removal of the chiral benzylic ether auxiliary.

Chapter 6 References

1. O. De Lucchi and L. Pasquato, *Tetrahedron*, **1988**, *44*, 6755.
2. L. A. Paquette, R. W. Houser, and M. Rosen, *J. Org. Chem.*, **1970**, *35*, 905.
3. J. J. Eisch, J. E. Galle, and L. E. Hallenbeck, *J. Org. Chem.*, **1982**, *47*, 1608; see also J. J. Eisch, L. E. Hallenbeck, and M. A. Lucarelli, *J. Org. Chem.*, **1991**, *56*, 4095.
4. O. De Lucchi, L. Pasquato, and G. Modena, *Tetrahedron Lett.*, **1984**, *25*, 3643.
5. M. A. A. M. El Tabei, N. V. Kirby, and S. T. Reid, *Tetrahedron Lett.*, **1980**, *21*, 565.
6. N. V. Kirby and S. T. Reid, *J. Chem. Soc., Chem. Commun.*, **1980**, 150; see also M. S. El Faghi El Amoudi, P. Geneste, and J. L. Olive, *J. Org. Chem.*, **1981**, *46*, 4258; M. S. El Faghi El Amoudi, P. Geneste, and J. L. Olive, *Nouv. J. Chim.*, **1981**, 251.
7. J. Delaunay, G. Mabon, A. Orliac, and J. Simonet, *Tetrahedron Lett.*, **1990**, *31*, 667; see also D. N. Harpp and C. Heitner, *J. Org. Chem.*, **1970**, *35*, 3256.
8. K. Kanematsu, N. Sugimoto, M. Kawaoka, S. Yeo, and M. Shiro, *Tetrahedron Lett.*, **1991**, *32*, 1351.
9. A. K. Musser and P. L. Fuchs, *J. Org. Chem.*, **1982**, *47*, 3121.
10. I. W. J. Still and T. S. Leong, *Tetrahedron Lett.*, **1979**, 1097; L. A. Paquette and L. D. Wise, *J. Am. Chem. Soc.*, **1967**, *89*, 6659; see also N. Ishibe, K. Hashimoto, and M. Sunami, *J. Org. Chem.*, **1974**, *39*, 103.
11. A. Padwa and M. W. Wannamaker, *J. Chem. Soc., Chem. Commun.*, **1987**, 1742; see also P. G. De Benedetti, C. De Micheli, R. Gandolfi, P. Gariboldi, and A. Rastelli, *J. Org. Chem.*, **1980**, *45*, 3646.
12. A. Padwa and M. W. Wannamaker, *Tetrahedron*, **1990**, *46*, 1145.
13. A. Padwa, S. P. Craig, U. Chiacchio, and D. N. Kline, *J. Org. Chem.*, **1988**, *53*, 2232.
14. A. Padwa, U. Chiacchio, D. N. Kline, and J. Perumattam, *J. Org. Chem.*, **1988**, *53*, 2238; A. Padwa, S. P. Carter, U. Chiacchio, and D. N. Kline, *Tetrahedron Lett.*, **1986**, *27*, 2683.
15. A. Padwa, D. N. Kline, and B. H. Norman, *Tetrahedron Lett.*, **1988**, *29*, 265.
16. A. Padwa and P. E. Yeske, *J. Am. Chem. Soc.*, **1988**, *110*, 1617.
17. P. Armstrong, R. Grigg, and W. J. Warnock, *J. Chem. Soc., Chem. Commun.*, **1987**, 1325; R. Grigg, M. J. Dorrity, F. Heaney, J. F. Malone, S. Rajviroongit, V. Sridharan, and S. Surendrakumar, *Tetrahedron*, **1991**, *47*, 8297; see also A. Padwa and B. H. Norman, *Tetrahedron Lett.*, **1988**, *29*, 2417.
18. M. Barzaghi, P. L. Beltrame, P. D. Croce, P. D. Buttero, E. Licandro, S. Maiorana, and G. Zecchi, *J. Org. Chem.*, **1983**, *48*, 3807; see also A. Bened, R. Durand, D. Pioch, P. Geneste, J-P. Declercq, G. Germain, J. Rambaud, D. Roques, C. Guimon, and G. P. Guillouzo, *J. Org. Chem.*, **1982**, *47*, 2461.

19. P. D. Croce, C. La Rosa, and G. Zecchi, *J. Chem. Soc., Perkin Trans. 1*, **1985**, 2621.

20. P. G. De Benedetti, S. Quartieri, A. Rastelli, M. De Amici, C. De Micheli, R. Gandolfi, and P. Gariboldi, *J. Chem. Soc., Perkin Trans. 2*, **1982**, 95; see also R. Annunziata, M. Cinquini, F. Cozzi, P. Giaroni, and L. Raimondi, *Tetrahedron:Asymmetry*, **1990**, *1*, 251.

21. D. A. Barr, G. Donegan, and R. Grigg, *J. Chem. Soc., Perkin Trans. 1*, **1989**, 1550.

22. T. Takahashi, T. Hagi, K. Kitano, Y. Takeuchi, and T. Koizumi, *Chem. Lett.*, **1989**, 593.

23. B. M. Trost, P. Seoane, S. Mignani, and M. Acemoglu, *J. Am. Chem. Soc.*, **1989**, *111*, 7487; B. M. Trost and D. M. T. Chan, *J. Am. Chem. Soc.*, **1982**, *104*, 3733.

24. R. D. Little and S. O. Myong, *Tetrahedron Lett.*, **1980**, *21*, 3339.

25. K. Alder, H. F. Rickert, and E. Windemuth, *Chem. Ber.*, **1938**, *71*, 2451.

26. R. V. C. Carr and L. A. Paquette, *J. Am. Chem. Soc.*, **1980**, *102*, 853.

27. L. A. Paquette and M. Rosen, *J. Org. Chem.*, **1968**, *33*, 3027.

28. W. A. Kinney, G. D. Crouse, and L. A. Paquette, *J. Org. Chem.*, **1983**, *48*, 4986.

29. R. Hirsenkorn and R. R. Schmidt, *Liebigs Ann. Chem.*, **1990**, 883.

30. L. A. Paquette, H-S. Lin, B. P. Gunn, and M. J. Coghlan, *J. Am. Chem. Soc.*, **1988**, *110*, 5818; K. J. Moriarty, C-C. Shen, and L. A. Paquette, *Synlett*, **1990**, *1*, 263.

31. J. R. Bull and R. I. Thomson, *J. Chem. Soc., Chem. Commun.*, **1986**, 451.

32. L. A. Paquette and R. V. Williams, *Tetrahedron Lett.*, **1981**, *22*, 4643; R. V. C. Carr, R. V. Williams, and L. A. Paquette, *J. Org. Chem.*, **1983**, *48*, 4976.

33. O. De Lucchi, V. Lucchini, L. Pasquato, and G. Modena, *J. Org. Chem.*, **1984**, *49*, 596; O. De Lucchi, V. Lucchini, M. Zamai, G. Modena, and G. Valle, *Can. J. Chem.*, **1984**, *62*, 2487; L. A. Paquette, H. Kunzer, K. E. Green, O. De Lucchi, G. Licini, L. Pasquato, and G. Valle, *J. Am. Chem. Soc.*, **1986**, *108*, 3453; see also O. De Lucchi, D. Fabbri, and V. Lucchini, *Tetrahedron*, **1992**, *48*, 1485, and references therein.

34. N. Ono, A. Kamimura, and A. Kaji, *J. Org. Chem.*, **1986**, *51*, 2139; N. Ono, A. Kamimura, and A. Kaji, *Tetrahedron Lett.*, **1986**, *27*, 1595.

35. A. D. Buss, G. C. Hirst, and P. J. Parsons, *J. Chem. Soc., Chem. Commun.*, **1987**, 1836.

36. D. Craig, D. A. Fischer, O. Kemal, and T. Plessner, *Tetrahedron Lett.*, **1988**, *29*, 6369; D. Craig, D. A. Fischer, O. Kemal, A. Marsh, T. Plessner, A. M. Z. Slawin, and D. J. Williams, *Tetrahedron*, **1991**, *47*, 3095.

37. T. Kametani, M. Aizawa, and H. Nemoto, *Tetrahedron*, **1981**, *37*, 2547.

38. M. E. Jung and V. C. Truc, *Tetrahedron Lett.*, **1988**, *29*, 6059; see also L. Strekowski, S. Kong, and M. A. Battiste, *J. Org. Chem.*, **1988**, *53*, 901.

39. O. De Lucchi, C. Marchioro, G. Valle, and G. Modena, *J. Chem. Soc., Chem. Commun.*, **1985**, 878.

40. R. Lopez and J. C. Carretero, *Tetrahedron:Asymmetry*, **1991**, *2*, 93.

41. K. Hayakawa, H. Nishiyama, and K. Kanematsu, *J. Org. Chem.*, **1985**, *50*, 512.

42. A. J. Guildford and R. W. Turner, *J. Chem. Soc., Chem. Commun.*, **1983**, 466; see also O. Arjona, R. Fernandez de la Pradilla, A. Mallo, J. Plumet, and A. Viso, *Tetrahedron Lett.*, **1990**, *31*, 1475.

43. G. Barbarella, M. Cinquini, and S. Colonna, *J. Chem. Soc., Perkin Trans. 1*, **1980**, 1646.

44. L. A. Paquette, J. W. Fischer, A. R. Browne, and C. W. Doecke, *J. Am. Chem. Soc.*, **1985**, *107*, 686.

45. H-J. Altenbach, B. Blech, J. A. Marco, and E. Vogel, *Angew. Chem., Int. Ed. Engl.*, **1982**, *21*, 778.

46. M. Hanack, B. Wilhelm, and L. R. Subramanian, *Synthesis*, **1988**, 592; R. S. Glass and D. L. Smith, *J. Org. Chem.*, **1974**, *39*, 3712.

47. M. Hanack and F. Massa, *Tetrahedron Lett.*, **1977**, 661.

48. M. Shen and A. G. Shultz, *Tetrahedron Lett.*, **1981**, *22*, 3347; see also P. Rajakumar and A. Kannan, *J. Chem. Soc., Chem. Commun.*, **1989**, 154.

49. A. Riera, M. Marti, A. Moyano, M. A. Pericas, and J. Santamaria, *Tetrahedron Lett.*, **1990**, *31*, 2173.

50. J. A. Kloek, *J. Org. Chem.*, **1981**, *46*, 1951; R. V. Williams and C-L. A. Sung, *J. Chem. Soc., Chem. Commun.*, **1987**, 590.

51. Z. Djeghaba, B. Jousseaume, M. Ratier, and J-G. Duboudin, *J. Organomet. Chem.*, **1986**, *304*, 115.

52. R. M. Acheson and P. J. Ansell, *J. Chem. Soc., Perkin Trans. 1*, **1987**, 1275.

53. O. De Lucchi, G. Licini, L. Pasquato, and M. Senta, *Tetrahedron Lett.*, **1988**, *29*, 831; see also O. De Lucchi, G. Licini, L. Pasquato, and M. Senta, *J. Chem. Soc., Chem. Commun.*, **1985**, 1597.

54. C. Herdeis and C. Hartke, *Synthesis*, **1988**, 76; C. Herdeis and C. Hartke-Karger, *Arch. Pharm. (Weinheim)*, **1990**, *323*, 937; C. Herdeis and C. Hartke, *Heterocycles*, **1989**, *29*, 287.

55. U. Pindur and H. Erfanian-Abdoust, *Heterocycles*, **1990**, *31*, 1751.

56. J. C. Jagt and A. M. van Leusen, *J. Org. Chem.*, **1974**, *39*, 564; see also A. M. van Leusen and J. C. Jagt, *Tetrahedron Lett.*, **1970**, 971; P. Bayard, F. Sainte, R. Beaudegnies, and L. Ghosez, *Tetrahedron Lett.*, **1988**, *29*, 3799.

57. K. Inomata, M. Kinoshita, H. Takemoto, Y. Murata, and H. Kotake, *Bull. Chem. Soc. Jpn.*, **1978**, *51*, 3341; see also T. Cuvigny, C. Herve du Penhoat, and M. Julia, *Tetrahedron*, **1986**, *42*, 5329.

58. H. M. R. Hoffmann, A. Weichert, A. M. Z. Slawin, and D. J. Williams, *Tetrahedron,* **1990**, *46*, 5591. The dimerisation of 2-phenylsulphonylbuta-1,3-diene is not so regioselective, see T. Chou, S. C. Hung, and H-H. Tso, *J. Org. Chem.,* **1987**, *52*, 3394.

59. T. Chou and S-C. Hung, *J. Org. Chem.,* **1988**, *53*, 3020.

60. S-S. P. Chou and D-J. Sun, *J. Chem. Soc., Chem. Commun.,* **1988**, 1176.

61. J. E. Bäckvall and S. K. Juntunen, *J. Am. Chem. Soc.,* **1987**, *109*, 6396.

62. J. E. Bäckvall, N. A. Plobeck, and S. K. Juntunen, *Tetrahedron Lett.,* **1989**, *30*, 2589.

63. J. E. Bäckvall and N. A. Plobeck, *J. Org. Chem.,* **1990**, *55*, 4528.

64. A. Padwa, Y. Gareau, B. Harrison, and B. H. Norman, *J. Org. Chem.,* **1991**, *56*, 2713; A. Padwa, B. Harrison, S. S. Murphree, and P. E. Yeske, *J. Org. Chem.,* **1989**, *54*, 4232.

65. L. E. Overman, C. B. Petty, T. Ban, and G. T. Huang, *J. Am. Chem. Soc.,* **1983**, *105*, 6335.

66. J. A. Miller and G. Mustafa Ullah, *J. Chem. Res. (S),* **1988**, 350.

67. T. G. Tolstikova, V. A. Davydova, E. E. Shults, G. F. Vafina, G. M. Safarova, F. A. Zarudii, D. N. Lazareva, and G. A. Tolstikov, *Khim.-Farm. Zh. SSSR,* **1990**, *24*, 27.

68. S-S. P. Chou and S-J. Wey, *J. Org. Chem.,* **1990**, *55*, 1270.

69. A. Weichert and H. M. R. Hoffmann, *J. Chem. Soc., Perkin Trans. 1,* **1990**, 2154.

70. J-E. Bäckvall and F. Rise, *Tetrahedron Lett.,* **1989**, *30*, 5347.

71. G. H. Posner and C. M. Kinter, *J. Org. Chem.,* **1990**, *55*, 3967; G. M. Posner and D. G. Wettlaufer, *Tetrahedron Lett.,* **1986**, *27*, 667. For related cycloadditions of sulphonylpyridones, see G. H. Posner and C. Switzer, *J. Org. Chem.,* **1987**, *52*, 1642.

CHAPTER 7

Carbon–Carbon Double-Bond Formation
by Sulphone Elimination

This chapter groups together three important types of sulphone reaction which involve desulphonylation with concomitant carbon–carbon double-bond formation. These are the Julia olefination process (i.e. the reductive elimination of derivatives of β-hydroxy sulphones), the eliminations of other types of β-substituted sulphones and the Ramberg–Bäcklund reaction.

The Julia reaction is by far the most important type of sulphone elimination process. The synthetic significance of this reaction is reflected by its widespread use in the total synthesis of natural products. The description of the Julia reaction, including its well-studied stereochemical characteristics, synthetic applications and operational variants thus warrants a separate section of this chapter. The less-well explored eliminations of other β-substituted sulphones comprise Section 7.2, whilst the last section of this chapter discusses the Ramberg–Bäcklund reaction. This last process differs greatly from the other chemistry in this chapter but is included at this point in order to group material under the broad banner of sulphone-mediated olefinations.

7.1 The Julia Olefination Reaction

The reductive elimination of β-hydroxy sulphones and derivatives thereof, first reported by Julia and Paris, has come to be known as the Julia reaction.[1] The Julia olefination procedure is outlined in general form in Scheme 1. Firstly, an α-sulphonyl carbanion is reacted with an aldehyde or ketone. The metal alkoxide adduct (1) is then usually derivatised *in situ* to give (3), where R[4] is COMe, COPh or SO_2Me. Treatment of the derivative (3) with sodium amalgam then produces the alkene product (4). As indicated, alternative protocols may be employed to achieve the same overall result. Thus, in exceptional cases, direct reductive elimination of the β-hydroxy sulphone (2) is effective (see for example Scheme 10).

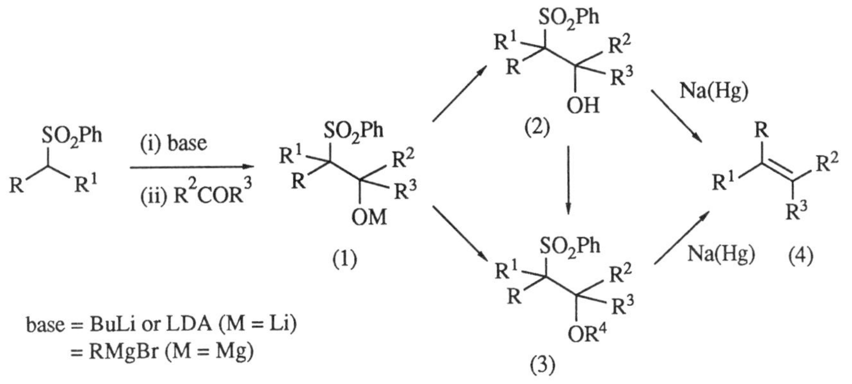

$$
R \overset{SO_2Ph}{\underset{R^1}{\bigtriangleup}} \quad \xrightarrow[\text{(ii) } R^2COR^3]{\text{(i) base}} \quad R \overset{SO_2Ph}{\underset{R}{\underset{OM}{\bigtriangleup}}} \overset{R^1}{\underset{R^3}{\bigtriangleup}} R^2
$$

(1)

$$
R \overset{SO_2Ph}{\underset{R}{\underset{OH}{\bigtriangleup}}} \overset{R^1}{\underset{R^3}{\bigtriangleup}} R^2
$$

(2)

$$
R \overset{SO_2Ph}{\underset{R}{\underset{OR^4}{\bigtriangleup}}} \overset{R^1}{\underset{R^3}{\bigtriangleup}} R^2
$$

(3)

Na(Hg)

$$
R^1 \overset{R}{\diagup} = \overset{R^2}{\underset{R^3}{\diagdown}}
$$

Na(Hg) (4)

base = BuLi or LDA (M = Li)
     = RMgBr (M = Mg)

Scheme 1

This direct method is rarely as efficient as elimination of the derivative (**3**). Another alternative is to isolate the β-hydroxy sulphone and then derivatise it in a separate step to give (**3**). This method sometimes gives improved overall yields. Examples of these options are discussed below.

In essence the Julia olefination resembles the Wittig and Peterson reactions, in that an α-heterosubstituted carbanion and a carbonyl compound are combined, and a vicinal elimination of two functional groups then occurs to give a double bond. Both types of reaction are therefore connective and regiospecific. The key difference is the need for subsequent manipulations in the Julia procedure, whereas the Wittig elimination occurs *in situ*. In practical terms the disadvantage of the two-stage Julia olefination is somewhat offset by the relative ease of removal of elimination by-products during work-up. As will become evident, an organic chemist will rarely be faced with a choice between a 'normal' non-stabilised Wittig and a Julia process, since the two reactions produce stereochemically complementary results.

Issues of key importance are the scope of the Julia reaction and, most importantly, the overall stereochemical outcome. These features of the chemistry have been addressed by Kocienski and Lythgoe. Indeed, two excellent reviews by Kocienski are available on this subject.[2,3] A number of important features of the reaction emerged from the examples carried out using sulphones (**5**) and (**6**), Scheme 2.[4] Reaction of the lithiosulphone derived from (**5**) with (Z)-dienal (**7**) gives (**8**), and then, after elimination, the triene (**9**), with retention of the original (Z)-alkene geometry. Reductive elimination is effected under conditions modified from the original report of Julia. The use of methanol as solvent at a temperature of -20°C, with ethyl acetate added to improve substrate solubility, gives good results. In later reports the ethyl acetate cosolvent is often replaced by THF.

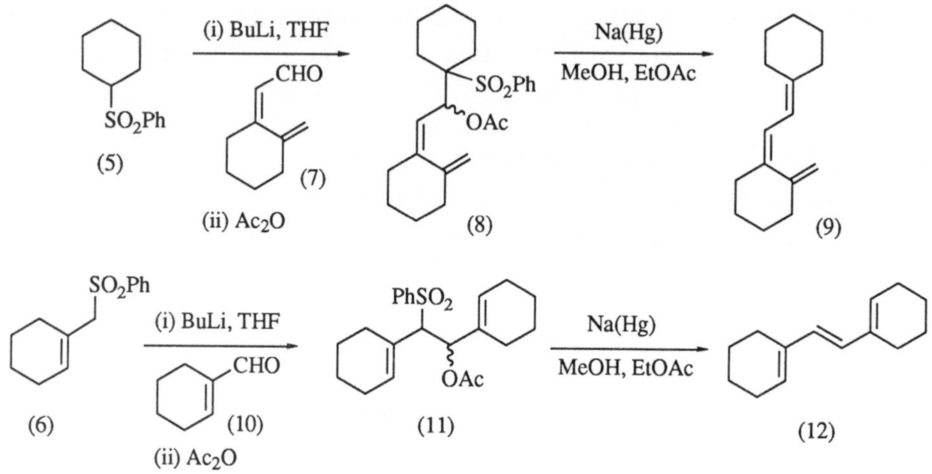

Scheme 2

These conditions minimise unwanted side reactions such as hydrolysis of the acetate or benzoate derivative and elimination of acetic or benzoic acid to form vinyl sulphones. The Kocienski–Lythgoe modification of the Julia process has been very widely accepted by the organic synthesis community, as will be illustrated later in this chapter.

Reaction of the carbanion derived from (6) with enal (10) followed by quenching with acetic anhydride gives (11), which on reductive elimination gives triene (12). Significantly, even though the intermediate acetoxy sulphones are formed as a diastereoisomeric mixture, the final product is formed exclusively as the (E)-isomer (12). It was also confirmed that the separated *threo-* and *erythro-*intermediates (11) each give only (E)-(12). The independence of the product stereochemistry from that of the initial 'aldol' adducts and the pronounced overall (E)-selectivity both contrast dramatically with simple Wittig chemistry. A mechanistic rationale involves electron transfer to the sulphone group, resulting in expulsion of benzenesulphinate anion (or possibly the corresponding $PhSO_2$ radical), to give an intermediate carbanion (13). This can then undergo elimination in an *anti*-periplanar fashion to give the alkene product, Scheme 3.

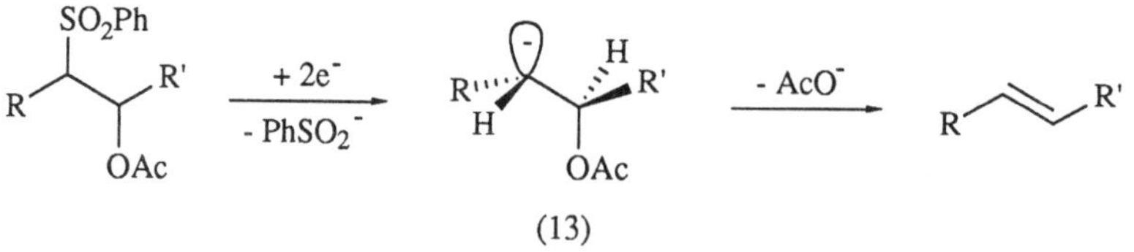

Scheme 3

The stereochemical results are in accord with the proposal that (13) has sufficient time to equilibrate and that elimination occurs with the bulky substituents as far apart as possible. The product isomer ratio therefore reflects the thermodynamic ratio, which in turn is governed by interactions between the R and R' groups in (13). It was to be expected therefore that the isomer ratio of the alkene products should be strongly influenced by the bulk of the two groups of R and R'.[5] From this point of view the very highly stereoselective formation of (12) can be understood as the result of substantial interactions between the two cyclohexenyl groups. Clearly this is an exceptionally favourable case and in general the stereoselectivity is not so high, as indicated by the $(E):(Z)$ ratios for the formation of (14)-(17).

The effect of increased branching is clear to see. The union of simple unbranched partners gives rather modest selectivity of about 4:1. With branching on one of the reaction partners the ratio is 9:1, and with branching on both partners effectively only the (E)-isomer is formed. The formation of conjugated systems, such as (17), using this method follows the same pattern. The rather severe steric interactions in the alternative (Z)-isomer (18), and hence the intermediate leading to it, explains this type of result. For the selective formation of disubstituted (E)-double bonds from such branched partners the Julia reaction is clearly the method of choice.

		$(E) : (Z)$
nC_6H_{13} —— nC_7H_{15}	(14)	4 : 1
nC_6H_{13} ——	(15)	9 : 1
	(16)	100 : 0
	(17)	97 : 3 (central C=C formed)

(18)

Some limitations of the method are evident. Firstly, with more substituted systems, i.e. those leading to trisubstituted double bonds, the retro-aldol reaction of the intermediate β-alkoxy sulphones can be a serious problem (see Section 3.2.3). Whilst this can usually be overcome and trisubstituted alkenes accessed (e.g. (9)), the stereochemical outcome of such reactions has not been firmly established.

Another problem can arise when eliminations are attempted in cyclic systems. If the β-leaving group (AcO⁻, PhCO₂⁻, etc.) cannot easily adopt an *anti*-periplanar orientation with the intermediate carbanionic lone pair then elimination is retarded, e.g. Scheme 4.[3] Thus in cholestane derivative (19) the Julia elimination product is obtained in only low yield and the alternative product of carbanion protonation is also obtained.

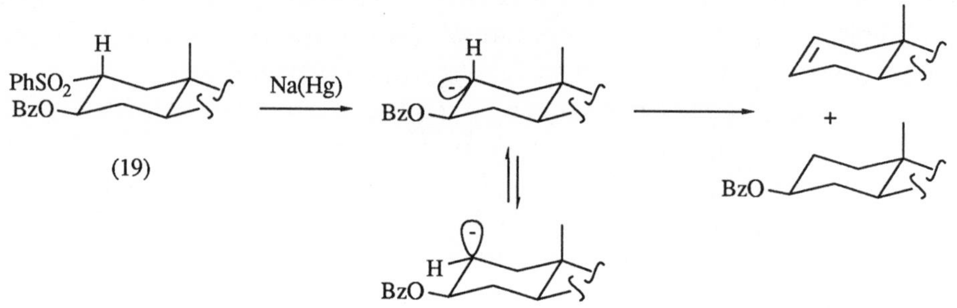

Scheme 4

Since the development of the Julia sequence as an effective means of alkene formation it has been widely used in synthesis. An early example from Lythgoe's group is the synthesis of 1α-hydroxytachysterol, achieved by coupling of (**20**) and (**21**), Scheme 5.[6]

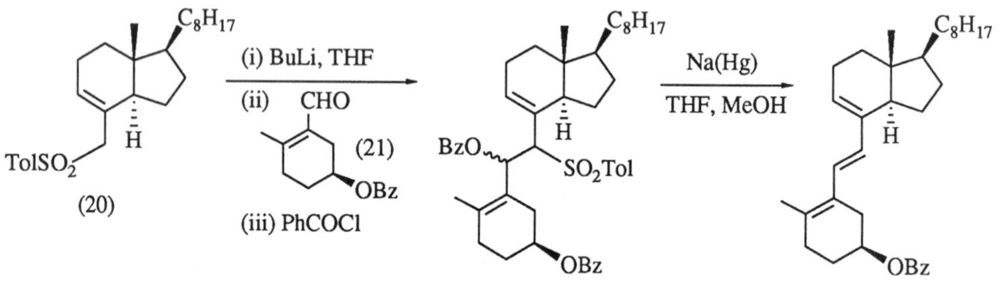

Scheme 5

In this case one diastereoisomeric β-benzoyloxy sulphone crystallised from the mixture and was taken on to the (*E*)-alkene. Another example, described by Kocienski and Todd, enabled the coupling of fragments required for a total synthesis of moenocinol, Scheme 6.[7]

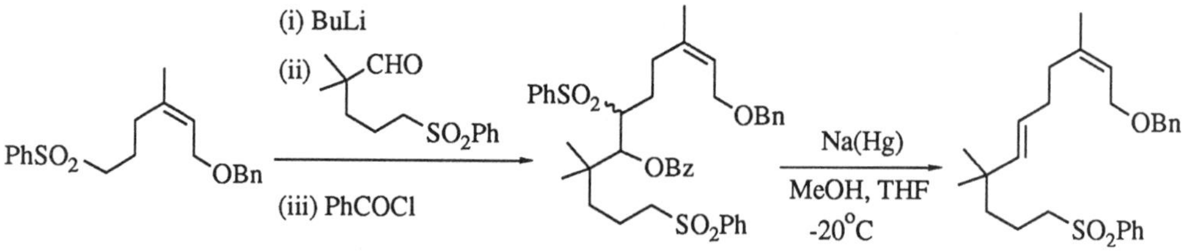

Scheme 6

Remarkably, the sulphone group remote from the benzoyloxy group is retained in the reductive elimination step. The presence of the β-oxygenation thus appears to actually facilitate selective reaction at the neighbouring sulphone group.

In one of a series of papers concerning calciferol and its relatives, Kocienski *et al.* described the silylation of a β-alkoxy sulphone intermediate, Scheme 7.[8]

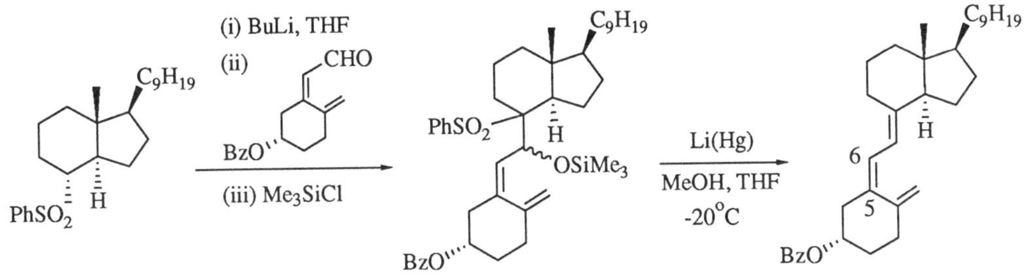

Scheme 7

Whereas the β-silyloxy or acetyloxy sulphones undergo smooth elimination to give the desired product, the corresponding benzoyloxy sulphones surprisingly give a mixture of stereoisomers about the C-5–C-6 double bond. The reasons for this anomalous behaviour are not clear.

More recently the Julia reaction has been used extensively in the total synthesis of complex natural products such as ionophores, and members of the milbemycin and avermectin family.[9] The two examples shown in Scheme 8 demonstrate how the coupling and reductive elimination can be carried out on complex systems bearing varied (and potentially sensitive) functionality.[10,11]

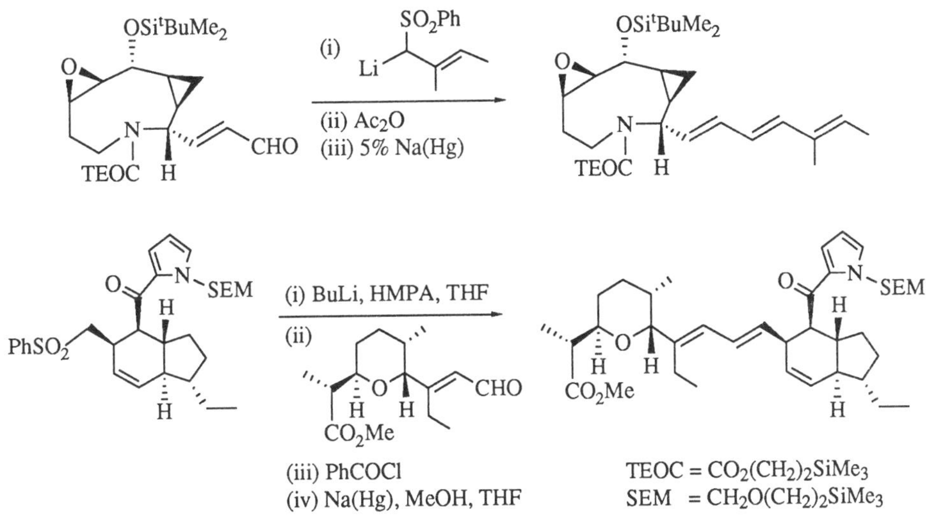

Scheme 8

Both targets are particularly appropriate for the Julia construction, since branching is present on the sulphone partner in each case.

Problems may be encountered in reactions of ketones and some aldehydes due to deprotonation by the sulphonyl carbanion.[12] This problem may be overcome by using the magnesium derivative of the sulphone rather than the more usual lithium derivative. Whilst the magnesium derivatives are usually prepared by brief warming of the sulphone with EtMgBr, similar beneficial effects have been observed by simply adding MgBr$_2$ to the lithiosulphone.[12] Another slight modification to the usual procedure was used in a synthesis of FK-506 and involves reductive elimination of β-trifluoroacetoxy sulphones using lithium naphthalenide.[13] In this case increasing the leaving group ability of the acyloxy group had a beneficial effect on an elimination otherwise plagued by side reactions.

Other reports have described Julia-type reductive elimination of β-acyloxy sulphones obtained by a slightly different tactic than usual. In these examples the required β-hydroxy sulphone intermediates are obtained by reduction of the corresponding ketosulphones, themselves obtained by sulphonyl carbanion acylation, e.g. Scheme 9.[14]

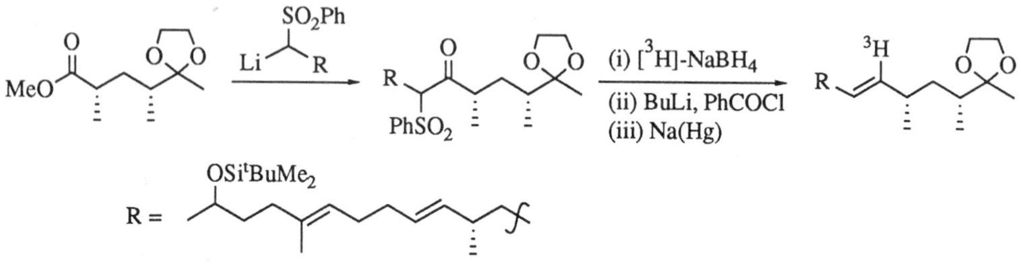

Scheme 9

In this particular example a tritiated product was prepared, leading ultimately to a labelled form of a proposed intermediate in monensin biosynthesis.

Recently, Kende and Mendoza have reported a modification of the Julia olefination which employs imidazolyl sulphones, Scheme 10.[15]

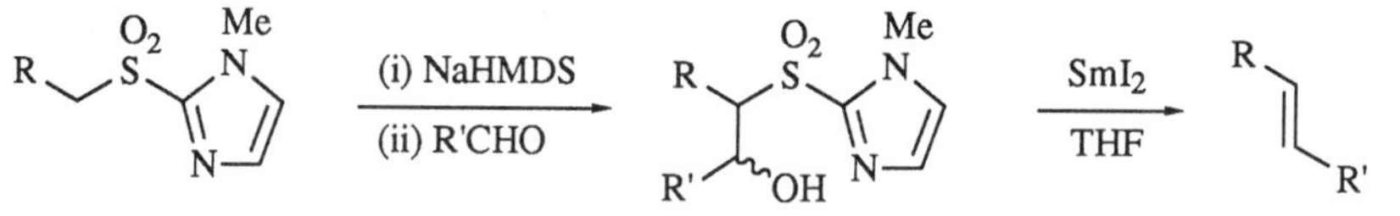

Scheme 10

This method offers the advantage of a smooth, high-yielding elimination, without the need for hydroxyl group derivatisation. That this is not simply a feature of the SmI$_2$ elimination in general is shown by the failure of a β-hydroxy *phenyl* sulphone to undergo elimination under these conditions. Unexpectedly, the elimination of imidazolyl sulphone (**22**) gives only minor amounts of the expected product (**23**), Scheme 11.

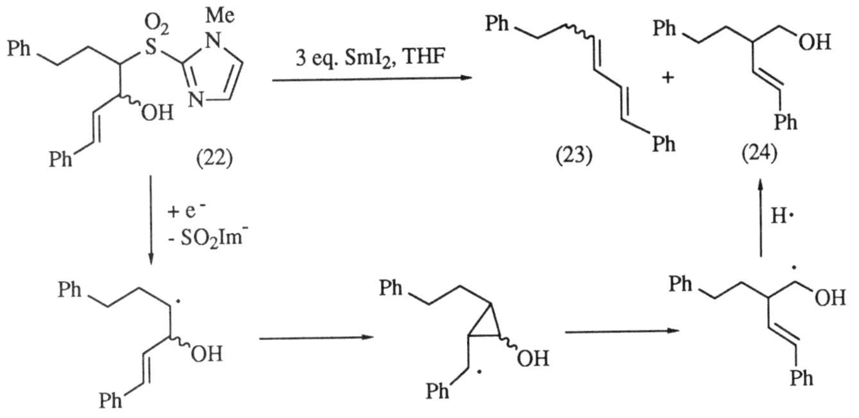

Scheme 11

The rearranged homoallylic alcohol (**24**) is isolated as the major product in 48% yield. This is proposed to be formed by a 3-*exo-trig* cyclisation of an intermediate radical to give a cyclopropanol, which then reopens to give an hydroxyl-stabilised radical, as shown. This type of behaviour may account for the anomalous alkene isomerisation problems described earlier (i.e. Scheme 7). Direct reductive elimination of β-hydroxy sulphones can also be achieved electrochemically.[16] This method is most suitable for the preparation of terminal alkenes, the preparation of internal alkenes proceeding in noticeably lower yield.

A report by Barton *et al.* describes a variant of the Julia reaction involving a free-radical method for the vicinal elimination, Scheme 12.[17]

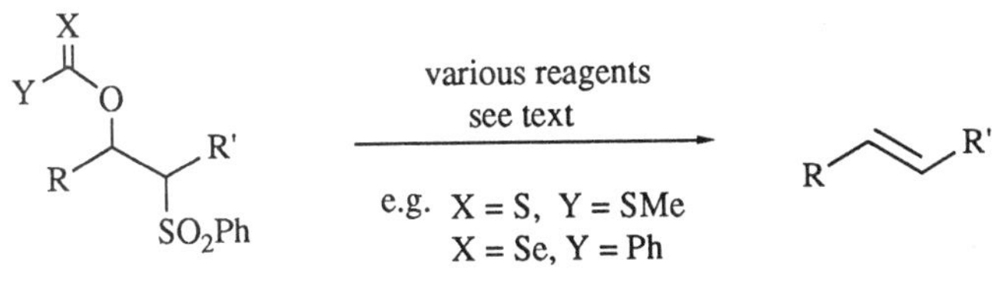

Scheme 12

Transformation of the intermediate β-hydroxy sulphone, formed as usual, into either a xanthate, thionocarbonate, or selenobenzoate is then followed by elimination under radical generating conditions. Four different types of conditions were examined, typically using either an acyl derivative of *N*-hydroxy-2-thiopyridone, or diphenylsilane in combination with Et_3B/O_2, $(PhCO_2)_2$ or AIBN, to generate the radicals required to effect elimination.

Finally, Hsiao and Shechter have utilised a more conventional Julia reaction as a means of obtaining allylsilanes of general structure (**25**), Scheme 13.[18]

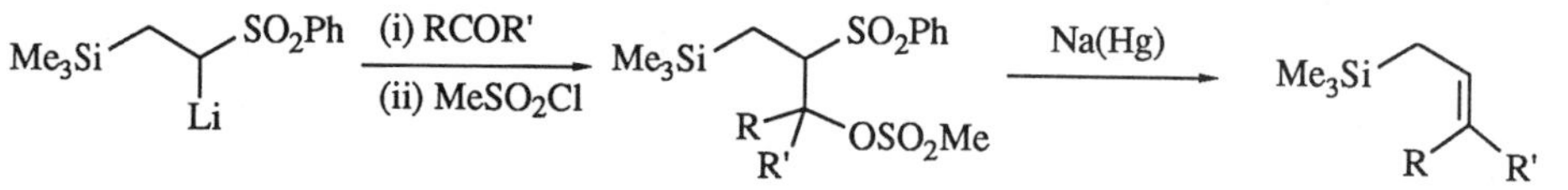

Scheme 13

Overall yields for the olefination are excellent (typically >90%) despite opportunities for silyl migration, or alternative elimination triggered by the β-silyl group. This latter mode of elimination is the subject of the next section.

7.2 Alternative β-Eliminations of Sulphones

7.2.1 Elimination of RSO_2H

Important studies of the base-induced eliminations of sulphones include the pioneering work of Fenton and Ingold[19] and the later reports of Hofmann and co-workers[20] and Colter and Miller.[21] Fenton and Ingold first reported that the elimination of simple sulphones is more facile using alcoholic sodium ethoxide rather than aqueous potassium hydroxide. Later work showed that the combination of potassium *tert*-butoxide as base in DMSO gives excellent yields of elimination products at only 55°C.[20] This study also showed that potassium alkoxides are superior to sodium or lithium alkoxides and that alcoholic solvents are inferior to DMSO. The report of Colter and Miller describes the elimination of various aryl 2-pentyl sulphones using bases in either ethylene glycol or pyridine, Scheme 14.[21]

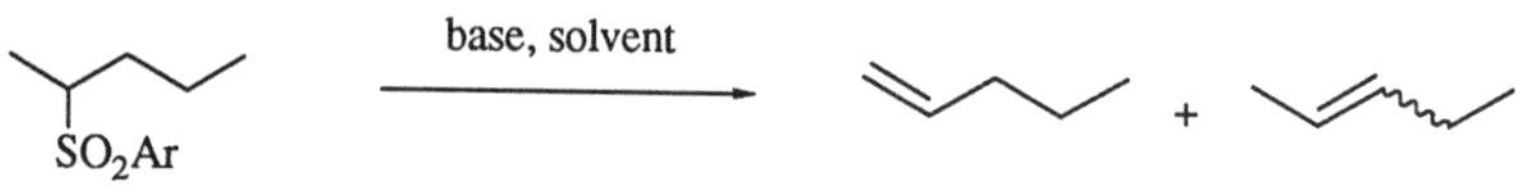

$$HO(CH_2)_2ONa/HO(CH_2)_2OH \quad ca.\ 75 : 25$$

$$^tBuOK/pyridine \quad ca.\ 95 : 5$$

Scheme 14

The base/solvent combinations were chosen in order to minimise isomerisation of the product alkenes. Immediately apparent is the strong preference for Hofmann-type elimination to give the less-substituted alkene product. This had been noted previously by Brown and Wheeler, who showed a clear correlation between the steric requirements of various leaving groups and their tendency toward Hofmann elimination.[22] In this work the methyl sulphone group was found to be second only to the trimethylammonium group in its preference for Hofmann elimination. As can be seen in Scheme 14 the more sterically demanding base ^{t}BuOK increases the proportion of Hofmann product.

In general the results for sulphone elimination are in accord with a β-elimination in which an E2 or E1cB mechanism operates. Certainly the Hofmann-type product distribution points to a transition state towards the E1cB extreme, with relatively little C–S bond breaking. This is certainly the case when the β-hydrogen is being removed from an activated position.[23] This includes the vast majority of synthetically significant eliminations, which often involve removal of allylic or benzylic hydrogens, or those activated by carbonyl groups.

In elimination reactions carried out in DMSO at 55°C using ^{t}BuOK as base, the ease of elimination of a series of isopropyl derivatives, i.e. CH_3CHXCH_3, was found to be Br$^-$ ≈ RSO_2^- ≈ $RSO^- > NO_3^- > SCN^- > RS^- > NO_2^- > CN^-$.[20] Certainly this, and other studies,[23] point to β-elimination of a sulphone being easier than that of most other uncharged sulphur and nitrogen groups. That sulphonyl groups serve so well in eliminations is remarkable considering that the alkanesulphonyl group has been shown to be an extremely poor leaving group (ca. 10^9 less reactive than comparable chlorides).[24] Again this supports an E1cB-type of elimination in which activation of the β-hydrogen by the strongly electron-withdrawing sulphone group is important.

Although eliminations of simple sulphones have not been utilised in synthesis the corresponding reactions of allylic, homoallylic and other activated sulphones have received significant attention. Of great interest in this regard has been the synthesis of retinoic acids, including vitamin A acid, first examined by Julia and Arnould.[25] In this procedure sulphone (26) is alkylated with bromoester (27) to give an intermediate which then undergoes elimination of benzenesulphinic acid on base treatment, Scheme 15.

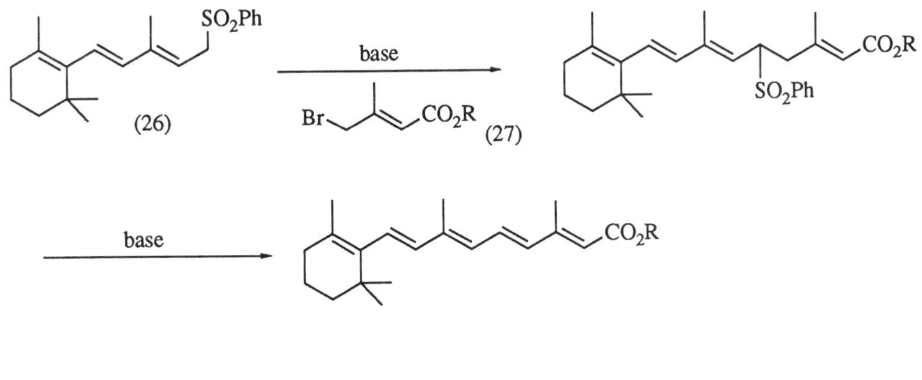

Scheme 15

In this case the sulphone alkylation is carried out using ᵗBuOK in THF, whilst the elimination is effected most efficiently using NaOMe. A number of variants on the same theme have appeared, each involving sulphone alkylation and then elimination, to construct the polyenic system required for vitamin A, e.g. Scheme 16.[26]

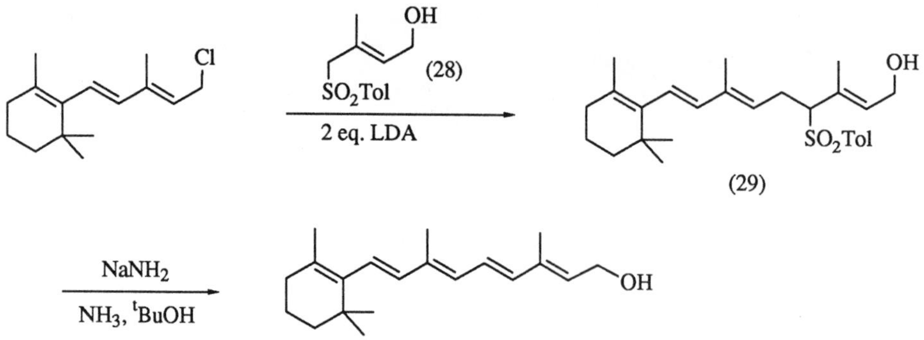

Scheme 16

In this example the nucleophilic and electrophilic partners are reversed compared to the situation shown in Scheme 15. Alkylation of the dianion derived from (**28**), using LDA as base, occurs smoothly in regioselective fashion. Elimination from the hydroxy sulphone (**29**) is then carried out using sodamide in ammonia in the presence of *tert*-butanol. Inclusion of the *tert*-butanol is essential to the elimination and is probably required to reprotonate the α-sulphonyl carbanion formed.

Herve du Penhoat and Julia have examined the elimination reactions of a range of homoallylic sulphones such as (**30**), Scheme 17.[27]

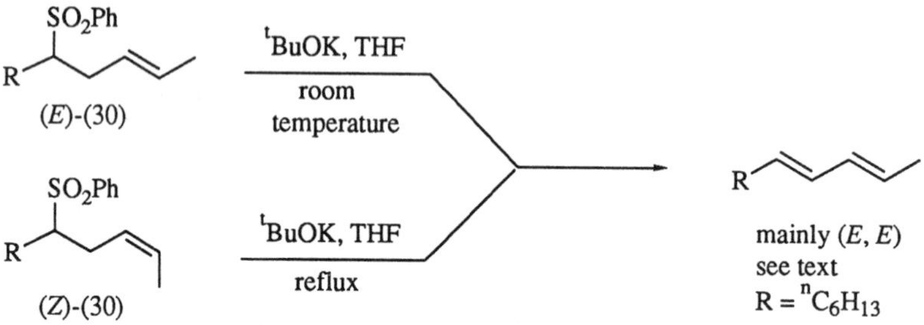

Scheme 17

On treatment with ᵗBuOK in THF (*E*)-(**30**) undergoes a remarkably stereoselective elimination of benzenesulphinic acid to give almost solely the (*E,E*)-2,4-undecadiene product. The elimination of the corresponding (*Z*)-isomer is somewhat more sluggish and produces a mixture of isomers in which the (*E,E*)-diene is the major component (84%). The elimination was then applied to the synthesis of a simple pheromone component of the codling moth. A more significant synthetic application of the reaction was reported by Nicolaou *et al.* in a total synthesis of the ionophore antibiotic X-14547A (indanomycin) Scheme 18.[28]

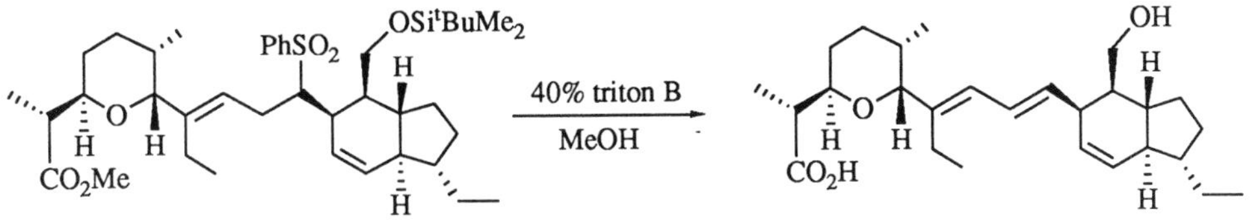

Scheme 18

Under the carefully defined conditions used, the elimination was accompanied by ester hydrolysis and removal of the silicon protecting group.

Otera and co-workers have examined the elimination and double elimination reactions of β-acetoxy and β-alkoxy sulphones, e.g. Scheme 19.[29] A double elimination occurs by initial formation of a vinyl sulphone, isomerisation to an allylic sulphone, and finally a 1,4-elimination to give a mixture of diene stereoisomers. Whilst the first two steps occur at room temperature, the elimination step requires heating in *tert*-butanol.

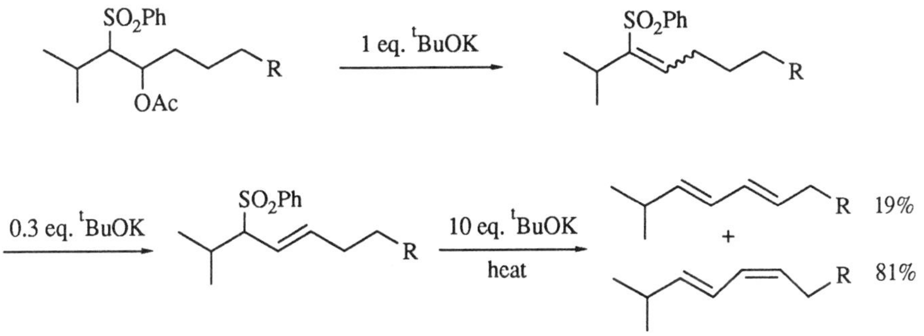

Scheme 19

Depending on the structure of the starting hydroxy sulphone derivative the elimination can produce dienes or alkynes. This approach provides a route to a range of conjugated products, including ene–yne, ene–yne–ene, yne–yne and ene–yne–yne types, e.g. Scheme 20.

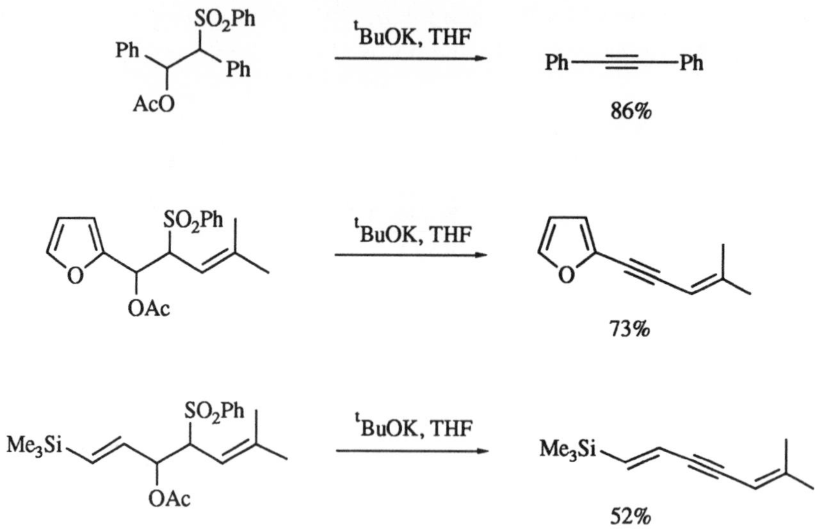

Scheme 20

The method was applied to the synthesis of muscone, and also vitamin A.[30] Surprisingly, in the latter case, the double elimination takes a mechanistically different course to that outlined in Scheme 19, and does not involve the initial formation of a vinyl sulphone.

Exactly the same type of double elimination has been applied to the synthesis of dienamides and dienoates, e.g. Scheme 21.[31]

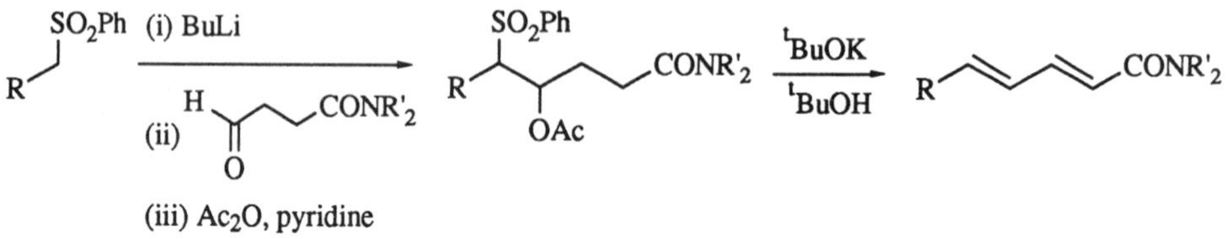

Scheme 21

The product diene is mainly the (*E,E*)-isomer (*ca.* 90%). Simple applications of the method to natural product synthesis were explored.

Many examples of elimination of sulphinic acids from β-sulphonyl carbonyl compounds have been reported, e.g. Scheme 22.[32]

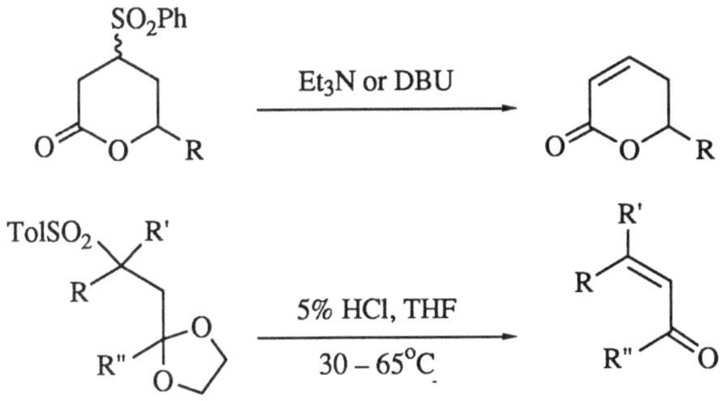

Scheme 22

In such cases elimination is facile, and can be conducted under very mild conditions. The eliminations of sulphones having geminal sulphur, nitrogen or oxygen substitution have also been studied, e.g. Scheme 23.[33–35]

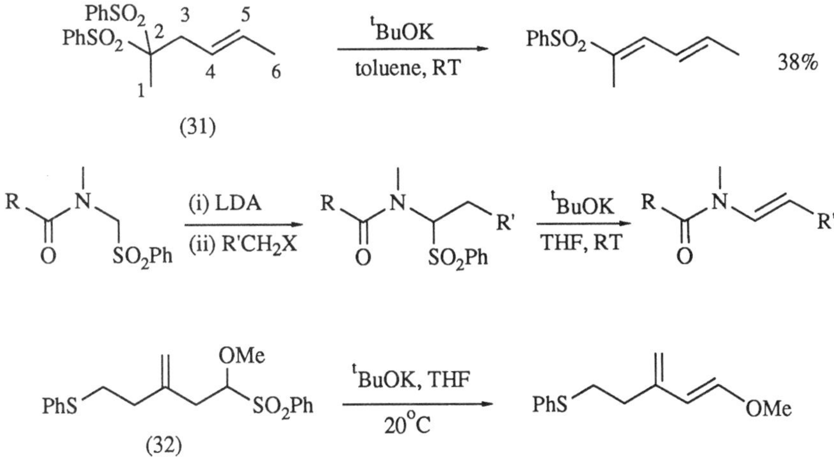

Scheme 23

In each case the α-heterofunctionality appears to facilitate the elimination, each of the examples shown being conducted at room temperature. Elimination from (**31**) is highly stereoselective but fairly low-yielding, due to polymerisation problems.[33] 1,4-Eliminations from the related allylic disulphones (i.e. having the C-3–C-4 double bond) were also explored but were found to be less stereoselective. Elimination from homoallylic ethers such as (**32**) occurs to give synthetically useful dienes; no comment was made on competing thiolate or methoxide elimination.

A number of special protocols can be employed to effect sulphone elimination in certain situations. For example, Trost *et al.* found that simple base treatment of sulphone (**33**) failed to effect elimination, Scheme 24.[36]

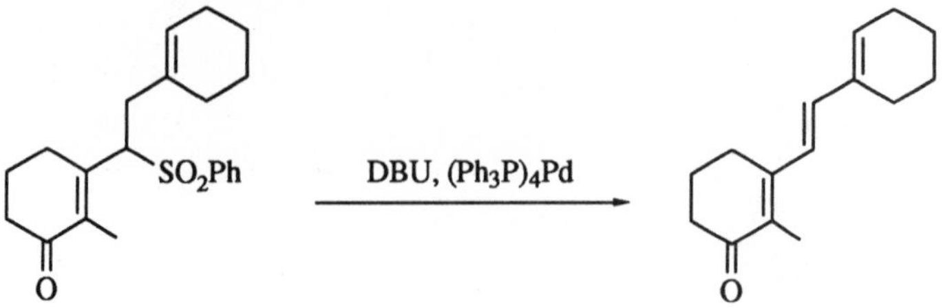

(33)   (34) 71%

Scheme 24

However, reaction with base in the presence of a palladium catalyst enables clean elimination to the desired trienone (**34**). A second special type of elimination procedure involves initial *ortho*-metallation of certain types of unsaturated sulphone, Scheme 25.[37]

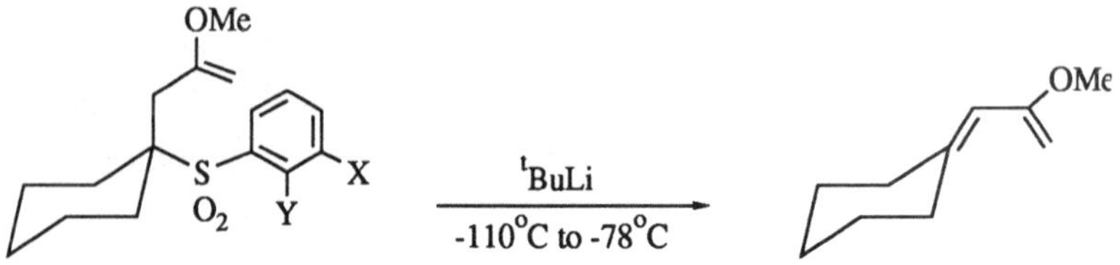

(35) X = OMe, Y = H
(36) X = OMe, Y = D

ᵗBuLi; D₂O (-90°C)

(37) 95%

Scheme 25

Treatment of (**35**) with ᵗBuLi at low temperature results in *ortho*-metallation, as revealed by labelling studies, i.e. (**35**) → (**36**). On warming to -78°C clean elimination to give (**37**) occurs. The reaction is limited to sulphones having no α-hydrogens, and which can undergo eliminative rearrangement by abstraction of an activated (i.e. allylic or benzylic) β-hydrogen.

7.2.2 Elimination of RSO₂X: X = SiR′₃, SnR′₃, NO₂, SO₂R′

Fluoride-initiated eliminations of β-silyl-substituted sulphones offer a very mild entry to a range of unsaturated products, as first shown by Kocienski, Scheme 26.[38] Substituted β-silyl sulphones (**38**) can be constructed either by alkylation of the parent compound (**39**) or by reaction of a sulphonyl carbanion with iodomethyl-trimethylsilane. Elimination using TBAF in refluxing THF then brings about conversion to the alkene in high yield. This type of chemistry has also been studied in some detail by Hsiao and Shechter.[39]

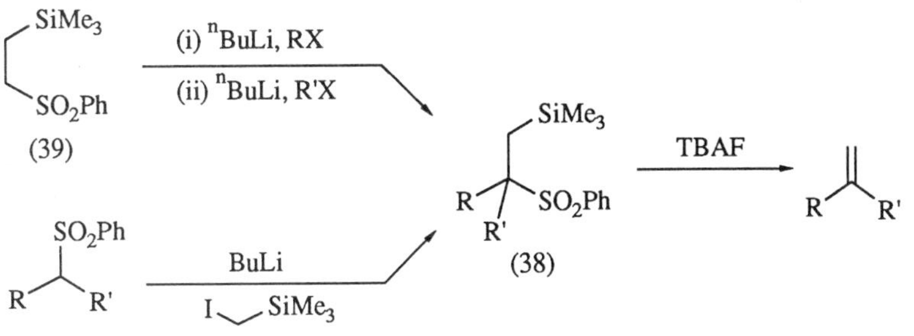

Scheme 26

One significant extension of the approach shown in Scheme 26 is the use of vinylogous systems which undergo 1,4-elimination, Scheme 27.[40]

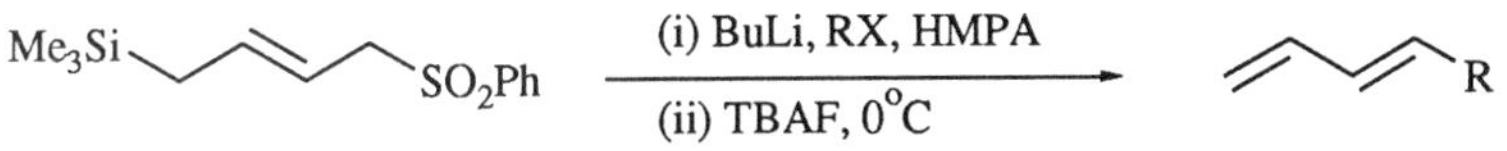

Scheme 27

The method furnishes (*E*)-dienes in good overall yield, the starting silyl sulphones acting effectively as a butadiene anion or 1,1-dianion equivalent. An interesting epoxide coupling reaction which incorporates an elimination of a β-silyl sulphone has been described, e.g. Scheme 28.[41]

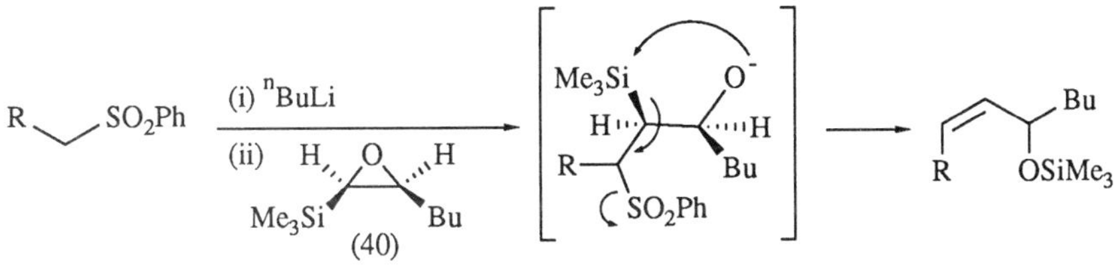

Scheme 28

In this sequence regioselective opening of the α–silyl epoxide (40) is followed by silyl group migration and then elimination. In the reaction involving simple sulphones the (*Z*)-allylic alcohol products are obtained predominantly in the form of their silyl ether derivatives. With more hindered sulphones a two-stage procedure involving $BF_3 \cdot OEt_2$-assisted attack on the epoxide is preferable; however, the overall stereoselectivity of this latter process is poor.

The elimination reaction of β-stannyl sulphones is even more facile than that of their silyl counterparts. Thus, methylenation of sulphones can be carried out by reaction of the derived carbanion with an iodomethyl trialkylstannane, e.g. Scheme 29.[42]

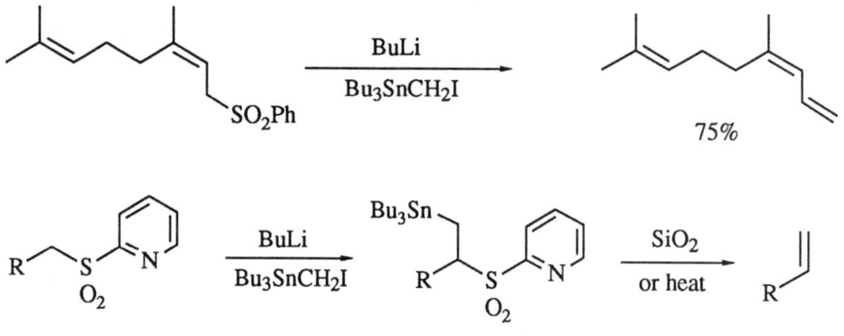

(41)

Scheme 29

In most cases the somewhat unstable stannyl sulphone intermediates are not isolated, and the alkene product is obtained directly, or after treatment of the crude mixture with silica gel. In the case of 2-pyridyl sulphone (41) isolation is possible by flash chromatography, and elimination occurs only on further treatment with silica gel, or on pyrolysis.

The same types of β-stannyl sulphones can also be accessed by the conjugate addition of stannyllithium reagents to vinyl sulphones, Scheme 30.[43]

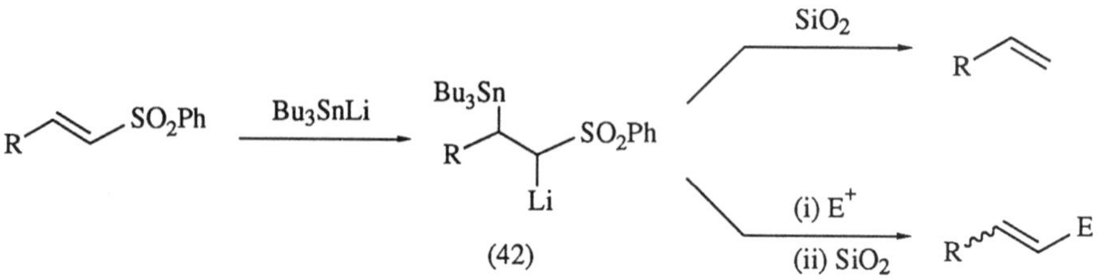

(42)

Scheme 30

In this case the intermediate sulphonyl carbanion (42) can be further elaborated by quenching with alkylating agents or aldehydes prior to elimination. Reactions carried out using Me_3SiCl as the electrophilic trapping agent lead to (Z)-vinyl silanes stereoselectively. The elimination can also be carried out under conditions in which the initially formed (Z)-isomers are isomerised to give the (E)-vinyl silanes. Another variant on this theme involves starting with a vinyl sulphone which already has a β-stannyl or silyl substituent, Scheme 31.[44,45]

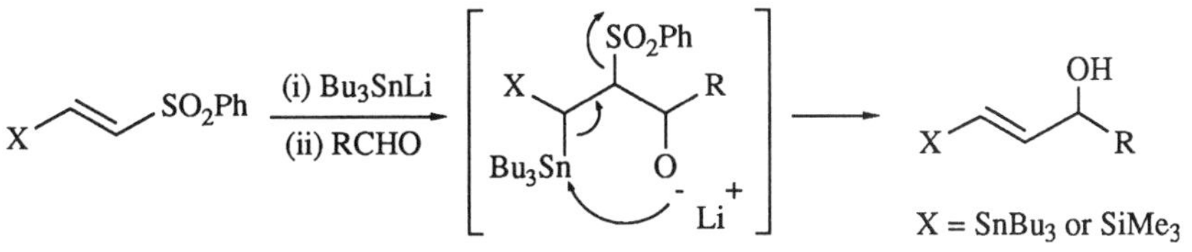

Scheme 31

The product hydroxy vinylstannanes or silanes are usually produced as the (*E*)-isomers, although some solvent dependency of the stereochemical outcome is observed. In the intermediate (**43**), where X = SiMe$_3$, loss of either a silicon or tin group could result in elimination. In the event, only loss of the tin grouping is observed.

Ono and co-workers have examined the elimination reactions of β-nitro-sulphones.[46] This reaction proceeds in high yield to give a range of alkenyl products, e.g. Scheme 32.

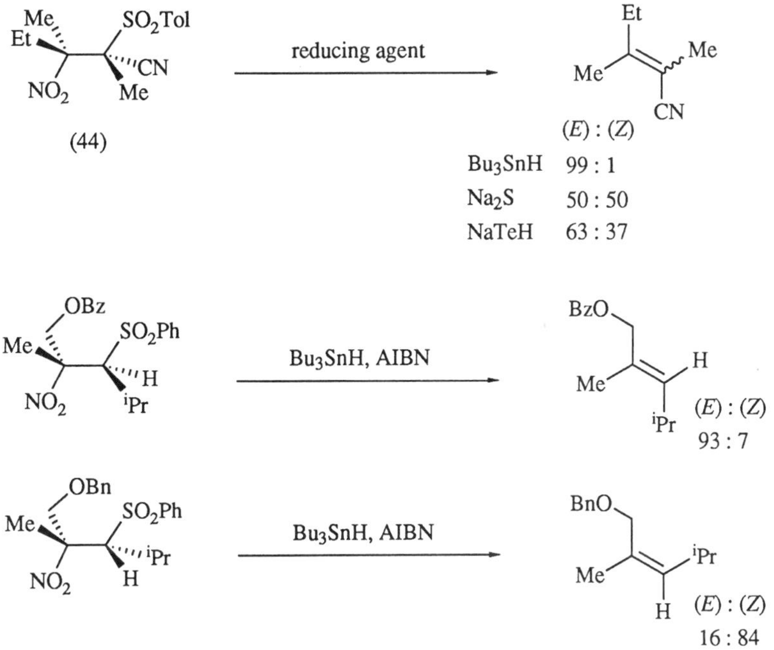

Scheme 32

In the case of eliminations leading to conjugated alkenes, reducing agents such as Na$_2$S, NaTeH and Bu$_3$SnH are effective. Most importantly, the latter reagent effects a highly stereospecific *anti*-elimination of the nitro and sulphone groups. The diastereoisomer of nitrosulphone (**44**) shown thus gives the (*E*)-alkene product, whereas the other diastereoisomeric nitrosulphone gives the (*Z*)-product. The stereospecific nature of the elimination is not retained when Na_2S or NaTeH are employed. In eliminations not leading to conjugated products only Bu_3SnH gives satisfactory results, and again high stereospecificity is observed.

The stereospecific nature of this radical-mediated elimination is rather puzzling, since analogous eliminations of β-bromosulphones are non-stereospecific.[47] A mechanism involving the formation of a free β-sulphonyl radical must therefore be ruled out. The authors have proposed a synchronous elimination pathway as shown in Scheme 33.

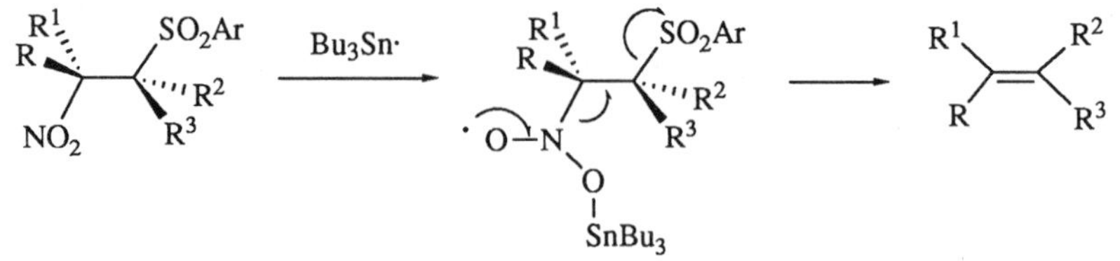

Scheme 33

The lack of stereospecificity in the reactions carried out with Na_2S was explained by invoking an alternative mechanism involving a loose ion radical pair.

De Lucchi *et al.* first examined the reductive elimination of 1,2-disulphones to give alkenes, Scheme 34.[48]

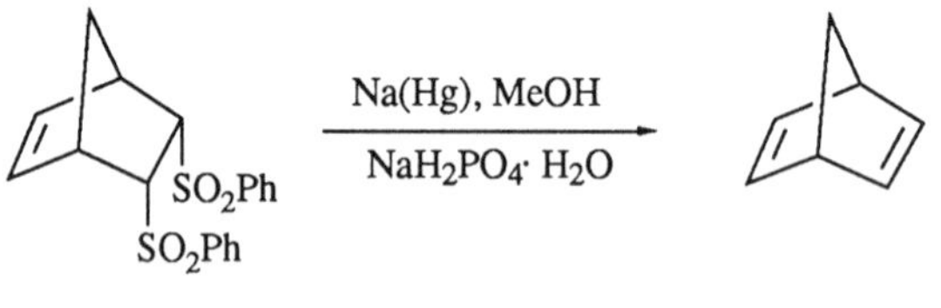

Scheme 34

The use of sodium amalgam in methanol under buffered conditions gives very good results, starting with either diastereoisomeric disulphone (obtained stereospecifically by Diels–Alder cycloadditions of (*E*)- or (*Z*)-1,2-bis(phenylsulphonyl)ethylenes). Whilst sodium in refluxing toluene, or lithium amalgam in toluene, also give acceptable results, a wide range of other reducing agents do not react with the disulphone, or give mixtures of unidentified products. A later report by Brown and Carpino describes the use of magnesium in methanol for the same type of double desulphonylation.[49]

Treatment of the unusual disulphone (**45**) with lithium dimethylcuprate furnishes diphenylacetylene in 83% yield, Scheme 35.[50]

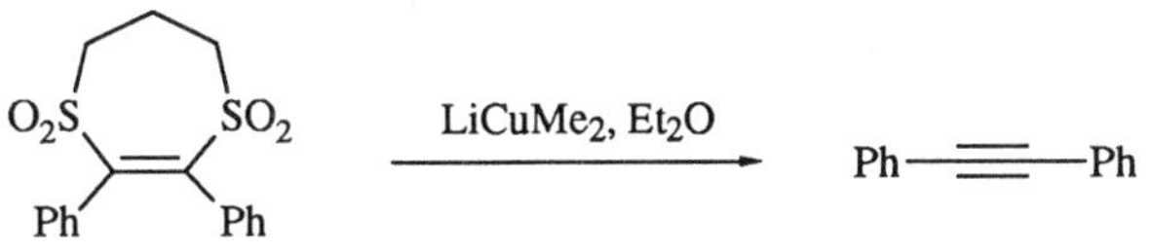

(45)

Scheme 35

This process presumably involves sequential electron transfers to the unsaturated system resulting in two successive sulphinate eliminations. Another elimination leading to alkynes and which involves sulphone intermediates was examined by the groups of Bartlett[51] and Lythgoe.[52] In this process a β-ketosulphone is converted to an enol phosphate derivative such as (46) and subsequently subjected to reductive elimination, Scheme 36.

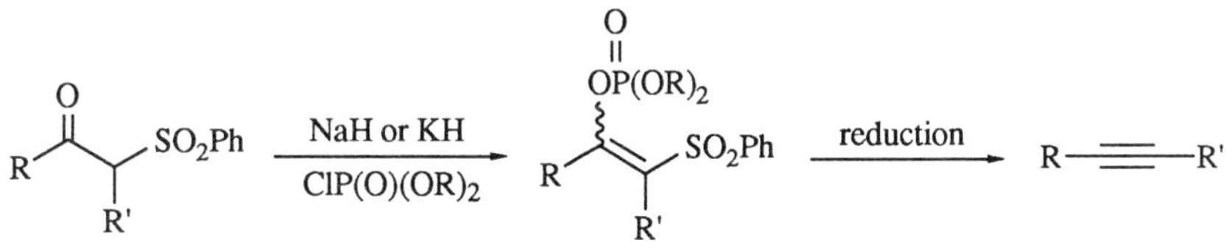

(46)

Scheme 36

Reductive elimination can be carried out using sodium amalgam in a mixture of THF and DMSO, or by using sodium in ammonia. The method can be applied to functionalised alkyne products, and also to the preparation of conjugated enyne systems.

7.3 The Ramberg–Bäcklund Reaction

The Ramberg–Bäcklund reaction is a rearrangement of certain types of sulphone having a leaving group at the α-position, which results in the regiospecific formation of a new carbon–carbon double bond with concomitant loss of SO_2, Scheme 37.

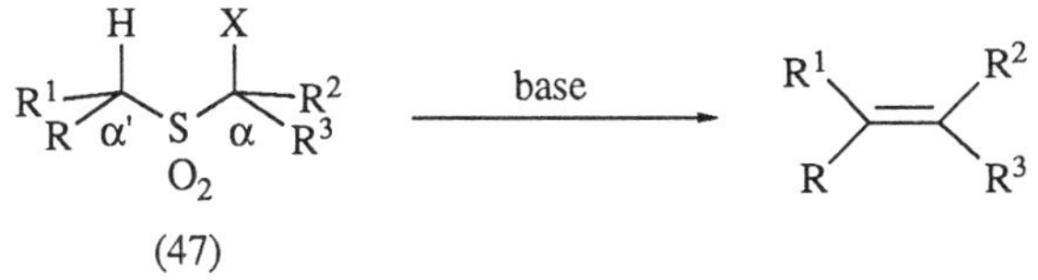

(47)

Scheme 37

Since its initial report in 1940 this reaction has been the focus of considerable attention, particularly concerning the mechanism of the process and its synthetic scope and utility.[53]   The reaction is most usually conducted using α-chlorosulphones, although other leaving groups, including other halogens or sulphinate, are also effective.[54]   The reaction is quite general for sulphones of formula (47) having an α-halogen and at least one α'-hydrogen. Extensive studies, primarily by the group of Bordwell, have provided convincing evidence that the mechanism of the Ramberg–Bäcklund reaction is as shown in Scheme 38.[55]

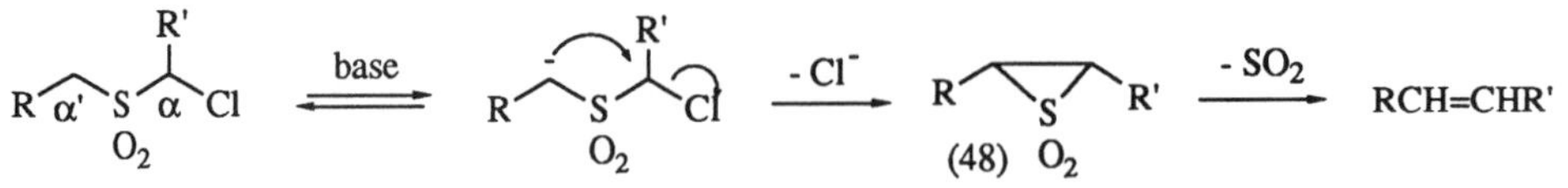

Scheme 38

The sequence involves deprotonation of the sulphone to form the corresponding carbanion which then undergoes intramolecular nucleophilic attack on the α-carbon centre to displace chloride and form the episulphone intermediate (48). Under the basic reaction conditions the episulphone is unstable, and loss of SO_2 occurs to complete the alkene formation. The significant features of each step of the reaction are discussed and illustrated below.

Firstly, deuteration studies using a variety of sulphones have shown that the first step in the sequence, sulphonyl carbanion formation, is reversible.[56] Thus, in reactions involving sulphones such as (49) and (50), when the rearrangement is carried out to partial completion in MeOD the starting sulphones are recovered with high levels of deuterium incorporation at the α'-centre.[57]

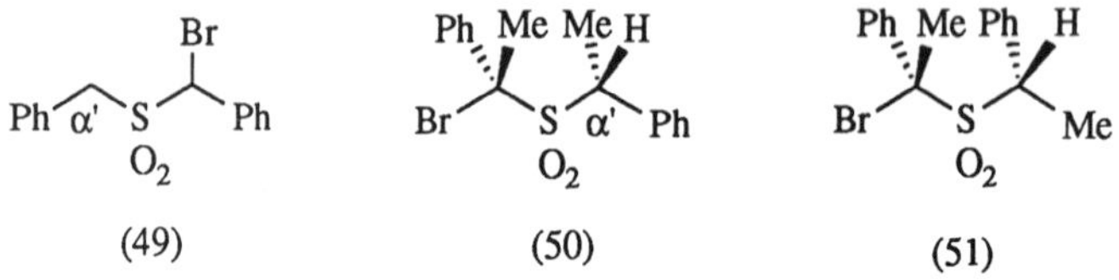

(49)   (50)   (51)

Pre-equilibrium of the sulphone and the corresponding carbanion may be partial or complete, depending on the substrate in question. In systems such as (50) the rate of exchange is much faster than the rate of epimerisation (about (50:1)) as discussed in Chapter 3. Compound (50) and its diastereoisomer (51) therefore serve as good probes of the stereochemical course of the Ramberg–Bäcklund reaction. Thus on treatment with NaOMe in methanol, sulphone (50) gives the (Z)-dimethylstilbene (52) highly selectively, whereas (51) gives only the (E)-isomer (53), Scheme 39.

Scheme 39

The stereochemical course of the reaction can thus be deduced to involve formation of the intermediate episulphone with inversion at both the α- and α'-centres, Scheme 40.

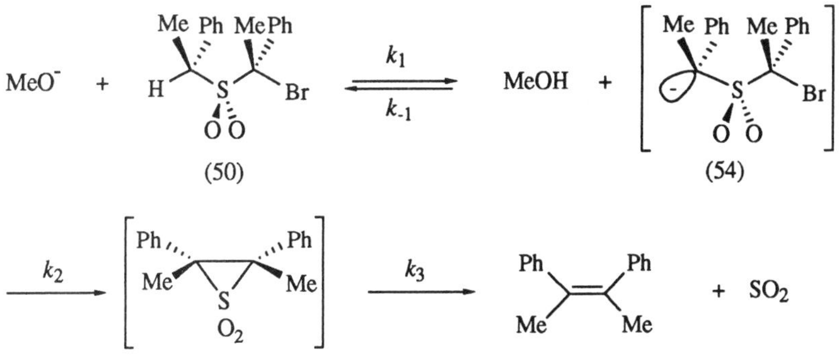

Scheme 40

As discussed in Chapter 3 initial sulphone deprotonation would be expected to occur with the hydrogen involved aligned along the bisector of the O–S–O bond angle. This gives carbanion (54), which in this case undergoes more rapid intramolecular alkylation than epimerisation. In the particular example shown, the stereochemistry of the final alkene must reflect that of the intermediate episulphone. This is not always the case if the intermediate episulphone has an acidic α-hydrogen, in which case epimerisation may result in formation of the less hindered episulphone, and consequently the (E)-alkene product.

Whilst for the reaction of (50) the intramolecular alkylation of (54) is the rate limiting step, in the case of (51) decomposition of the episulphone is rate-limiting. The former situation is more general, as indicated by the pronounced leaving group effect usually observed, i.e. k^{Br}/k^{Cl} = *ca.* 200.[58]

A remarkable observation is that the presence of the halogen at the α-position greatly enhances the rate of hydrogen exchange at the α' position (i.e. *ca.* 1000 times!).[59] This effect has been attributed to halogen-promoted solvent exchange, and is presumably operative in most substrates. The actual rate of ring closure depends on a balance of several effects involving the substituents at both the α- and α'-positions.[60] Thus alkyl substitution at the α'-centre (undergoing deprotonation) decreases k_1, whereas phenyl substitution increases k_1. This effect may be offset by the greater nucleophilicity of the sulphonyl carbanion with increasing alkyl substitution. Not surprisingly, activation of the leaving halogen by making the α-centre benzylic or allylic increases the rate of intramolecular displacement.

Decomposition of the presumed episulphone intermediates in the Ramberg–Bäcklund reaction has been shown to be stereospecific.[61] For example, reaction of diazoethane with sulphur dioxide allows isolation of pure crystals of *cis*-2-butene episulphone which on heating gives (Z)-2-butene quantitatively. Such studies were conducted using independently synthesised episulphones, and only very recently has one such intermediate been isolated from a Ramberg–Bäcklund reaction.[62] Thus treatment of iodosulphone (55) with 1.2 equivalents of ^{t}BuOK at temperatures up to 0°C gives good yields of episulphone (56), Scheme 41.

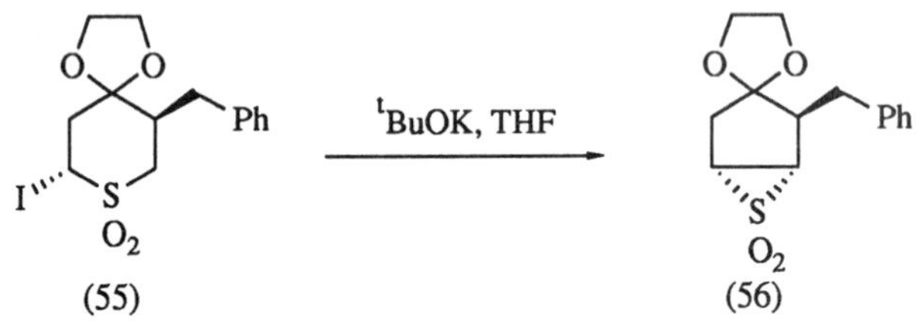

Scheme 41

The structure of (56) was unequivocally demonstrated from spectral and analytical evidence, and by its conversion to the corresponding alkene.

A very important general feature of the Ramberg–Bäcklund reaction is its tendency to furnish the (Z)-alkene predominantly from acyclic starting sulphones, e.g. Scheme 42.[61,63]

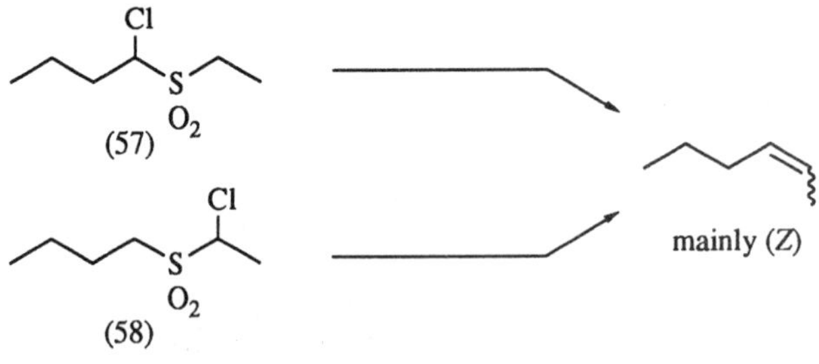

Scheme 42

However, the stereochemical results from such reactions appear to be rather dependent on the type of base used. Whilst weak bases such as aqueous NaOH give modest (Z)-selectivity, stronger bases such as ^{t}BuOK in DMSO have been shown to give unexpectedly high levels of (E)-selectivity in certain cases.[61]

Even more intriguing is that the two isomers (57) and (58) give consistently different (E):(Z) ratios (30.6 : 69.4 for (57) and 33.8 : 66.2 for (58)) when treated with NaOH. That the *cis*-selectivity is greater when the leaving group is attached to the more bulky of the two alkyl groups was shown for a number of isomeric pairs of sulphones. The so-called *cis*-effect has been the subject of much speculation, with the most convincing explanations being offered by Paquette and Wittenbrook.[63]

The most usual conditions for conducting the Ramberg–Bäcklund reaction involve the use of aqueous sodium hydroxide, or sodium methoxide in methanol. Other metal alkoxides, particularly potassium *tert*-butoxide, have also been used, as have solvents such as THF and DMSO. A phase-transfer method has been reported which involves the use of aqueous sodium hydroxide and dichloromethane in the presence of catalysts such as Aliquat-336 and TEBA.[64]

The above conditions rely on the use of preformed halogenosulphones, which are usually readily available, as described in Chapter 2. Alternatively, Meyers' method, which uses potassium hydroxide, *tert*-butanol and carbon tetrachloride, allows *in situ* chlorination and Ramberg–Bäcklund reaction.[65] In the original report a range of sulphone chlorinations and Ramberg–Bäcklund reactions was reported, e.g. Scheme 43.

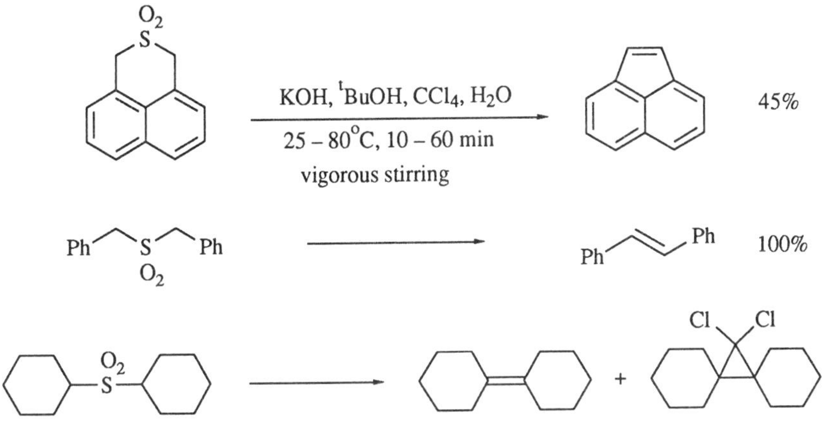

KOH, ^{t}BuOH, CCl$_4$, H$_2$O
25 – 80°C, 10 – 60 min
vigorous stirring

45%

100%

(59)   (60) 32%   (61) 60%

Scheme 43

The process gives the expected Ramberg–Bäcklund products with dibenzylic sulphones, whilst secondary dialkyl sulphones such as (59) are converted to alkenes, e.g. (60), and dichlorocyclopropanes such as (61). The latter arise by attack of dichlorocarbene on the initial Ramberg–Bäcklund product. A subsequent report by Meyers *et al.* describes the successful application of the CCl_4–KOH system to the Ramberg–Bäcklund reaction of benzhydryl sulphones to give 1,1-diarylalkenes.[66] Other workers have demonstrated that the system usually works well in conversion of diallylic sulphones to conjugated trienes, e.g. Scheme 44.[67]

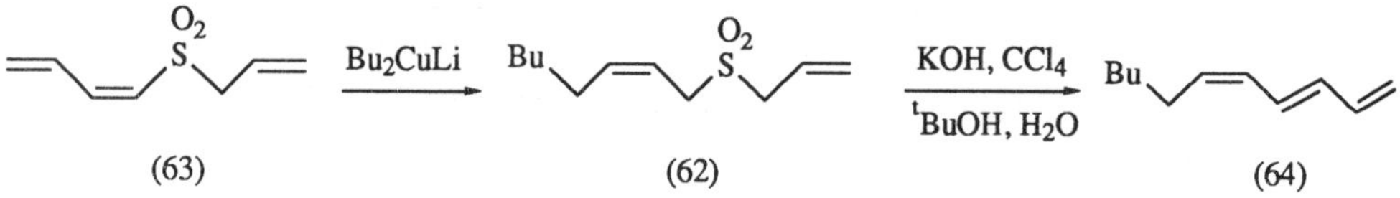

(63)   (62)   (64)

Scheme 44

The diallylic sulphone (62) was first prepared by cuprate addition to (63). Ramberg–Bäcklund reaction using Meyers' modification then occurs in moderate yield to give mainly the trienyl product (64) in which the (Z)-double-bond geometry present in (62) is retained, and in which the new double bond has mainly the (E)-geometry. Unfortunately the level of stereocontrol in each step is only moderate, resulting in a cocktail of minor isomers in the final triene (64). The (E)-selectivity observed in this example appears general for reactions involving benzylic or allylic sulphones, due to epimerisation in the intermediate episulphones.

Simple primary dialkyl sulphones undergo conversion to *cis*-dialkylvinyl-sulphonic acid salts on treatment with the CCl_4–KOH system, Scheme 45.[68]

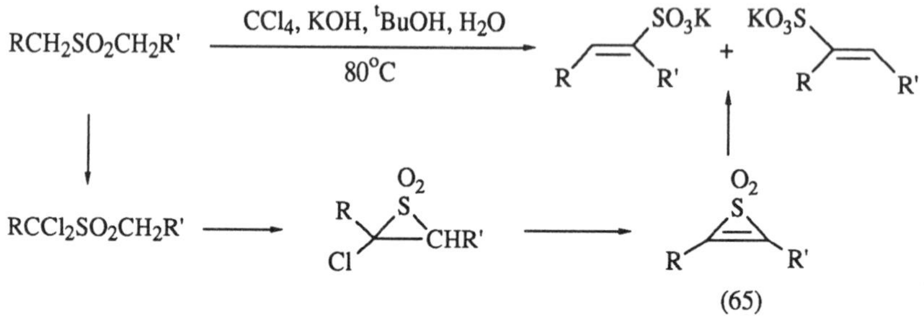

(65)

Scheme 45

This transformation was shown by Meyers to occur via the sequence shown, involving formation of a thiirene dioxide intermediate (65).[69] The reason for the different reaction course, compared with the examples in Schemes 43 and 44, is the failure in simple dialkyl sulphones of episulphone formation to compete with α,α-dichlorination.

Although thiirene dioxide intermediates such as (**65**) had been observed spectroscopically in Ramberg–Bäcklund reactions,[70] it was not until 1971 that this type of compound was properly synthesised and characterised.[71] The report of Carpino *et al.* describes the synthesis of thiirene dioxides by either a modified Ramberg–Bäcklund process, or by dehydrobromination of independently synthesised α-bromo-episulphones. The chemistry of such heterocyclic sulphones is discussed in more detail in Chapter 8.

A significant area of application of the Ramberg–Bäcklund reaction has been in the synthesis of strained alkenes. For example Gassman and Mlinaric-Makerski have reported the synthesis of 7-methyl-*trans*-bicyclo[4.1.0]hept-3-ene (**66**),[72] whilst Becker and Labhart succeeded in preparing anti-Bredt alkene (**67**), Scheme 46.[73]

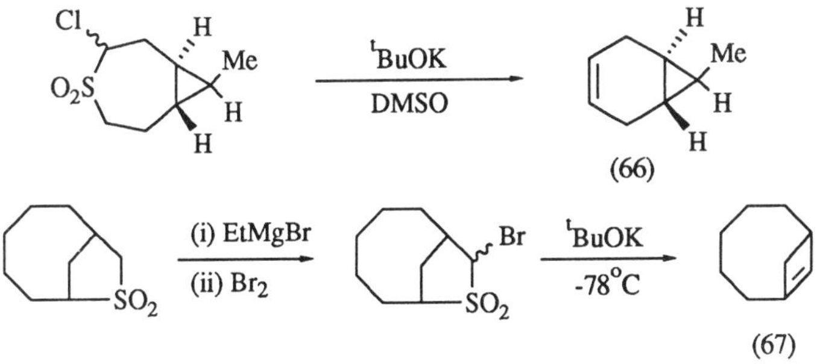

Scheme 46

The latter report also contains some useful results concerning the importance of the favourable W-type arrangement of atoms H–C–S–C–X involved in the Ramberg–Bäcklund rearrangement.[74]   Thus, of the three decalin-type sulphones (**68**)-(**70**) only the first was found to undergo clean Ramberg–Bäcklund rearrangement to give (**71**), Scheme 47.

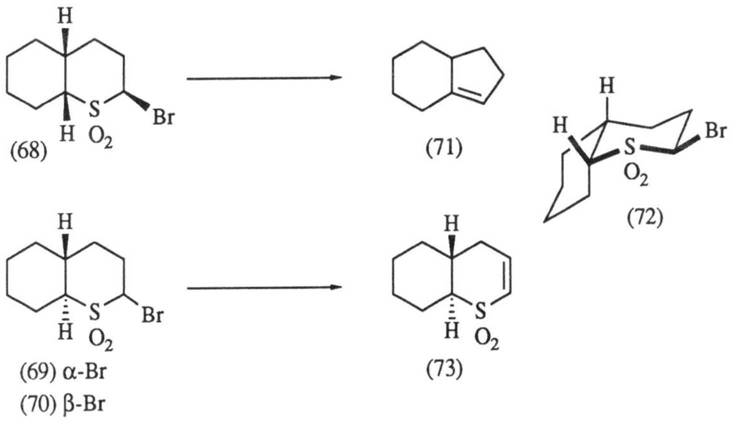

Scheme 47

The W-type arrangement of the groups involved in the rearrangement of (68) can be clearly seen when the compound is drawn in the chair–chair conformation (72). Such arrangements cannot be attained by (69) and (70) and predominant 1,2-elimination occurs to give (73). The proper W-type arrangement cannot be attained in sulphones such as that leading to (67); however, in this case, 1,2-elimination is highly unfavourable.

The Ramberg–Bäcklund reaction also fails to compete with 1,2-elimination when α-chlorosulphone (74) is reacted under a variety of basic conditions.[75] Instead of the desired alkene (75) the reaction using ${}^tBuOK$ as base produces sulphone (76). This comes about by dehydrochlorination and Michael addition of *tert*-butanol, Scheme 48.

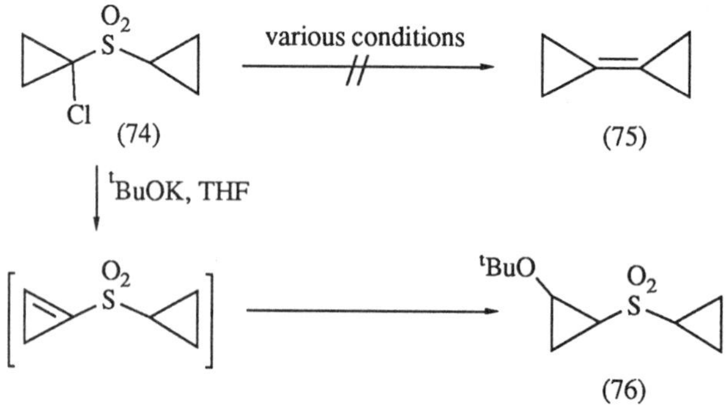

Scheme 48

Thus in assessing the possible application of the Ramberg–Bäcklund reaction to a particular synthetic problem it is important to take into consideration the geometric requirement for successful reaction, and to assess the likelihood of competing reactions such as 1,2-elimination.

Block *et al.* have described a broad range of useful Ramberg–Bäcklund reactions in which the α-halogenosulphonyl group is introduced directly into an unsaturated starting material by means of free radical halosulphonylation.[76] Thus addition of bromomethanesulphonyl bromide to a simple alkene gives a 1:1 adduct which can then be treated with base to give a bromomethyl alkenyl sulphone, e.g. (77), Scheme 49.

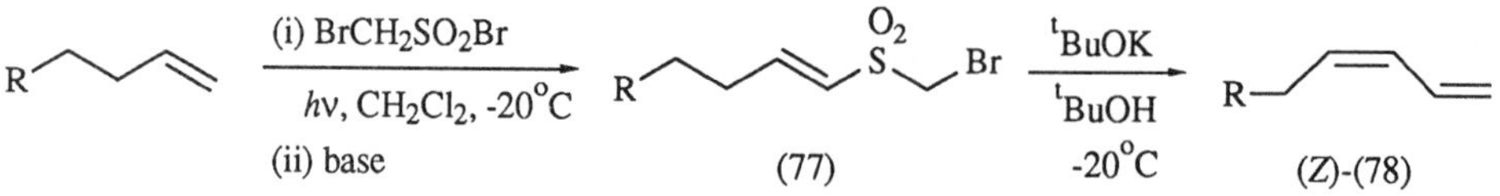

Scheme 49

Subsequent vinylogous Ramberg–Bäcklund reaction then gives a diene product, thereby completing a three-step homologation of alkene to diene. Significantly, in simple cases (e.g. R = ⁿBu) the formation of both (77) and (78) is reasonably stereocontrolled. When the base used for elimination is DBN the vinyl sulphone (77) is formed almost exclusively (>97%) in the (E)-form. Ramberg–Bäcklund elimination proceeds to give a 83:17 mixture of (Z)- and (E)-1,3-nonadiene. The stereoselectivity in this latter step has been attributed to an attractive interaction between the developing negative charge α to the sulphone and the methylene at the δ-position, i.e. (79) is favoured over (80), Scheme 50.[77]

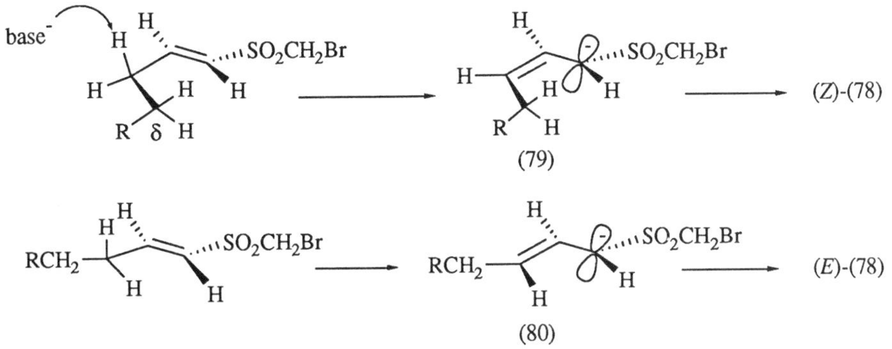

Scheme 50

This effect is not general, and variable results are seen when the δ-methylene is replaced by other groups such as PhO and phenyl. One particularly interesting application of the method is the reaction with silyl enol ethers. Depending on the reaction conditions used, either Ramberg–Bäcklund-type elimination to give an enone, or intramolecular O-alkylation to give novel 1,3-oxathiolane dioxides such as (81) can be achieved, Scheme 51.

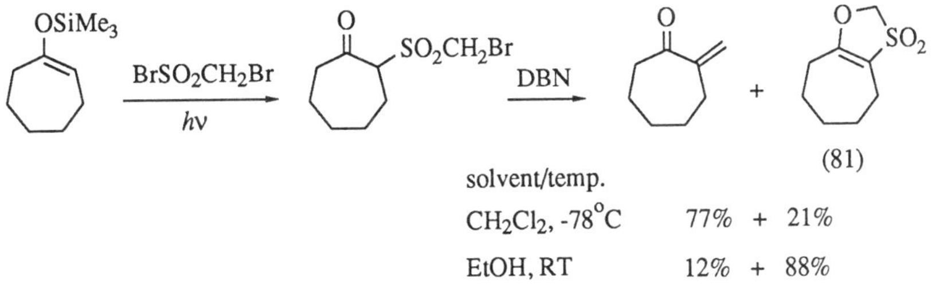

solvent/temp.		
CH_2Cl_2, -78°C	77%	21%
EtOH, RT	12%	88%

Scheme 51

Other reactions, starting from non-conjugated dienes and alkynes, also give interesting results. Overall the methodology offers a concise way of preparing a range of diene, polyene and enyne products, although the yields and degree of stereocontrol available are somewhat substrate dependent.

A more recent report from Block's group describes the use of the Ramberg–Bäcklund reaction as part of an iterative 'ring-growing' procedure, e.g. Scheme 52.[78]

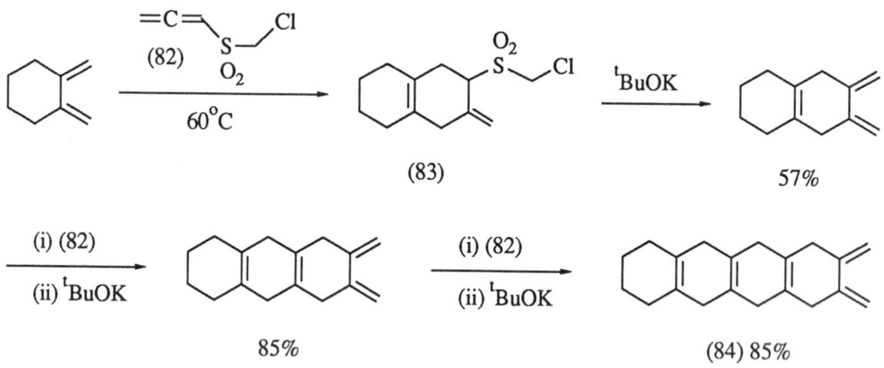

Scheme 52

The allenyl chloromethyl sulphone (82) participates in Diels–Alder cycloadditions to give intermediates such as (83). These can then be converted to a conjugated diene product ready for further reaction with (82). In the case shown in Scheme 52 two subsequent iterations were performed in highly efficient fashion to give tetracyclic system (84).

The Ramberg–Bäcklund reaction is readily applicable to the synthesis of unsaturated acid or ester products as indicated by the examples in Scheme 53.[79,80]

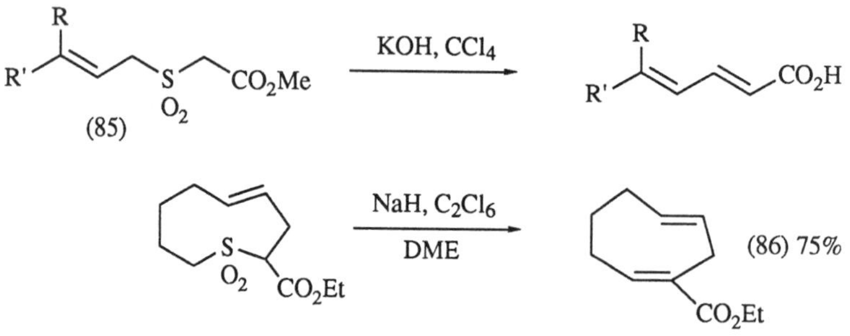

Scheme 53

Using Meyers-type conditions the reaction of sulphonyl esters such as (85) occurs to give dienoic acids.[79] Another one-pot sulphone chlorination/Ramberg–Bäcklund reaction procedure allows the conversion of certain cyclic and acyclic sulphones to α,β-unsaturated esters such as (86).[80] Analogous conversions of sulphones having other ring sizes were also examined, the results being rather variable.

A number of reports have described the use of sulphinate as the leaving group in modifications of the Ramberg–Bäcklund reaction, e.g. Scheme 54.[81-83]

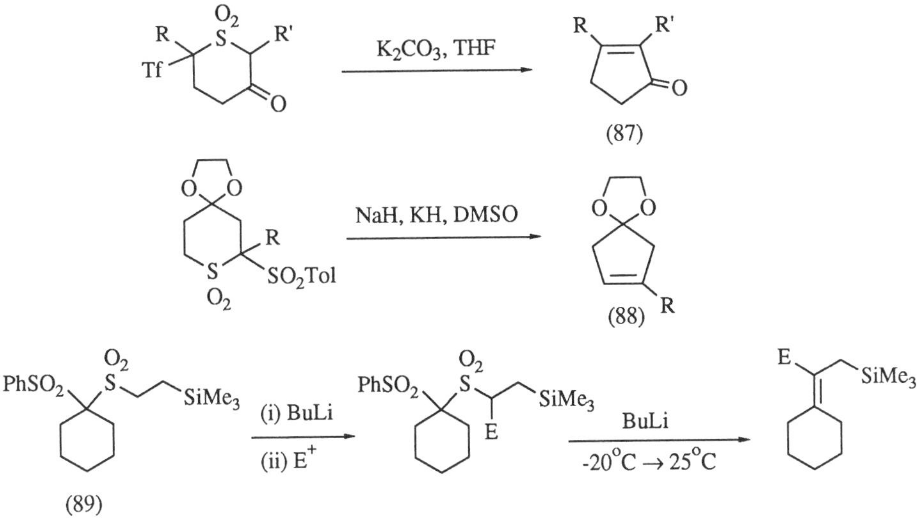

Scheme 54

The preparations of (87) and (88) are just two examples of a significant number of related cyclopentenone syntheses achieved using the Ramberg–Bäcklund approach.[84] Thus use of the additional sulphone group in place of a halogen provides the opportunity for regioselective alkylation prior to the alkene-forming elimination. The use of triflones allows the use of rather mild elimination conditions compared to those normally required. Disulphone (89) and related compounds provide an attractive and versatile route to allylsilanes. Sulphonyl carbanion alkylation is possible by maintaining a low temperature, and subsequent sulphonyl carbanion formation and warming to room temperature then brings about the Ramberg–Bäcklund elimination with formation of a substituted allylsilane.

Recently the Ramberg–Bäcklund reaction has been utilised for the preparation of a series of cyclic conjugated enediynes related to the calicheamicin and esperamicin natural products, Scheme 55.[85]

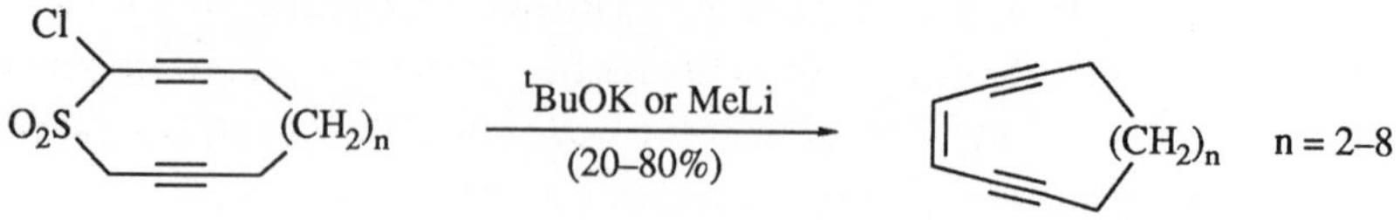

Scheme 55

Boeckman *et al.* have also used this approach to effect a crucial ring-contraction in their synthesis of (+)-eremantholide A, Scheme 56.[86]

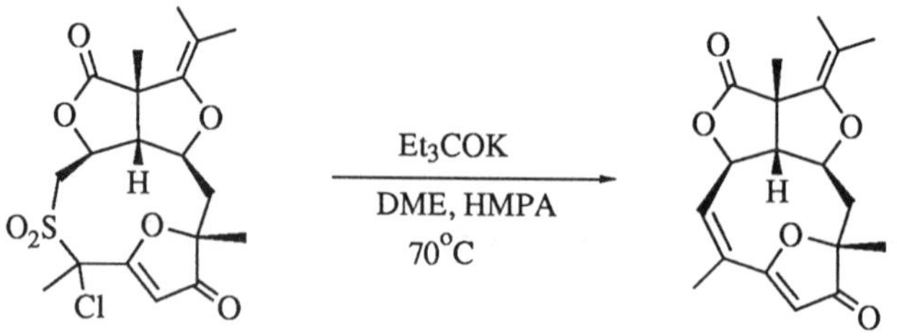

Scheme 56

Finally, a modification of the Ramberg–Bäcklund reaction uses a Michael addition onto an unsaturated sulphone, rather than proton abstraction, to initiate rearrangement, e.g. Scheme 57.[87]

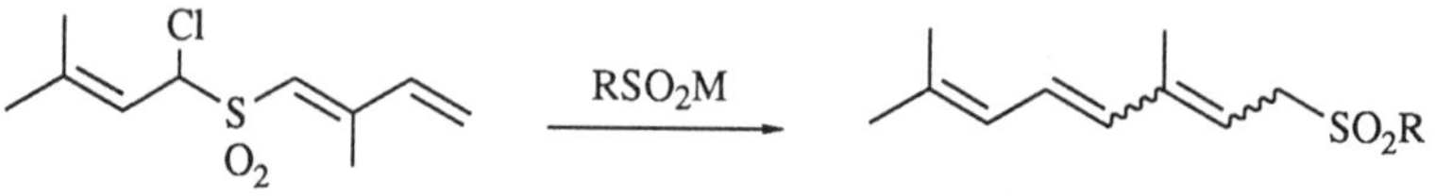

Scheme 57

Unfortunately the reaction usually gives mixtures of stereoisomers. Although the study was limited to the use of sulphinate as nucleophile, Block *et al.* have also reported a single example using sodium isopropoxide to initiate the reaction.[76]

Chapter 7 References

1. M. Julia and J-M. Paris, *Tetrahedron Lett.*, **1973**, 4833.
2. P. Kocienski, *Phosphorus and Sulphur*, **1985**, *24*, 97.
3. P. J. Kocienski, *Chem. Ind. (London)*, **1981**, 548.
4. P. J. Kocienski, B. Lythgoe, and S. Ruston, *J. Chem. Soc., Perkin Trans. 1*, **1978**, 829.
5. P. J. Kocienski, B. Lythgoe, and I. Waterhouse, *J. Chem. Soc., Perkin Trans. 1*, **1980**, 1045.
6. P. J. Kocienski and B. Lythgoe, *J. Chem. Soc., Perkin Trans. 1*, **1980**, 1400; see also P. J. Kocienski, B. Lythgoe, and D. A. Roberts, *J. Chem. Soc., Perkin Trans. 1*, **1978**, 834; K. Okada and K. Mori, *Agric. Biol. Chem.*, **1983**, *47*, 89.
7. P. Kocienski and M. Todd, *J. Chem. Soc., Perkin Trans. 1*, **1983**, 1777; see also P. Kocienski and M. Todd, *J. Chem. Soc., Chem. Commun.*, **1982**, 1078.
8. P. J. Kocienski, B. Lythgoe, and S. Ruston, *J. Chem. Soc., Perkin Trans. 1*, **1979**, 1290.
9. See, *inter alia*, R. Baker, M. J. O'Mahony, and C. J. Swain, *J. Chem. Soc., Chem. Commun.*, **1985**, 1326; F. Matsuda, N. Tomiyosi, M. Yanagiya, and T. Matsumoto, *Chem. Lett.*, **1987**, 2097; A. G. M. Barrett, R. A. E. Carr, S. V. Attwood, G. Richardson, and N. D. A. Walshe, *J. Org. Chem.*, **1986**, *51*, 4840; J. D. White and G. L. Bolton, *J. Am. Chem. Soc.*, **1990**, *112*, 1626; D. A. Evans, R. L. Dow, T. L. Shih, J. M. Takacs, and R. Zahler, *J. Am. Chem. Soc.*, **1990**, *112*, 5290; S. E. de Laszlo, M. J. Ford, S. V. Ley, and G. N. Maw, *Tetrahedron Lett.*, **1990**, *31*, 5525.
10. G. Kim, M. Y. Chu-Moyer, and S. J. Danishefsky, *J. Am. Chem. Soc.*, **1990**, *112*, 2003.
11. M. P. Edwards, S. V. Ley, S. G. Lister, and B. D. Palmer, *J. Chem. Soc., Chem. Commun.*, **1983**, 630.
12. S. J. Danishefsky, H. G. Selnick, M. P. DeNinno, and R. E. Zelle, *J. Am. Chem. Soc.*, **1987**, *109*, 1572.
13. A. B. Jones, A. Villalobos, R. G. Linde II, and S. J. Danishefsky, *J. Org. Chem.*, **1990**, *55*, 2786.
14. D. S. Holmes, U. C. Dyer, S. Russell, J. A. Sherringham, and J. A. Robinson, *Tetrahedron Lett.*, **1988**, *29*, 6357; U. C. Dyer and J. A. Robinson, *J. Chem. Soc., Perkin Trans. 1*, **1988**, 53. For related work see D. V. Patel, F. VanMiddlesworth, J. Donaubauer, P. Gannett, and C. J. Sih, *J. Am. Chem. Soc.*, **1986**, *108*, 4603. For other examples of this type of Julia strategy see B. M. Trost, J. Lynch, P. Renaut, and D. H. Steinman, *J. Am. Chem. Soc.*, **1986**, *108*, 284; B. M. Trost and Y. Matsumura, *J. Org. Chem.*, **1977**, *42*, 2036.
15. A. S. Kende and J. S. Mendoza, *Tetrahedron Lett.*, **1990**, *31*, 7105.

16. T. Shono, Y. Matsumura, and S. Kashimura, *Chem. Lett.*, **1978**, 69; S. Gambino, P. Martigny, G. Mousset, and J. Simonet, *J. Electroanal. Chem.*, **1978**, *90*, 105.

17. D. H. R. Barton, J. C. Jaszberenyi, and C. Tachdjian, *Tetrahedron Lett.*, **1991**, *32*, 2703. An example of such an elimination had been reported previously, see B. Lythgoe and I. Waterhouse, *Tetrahedron Lett.*, **1977**, 4223.

18. C-N. Hsiao and H. Shechter, *Tetrahedron Lett.*, **1982**, *23*, 1963.

19. G. W. Fenton and C. K. Ingold, *J. Chem. Soc.*, **1930**, 705.

20. J. E. Hofmann, T. J. Wallace, and A. Schriesheim, *J. Am. Chem. Soc.*, **1964**, *86*, 1561; T. J. Wallace, J. E. Hofmann, and A. Schriesheim, *J. Am. Chem. Soc.*, **1963**, *85*, 2739.

21. A. K. Colter and R. E. Miller, Jr., *J. Org. Chem.*, **1971**, *36*, 1898.

22. H. C. Brown and O. H. Wheeler, *J. Am. Chem. Soc.*, **1956**, *78*, 2199.

23. For an example in which the hydrogen removed is activated by a sulphone, see D. R. Marshall, P. J. Thomas, and C. J. M. Stirling, *J. Chem. Soc., Chem. Commun.*, **1975**, 940; see also C. J. M. Stirling, *Int. J. Sulphur Chem., C*, **1971**, *6*, 41.

24. X. Creary, *J. Org. Chem.*, **1985**, *50*, 5080. Triflones are somewhat more reactive, see J. B. Hendrickson, A. Giga, and J. Wareing, *J. Am. Chem. Soc.*, **1974**, *96*, 2275.

25. M. Julia and D. Arnould, *Bull. Soc. Chim. Fr.*, **1973**, 746.

26. G. L. Olson, H-C. Cheung, K. D. Morgan, C. Neukom, and G. Saucy, *J. Org. Chem.*, **1976**, *41*, 3287; see also S. C. Welch and J. M. Gruber, *J. Org. Chem.*, **1982**, *47*, 385; P. S. Manchand, M. Rosenberger, G. Saucy, P. A. Wehrli, H. Wong, L. Chambers, M. P. Ferro, and W. Jackson, *Helv. Chim. Acta,* **1976**, *59*, 387.

27. C. Herve du Penhoat and M. Julia, *Tetrahedron*, **1986**, *42*, 4807.

28. K. C. Nicolaou, D. A. Claremon, D. P. Papahatjis, and R. L. Magolda, *J. Am. Chem. Soc.*, **1981**, *103*, 6969.

29. T. Mandai, T. Yanagi, K. Araki, Y. Morisaki, M. Kawada, and J. Otera, *J. Am. Chem. Soc.*, **1984**, *106*, 3670; J. Otera, H. Misawa, and K. Sugimoto, *J. Org. Chem.*, **1986**, *51*, 3830.

30. J. Otera, H. Misawa, T. Onishi, S. Suzuki, and Y. Fujita, *J. Org. Chem.*, **1986**, *51*, 3834.

31. T. Moriyama, T. Mandai, M. Kawada, J. Otera, and B. M. Trost, *J. Org. Chem.*, **1986**, *51*, 3896; T. Mandai, T. Moriyama, K. Tsujimoto, M. Kawada, and J. Otera, *Tetrahedron Lett.*, **1986**, *27*, 603; see also N. A. Plobeck and J-E. Bäckvall, *J. Org. Chem.*, **1991**, *56*, 4508.

32. J. C. Carretero and L. Ghosez, *Tetrahedron Lett.*, **1988**, *29*, 2059; T. Yoshida and S. Saito, *Chem. Lett.*, **1982**, 165. For additional examples see H. Ishibashi, H. Nakatani, T. S. So, T. Fujita, and M. Ikeda, *Heterocycles*, **1990**, *31*, 215; D. Orr, *Synthesis*, **1979**, 139; B. Corbel, J. M. Decesare, and T. Durst, *Can. J. Chem.*, **1978**, *56*, 505; S. Torii, K. Uneyama, and I. Kawahara, *Bull. Chem. Soc. Jpn.*, **1978**, *51*, 949.

33. T. Cuvigny, C. Herve du Penhoat, and M. Julia, *Tetrahedron*, **1986**, *42*, 5321.

34. L. Berthon and D. Uguen, *Tetrahedron Lett.*, **1985**, *26*, 3975.

35. T. Mandai, K. Hara, T. Nakajima, M. Kawada, and J. Otera, *Tetrahedron Lett.*, **1983**, *24*, 4993; see also S. V. Ley, B. Lygo, F. Sternfeld, and A. Wonnacott, *Tetrahedron*, **1986**, *42*, 4333; C. Greck, P. Grice, S. V. Ley, and A. Wonnacott, *Tetrahedron Lett.*, **1986**, *27*, 5277.

36. B. M. Trost, N. R. Schmuff, and M. J. Miller, *J. Am. Chem. Soc.*, **1980**, *102*, 5979.

37. X. Radisson, P. L. Kwiatkowski, and P. L. Fuchs, *Synth. Commun.*, **1987**, 39.

38. P. J. Kocienski, *Tetrahedron Lett.*, **1979**, 2649; see also reference 2.

39. C-N. Hsiao and H. Shechter, *J. Org. Chem.*, **1988**, *53*, 2688.

40. C-N. Hsiao and H. Shechter, *Tetrahedron Lett.*, **1984**, *25*, 1219.

41. M. Masnyk and J. Wicha, *Tetrahedron Lett.*, **1988**, *29*, 2497; S. Marczak, M. Masnyk, and J. Wicha, *Liebigs Ann. Chem.*, **1990**, 345.

42. M. Ochiai, S. Tada, K. Sumi, and E. Fujita, *Tetrahedron Lett.*, **1982**, *23*, 2205; M. Ochiai, T. Ukita, E. Fujita, and S. Tada, *Chem. Pharm. Bull.*, **1984**, *32*, 1829; see also B. A. Pearlman, S. R. Putt, and J. A. Fleming, *J. Org. Chem.*, **1985**, *50*, 3622.

43. M. Ochiai, T. Ukita, and E. Fujita, *J. Chem. Soc., Chem. Commun.*, **1983**, 619.

44. M. Ochiai, T. Ukita, and E. Fujita, *Chem. Lett.*, **1983**, 1457.

45. M. Ochiai, T. Ukita, and E. Fujita, *Tetrahedron Lett.*, **1983**, *24*, 4025.

46. N. Ono, R. Tamura, J. Hayami, and A. Kaji, *Tetrahedron Lett.*, **1978**, 763; N. Ono, H. Miyake, R. Tamura, I. Hamamoto, and A. Kaji, *Chem. Lett.*, **1981**, 1139; N. Ono, A. Kamimura, and A. Kaji, *J. Org. Chem.*, **1987**, *52*, 5111.

47. T. E. Boothe, J. L. Greene, Jr., and P. B. Shevlin, *J. Org. Chem.*, **1980**, *45*, 794.

48. O. De Lucchi, V. Lucchini, L. Pasquato, and G. Modena, *J. Org. Chem.*, **1984**, *49*, 596.

49. A. C. Brown and L. A. Carpino, *J. Org. Chem.*, **1985**, *50*, 1749.

50. K. Saigo, Y. Hashimoto, L. Fang, and M. Hasegawa, *Heterocycles*, **1989**, *29*, 2079.

51. P. A. Bartlett, F. R. Green III, and E. H. Rose, *J. Am. Chem. Soc.*, **1978**, *100*, 4852.

52. B. Lythgoe and I. Waterhouse, *J. Chem. Soc., Perkin Trans. 1*, **1979**, 2429.

53. L. Ramberg and B. Bäcklund, *Arkiv Kemi, Mineral. Geol.*, **1940**, *13A*, N° 27; L. A. Paquette, *Org. React.*, **1977**, *25*, 1; F. G. Bordwell, *Acc. Chem. Res.*, **1970**, *3*, 281; F. G. Bordwell, in *Organosulphur Chemistry*, Ed. M. J. Janssen, Interscience, New York, **1967**, p.271; L. A. Paquette, *Acc.Chem. Res.*, **1968**, *1*, 209; L. A. Paquette, *Mech. Mol. Migr.*, **1968**, *1*, 121; E. Block, in *React. Organosulphur Comp.*, Academic Press, New York, **1978**, p.77; E. Vedejs and G. A. Krafft, *Tetrahedron*, **1982**, *38*, 2857; J. M. Clough, in *Comprehensive Organic Synthesis*, Ed. B. M. Trost and I. Fleming, Pergamon Press, Oxford, **1991**, *3*, 861.

54. C. Y. Meyers, D. H. Hua, and N. J. Peacock, *J. Org. Chem.*, **1980**, *45*, 1719.

55. F. G. Bordwell and G. D. Cooper, *J. Am. Chem. Soc.*, **1951**, *73*, 5187.

56. N. P. Neureiter and F. G. Bordwell, *J. Am. Chem. Soc.*, **1963**, *85*, 1209.

57. F. G. Bordwell and E. Doomes, *J. Org. Chem.*, **1974**, *39*, 2526.

58. F. G. Bordwell and J. M. Williams, Jr., *J. Am. Chem. Soc.*, **1968**, *90*, 435.

59. F. G. Bordwell and J. B. O'Dwyer, *J. Org. Chem.*, **1974**, *39*, 2519.

60. F. G. Bordwell and M. D. Wolfinger, *J. Org. Chem.*, **1974**, *39*, 2521.

61. N. P. Neureiter, *J. Am. Chem. Soc.*, **1966**, *88*, 558; see also D. Scholz and P. Burtscher, *Liebigs Ann. Chem.*, **1985**, 517; D. Scholz, *Chem. Ber.*, **1981**, *114*, 909.

62. A. G. Sutherland and R. J. K. Taylor, *Tetrahedron Lett.*, **1989**, *30*, 3267.

63. L. A. Paquette and L. S. Wittenbrook, *J. Am. Chem. Soc.*, **1968**, *90*, 6783.

64. G. D. Hartman and R. D. Hartman, *Synthesis*, **1982**, 504.

65. C. Y. Meyers, A. M. Malte, and W. S. Matthews, *J. Am. Chem. Soc.*, **1969**, *91*, 7510; C. Y. Meyers, A. M. Malte, and W. S. Matthews, *Quart. Rep. Sulphur Chem.*, **1970**, *5*, 229; see also C. Y. Meyers, in *Topics in Organic Sulphur Chemistry*, Ed. M, Tislev, University Press, Ljubljana, **1978**, p.207.

66. C. Y. Meyers, W. S. Matthews, G. J. McCollum, and J. C. Branca, *Tetrahedron Lett.*, **1974**, 1105.

67. F. Naf, R. Decorzant, and S. D. Escher, *Tetrahedron Lett.*, **1982**, *23*, 5043; see also M. Julia, D. Lave, M. Mulhauser, M. Ramirez-Munoz, and D. Uguen, *Tetrahedron Lett.*, **1983**, *24*, 1783; G. Buchi and R. M. Freidinger, *J. Am. Chem. Soc.*, **1974**, *96*, 3332.

68. C. Y. Meyers and L. L. Ho, *Tetrahedron Lett.*, **1972**, 4319; see also L. A. Paquette, L. S. Wittenbrook, and V. V. Kane, *J. Am. Chem. Soc.*, **1967**, *89*, 4487.

69. C. Y. Meyers, L. L. Ho, G. J. McCollum, and J. Branca, *Tetrahedron Lett.*, **1973**, 1843.

70. F. G. Bordwell, J. M. Williams, Jr., and B. B. Jarvis, *J. Org. Chem.*, **1968**, *33*, 2026.

71. L. A. Carpino, L. V. McAdams III, R. H. Rynbrandt, and J. W. Spiewak, *J. Am. Chem. Soc.*, **1971**, *93*, 476; see also J. C. Philips, J. V. Swisher, D. Haidukewych, and O. Morales, *J. Chem. Soc., Chem. Commun.*, **1971**, 22.

72. P. G. Gassman and K. Mlinaric-Makerski, *Tetrahedron Lett.*, **1988**, *29*, 4803; see also H-D. Martin, B. Mayer, M. Pütter, and H. Höchstetter, *Angew. Chem., Int. Ed. Engl.*, **1981**, *20*, 677; L. A. Paquette and M. P. Trova, *J. Am. Chem. Soc.*, **1988**, *110*, 8197.

73. K. B. Becker and M. P. Labhart, *Helv. Chim. Acta.*, **1983**, *66*, 1090; see also J. Kattenburg, E. R. De Waard, and H. O. Huisman, *Tetrahedron*, **1974**, *30*, 463; J. Kattenburg, E. R. De Waard, and H. O. Huisman, *Tetrahedron*, **1974**, *30*, 3177. For another transannular example see E. J. Corey and E. Block, *J. Org. Chem.*, **1969**, *34*, 1233.

74. For previous discussions on this topic see A. Nickou and N. H. Werstiunk, *J. Am. Chem. Soc.*, **1967**, *89*, 3914; L. A. Paquette and R. W. Houser, *J. Am. Chem. Soc.*, **1969**, *91*, 3870; F. G. Bordwell and E. Doomes, *J. Org. Chem.*, **1974**, *39*, 2531.

75. L. A. Paquette and R. W. Houser, *J. Org. Chem.*, **1971**, *36*, 1015.

76. See E. Block, M. Aslam, V. Eswarakrishnan, K. Gebreyes, J. Hutchinson, R. Iyer, J-A. Laffitte, and A. Wall, *J. Am. Chem. Soc.*, **1986**, *108*, 4568, and references therein to previous papers in this series.

77. K. Inomata, T. Hirata, H. Suhara, H. Kinoshita, H. Kotake, and H. Seuda, *Chem. Lett.*, **1988**, 2009.

78. E. Block and D. Putman, *J. Am. Chem. Soc.*, **1990**, *112*, 4072.

79. P. A. Grieco and D. Boxler, *Synth. Commun.*, **1975**, *5*, 315.

80. E. Vedejs and S. P. Singer, *J. Org. Chem.*, **1978**, *43*, 4884; see also T. Fujisawa, B. I. Mobele, and M. Shimizu, *Tetrahedron Lett.*, **1991**, *32*, 7055.

81. J. B. Hendrickson and P. S. Palumbo, *J. Org. Chem.*, **1985**, *50*, 2110.

82. H. Matsuyama, Y. Miyazawa, and M. Kobayashi, *Chem. Lett.*, **1986**, 433.

83. M. G. Ranasinghe and P. L. Fuchs, *J. Am. Chem. Soc.*, **1989**, *111*, 779; D. Scarpetti and P. L. Fuchs, *J. Am. Chem. Soc.*, **1990**, *112*, 8084.

84. See H. Matsuyama, Y. Miyazawa, Y. Takei, and M. Kobayashi, *J. Org. Chem.*, **1987**, *52*, 1703; H. Matsuyama, Y. Ebisawa, M. Kobayashi, and N. Kamigata, *Heterocycles*, **1989**, *29*, 449; G. Casy and R. J. K. Taylor, *J. Chem. Soc., Chem. Commun.*, **1988**, 454; G. Casy and R. J. K. Taylor, *Tetrahedron*, **1989**, *45*, 455; J. Kattenberg, E. R. de Waard, and H. O. Huisman, *Tetrahedron Lett.*, **1977**, 1173.

85. K. C. Nicolaou, G. Zuccarello, Y. Ogawa, E. J. Schweiger, and T. Kumazawa, *J. Am. Chem. Soc.*, **1988**, *110*, 4866.

86. R. K. Boeckman, Jr., S. K. Yoon, and D. K. Heckendorn, *J. Am. Chem. Soc.*, **1991**, *113*, 9682.

87. T. B. R. A. Chen, J. J. Burger, and E. R. de Waard, *Tetrahedron Lett.*, **1977**, 4527; J. J. Burger, T. B. R. A. Chen, E. R. de Waard, and H. O. Huisman, *Tetrahedron*, **1981**, *37*, 417.

CHAPTER 8

Chemistry of Cyclic Sulphones

The synthesis and reactivity of some cyclic sulphones follows closely the chemistry described, largely for acyclic derivatives, in previous chapters. However, the chemistry of some cyclic sulphones, particularly three-, four- and some five-membered ring sulphones, is often unique to that particular ring system. For this reason the chemistry of such cyclic sulphones merits special attention in a separate chapter.

The important cyclic sulphone structures (1)-(8), along with the accepted nomenclature, are shown below. The compound names follow from those of the parent sulphur heterocycle with dioxide indicating the sulphone oxidation state (the monooxides, i.e. sulphoxides, are also well known).

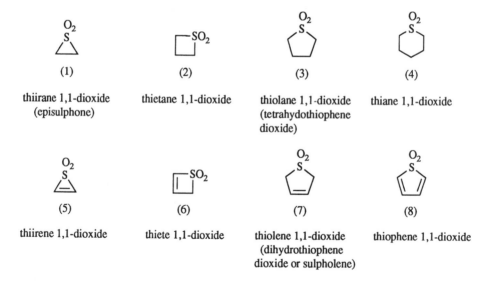

(1)	(2)	(3)	(4)
thiirane 1,1-dioxide (episulphone)	thietane 1,1-dioxide	thiolane 1,1-dioxide (tetrahydothiophene dioxide)	thiane 1,1-dioxide

(5)	(6)	(7)	(8)
thiirene 1,1-dioxide	thiete 1,1-dioxide	thiolene 1,1-dioxide (dihydrothiophene dioxide or sulpholene)	thiophene 1,1-dioxide

This chapter concentrates on the unique chemistry of the small ring systems, particularly (1) and (5)-(8), although some larger ring compounds and some polyfunctionalised normal-sized cyclic sulphones are also included.

8.1 Three-membered Ring Sulphones

Thiirane 1,1-dioxides, or episulphones, have been described in Chapter 7 in the discussion of the Ramberg–Bäcklund reaction. Under typical Ramberg–Bäcklund reaction conditions, i.e. strong base, the intermediate episulphone is almost never isolable. Syntheses of episulphones have therefore relied heavily on three alternative methods of preparation.

The earliest of these involves the reaction of a diazoalkane with SO_2, for example to give (9) and (10), Scheme 1.

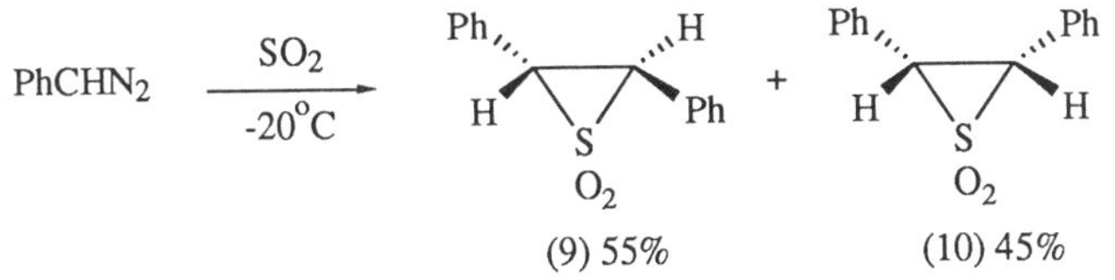

(9) 55%   (10) 45%

Scheme 1

This method, originally reported by Staudinger and Pfenninger,[1] has subsequently been used by the groups of Bordwell,[2] Neureiter,[3] and Matsumura[4] for the synthesis of episulphones. The products are thermally sensitive, and on warming decompose stereospecifically to give the corresponding alkene and SO_2. Single stereoisomers of the episulphones have been isolated by either fractional crystallisation at low temperature, or by warming to effect selective destruction of the more labile of the two isomers. In this way cis-episulphone (10) can be obtained in high purity, and converted to (Z)-stilbene. The main impetus for the synthesis of episulphones has been the study of their base-accelerated conversion to alkenes, particularly the stereochemical aspects. Thus, in contrast to the stereospecific loss of SO_2 from (10) observed simply on warming, treatment with dilute alkali gives mixtures of (Z)- and (E)-stilbene. This is due to the formation of the derived sulphonyl carbanion which allows episulphone epimerisation. The strength of the base needed to effect this epimerisation depends on the structure of the episulphone. Thus, whilst loss of stereospecificity in the conversion of (10) to (Z)-stilbene is observed using dilute NaOH,[4] the decomposition of the corresponding butene episulphone (Ph replaced by Me) maintains stereospecificity unless a stronger base such as ^tBuOK is employed.[3]

The mechanism of these reactions has been discussed in detail by Bordwell,[2] the evidence pointing to a non-concerted process. Of the various dipolar ion and diradical mechanisms possible, that shown in Scheme 2 is considered the most consistent with experimental evidence.

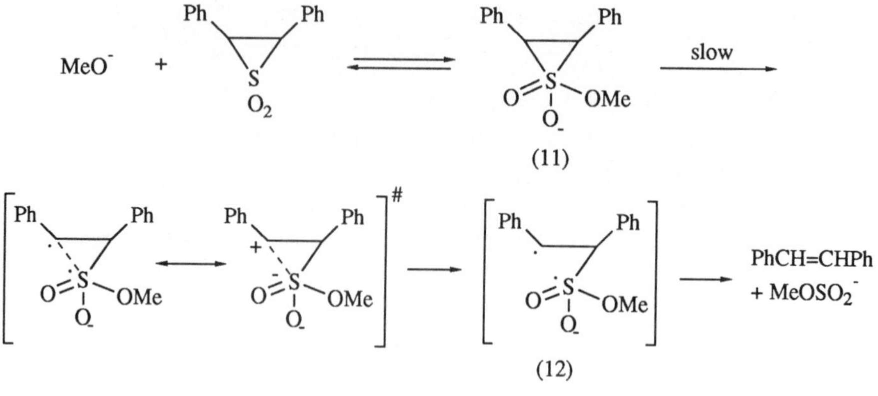

Scheme 2

Thus addition of alkoxide to the sulphone group leads to (**11**) and then to diradical (**12**) via a transition state having dipolar character. Decomposition of (**12**) must be faster than rotation in order to maintain the overall stereospecificity of the SO_2 extrusion. Evidence in favour of diradical intermediates includes the lack of by-products which could reasonably be expected from a purely dipolar process, and the observation that stored samples of episulphones always give some polysulphone. The rather unusual attack at the sulphone sulphur atom is lent some support by findings that some organometallic reagents attack episulphones in this way (e.g. Scheme 6).

The second method for the preparation of episulphones involves the treatment of a sulphonyl chloride with a diazoalkane in the presence of a tertiary amine.[5]   A notable example is the preparation of episulphone (**13**) from D-camphorsulphonyl chloride, Scheme 3.[6]

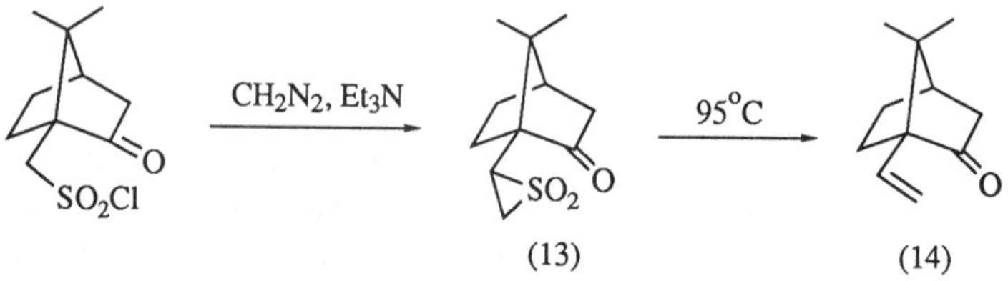

Scheme 3

The episulphone (**13**), of undefined stereochemistry at C-10, can be isolated as colourless crystals. On heating, this sulphone gives the vinyl ketone (**14**) in good yield. The mechanism of this episulphone preparation is actually closely related to that of the Staudinger method described in Scheme 1. Both involve the generation of sulphene intermediates, Scheme 4.

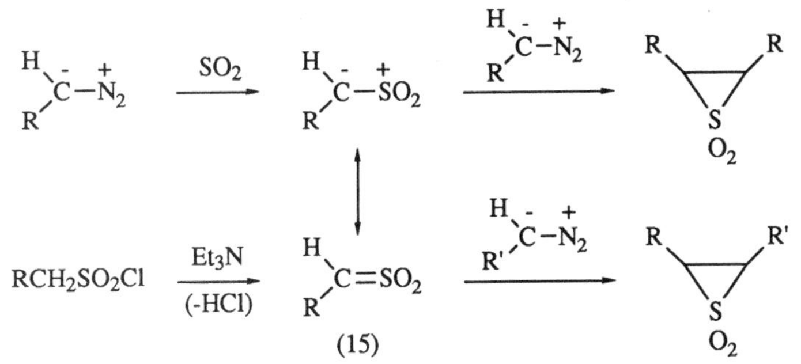

Scheme 4

In the first case the diazoalkane serves to form the sulphene (**15**) and subsequently as its reaction partner, thus producing a symmetrical episulphone product. By using a sulphonyl chloride to generate the sulphene partner, unsymmetrical episulphones such as (**13**) become available.

The final method for generating episulphones involves the reaction of a sulphene with another molecule of sulphene in the form of an amine complex, Scheme 5.[7,8]

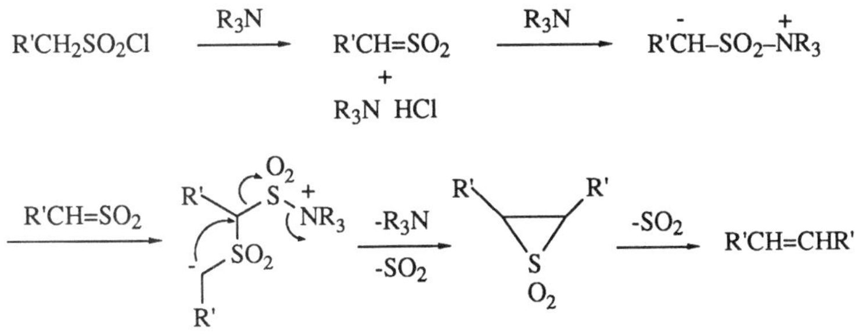

Scheme 5

Depending upon the reaction conditions, and particularly the temperature, either the episulphone or derived symmetrical alkenes can be isolated from this reaction. The detailed studies by Opitz and co-workers have shown that both the stereochemical outcome of this sequence, and indeed the chemical fate of sulphene intermediates, can be directed by the nature of the amine base employed.[8]

Due mainly to their instability, and notwithstanding the base-mediated SO_2 eliminations already described, hardly any chemical transformations of episulphones have been explored. Two notable exceptions are the reaction of episulphones with organometallics, and with complex metal hydrides, described by the groups of Vilsmaier[9] and Matsumura[10] respectively. The first type of reaction provides some rather significant results in that it enables easy distinction between two possible reaction pathways, Scheme 6.[9]

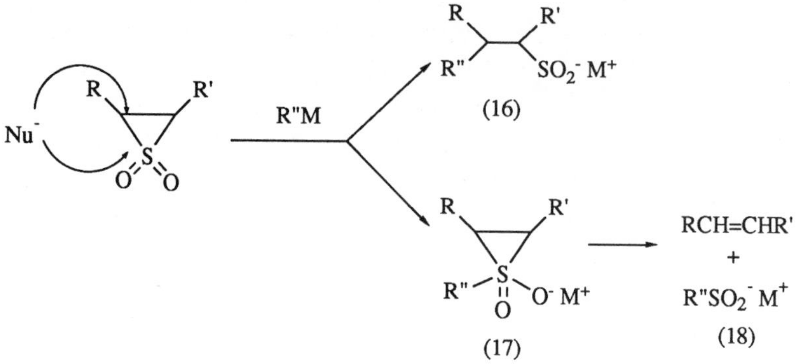

Scheme 6

Thus initial attack at carbon should result in the formation of an extended sulphinate (16) by C–S cleavage. Alternatively, attack on the sulphur atom of the episulphone would lead to (17) and then to the alkene and a sulphinic acid (18). In the event, only the latter outcome is observed, with the reaction of episulphone with RLi or R_2Mg giving good yields of RSO_2Li or $((RSO_2)_2Mg$. This result lends considerable support to the mechanism outlined in Scheme 2 for the base-mediated decomposition of episulphones. Nucleophilic opening of episulphone by halogen is also observed on reaction with Grignard reagents (RMgX as opposed to R_2Mg), furnishing mixtures of $(RSO_2)_2Mg$ and $(HalCH_2CH_2SO_2)_2Mg$ sulphinate salts.

In the reactions of episulphones with reducing agents such as $LiAlH_4$, $NaBH_4$ and $LiBH_4$ three types of product can be obtained, Scheme 7.[10]

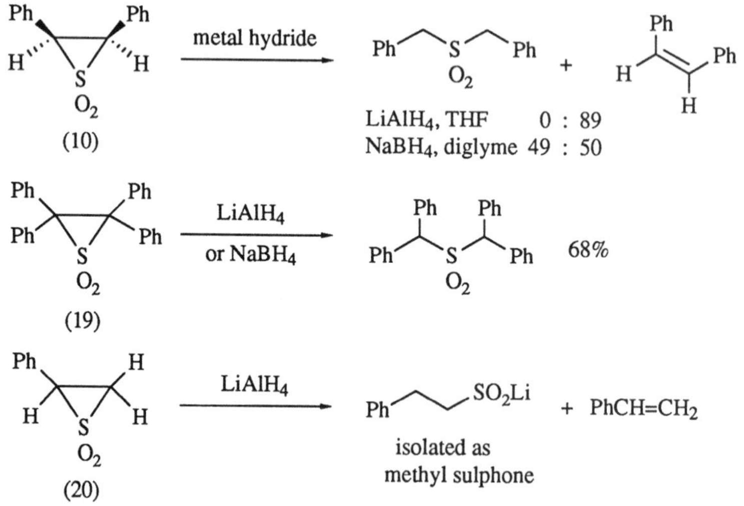

Scheme 7

The *cis*-disubstituted episulphone (**10**) follows exclusively the SO_2 elimination pathway with $LiAlH_4$ in THF, giving (Z)-stilbene. In other solvents, however, $LiAlH_4$ also gives some of the C–C cleavage product dibenzyl sulphone. This is produced in higher yield by reaction with other reducing agents such as $LiBH_4$ and $NaBH_4$. Tetraphenylethylene episulphone (**19**) gives the C–C cleavage product in good yield, whereas styrene episulphone (**20**) gives β-phenylethylsulphinic acid along with styrene-derived products.

The SO_2 elimination to give alkene products presumably occurs along the lines shown in Schemes 2 and 6. The alternative C–C cleavage reaction seems to be favoured in episulphones having at least one phenyl group on each ring carbon. The main driving force for this reaction appears to be the relief of steric interaction between the neighbouring phenyl groups, C–C fission also being aided by the additional carbanion stabilisation offered by aromatic conjugation. Such C–C cleavage was apparently not encountered by Vilsmaier *et al.* in the organometallic reactions already described, possibly because no phenyl-substituted systems were employed.

Carpino and co-workers first described the synthesis and some characteristic reactions of simple thiirene 1,1-dioxides.[11] Two different approaches to the thiirene structure were investigated, Scheme 8.

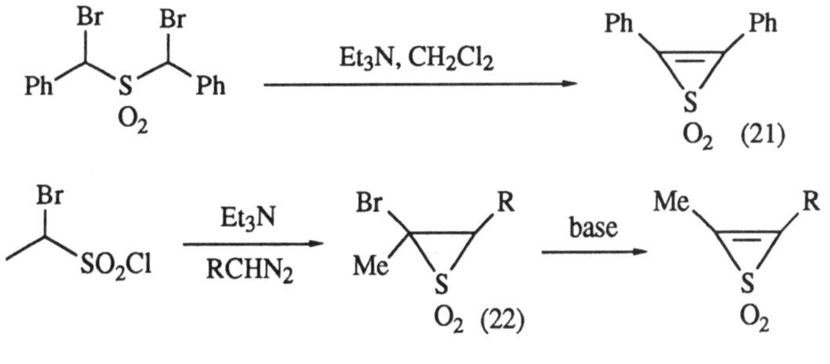

Scheme 8

Whilst a modified Ramberg–Bäcklund procedure serves to prepare the diphenyl-substituted compound (**21**) this procedure was found not to be suitable for the preparation of alkyl-substituted analogues. However, these compounds can be prepared by dehydrobromination of episulphones of structure (**22**), themselves available by the diazoalkane–sulphene reaction described earlier (Scheme 4).

Remarkably, the thiirene 1,1-dioxides proved rather more stable than their saturated episulphone counterparts. Carpino was unambiguously able to demonstrate the structure of these unusual compounds by spectroscopic examination, and from their chemical transformations, Scheme 9.

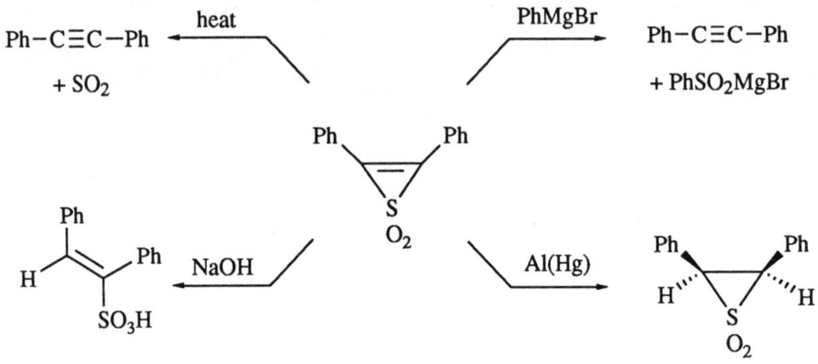

Scheme 9

This chemistry largely parallels that observed for the saturated analogues, attack of NaOH, PhMgBr and LiAlH$_4$ occurring on the sulphone group. By contrast, the methyl-substituted compound (**23**) reacts with base to give an unstable alkynyl sulphinate (**24**) which can be isolated as the corresponding alkynylsulphonyl chloride (**25**), Scheme 10.

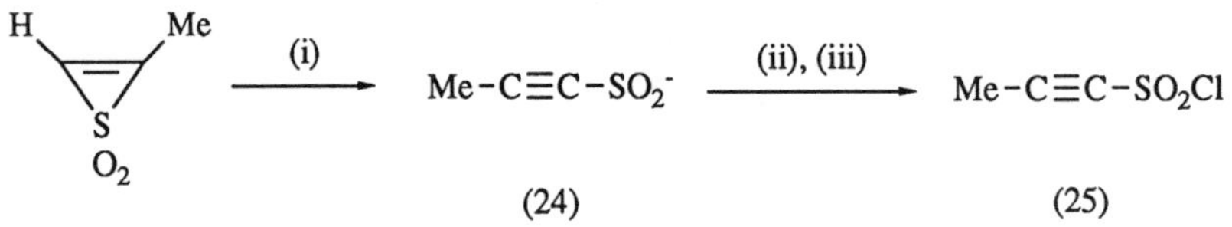

reagents:
(i) NaOH (ii) NaOCl; HCl (iii) NaHCO$_3$; PCl$_5$

Scheme 10

The synthesis of diaryl-substituted thiirene 1,1-dioxides such as (**21**) was also described by Philips *et al.*[12] Both this group, and that of Bordwell, proposed a stepwise mechanism for the SO$_2$ extrusion reactions of these compounds.[13]

Other reactions of thiirene 1,1-dioxides which have attracted attention include cycloadditions and nucleophilic additions. Examples in the former category include those with phenyldiazomethane,[11] benzonitrilium *N*-phenylimide (**26**),[14] mesoionic oxazolone (**27**),[15] and dienamine (**28**),[14] Scheme 11.

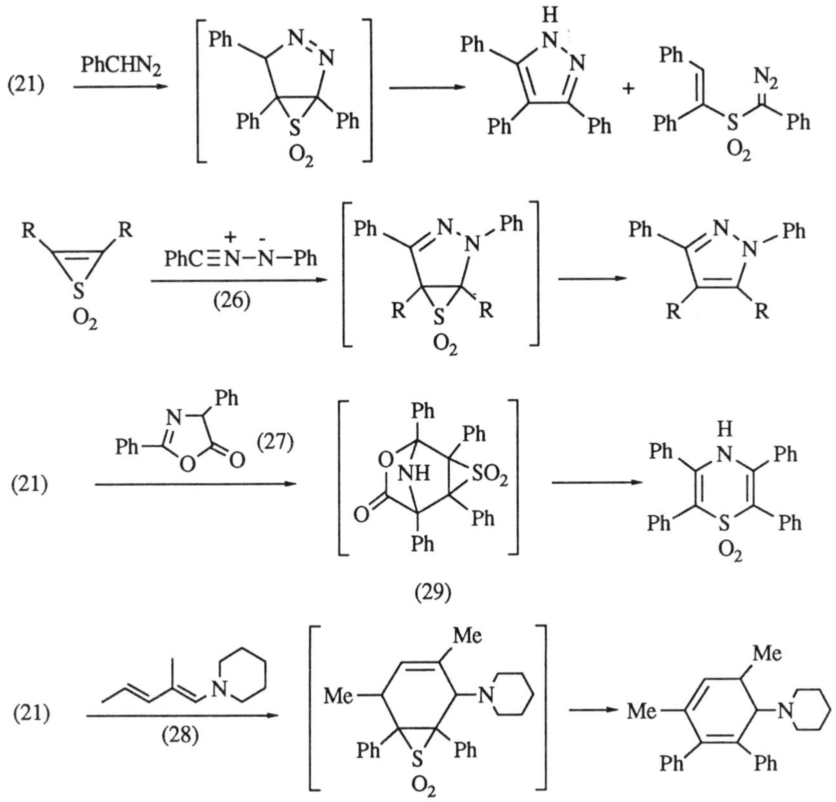

Scheme 11

In this type of reaction the cycloaddition results in the formation of an episulphone intermediate, which most usually decomposes to give a sulphur-free product. Interestingly (29) extrudes CO_2 rather than SO_2 to give the observed product.

A limited range of nucleophilic additions to the thiirene 1,1-dioxide system has been examined, most notably by Jarvis and co-workers, e.g. Scheme 12.[16]

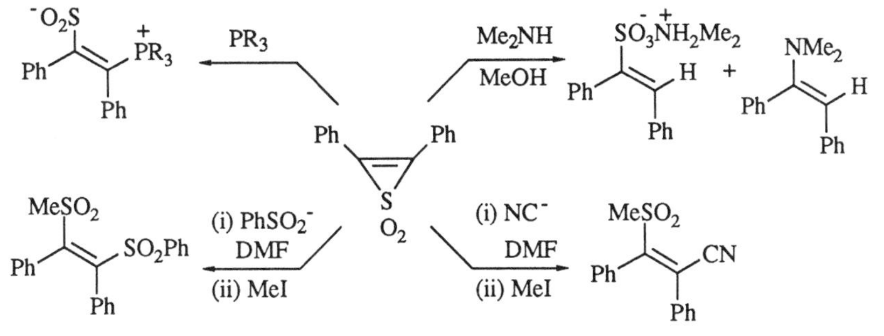

Scheme 12

Thus for relatively soft nucleophiles addition to the vinyl sulphone-type double bond, followed by either ring-opening to give a vinyl sulphinate, or protonation and SO_2 extrusion, appear to be common pathways.

A remarkable array of products is accessible from the reaction of thiirene 1,1-dioxides with α-metallated nitriles.[17] The reaction outcome is strongly dependent on the substituents present on the nitrile. Products arising from initial Michael addition or attack on the sulphone sulphur centre can be obtained, e.g. Scheme 13.

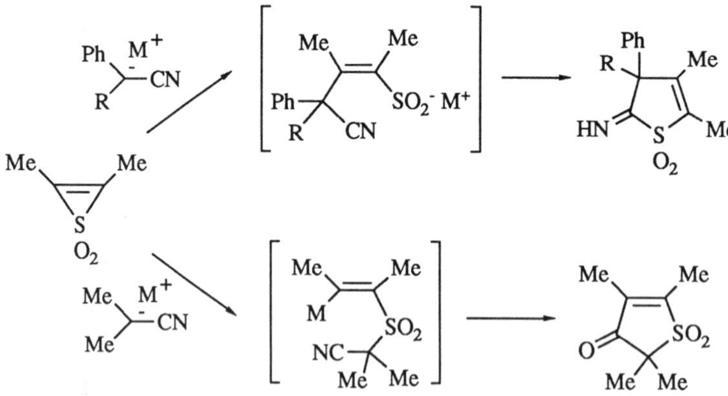

Scheme 13

It is clear that the reactions of thiirenes with nucleophiles offer a largely untapped entry into an interesting range of heterocyclic systems.

8.2 Four-membered Ring Sulphones

Thietane 1,1-dioxides can be accessed by the classical approach involving oxidation of thietanes, e.g. Scheme 14.[18]

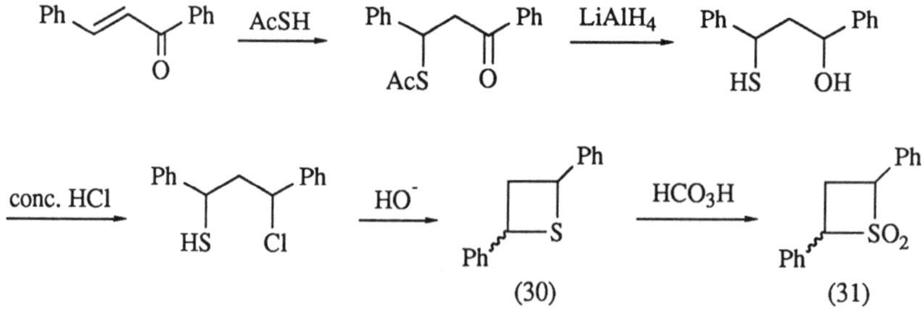

Scheme 14

Although thietane (**30**) is formed as a mixture of diastereoisomers it is possible to effect some degree of separation at this stage, and at the thietane 1-oxide stage, in order to access the dioxide (**31**) in either the *cis*- or *trans*-form.

By far the most important and direct route to thietane dioxides involves the use of sulphenes in cycloadditions with electron-rich alkenes. The chemistry of sulphenes has been reviewed,[19] and their behaviour in solution studied extensively.[20] A typical example is the reaction of sulphene (**32**), generated by base-induced dehydrohalogenation of a sulphonyl chloride, with an enamine, Scheme 15.[21]

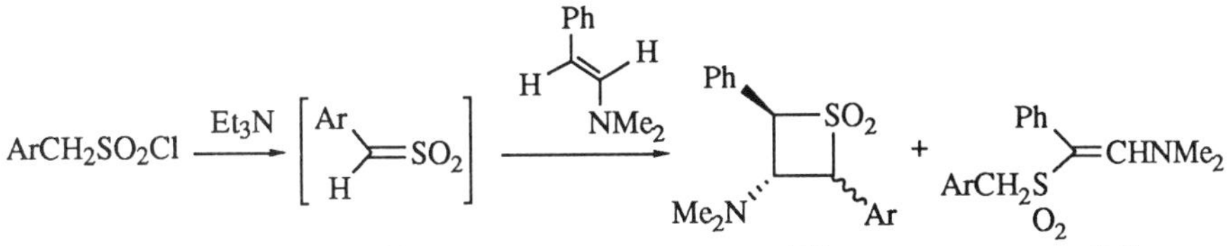

(32) (33) (34)

Scheme 15

Although the yields of cycloadducts such as (**33**) are usually high, some acyclic enamine substitution products such as (**34**) can also be formed. It should be noted that the intermediate involved in this reaction may be a sulphene–amine complex rather than a free sulphene.[20] The *trans*-stereochemistry of the starting enamine is translated into the cyclic product, the relative stereochemistry at the remaining centre being dependent on both the reaction conditions and the nature of the aryl group (Ar). The stabilities of the stereoisomeric products were found to be rather different, the isomer with the two aromatic groups *cis* being very much more stable than the *trans*-compound. In fact, thermolysis of the *trans*-compound leads to extensive decomposition by several pathways, including cycloreversion.

Paquette and Rosen probed the question of a concerted versus a non-concerted mechanism for the cycloadditions of sulphenes to enamines.[22] In a series of experiments which furnished a range of interesting products, including sulphones (**35**)-(**37**), good evidence for non-concerted reaction was obtained.

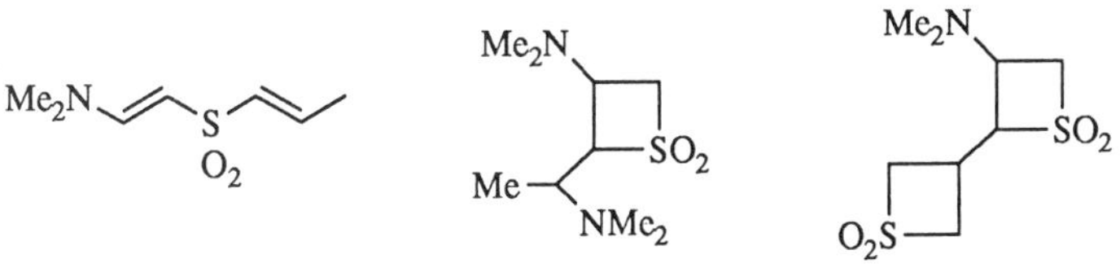

(35) (36) (37)

In many reports of sulphene cycloadditions stereochemical detail is lacking. The report of Stephen and Marcus indicates that such reactions can give products in stereoselective fashion, e.g. Scheme 16.[23]

Scheme 16

Thus in reactions of bicyclic and tricyclic enamines single cycloadducts are obtained. The stereochemical assignments were made on the basis of the expected least hindered (*exo*) approach of the sulphene to the enamine. The somewhat modest isolated yields of adduct do not preclude the possibility that the *endo*-product is also produced, but is lost or decomposed at some stage. Later studies have examined the addition of sulphene to a conformationally locked cyclohexanone-derived enamine, and negligible levels of stereoselectivity have been found.[24]

Sulphene cycloaddition with other types of electron-rich alkenes have also been described, Scheme 17.[25–27]

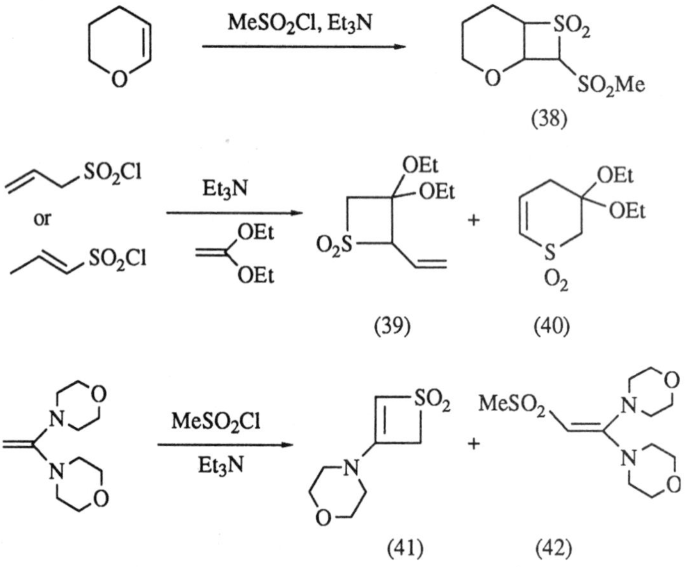

Scheme 17

In the reactions of various vinyl ethers, including tetrahydropyran, with CH_3SO_2Cl and Et_3N, no adduct of the simple parent sulphene is isolated.[25] Instead good yields of adducts such as (**38**) are obtained. This is due to the formation of mesyl sulphene (or its amine complex) under the reaction conditions, this being more reactive than sulphene itself.[20] The use of vinyl sulphenes allows an alternative [4+2]-type of cycloadduct, e.g. (**40**), to be formed in competition with the usual [2+2] adduct (**39**), although only to a small extent.[26] Both ketene O,N-acetals and ketene N,N-acetals have also been employed.[27] In the case of the bis(morpholine)enamine shown, the thiete 1,1-dioxide system (**41**) is formed in good yield when the sulphene reaction is conducted in solvents such as benzene. The product presumably arises by elimination from an intermediate aminal. In more polar solvents, and with groups other than morpholine, the formation of acyclic products analogous to (**42**) is favoured.[28]

Cycloadditions of bis(trifluoromethyl)sulphene (**43**) have also been described, e.g. Scheme 18.[29]

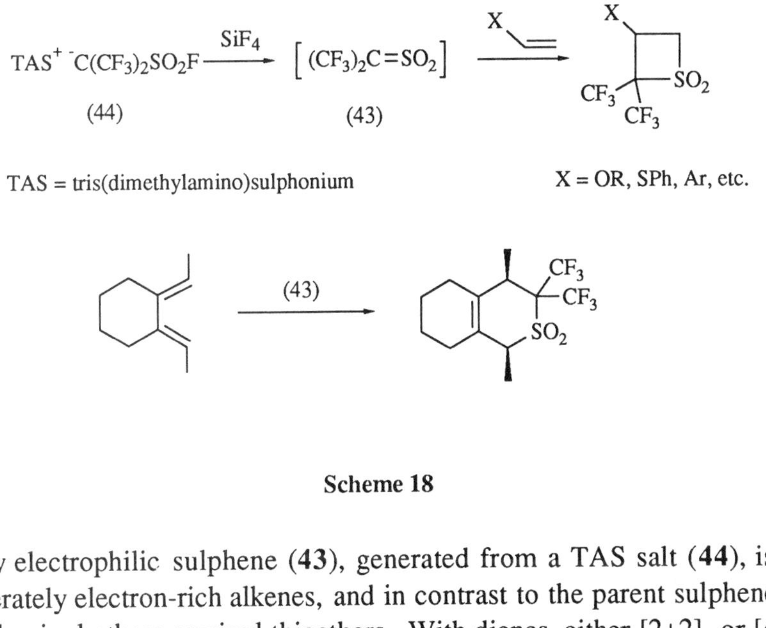

Scheme 18

The highly electrophilic sulphene (**43**), generated from a TAS salt (**44**), is reactive with moderately electron-rich alkenes, and in contrast to the parent sulphene it reacts with simple vinyl ethers or vinyl thioethers. With dienes, either [2+2]- or [4+2]-type cycloadducts are obtained. Ordinary alkenes such as cyclohexene and 1-phenyl-cyclohexene, or electron-poor alkenes such as vinyl acetate or acrylonitrile, do not give sulphene adducts. Although the simplest way to generate sulphenes is undoubtedly the treatment of alkanesulphonyl chlorides with triethylamine, this method has some disadvantages. The presence of an amine (or its hydrochloride salt) may be detrimental either to the sulphene partner or the reaction product, or cause other complications such as oligomerisation of the sulphene.[20,25] To overcome these problems Block and co-workers have developed a different synthesis of sulphenes which involves treatment of an α-silylsulphonyl chloride (e.g. $Me_3SiCH_2SO_2Cl$) with fluoride.[30] The method is rather mild and appears to give superior yields of sulphene adducts in many cases.

As in the case of the corresponding three-membered sulphones, much of the chemistry of thietane 1,1-dioxide systems has involved desulphurisation. Trost *et al.* carried out one such study in which stereochemical aspects were considered, Scheme 19.[31]

33.8 : 40.1 from *cis*-(45)
39.2 : 32.9 from *trans*-(45)

Scheme 19

The reaction was found to give yields of 1,2-dimethylcyclopropane of up to 50%, along with up to 12% of 2-pentenes. Thermolysis of each of the diastereoisomers of (45) gives slightly different ratios of the *cis*- and *trans*-cyclopropane products, indicating a low level of 'memory' in the reaction. Such SO_2 extrusions have also been carried out photolytically.[32]

In the case of thietane 1,1-dioxide systems having an appropriately positioned hydroxyl function, ring opening can occur by a retro-aldol process, e.g. Scheme 20.[33]

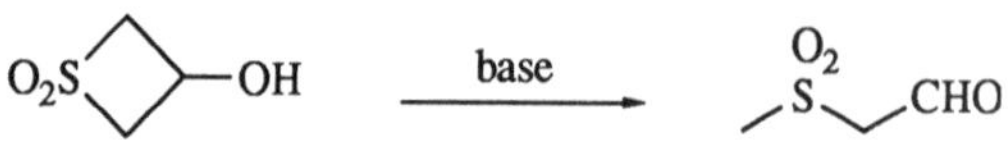

Scheme 20

Comparison of the rates of such cleavage processes with those of similar acyclic hydroxy sulphones allows estimation of the strain energy of the four-membered ring.

In a series of papers Dodson and co-workers demonstrated several interesting transformations of 2,4-diphenylthietane 1,1-dioxide, Scheme 21.[18,34]

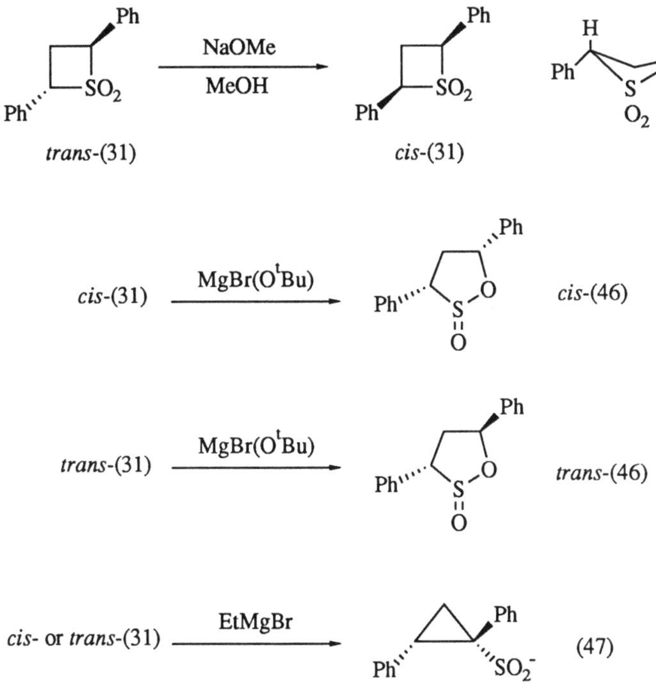

Scheme 21

Firstly, it was found that treatment of *trans*-(**31**) with base results in complete epimerisation to the *cis*-isomer. This is explained by the sulphone ring adopting a puckered conformation, thus allowing both phenyl substituents to occupy pseudoequatorial positions in the *cis*-isomer, as shown.[18]   A stereospecific transformation of either *cis*- or *trans*-(**31**) to the corresponding *cis*- or *trans*-(**46**) respectively, occurs on treatment with MgBr(O^{t}Bu). The reaction clearly involves a 1,2-bond migration, with the driving force being relief of ring strain. Both concerted and anion–diradical mechanisms are possible. Finally, the authors described the stereoselective transformation of either stereoisomer of (**31**) to the cyclopropanesulphinate (**47**) by a ring-contraction resembling a Stevens rearrangement.

A few scattered reports have shown that substitution of the thietane 1,1-dioxide ring can be effected by sulphonyl carbanion formation, Scheme 22.[35–37]

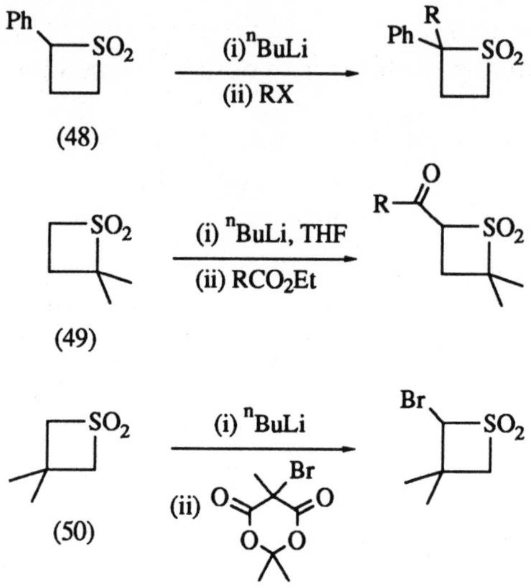

Scheme 22

Thus alkylation, acylation and bromination have been demonstrated with the simple systems (48)-(50).

The cycloaddition of sulphenes with enamines discussed earlier also provides a good route to thiete 1,1-dioxides along the lines generalised in Scheme 23.[38-40]

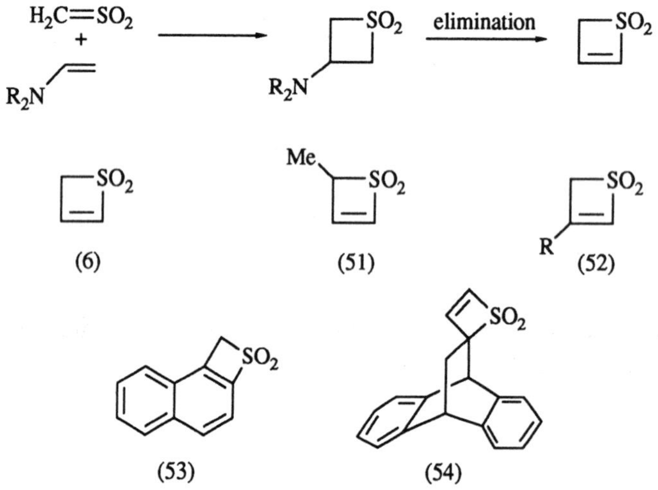

Scheme 23

This approach has been used to access the compounds (6) and (51)-(54) shown, including (51) in non-racemic form[39] and the naphtho-fused sulphone (53).[40] Either

Cope- or Hofmann-type elimination is usually employed to effect the thietane to thiete conversion. A different approach has also been examined by Dittmer and co-workers, Scheme 24.[41,42]

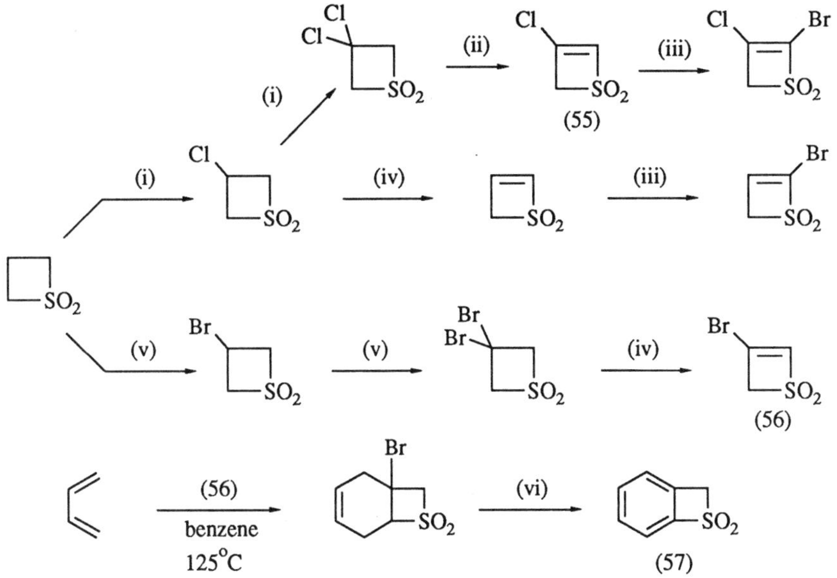

reagents:
(i) Cl₂, hv, CCl₄   (ii) Et₃N, toluene
(iii) Br₂, CCl₄, hv; Et₃N, toluene   (iv) Et₃N, benzene
(v) Br₂–Cl₂, hv, CCl₄   (vi) Br₂, CCl₄; DBN, benzene

Scheme 24

As can be seen, the radical halogenation of thietane 1,1-dioxide provides access to a range of chlorine- and bromine-containing four-membered ring sulphones. In particular the chlorothiete 1,1-dioxide (**55**) participates in useful addition–elimination chemistry[41] and the bromo derivative (**56**) allows the preparation of (**57**) by a [4+2] cycloaddition.[42]

Thermolysis of thiete 1,1-dioxide provides the cyclic sulphinic ester (sultine) (**58**) in good yield, Scheme 25.[43]

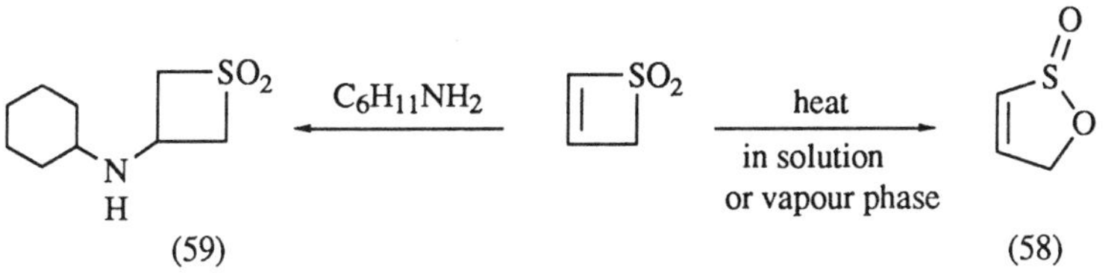

Scheme 25

The same paper describes attempts to establish the mechanism of this transformation, suspected to involve vinyl sulphene as an intermediate, and also includes the preparation of (59) by simple conjugate addition.

Dittmer and Davis examined the reactions of sulphone (60), Scheme 26.[44]

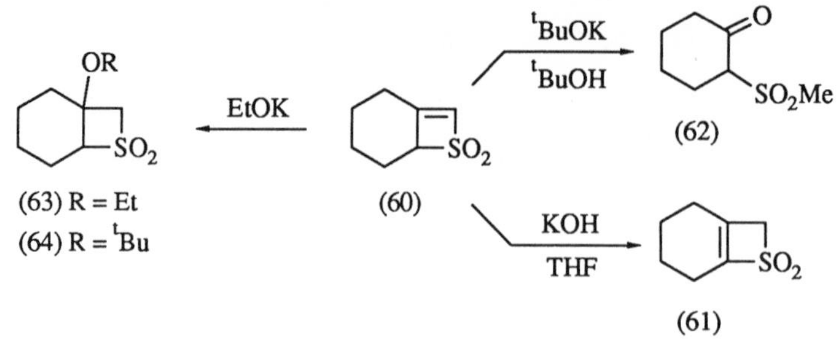

(63) R = Et
(64) R = ^{t}Bu

(60)

(62)

(61)

Scheme 26

Depending upon the base used this sulphone can be isomerised to (61), ring opened to (62) or converted to adduct (63). The formation of (62) was rationalised in terms of initial conjugate addition of *tert*-butoxide to give an adduct (64) which rearranges to (62) with loss of isobutylene. Reactions of (61) were also examined, including hydrogenation to give the thietane analogue, and dehydrogenation to give (57). A range of addition and addition–elimination chemistry of thiete 1,1-dioxides has been examined, e.g. Scheme 27.[41,45]

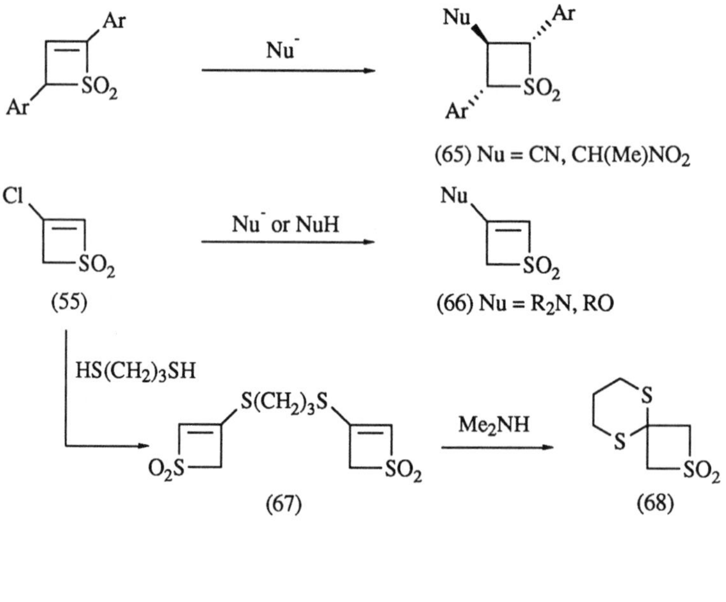

(65) Nu = CN, CH(Me)NO$_2$

(55)

(66) Nu = R$_2$N, RO

(67)

(68)

Scheme 27

Not surprisingly, addition reactions result in the formation of the more stable *cis*-diaryl arrangement in (**65**).[45] With (**55**), either addition–elimination to give (**66**) or double addition can occur, depending upon the nucleophile.[41] Treatment of the interesting bis(thiete 1,1-dioxide) (**67**) with base results in the formation of (**68**), presumably triggered by amine addition. Another contribution from Dittmer's group describes the aldol-type reaction of naphthothiete dioxide (**69**), Scheme 28.[46]

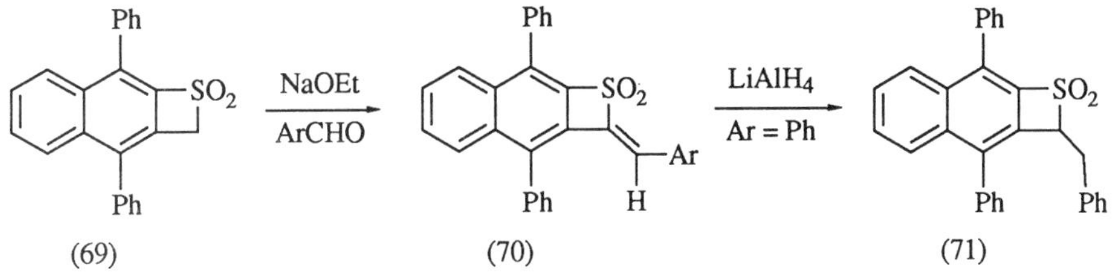

Scheme 28

Reaction with aromatic aldehydes proceeds in predictable fashion to give benzylidene products (**70**). Reaction of one such product (Ar = Ph) with LiAlH$_4$ results mainly in conjugate reduction to the benzyl-substituted naphthothiete dioxide (**71**).

Finally, it is worthy of note that additional heteroatoms can be incorporated into the ring of thiete 1,1-dioxides. For example, Block *et al.* described the synthesis of 1,3-dithietanes at various oxidation states, Scheme 29.[47]

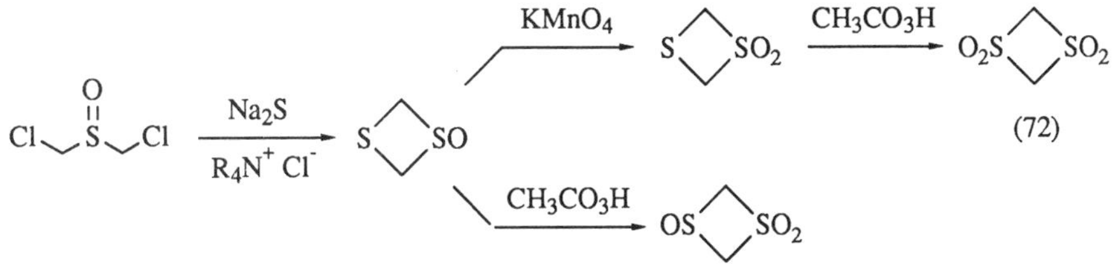

Scheme 29

In contrast to the puckered systems described above, 1,3-dithietane 1,1,3,3-tetraoxide (**72**) is nearly planar and almost square, as revealed by X-ray diffraction methods.

8.3 Five-membered Ring Sulphones

As mentioned earlier, the chemistry of many types of five-membered ring sulphones follows closely the usual types of sulphone reactivity described in earlier chapters. This chapter focuses on the atypical chemistry associated mainly with unsaturated five-membered ring sulphones.

8.3.1 Dihydrothiophene 1,1-dioxides (Sulpholenes)

Most of the chemistry of sulpholenes has been focused on methods to generate substituted 3-sulpholenes of general structure (**73**), which can then be thermolysed to effect conversion to a diene, Scheme 30.

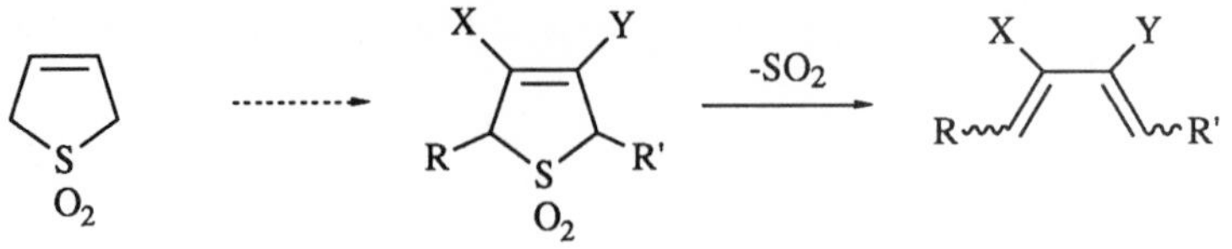

(7)   (73)

Scheme 30

Efforts have been concentrated on the development of methods for the conversion of simple starting materials such as the parent 2,5-dihydrothiophene 1,1-dioxide (or 3-sulpholene) (**7**) into more complex compounds (**73**) bearing alkyl and heteroatom substitution. Of key importance to the strategy outlined in Scheme 30 is the stereocontrol possible in the sulpholene elaboration and the SO_2 extrusion. The extrusion of SO_2 from sulpholenes (**7**), episulphones (**1**) and higher homologues (**74**) and (**75**) has been studied in detail by Mock.[48]

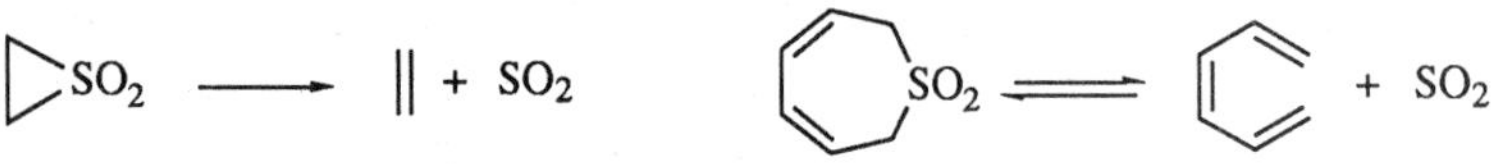

(1)   (74)

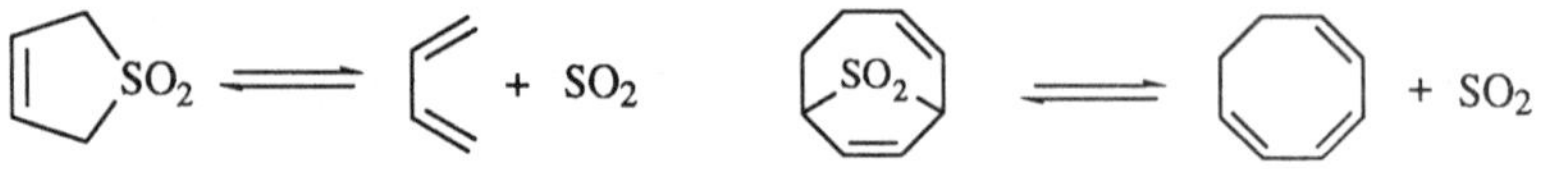

(7)   (75)

These reactions (and their reverse: SO_2 addition to dienes, trienes, etc.[49]) are found to proceed by concerted, cheletropic mechanisms. The orbital symmetry control in such reactions has important stereochemical consequences, and hence implications for their synthetic utility.

As described above, episulphones decompose stereospecifically by a concerted suprafacial process. This apparently 'disallowed' transformation has been rationalised as a 'nonlinear' concerted cycloregression, although in some cases a stereospecific diradical mechanism may be involved. The stereochemical results for the sulpholene and dihydrothiepin 1,1-dioxide SO_2 extrusions demonstrate that these reactions are also highly stereospecific, Scheme 31.

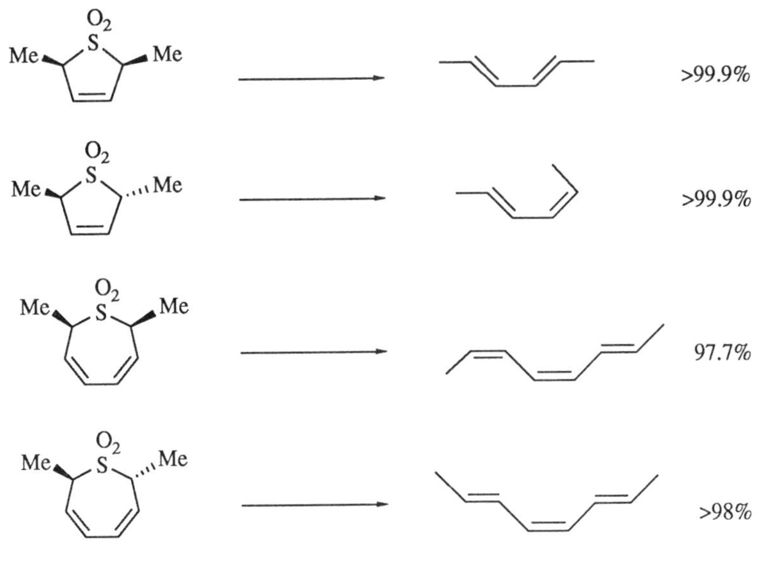

Scheme 31

The extrusion proceeds in a suprafacial fashion for the sulpholene system, and by an antarafacial process for the seven-membered ring sulphones.

Most synthetic chemistry involving 3-sulpholenes has centred on the stereoselective introduction of substituents at the 2- and 5-positions. Stereochemical information in the sulpholene is then translated into alkene geometry on thermolysis and SO_2 extrusion.

A problem in the manipulation of 3-sulpholenes is their tendency to undergo eliminative ring-opening on treatment with basic reagents, Scheme 32.[50,51]

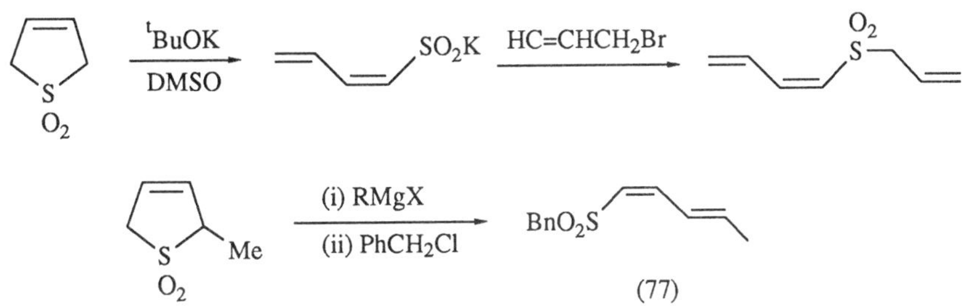

Scheme 32

This type of reaction is an effective (and largely overlooked) method for the preparation of dienyl sulphinic acids and derived dienyl sulphones. The opening reaction of unsymmetrical sulpholenes such as (**76**) with Grignard reagents is regioselective, the product presumably being the (*Z,E*)-isomer (**77**).[51] An extension of this reaction has been described which allows the initially formed sulphinate salts to be converted into dienyl sulphoxides, Scheme 33.[52]

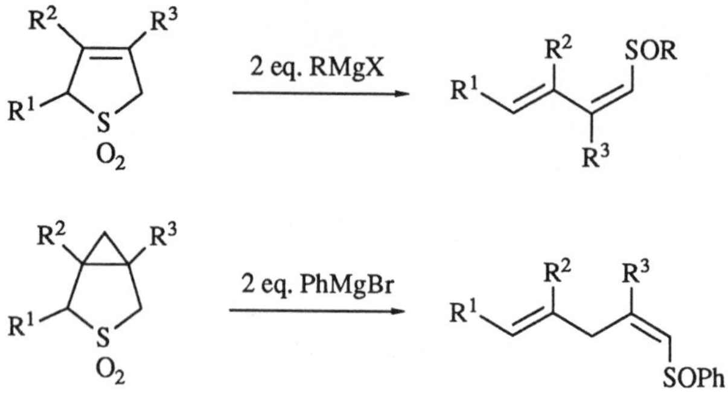

(78)

Scheme 33

Although the yields are only moderate, and rather variable, this is a rather direct route to 1,3-dienyl sulphoxides. The method can also be applied to the synthesis of 1,4-dienyl sulphoxides by employing the cyclopropyl sulphone (78) as the starting material.[53] Whilst most synthetic work has involved the manipulation of simple preformed sulpholenes, a wide range of *de novo* syntheses of specific systems are available, e.g. Scheme 34.[54–56]

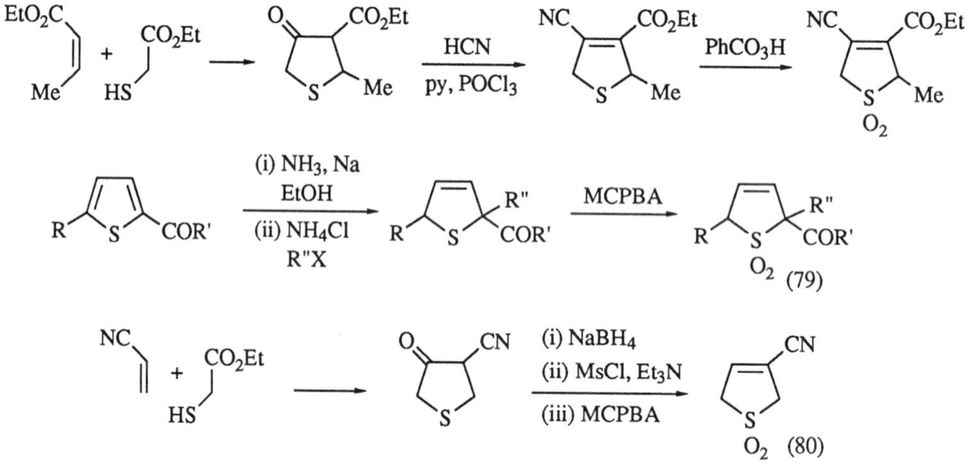

(79)

(80)

Scheme 34

Further detailed discussion of the availability of substituted thiophenes and their derived oxides is beyond the scope of this book. Suffice to say that these compounds are readily available. Both sulpholenes (79) and (80) were submitted to thermolysis to produce the corresponding diene, Scheme 35.

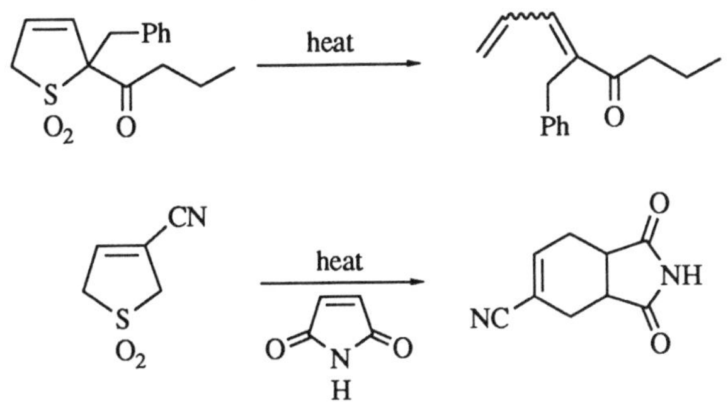

Scheme 35

These examples illustrate the two main synthetic uses for sulpholene SO$_2$ extrusion. Thus, whilst stereocontrolled diene synthesis is often the ultimate aim, equally significant is the use of the generated diene for *in situ* cycloaddition chemistry. Examples of the latter type of strategy are discussed in Chapter 6.

Some examples of sulpholene manipulation involving the introduction of heteroatoms with no change in the carbon skeleton are outlined in Scheme 36.[57–59]

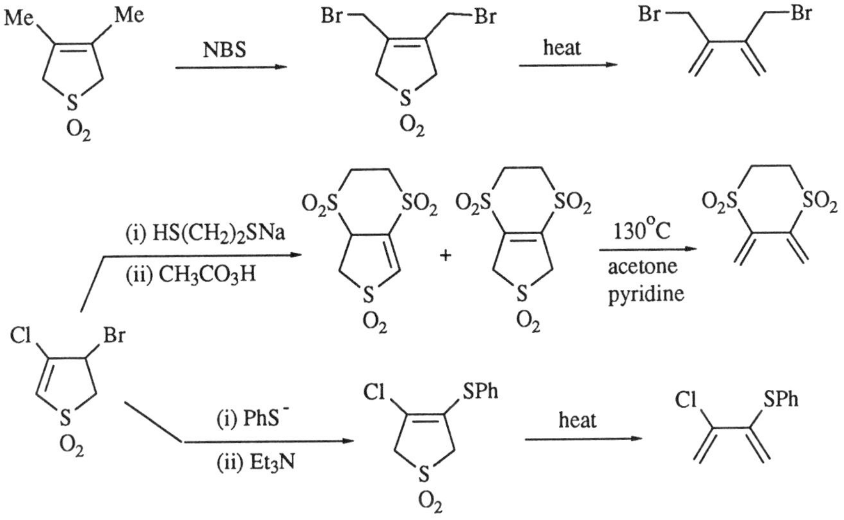

Scheme 36

Such transformations make available a host of dienes, both electron-rich and electron-poor, for cycloadditions.

The introduction of aromatic and vinylic substituents onto a sulpholene nucleus using palladium chemistry has also been examined, Scheme 37.[60,61]

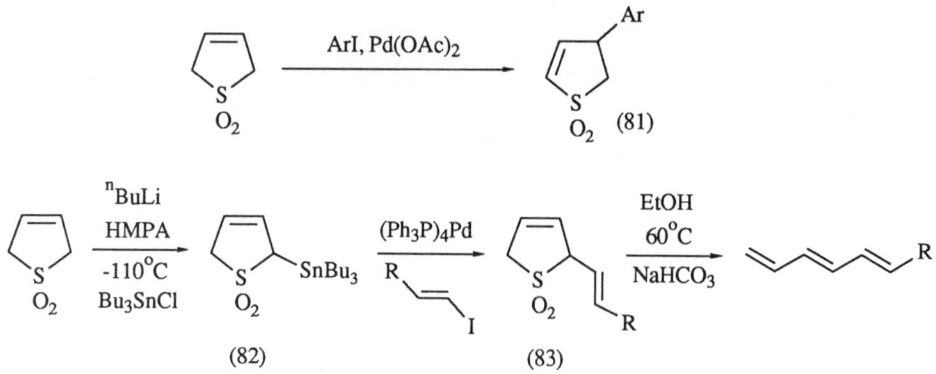

Scheme 37

The synthesis of (**81**) represents the straightforward application of a Heck reaction to the synthesis of a 4-aryl 2-sulpholene.[60] Palladium-catalysed coupling of 2-stannyl sulpholenes (**82**) with vinyl iodides gives substituted products (**83**), which on heating furnish trienes in a stereoselective fashion.[61]

The SO_2 extrusion reactions are usually carried out under relatively mild conditions, *ca.* 60–200°C, although on occasion higher temperatures are required. The reaction can also be carried out at lower temperatures by the use of lithium aluminium hydride,[62] or ultrasonically dispersed potassium[63] to promote the extrusion.

The very low reaction temperature utilised in preparing stannyl sulpholene (**82**) is typical if unwanted ring opening along the lines shown in Schemes 32 and 33 is to be avoided. Takayama and co-workers first described methods for the direct alkylation of 3-sulpholenes, Scheme 38.[64]

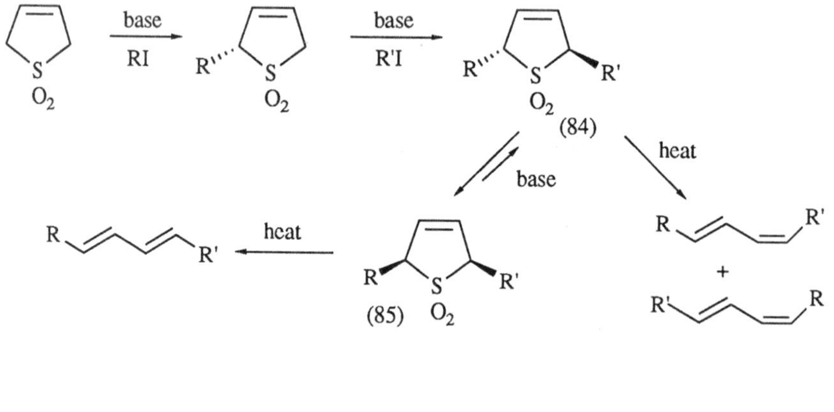

Scheme 38

Sequential alkylations can best be accomplished by employing $(MeSi)_2NLi$ and an alkyl iodide at -78°C. The success of the procedure relies on the generation of the unstable sulpholene anion in the presence of the electrophile. Alternative conditions include the use of ^nBuLi at -105°C in a mixture of THF and HMPA (other higher temperature alkylations have also been developed; *vide infra*).[65] Of great significance is the high regio- and stereoselectivity observed in the formation of *trans*-(84). Under neutral conditions these products undergo SO_2 extrusion to give the expected (E,Z)-dienes. However, under basic conditions $(K_2CO_3, EtOH, 125°C)$ initial sulpholene epimerisation occurs to give the *cis*-compounds (85) which then give (E,E)-diene products. This finding adds versatility to the method since either kind of diene can be accessed at will.

Chou *et al.* have further developed the idea of sulpholene alkylation by anion generation in the presence of an electrophile.[66] Their procedure utilises NaH in DMF at -4°C, Scheme 39.

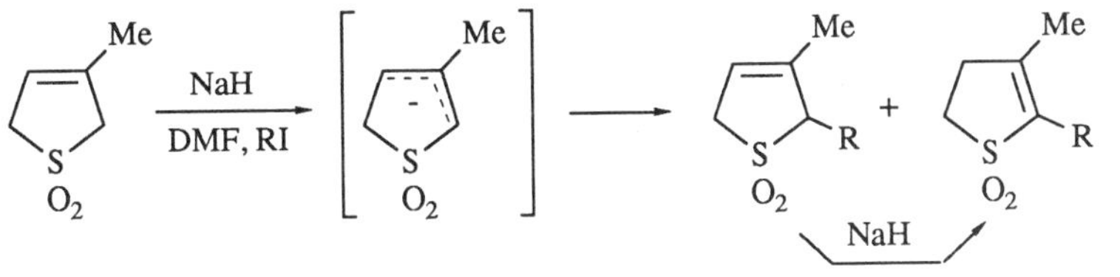

Scheme 39

The use of alkyl iodides is again crucial for optimal results, alkyl bromides giving lower yields. Remarkably, in the case shown, anion formation is highly regioselective, giving the product alkylated only at the position adjacent to the methyl substituent. This effect was also observed previously in the Grignard-promoted sulpholene ring-opening reactions described in Scheme 32.[51] Clean isomerisation of the 3-sulpholene products to the corresponding 2-sulpholenes is also possible using NaH in DMF.

Further reports from Chou's group describe interesting extensions of this chemistry for the preparation of polyalkylated and polycyclic sulpholenes, e.g. Scheme 40.[67,68] Thus di-, tri- and tetramethylation of 3-sulpholene can be controlled simply by adjusting the number of equivalents of base/MeI mixture employed. The formation of bis-spiroannulated sulpholene (86) and the bridged product (87) are variants on the same theme, the latter offering a useful entry into cycloheptadienes by the $LiAlH_4$-promoted SO_2 extrusion.

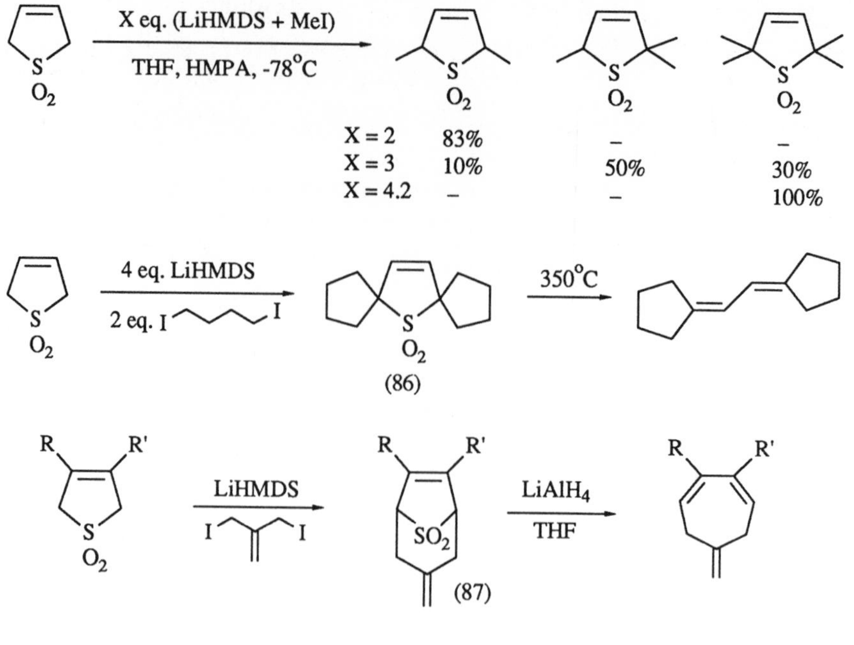

Scheme 40

Modification of the behaviour of sulpholene anions can be accomplished by introducing additional anion-stabilising groups such as trimethylsilyl, or additional sulphone groups, Scheme 41.[69,70]

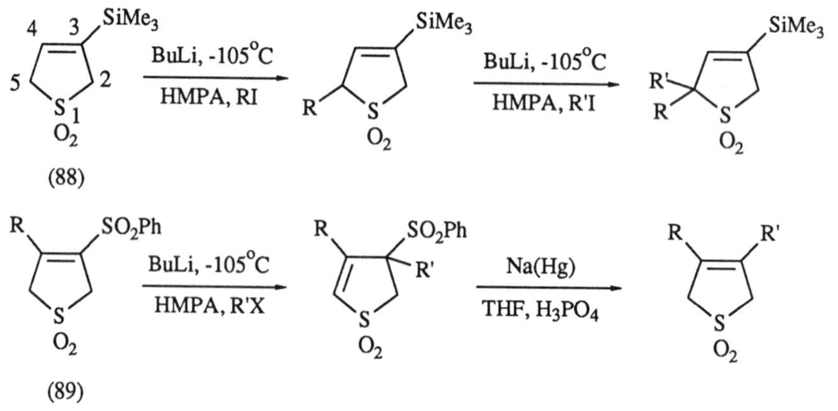

Scheme 41

3-Silylated sulpholene (**88**) (and the corresponding 2-silylated isomer) are readily available from the parent sulpholene.[71] Sequential alkylations of this compound occur at the 5-position, in sharp contrast to the 2,5-dialkylation shown in previous schemes.[69] Further regiocomplementary results are obtained using a dianion derived from (88). Along similar lines the disulphone (89) allows overall substitution at the 3-position.[70]

Another contribution from Chou *et al.* describes the acylation, and one-pot acylation–alkylation, of 3-sulpholenes, Scheme 42.[72]

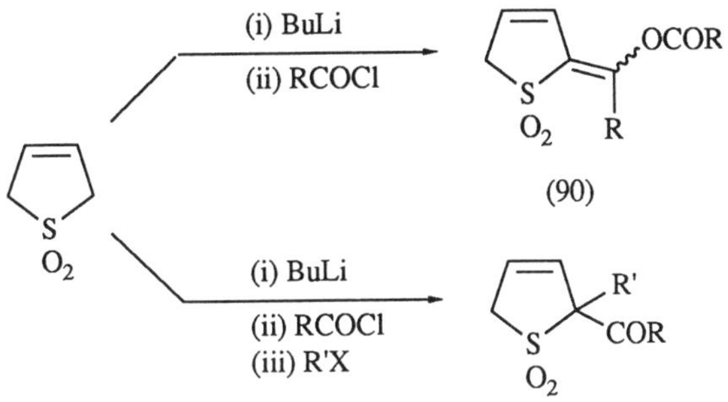

Scheme 42

With acid chlorides a double acylation occurs, resulting in the formation of products such as (90). By using a deficiency of acylating agent and then adding an excess of a reactive alkyl halide the second acylation can be avoided, resulting in the formation of (91). Interestingly this paper also reports that simple sulpholene carbanions can be generated and remain stable at -105°C for at least 15 minutes.

Reactions of 3-sulpholene carbanions with carbonyl compounds have been examined by Takayama's group, Scheme 43.[73]

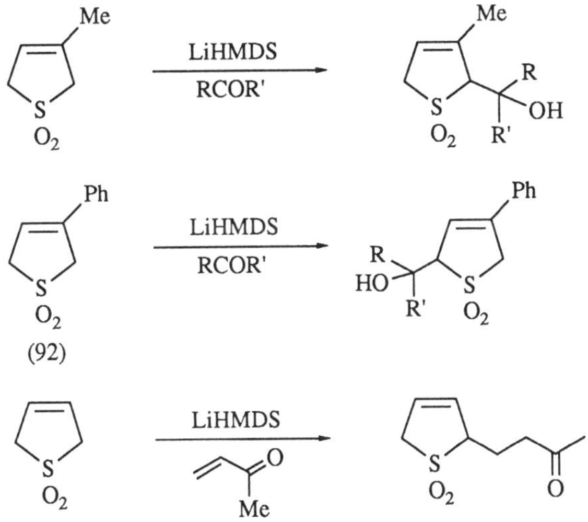

Scheme 43

These reactions were conducted at -78°C by adding the base in one portion to a mixture of the sulpholene and carbonyl compound. The reactions with 3-phenyl-sulpholene (**92**) show a reversal in regiochemical outcome compared to those for the 3-methyl series. With methyl vinyl ketone predominant Michael addition is observed to give ketosulphones such as (**93**).

An alternative method for achieving regioselective hydroxyalkylation of sulpholenes involves allylzincation with bromosulpholene (**94**), Scheme 44.[74]

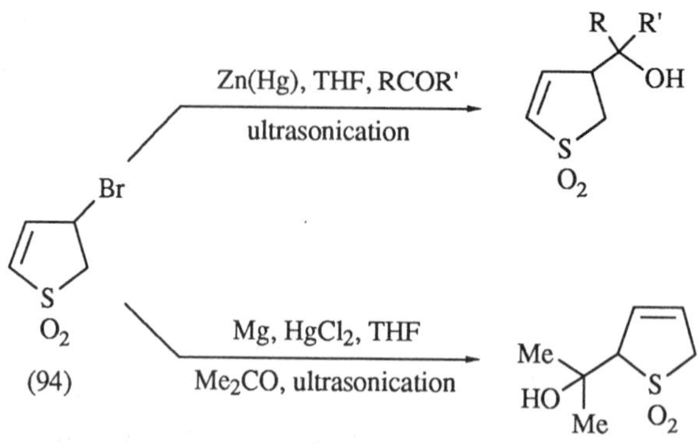

Scheme 44

Vital to the success of the reaction is the use of ultrasound, otherwise little or no product is obtained and (**94**) simply suffers debromination. Complementary regiocontrol appears possible by changing the metal used to magnesium. Oxidation of some of the secondary alcohol products obtained was also examined. The same chemistry can also be applied to the analogue of (**94**) having an additional phenylthio substituent at C-3.[75] This chemistry is particularly significant in allowing facile access to 3-substituted compounds, since the sulpholene carbanion chemistry usually only allows substitution α to the sulphone function.

All of the above chemistry has concentrated on using the sulpholene as a nucleophile. A few reports also describe efforts to develop useful electrophilic sulpholenes, e.g. Scheme 45.[76,77] Bromosulpholene (**94**) reacts with various heteronucleophiles and organocuprates to give substitution products in good yield.[76] The *exo*-methylene derivative (**95**) can react with nucleophiles at C-6, and undergoes substitution through carbanion chemistry at C-5. Nitronate anions proved the most effective carbon nucleophiles in the former type of reaction, one example allowing a synthesis of the simple naturally occurring substituted butadiene ipsenol.[77] Other simple diene-containing natural products which have been obtained using the SO_2 extrusion reaction include *trans*-β-ocimene and α-farnesene,[78] and components of the fire-ant trail pheromone.[79] The sulphone alkylation strategy has also proved useful in the synthesis of vitamin D_3 19-alkanoic acids.[80]

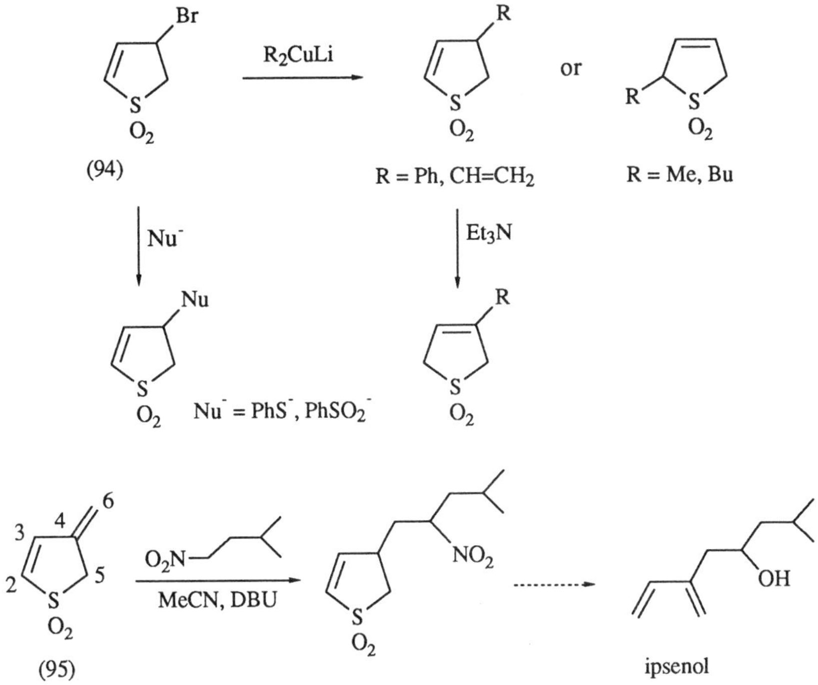

Scheme 45

The other main use of the SO_2 extrusion reaction has been in the generation of dienes for (usually *in situ*) Diels–Alder cycloadditions. This strategy can be particularly useful for assembling systems capable of undergoing intramolecular Diels–Alder (IMDA) reactions as exemplified by the synthesis of selina-3,7(11)-diene (**96**),[81] and elaeokanine A (**97**),[82] Scheme 46.

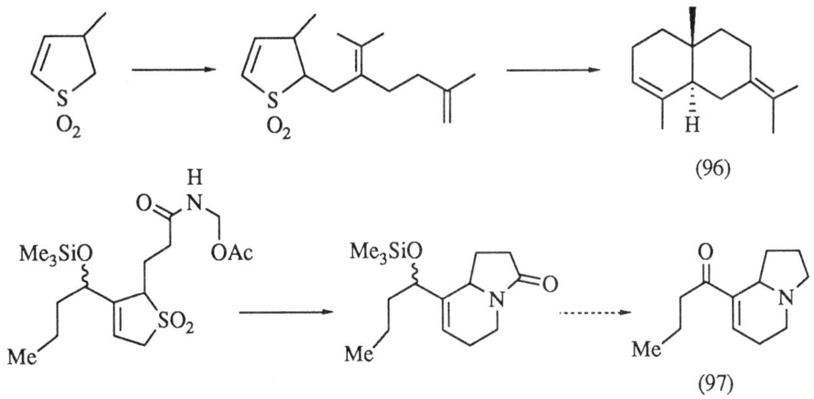

Scheme 46

SO_2 extrusion from benzo-fused thiophene 1,1-dioxides is a useful entry into *ortho*-quinodimethane intermediates. This chemistry has its origins in the report from Cava and Deana which described thermolysis of sulphone (**98**) in the presence of *N*-phenylmaleimide to give cycloadduct (**99**) in 78% yield, Scheme 47.[83]

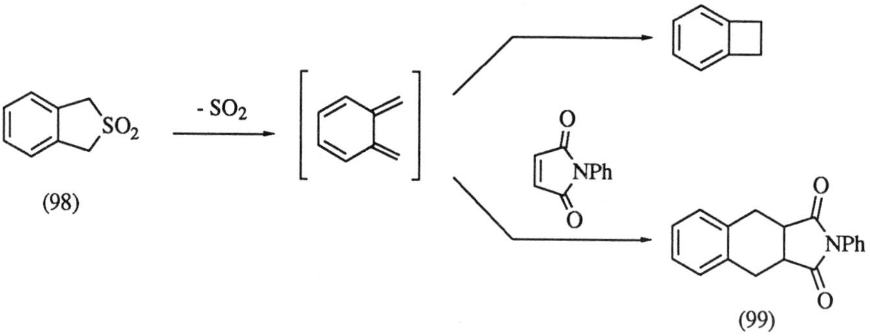

Scheme 47

In the absence of a trap for the presumed quinodimethane intermediate benzocyclobutene is produced in low yield. Durst and co-workers have recently examined analogous chemistry using systems of structure (**100**) incorporating an additional heteroatom.[84] The furan-annulated sulpholene (**101**) has also seen use in this type of Diels–Alder chemistry.[85]

(**100**) X = OR, SR, NR$_2$ (**101**)

Examples involving SO_2 extrusion followed by IMDA reaction are also known, e.g. Scheme 48.[86]

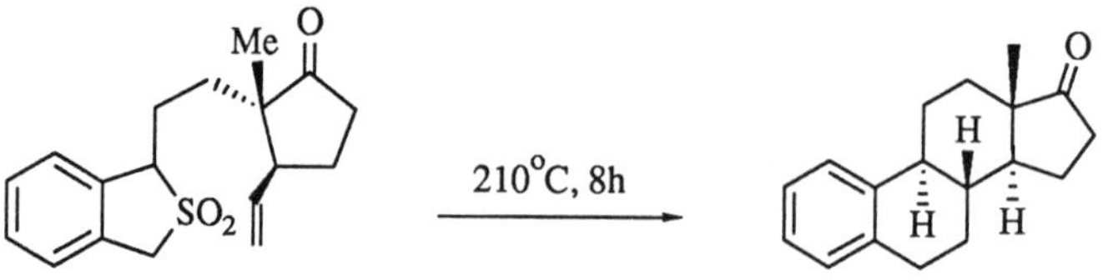

Scheme 48

In this case a remarkably efficient route to steroidal and other polycyclic systems was developed. In contrast to the troublesome sulpholene carbanion chemistry the alkylation of 1,3-dihydrobenzo[c]thiophene 2,2-dioxides such as (**98**) is straightforward. It has also proved possible to extrude SO_2 from five-membered ring sulphones fused to another appropriately reactive ring system. One such example provided the first, indirect, solution to the alkylations of sulpholenes, by using a cyclopentadiene adduct as a masked 3-sulpholene, Scheme 49.[87]

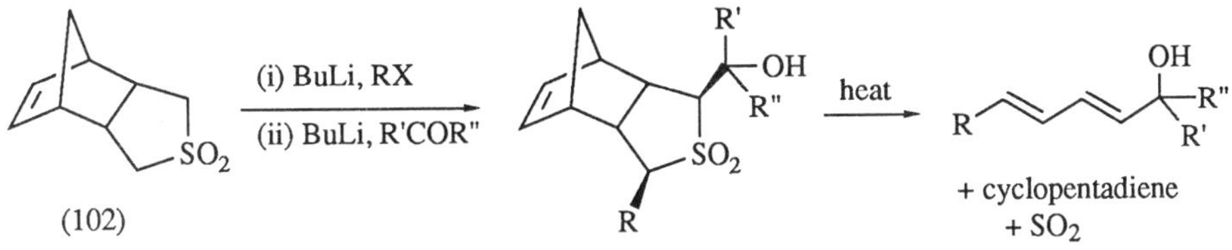

Scheme 49

Monoalkylation, 1,4-dialkylation and aldol-type substitution of sulphone (**102**) can be achieved stereoselectively, the electrophile attacking the more exposed *exo*-face of the sulphonyl carbanion. Pyrolysis under fairly severe conditions (650°C) then releases SO_2, cyclopentadiene, and the desired acyclic diene. The starting sulphone (**102**) is readily synthesised from the cycloadduct of maleic anhydride and cyclopentadiene. Finally, similar extrusion from five-membered ring sulphones fused to small rings can be achieved, e.g. Scheme 50.[88]

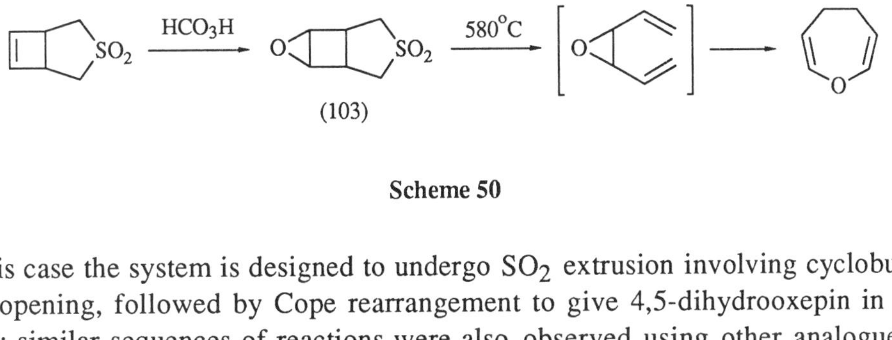

Scheme 50

In this case the system is designed to undergo SO_2 extrusion involving cyclobutane ring opening, followed by Cope rearrangement to give 4,5-dihydrooxepin in 55% yield; similar sequences of reactions were also observed using other analogues of epoxide (**103**), e.g. the corresponding aziridine. A similar strategy was reported by the same group for the preparation of 1,5-dienes in stereoselective fashion.[89]

8.3.2 Thiophene 1,1-dioxides
The parent thiophene 1,1-dioxide (**4**) is an unstable material. Its synthesis and reactivity, particularly its tendency towards dimerisation with evolution of SO_2, were described in a series of papers by Bailey and Cummins, Scheme 51.[90]

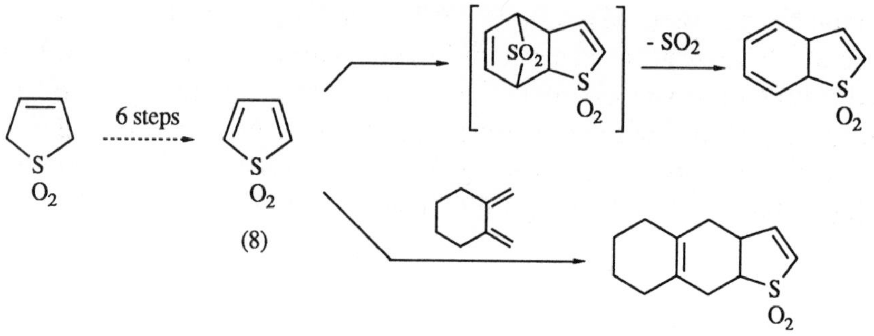

Scheme 51

A rather long-winded route was followed to obtain (8), which was found to be stable as a solution in chloroform if stored at 5°C. More substituted derivatives can be isolated as stable crystalline compounds. A particularly mild method for preparing these sulphones involves the oxidation of thiophenes with dimethyldioxirane, Scheme 52.[91]

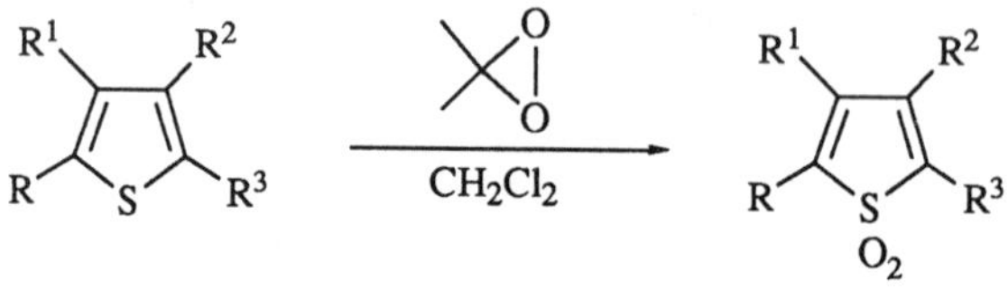

Scheme 52

This reagent was found superior to MCPBA and hydrogen peroxide–acetic acid. Substitution at both the 2- and 5-positions appears to improve the stability of the product thiophene 1,1-dioxide.

More extensive oxidation of a rather hindered thiophene 1,1-dioxide system (104) has been explored, Scheme 53.[92]

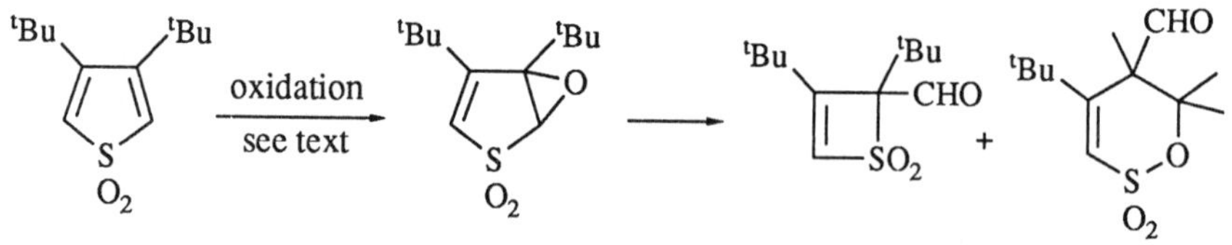

Scheme 53

Treatment of sulphone (**104**) with MCPBA under basic or buffered conditions allows the isolation of epoxide (**105**) in up to 77% yield. However, if (**104**) is reacted with an excess of MCPBA in the absence of base or buffer, the major products are thiete 1,1-dioxide (**106**) and sultone (**107**). A range of transformations of (**105**) was also examined, along with oxidation of (**104**) using alkaline hydrogen peroxide.[93]

The most remarkable synthesis of the thiophene 1,1-dioxide system involves the cyclisation of bis-allenic sulphone (**108**), Scheme 54.[94]

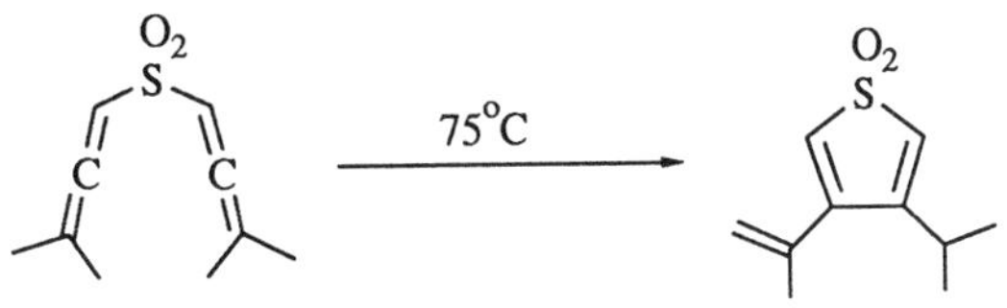

(108)   (109)

Scheme 54

Quantitative conversion to (**109**) is observed simply on heating at 75°C. The mechanism of this transformation is uncertain but an intramolecular ene process appears perhaps most reasonable.

The fairly sparse chemistry of thiophene 1,1-dioxides is dominated by their use in cycloaddition reactions. The tetrachloro derivative (**110**) has received significant attention in this respect, most notably by Raasch, Scheme 55.[95]

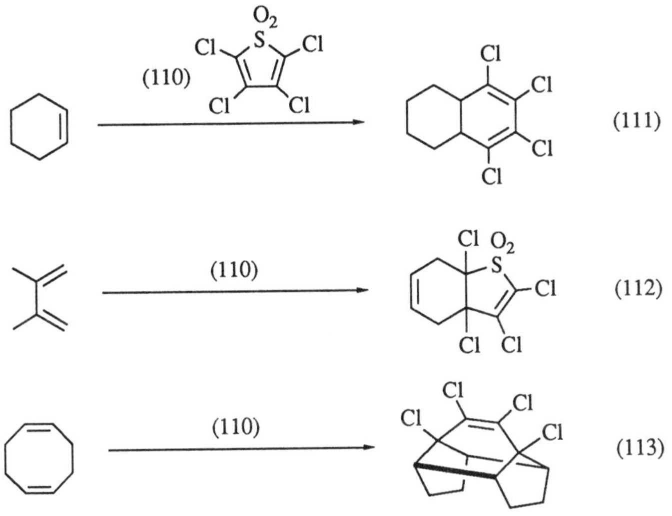

Scheme 55

Sulphone (**110**) is reasonably stable and yet it is reactive enough to undergo cycloaddition with alkenes, even at ambient temperature. A great range of adducts such as (111) was prepared, resulting from cycloaddition and SO_2 extrusion from the intermediate sulpholene. With dienes the same course can be followed, or (110) can act as the 2π-cycloaddition partner to give products such as (112). In the case of 1,5-cyclooctadiene the initial cycloadduct corresponding to (111) can undergo subsequent intramolecular Diels–Alder reaction to give the interesting caged structure (113).

The group of Nakayama has studied the cycloaddition chemistry of the hindered sulphone (104) mentioned above, and its relatives, Scheme 56.[96]

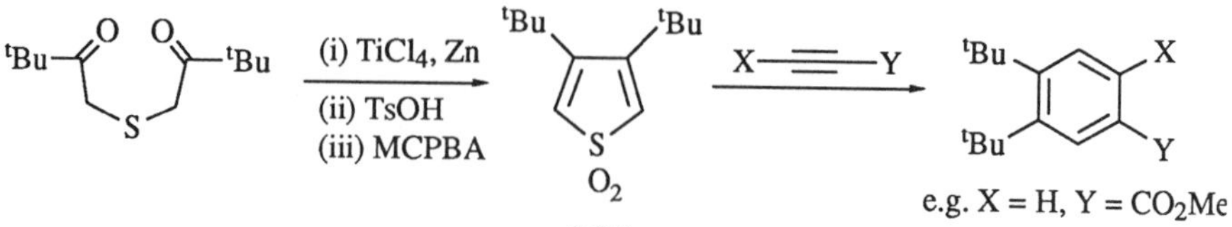

e.g. X = H, Y = CO₂Me
X = H, Y = Ph

Scheme 56

This is a good way to prepare a range of aromatic compounds incorporating the *ortho*-di-*tert*-butyl feature. Both electron-rich and electron-poor alkynes participate, and benzyne can be used in the reaction to furnish the related naphthalene product.

A synthesis of azulenes is facilitated by the finding that thiophene 1,1-dioxides will undergo [6+4] cycloadditions with 6-aminofulvenes, Scheme 57.[97]

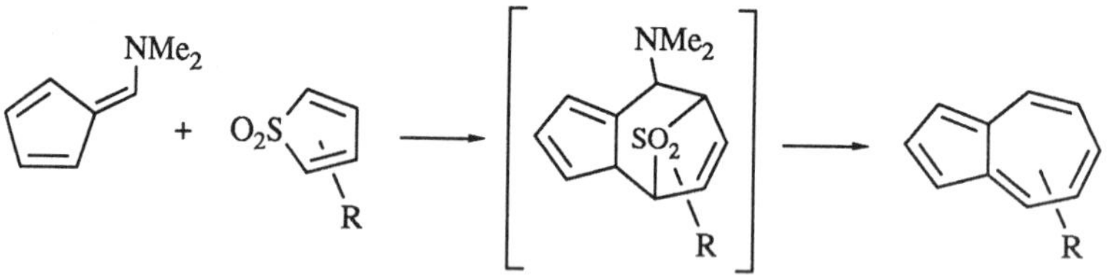

Scheme 57

Yields of azulenes are modest at best, although this is compensated for by the shortness of this approach. The characteristics of the reaction are consistent with the intermediacy of the [6+4] cycloadduct (114). Zwitterionic intermediates appear not to be involved, as suggested by the insensitivity of the process to solvent effects. Reactions of unsymmetrical thiophene 1,1-dioxides occur regioselectively.

Other types of cycloaddition of thiophene dioxides which have received attention include those with nitrones, and nitrile oxides.[98]

A further well-established method for the generation of thiophene 1,1-dioxide and its derivatives involves the double dehydrobromination of 3,4-dibromothiolane 1,1-dioxides such as (115) or (116), Scheme 58.[99,100]

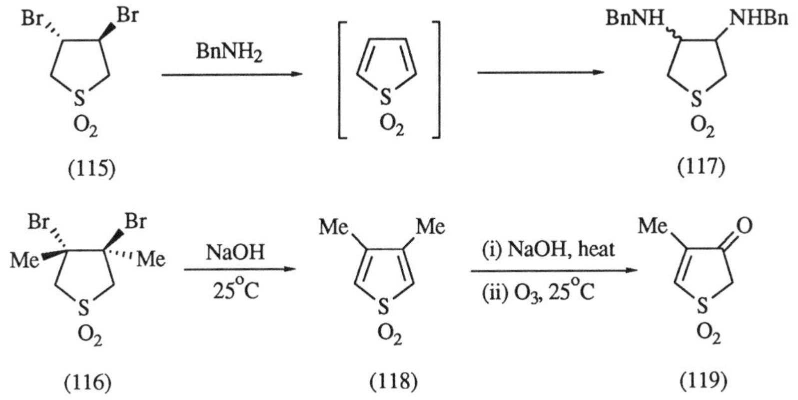

Scheme 58

Thus, on treatment with excess benzylamine the sulphone (**115**) undergoes elimination to form thiophene 1,1-dioxide, which then undergoes double addition of the amine to give adduct (**117**) as a mixture of stereoisomers.[99] In the case of (**116**) the double elimination allows (**118**) to be isolated in crystalline form in high yield.[100] Subsequent conversion to (**119**) is possible by isomerisation of (**118**) to form an *exo*-methylene isomer, followed by ozonolysis. Sulphone enone (**119**) was found to undergo a single electron transfer (SET) process involving head-to-tail coupling on treatment with LDA.

exo-Methylene isomers of thiophene 1,1-dioxides have also been obtained by reaction of the parent compound (formed *in situ*) with dihalogenocarbenes.[101] Finally, a range of products has been obtained by the group of Gronowitz by treatment of certain halogenated thiophene 1,1-dioxides with organometallics such as alkyllithiums and Grignard reagents.[102]

8.4 Functionalised Normal-sized Rings

Cyclisations of sulphonyl radicals can be carried out starting from unsaturated cobaloximes, Scheme 59.[103]

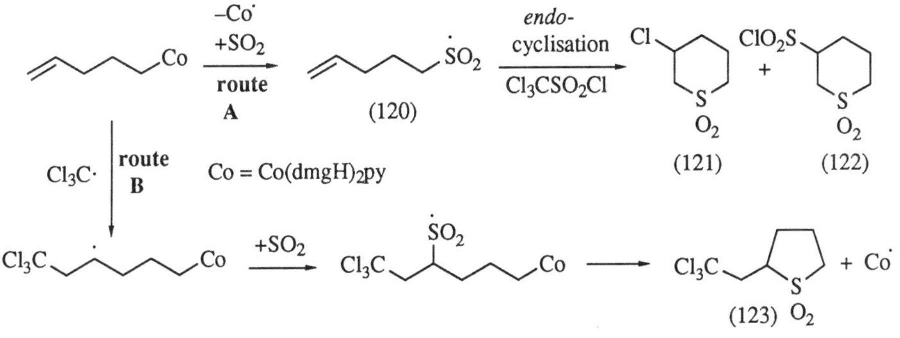

Scheme 59

Photolysis in the presence of trichloromethanesulphonyl chloride usually gives mixtures of five- or six-membered ring sulphone products, e.g. (121)–(123). The former arise via route B, involving an intramolecular homolytic displacement of cobaloxime (II). The six-membered products probably arise by *endo*-cyclisation of an intermediate sulphonyl radical (120). The preference for *endo*-cyclisation in such reactions has also been observed under alternative cyclisation conditions. [104]

Cyclic sulphones chlorinated in the β-position can also be obtained by direct ionic chlorination of the parent sulphones using sulphuryl chloride, Scheme 60. [105]

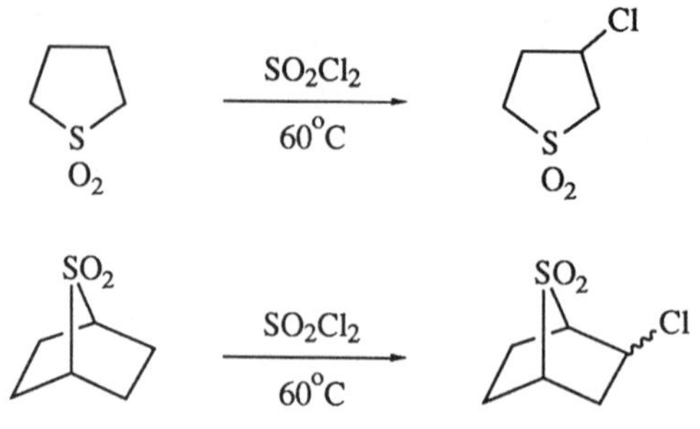

(124)   mainly *endo*

Scheme 60

The reaction is perhaps most suitable for symmetrical cyclic sulphones, since with most other types of substrate rather complex mixtures are possible. The reaction can also be conducted under radical conditions using di-*tert*-butyl peroxide as initiator. Under such conditions the product distribution is significantly different to that under thermal conditions, sulphones such as (124) giving practically no product due to homolytic C–SO$_2$ bond cleavage.

Sulphur dioxide can undergo reaction with some enol ethers to give sulpholane products, e.g. Scheme 61. [106]

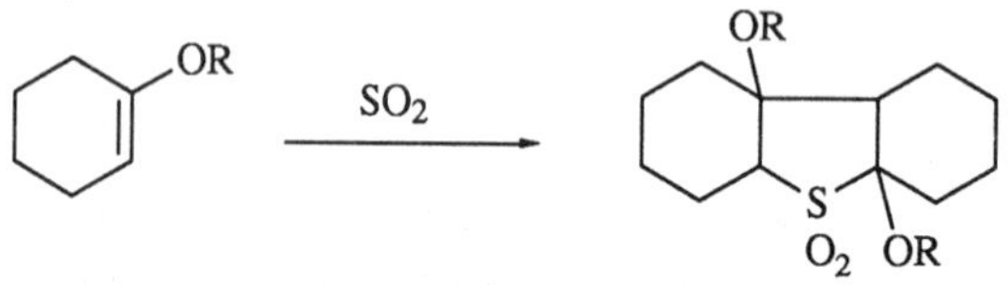

(125)

Scheme 61

This reaction is proposed to occur by addition of SO$_2$ to the vinyl ether to generate a dipolar intermediate which then reacts with a further equivalent of the alkene. Further transformation of adducts (125) by elimination to give a sulpholene, and loss of SO$_2$ to give diene products was also described.

The cycloaddition chemistry of sulphenes has already been described in some detail in this chapter. The cycloadditions of sulphines and thiocarbonyl compounds can also provide useful access to six-membered ring sulphones following oxidation. Examples of this chemistry, along with a further cycloaddition of a vinyl sulphene, are shown in Scheme 62.[107–109]

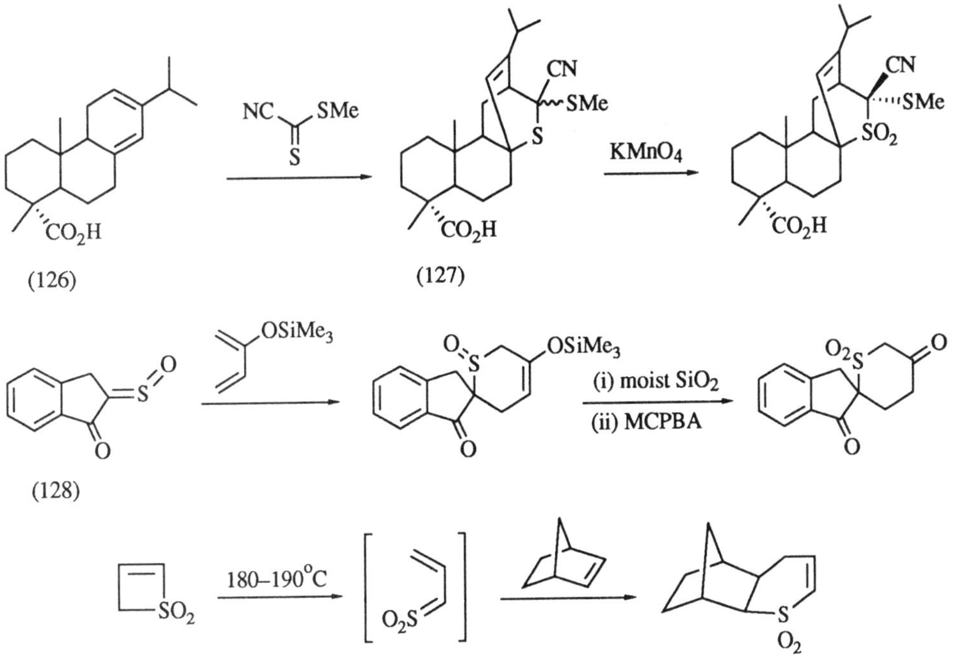

Scheme 62

Thus levopimaric acid (**126**) undergoes 1,4-cycloaddition with methyl cyano-dithioformate to give an almost quantitative yield of stereoisomeric adducts (**127**).[107] Subsequent oxidation of the ring sulphur atom to give a monosulphone is possible using KMnO₄. The α-oxosulphine (**128**) undergoes regioselective cycloaddition with electron-rich heterodienes to give adducts which can be oxidised to cyclic sulphones.[108] Cycloadditions of vinyl sulphene, generated *in situ* from thiete 1,1-dioxide, allow a direct synthesis of six-membered vinyl sulphones.[109]

Cyclisation of certain sulphone-containing alkynecarboxylic acids using HgO provides a route to cyclic sulphones such as (**129**), Scheme 63.[110]

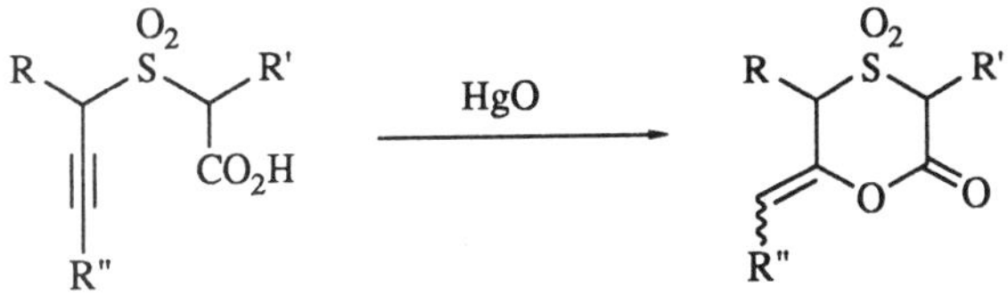

Scheme 63

The reaction occurs in toluene at reflux using only catalytic quantities of HgO, although yields are somewhat variable.

Many types of cyclic sulphone having additional functionality including additional heterocyclic rings have been synthesised using sulphide oxidations, e.g. pyrazolone (130),[111] carbohydrate sulphone (131),[112] and bicycloacetal (132).[113]

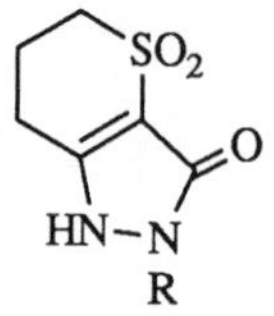

(130) 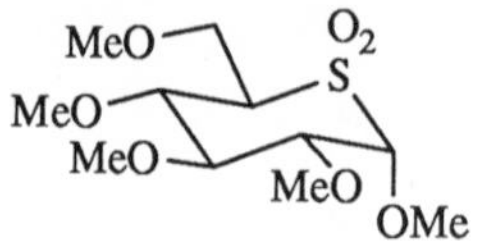   (131) 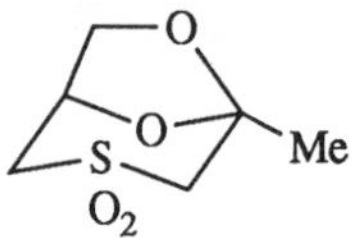   (132)

8.5 Larger Ring Sulphones

Some seven-, eight- and nine-membered ring ketosulphones have been prepared using ring expansion reactions, e.g. Scheme 64.[114]

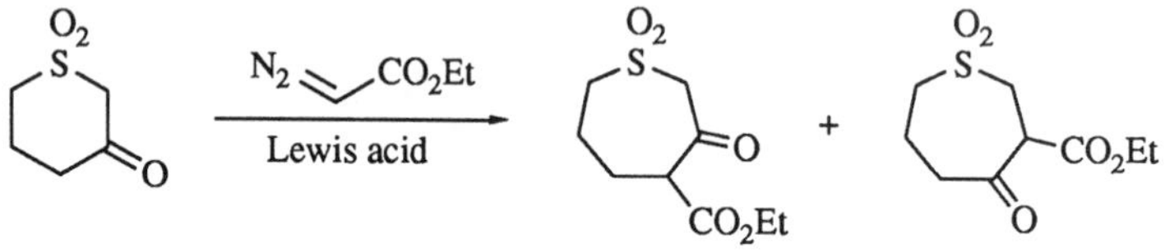

(133)

Scheme 64

The precise ratio of isomeric products is dependent on the Lewis acid used. The best compromise between speed of reaction and regioselectivity was found by using tin tetrachloride. Using this catalyst the 3-keto product (133) is formed with high selectivity.

An interesting nine-membered diketosulphone (134) is available by ozonolysis of sulpholene (135), Scheme 65.[115]

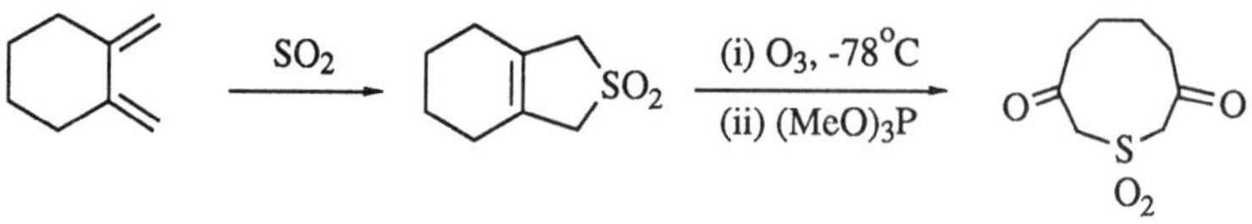

(135)   (134)

Scheme 65

The structure of this novel sulphone, which adopts a twist-chair–chair conformation, was elucidated by single-crystal X-ray analysis.

Thiepin 1,1-dioxide (**136**) and a number of derivatives can be prepared by a variant of the Staudinger reaction, Scheme 66.[116]

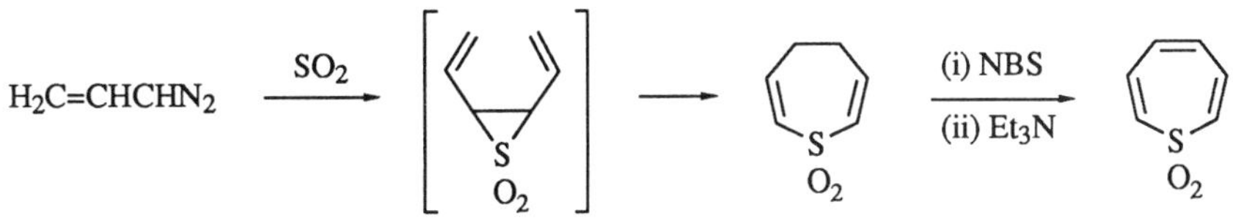

(137) (136)

Scheme 66

The transformation is proposed to occur by formation of a divinyl episulphone, which undergoes Cope rearrangement to give 4,5-dihydrothiepin 1,1-dioxide (**137**).

Larger ring sulphones and disulphones are most commonly prepared by thiolate displacement and sulphide oxidation, e.g. (**138**)-(**140**).[117–119]

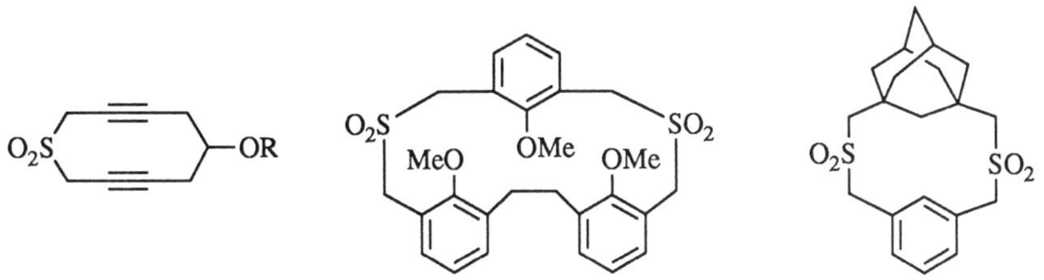

(138) (139) (140)

Large ring disulphones such as (**139**) and (**140**) are useful precursors for strained or unusual carbocycles by double SO_2 extrusion.[120]

Chapter 8 References

1. H. Staudinger and F. Pfenninger, *Chem. Ber.*, **1916**, *49*, 1941. Note that thiadiazoline intermediates are involved in reactions between hindered partners, see H. H. Inhoffen, R. Jonas, H. Krosche, and U. Eder, *Liebigs Ann. Chem.*, **1966**, *694*, 19.
2. F. G. Bordwell, J. M. Williams, Jr., E. B. Hoyt, Jr., and B. B. Jarvis, *J. Am. Chem. Soc.*, **1968**, *90*, 429.
3. N. P. Neureiter, *J. Am. Chem. Soc.*, **1966**, *88*, 558.
4. N. Tokura, T. Nagai, and S. Matsumura, *J. Org. Chem.*, **1966**, *31*, 349; S. Matsumura, T. Nagai, and N. Tokura, *Bull. Chem. Soc. Jpn.*, **1968**, *41*, 2672.
5. G. Opitz and K. Fischer, *Angew. Chem., Int. Ed. Engl.*, **1965**, *4*, 70.
6. N. Fischer and G. Opitz, *Org. Synth.*, **1968**, *48*, 106.
7. J. Nakayama, M. Tanuma, Y. Honda, and M. Hoshino, *Tetrahedron Lett.*, **1984**, *25*, 4553.
8. G. Opitz, T. Ehlis, and K. Rieth, *Tetrahedron Lett.*, **1989**, *30*, 3131; G. Opitz, T. Ehlis, and K. Rieth, *Chem. Ber.*, **1990**, *123*, 1989; see also reference 19.
9. E. Vilsmaier, R. Tropitzsch, and O. Vostrowsky, *Tetrahedron Lett.*, **1974**, 3987.
10. S. Matsumura, T. Nagai, and N. Tokura, *Tetrahedron Lett.*, **1966**, 3929; S. Matsumura, T. Nagai, and N. Tokura, *Bull. Chem. Soc. Jpn.*, **1968**, *41*, 635.
11. L. A. Carpino and R. H. Rynbrandt, *J. Am. Chem. Soc.*, **1966**, *88*, 5682; L. A. Carpino, L. V. McAdams III, R. H. Rynbrandt, and J. W. Spiewak, *J. Am. Chem. Soc.*, **1971**, *93*, 476.
12. J. C. Philips, J. V. Swisher, D. Haidukewych, and O. Morales, *J. Chem. Soc., Chem. Commun.*, **1971**, 22.
13. J. C. Philips and O. Morales, *J. Chem. Soc., Chem. Commun.*, **1977**, 713; F. G. Bordwell, J. M. Williams, Jr., and B. B. Jarvis, *J. Org. Chem.*, **1968**, *33*, 2026.
14. M. Komatsu, Y. Yoshida, M. Uesaka, Y. Ohshiro, and T. Agawa, *J. Org. Chem.*, **1984**, *49*, 1300.
15. H. Matsukubo, M. Kojima, and H. Kato, *Chem. Lett.*, **1975**, 1153.
16. B. B. Jarvis, W. P. Tong, and H. L. Ammon, *J. Org. Chem.*, **1975**, *40*, 3189; see also B. B. Jarvis and W. P. Tong, *Synthesis*, **1975**, 102.
17. Y. Yoshida, M. Komatsu, Y. Ohshiro, and T. Agawa, *J. Org. Chem.*, **1979**, *44*, 830.
18. R. M. Dodson, E. H. Francis, and G. Klose, *J. Org. Chem.*, **1970**, *35*, 2520; see also F. Lautenschlaeger, *J. Org. Chem.*, **1969**, *34*, 3998; F. S. Abbott and K. Haya, *Can. J. Chem.*, **1978**, *56*, 71, and reference 31.
19. J. F. King, *Acc. Chem. Res.*, **1975**, *8*, 10; G. Opitz, *Angew. Chem., Int. Ed. Engl.*, **1967**, *6*, 107; W. E. Truce and L. K. Liu, *Mech. React. Sulphur Comp.*, 145; N. H. Fischer, *Synthesis*, **1970**, 393.

20. H. Pritzkow, K. Rall, S. Reimann-Andersen, and W. Sundermeyer, *Angew. Chem., Int. Ed. Engl.*, **1990**, *29*, 60; W. Sundermeyer and J. Waldi, *Chem. Ber.*, **1990**, *123*, 1687; G. Opitz, K. Rieth, and T. Ehlis, *Chem. Ber.*, **1990**, *123*, 1563; J. F. King and D. R. K. Harding, *Can. J. Chem.*, **1976**, *54*, 2652; J. S. Grossert and M. M. Bharadwaj, *J. Chem. Soc., Chem. Commun.*, **1974**, 144.

21. F. S. Abbott, J. E. Coates, and K. Haya, *J. Org. Chem.*, **1977**, *42*, 3502.

22. L. A. Paquette and M. Rosen, *J. Am. Chem. Soc.*, **1967**, *89*, 4102.

23. J. F. Stephen and E. Marcus, *J. Org. Chem.*, **1969**, *34*, 2535; see also H. Marzarguil and A. Lattes, *Bull. Soc. Chim. Fr.*, **1969**, 3713.

24. P. Bradamante, M. Forchiassin, G. Pitacco, C. Russo, and E. Valentin, *J. Heterocyclic Chem.*, **1982**, *19*, 985.

25. G. Opitz, K. Rieth, and G. Walz, *Tetrahedron Lett.*, **1966**, 5269; G. Opitz and D. Bucher, *Tetrahedron Lett.*, **1966**, 5263.

26. W. E. Truce and J. R. Norell, *J. Am. Chem. Soc.*, **1963**, *85*, 3231.

27. R. H. Hasek, R. H. Meen, and J. C. Martin, *J. Org. Chem.*, **1965**, *30*, 1495.

28. W. E. Truce and P. N. Son, *J. Org. Chem.*, **1965**, *30*, 71.

29. B. E. Smart and W. J. Middleton, *J. Am. Chem. Soc.*, **1987**, *109*, 4982; see also U. Hartwig, K. Rall, and W. Sundermeyer, *Chem. Ber.*, **1990**, *123*, 595.

30. E. Block and M. Aslam, *Tetrahedron Lett.*, **1982**, *23*, 4203; E. Block and A. Wall, *J. Org. Chem.*, **1987**, *52*, 809; E. Block and A. Wall, *Tetrahedron Lett.*, **1985**, *26*, 1425.

31. B. M. Trost, W. L. Schinski, F. Chen, and I. B. Mantz, *J. Am. Chem. Soc.*, **1971**, *93*, 676; see also J. F. King, K. Piers, P. J. H. Smith, C. L. McIntosh, and P. de Mayo, *J. Chem. Soc., Chem. Commun.*, **1969**, 31; C. L. McIntosh and P. de Mayo, *J. Chem. Soc., Chem. Commun.*, **1969**, 32, and reference 43.

32. A. Padwa and R. Gruber, *J. Org. Chem.*, **1970**, *35*, 1781; see also reference 35.

33. D. J. Young and C. J. M. Stirling, *J. Chem. Soc., Chem. Commun.*, **1987**, 552.

34. R. M. Dodson, P. D. Hammen, and R. A. Davis, *J. Org. Chem.*, **1971**, *36*, 2693; R. M. Dodson, P. D. Hammen, E. H. Jancis, and G. Klose, *J. Org. Chem.*, **1971**, *36*, 2698; R. M. Dodson, P. D. Hammen, and J. Y. Fan, *J. Org. Chem.*, **1971**, *36*, 2703.

35. J. D. Finlay, D. J. H. Smith, and T. Durst, *Synthesis*, **1978**, 579.

36. P. Del Buttero and S. Maiorana, *Synthesis*, **1975**, 333.

37. J. P. Marino, *J. Chem. Soc., Chem. Commun.*, **1973**, 861.

38. P. L-F. Chang and D. C. Dittmer, *J. Org. Chem.*, **1969**, *34*, 2791; L. A. Paquette, M. Rosen, and H. Stucki, *J. Org. Chem.*, **1968**, *33*, 3020; W. O. Siegl and C. R. Johnson, *J. Org. Chem.*, **1970**, *35*, 3657.

39. L. A. Paquette and J. P. Freeman, *J. Am. Chem. Soc.*, **1969**, *91*, 7548.

40. L. A. Paquette and M. Rosen, *J. Org. Chem.*, **1968**, *33*, 2130.

41. T. C. Sedergran, M. Yokoyama, and D. C. Dittmer, *J. Org. Chem.*, **1984**, *49*, 2408.

42. D. C. Dittmer and T. R. Nelsen, *J. Org. Chem.*, **1976**, *41*, 3044.
43. J. F. King, P. De Mayo, C. L. McIntosh, K. Piers, and D. J. H. Smith, *Can. J. Chem.*, **1970**, *48*, 3704.
44. D. C. Dittmer and F. A. Davis, *J. Org. Chem.*, **1967**, *32*, 3872.
45. J. E. Coates and F. S. Abbott, *J. Org. Chem.*, **1977**, *42*, 3506.
46. D. C. Dittmer and J. M. Balquist, *J. Org. Chem.*, **1968**, *33*, 1364.
47. E. Block, E. R. Corey, R. E. Penn, T. L. Renken, P. F. Sherwin, H. Bock, T. Hirabayashi, S. Mohmand, and B. Solouki, *J. Am. Chem. Soc.*, **1982**, *104*, 3119.
48. W. L. Mock, *J. Am. Chem. Soc.*, **1975**, *97*, 3666; W. L. Mock, *J. Am. Chem. Soc.*, **1975**, *97*, 3673; see also S. D. McGregor and D. M. Lemal, *J. Am. Chem. Soc.*, **1966**, *88*, 2858; N. S. Isaacs and A. A. R. Laila, *J. Chem. Soc., Perkin Trans. 2*, **1976**, 1470; R. M. Kellogg and W. L. Prins, *J. Org. Chem.*, **1974**, *39*, 2366.
49. W. L. Mock, *J. Am. Chem. Soc.*, **1966**, *88*, 2857.
50. R. L. Crumbie and D. D. Ridley, *Aust. J. Chem.*, **1981**, *34*, 1017; F. Naf, R. Decorzant, and S. D. Escher, *Tetrahedron Lett.*, **1982**, *23*, 5043.
51. R. C. Krug, J. A. Rigney, and G. R. Tichelaar, *J. Org. Chem.*, **1962**, *27*, 1305.
52. Y. Gaoni, *Tetrahedron Lett.*, **1977**, 4521.
53. A. I. Meyers and T. Takaya, *Tetrahedron Lett.*, **1971**, 2609.
54. T. Takaya, S. Kosaka, Y. Otsuji, and E. Imoto, *Bull. Chem. Soc. Jpn.*, **1968**, *41*, 2086.
55. K. Kosugi, A. V. Anisimov, H. Yamamoto, R. Yamashiro, K. Shirai, and T. Kumamoto, *Chem. Lett.*, **1981**, 1341.
56. P. G. Baraldi, A. Barco, S. Benetti, S. Manfredini, G. P. Pollini, D. Simoni, and V. Zanirato, *Tetrahedron*, **1988**, *44*, 6451.
57. A. Commercon and G. Ponsinet, *Tetrahedron Lett.*, **1985**, *26*, 4093.
58. S-J. Lee, J-C. Lee, M-L. Peng, and T. Chou, *J. Chem. Soc., Chem. Commun.*, **1989**, 1020.
59. T. Chou, S-J. Lee, M-L. Peng, D-J. Sun, and S-S. P. Chou, *J. Org. Chem.*, **1988**, *53*, 3027.
60. P. J. Harrington and K. A. DiFiore, *Tetrahedron Lett.*, **1987**, *28*, 495.
61. H. Takayama and T. Suzuki, *J. Chem. Soc., Chem. Commun.*, **1988**, 1044.
62. Y. Gaoni, *Tetrahedron Lett.*, **1977**, 947.
63. T. Chou and M-L. You, *J. Org. Chem.*, **1987**, *52*, 2224.
64. S. Yamada, H. Ohsawa, T. Suzuki, and H. Takayama, *J. Org. Chem.*, **1986**, *51*, 4934; S. Yamada, H. Ohsawa, T. Suzuki, and H. Takayama, *Chem. Lett.*, **1983**, 1003; see also H. Takayama, H. Suzuki, T. Nomoto, and S. Yamada, *Heterocycles*, **1986**, *24*, 303; T. Nomoto and H. Takayama, *Heterocycles*, **1985**, *23*, 2913.
65. Y-T. Tao, C-L. Liu, S-J. Lee, and S-S. P. Chou, *J. Org. Chem.*, **1986**, *51*, 4718.

66. T. Chou, H-H. Tso, and L-J. Chang, *J. Chem. Soc., Perkin Trans. 1*, **1985**, 515.

67. T. Chou, L-J. Chang, and H-H. Tso, *J. Chem. Soc., Perkin Trans. 1*, **1986**, 1039.

68. T. Chou, S-J. Lee, H-H. Tso, and C-F. Yu, *J. Org. Chem.*, **1987**, *52*, 5082; T. Chou and C-Y. Chang, *J. Org. Chem.*, **1991**, *56*, 4560.

69. Y-T. Tao and M-L. Chen, *J. Org. Chem.*, **1988**, *53*, 69.

70. S-S. P. Chou and C-M. Sun, *Tetrahedron Lett.*, **1990**, *31*, 1035.

71. T. Chou, H-H. Tso, Y-T. Tao, and L. C. Lin, *J. Org. Chem.*, **1987**, *52*, 244.

72. T. Chou, H-H. Tso and L. C. Lin, *J. Org. Chem.*, **1986**, *51*, 1000.

73. S. Yamada, H. Suzuki, H. Naito, T. Nomoto, and H. Takayama, *J. Chem. Soc., Chem. Commun.*, **1987**, 332.

74. H-H. Tso, T. Chou, and S. C. Hung, *J. Chem. Soc., Chem. Commun.*, **1987**, 1552.

75. S-S. P. Chou, D-J. Sun, and S-J. Weh, *Synth. Commun.*, **1989**, *19*, 1593.

76. T. Chou, S. C. Hung, and H-H. Tso, *J. Org. Chem.*, **1987**, *52*, 3394.

77. T. Nomoto and H. Takayama, *J. Chem. Soc., Chem. Commun.*, **1989**, 295.

78. T. Chou, H-H. Tso, and L-J. Chang, *J. Chem. Soc., Chem. Commun.*, **1984**, 1323.

79. S-S. P. Chou and W-H. Lee, *Synthesis*, **1990**, 219.

80. S. Yamada, T. Suzuki, H. Takayama, K. Miyamoto, I. Matsunaga, and Y. Nawata, *J. Org. Chem.*, **1983**, *48*, 3483.

81. S-J. Lee and T. Chou, *J. Chem. Soc., Chem. Commun.*, **1988**, 1188.

82. H. F. Schmitthenner and S. M. Weinreb, *J. Org. Chem.*, **1980**, *45*, 3372; S. F. Martin, S. R. Desai, G. W. Phillips, and A. C. Miller, *J. Am. Chem. Soc.*, **1980**, *102*, 3294.

83. M. P. Cava and A. A. Deana, *J. Am. Chem. Soc.*, **1959**, *81*, 4266; see also M. P. Cava and J. A. Kuczkowski, *J. Am. Chem. Soc.*, **1970**, *92*, 5800.

84. J. L. Charlton and T. Durst, *Tetrahedron Lett.*, **1984**, *25*, 5287; J. L. Charlton, M. M. Alauddin, and G. H. Penner, *Can. J. Chem.*, **1986**, *64*, 793; K. Khan and T. Durst, *Can. J. Chem.*, **1987**, *65*, 482.

85. T. Suzuki, K. Kubomura, H. Fuchii, and H. Takayama, *J. Chem. Soc., Chem. Commun.*, **1990**, 1687.

86. K. C. Nicolaou, W. E. Barnette, and P. Ma, *J. Org. Chem.*, **1980**, *45*, 1463; see also W. Oppolzer, D. A. Roberts, and T. G. C. Bird, *Helv. Chim. Acta*, **1979**, *62*, 2017.

87. R. Bloch and J. Abecassis, *Tetrahedron Lett.*, **1982**, *23*, 3277; R. Bloch and J. Abecassis, *Tetrahedron Lett.*, **1983**, *24*, 1247; R. Bloch, D. Hassan, and X. Mandard, *Tetrahedron Lett.*, **1983**, *24*, 4691; R. Bloch, C. Benecou, and E. Guibe-Jampel, *Tetrahedron Lett.*, **1985**, *26*, 1301.

88. R. A. Aitken, J. I. G. Cadogan, I. Gosney, B. J. Hamill, and L. M. McLaughlin, *J. Chem. Soc., Chem. Commun.*, **1982**, 1164.

89. J. I. G. Cadogan, C. M. Buchan, I. Gosney, B. J. Hamill, and L. M. McLaughlin, *J. Chem. Soc., Chem. Commun.*, **1982**, 325.

90. W. J. Bailey and E. W. Cummins, *J. Am. Chem. Soc.*, **1954**, *76*, 1932; W. J. Bailey and E. W. Cummins, *J. Am. Chem. Soc.*, **1954**, *76*, 1936; W. J. Bailey and E. W. Cummins, *J. Am. Chem. Soc.*, **1954**, *76*, 1940.
91. Y. Miyahara and T. Inazu, *Tetrahedron Lett.*, **1990**, *31*, 5955.
92. J. Nakayama and Y. Sugihara, *J. Org. Chem.*, **1991**, *56*, 4001.
93. S. Marmor, *J. Org. Chem.*, **1977**, *42*, 2927.
94. S. Braverman and D. Segev, *J. Am. Chem. Soc.*, **1974**, *96*, 1245.
95. M. S. Raasch, *J. Org. Chem.*, **1980**, *45*, 856; see also K. Beck and S. Hunig, *Angew. Chem., Int. Ed. Engl.*, **1986**, *25*, 187; H. Bluestone, R. Bimber, R. Berkey, and Z. Mandel, *J. Org. Chem.*, **1961**, *26*, 346.
96. J. Nakayama, S. Yamaoka, T. Nakanishi, and M. Hoshino, *J. Am. Chem. Soc.*, **1988**, *110*, 6598; see also J. Nakayama and R. Hasemi, *J. Am. Chem. Soc.*, **1990**, *112*, 5654.
97. S. E. Reiter, L. C. Dunn, and K. N. Houk, *J. Am. Chem. Soc.*, **1977**, *99*, 4199; D. Copland, D. Leaver, and W. B. Menzies, *Tetrahedron Lett.*, **1977**, 639; D. Mukherjee, L. C. Dunn, and K. N. Houk, *J. Am. Chem. Soc.*, **1979**, *101*, 251.
98. F. Marinone Albini, P. Ceva, A. Mascherpa, E. Albini, and P. Caramella, *Tetrahedron*, **1982**, *38*, 3629; A. Bened, R. Durand, D. Pioch, P. Geneste, J. P. Declercq, G. Germain, J. Rambaud, and R. Roques, *J. Org. Chem.*, **1981**, *46*, 3502.
99. H. A. Bates, S. Smilowitz, and J. Lin, *J. Org. Chem.*, **1985**, *50*, 899; see also reference 76.
100. C. S. V. Houge-Frydrych, W. B. Motherwell, and D. M. O'Shea, *J. Chem. Soc., Chem. Commun.*, **1987**, 1819.
101. B. R. Dent and G. J. Gainsford, *Aust. J. Chem.*, **1989**, *42*, 1307.
102. G. Nikitidis, S. Gronowitz, A. Hallberg, and C. Stalhandske, *J. Org. Chem.*, **1991**, *56*, 4064.
103. M. R. Ashcroft, P. Bougeard, A. Bury, C. J. Cooksey, and M. D. Johnson, *J. Org. Chem.*, **1984**, *49*, 1751.
104. P. N. Culshaw and J. C. Walton, *Tetrahedron Lett.*, **1990**, *31*, 6433.
105. I. Tabushi, Y. Tamura, and Z. Yoshida, *Tetrahedron*, **1974**, *30*, 1457.
106. D. Masilamani, E. H. Manahan, J. Vitrone, and M. M. Rogic, *J. Org. Chem.*, **1983**, *48*, 4918.
107. J. D. Friedrich, *J. Org. Chem.*, **1987**, *52*, 2442.
108. J. B. M. Rewinkel and B. Zwanenburg, *Recl. Trav. Chim. Pays-Bas*, **1990**, *109*, 190.
109. D. C. Dittmer, J. E. McCaskie, J. E. Babiarz, and M. V. Ruggeri, *J. Org. Chem.*, **1977**, *42*, 1910.
110. M. Yamamoto, H. Munakata, M. Z. Hussein, S. Kohmoto, and K. Yamada, *J. Chem. Res. (S)*, **1990**, 12.
111. S. Fatutta, P. Nitti, G. Pitacco, and E. Valentin, *J. Heterocyclic Chem.*, **1989**, *26*, 183.

112. H. Yuasa, A. Takenaka, and H. Hashimoto, *Bull. Chem. Soc. Jpn.*, **1990**, *63*, 3473.
113. P. Calinaud and J. Gelas, *Can. J. Chem.*, **1983**, *61*, 2095; P. Calinaud and J. Gelas, *Can. J. Chem.*, **1983**, *61*, 2103.
114. J. H. Dodd, C. F. Schwender, and Y. Gray-Nunez, *J. Heterocyclic Chem.*, **1990**, *27*, 1453.
115. L. D. Quin, J. Leimert, E. D. Middlemas, R. W. Miller, and A. T. McPhail, *J. Org. Chem.*, **1979**, *44*, 3496.
116. L. A. Paquette and S. Maiorana, *J. Chem. Soc., Chem. Commun.*, **1971**, 313.
117. K. C. Nicolaou, G. Skokotas, P. Maligres, G. Zuccarello, E. J. Schweiger, K. Toshima, and S. Wendeborn, *Angew. Chem., Int. Ed. Engl.*, **1989**, *28*, 1272.
118. A. Tsuge, T. Sawada, S. Mataka, N. Nishiyama, H. Sakashita, and M. Tashiro, *J. Chem. Soc., Chem. Commun.*, **1990**, 1066.
119. F. Vogtle, J. Dohm, and K. Rissanen, *Angew. Chem., Int. Ed. Engl.*, **1990**, *29*, 902.
120. F. Vogtle and L. Rossa, *Tetrahedron Lett.*, **1977**, 3577; F. Vogtle, *Angew. Chem., Int. Ed. Engl.*, **1969**, *8*, 274; M. Nakazaki, K. Yamamoto, and Y. Miura, *J. Org. Chem.*, **1978**, *43*, 1041; F. S. Guziec, Jr. and L. J. Sanfilippo, *Tetrahedron*, **1988**, *44*, 6241. For a recent synthetic application, see P. A. Wender, M. Harmata, D. Jeffrey, C. Mukai, and J. Suffert, *Tetrahedron Lett.*, **1988**, *29*, 909.

CHAPTER 9

Desulphonylation

The sulphone group serves in synthesis as an activating group for C–C and C=C bond formation. However, the sulphone is almost never required in the final target molecule and so methods for sulphone removal (desulphonylation) are important. Methods have been devised for the removal of sulphones from all types of functionalised molecules, most usually involving reductive desulphonylation, i.e. replacement of the sulphone group by hydrogen. Recent investigations have focused on developing more constructive processes such as oxidative and alkylative desulphonylation, in which the removal of the sulphone allows concomitant introduction of useful oxygenation, or carbon–carbon bond formation. The ease of removal of a sulphone (typically, removal of an $ArSO_2$ group is required) from a particular substrate is critically dependent on other local functionality present; thus, special desulphonylation methods have been developed for allyl sulphones, vinyl sulphones, β-ketosulphones, etc., and are described in separate sections of this chapter.

9.1 Simple Alkyl Aryl and Dialkyl Sulphones

Some of the methods described in this section are fairly general, and may be applied to more complex functionalised sulphones. In particular the use of group (I)–(III) metals, or amalgams thereof, for the removal of sulphone functionality, is very widespread. Some examples which use such reagents are also mentioned in later sections of this chapter.

9.1.1 Reductive Desulphonylation

Truce and co-workers conducted an early survey of the utility of various alkali metals in liquid amines for the cleavage of sulphones.[1] With lithium in methylamine dialkyl sulphones, RSO_2R', yield the hydrocarbon and sulphinic acid, whereas diaryl sulphones, $ArSO_2Ar'$, yield the thiol and hydrocarbon, since the initially formed arenesulphinate undergoes rapid reduction under the reaction conditions. In contrast, sodium in liquid ammonia is unreactive towards dialkyl sulphones, and cleaves diaryl sulphones to the hydrocarbon and sulphinic acid.

334

The same group also described reactions using alkyl aryl sulphones, and here the results appear at variance with more recent reports. Firstly, alkyl aryl sulphones are described as unreactive with sodium in liquid ammonia. Secondly, the reaction of such sulphones with lithium in methylamine is reported to occur with high preference for aryl–S bond cleavage. That alkyl aryl sulphones can react with sodium in ammonia, and that selective alkyl–S bond cleavage can be achieved using lithium in ethylamine is demonstrated by the successful desulphonylations of (1) and (2), Scheme 1.[2,3]

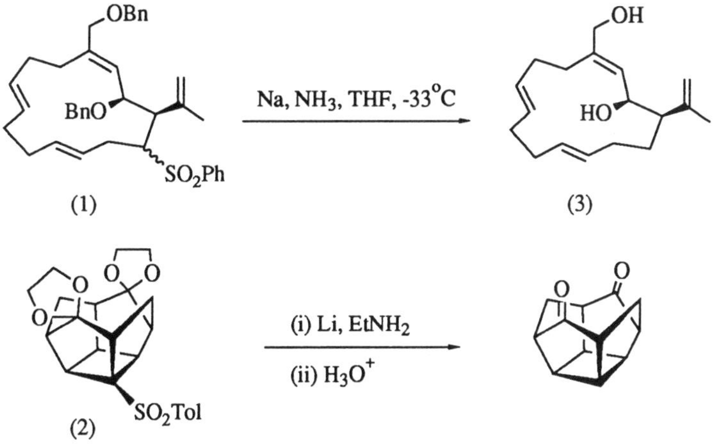

Scheme 1

It is possible that the precise structure of the sulphone or the addition of cosolvents such as THF significantly influences the outcome of such desulphonylations. In the reaction of (1) debenzylation is observed to yield diol (3). It was also found that hydrogenolysis of sulphone (1) could be effected without debenzylation by employing sodium amalgam in methanol.

Ultrasonically dispersed potassium in toluene is effective for the cleavage of a range of cyclic sulphones, e.g. (4) and (5), Scheme 2.[4,5]

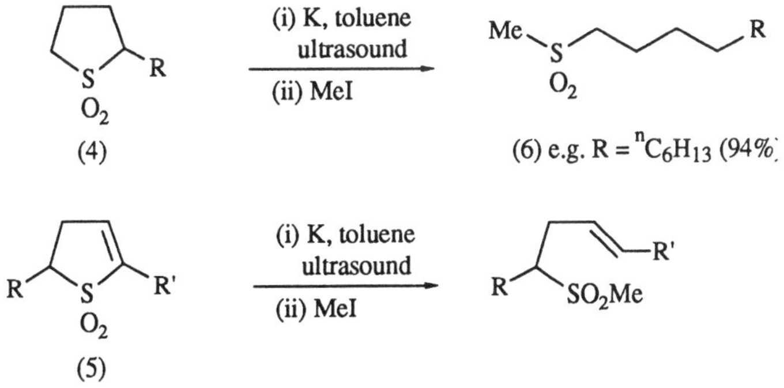

Scheme 2

The reaction proceeds with four- to six-membered ring sulphones at 0°C, to give acyclic sulphones in high yield after sulphinate alkylation using MeI. With unsymmetrical sulphones such as (4) the C–S cleavage occurs selectively at the more substituted carbon to give primary sulphone products, such as (6), after sulphinate methylation. The reagent gives unsaturated sulphones from 2-sulpholenes such as (5), and also promotes extrusion of sulphur dioxide from 3-sulpholenes.[5]

An early report on the use of sodium amalgam in refluxing ethanol revealed that diaryl and alkyl aryl, but not dialkyl sulphones, are reactive under such conditions.[6] The selective alkyl–S bond cleavage which occurs with alkyl aryl sulphones makes this reagent suitable for removing arenesulphonyl groups from synthetic intermediates. In a modification of this procedure, Trost *et al.* advocate the use of 6% Na(Hg) in methanol in the presence of four equivalents of disodium hydrogen phosphate at -20°C to room temperature.[7] This mild and chemoselective procedure has gained widespread acceptance as the first resort for common desulphonylations. Simple sulphones as well as those having a range of additional functionality (unprotected alcohol, alkene, ester, α,β-unsaturated nitrile) are smoothly desulphonylated. Examples of desulphonylation from Fuchs' group demonstrate the suitability of this approach for complex synthetic intermediates, Scheme 3.[8]

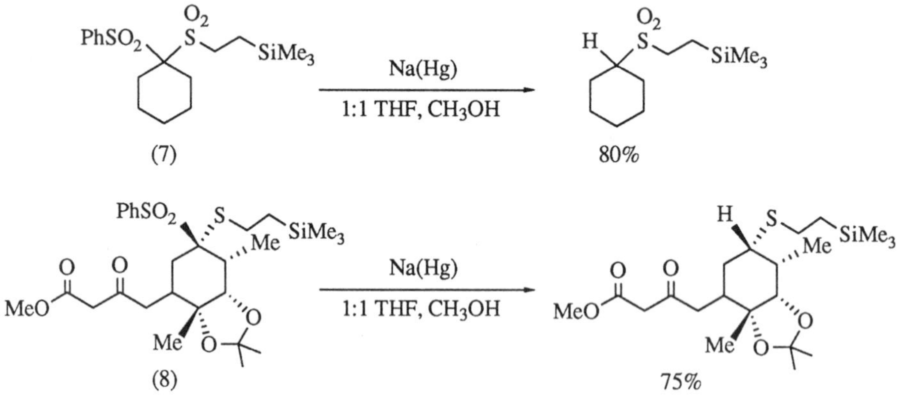

Scheme 3

Particularly impressive is the high yielding removal of $PhSO_2$ from (7) in the presence of the sensitive β-silylethyl sulphone group, and in the case of (8) the removal of the sulphone in the presence of a range of other functional groups, including a sulphide.

Another combination useful for the removal of phenylsulphonyl groups is that of SmI_2 in a mixture of THF and HMPA, the latter solvent additive being crucial to the reaction, e.g. Scheme 4.[9]

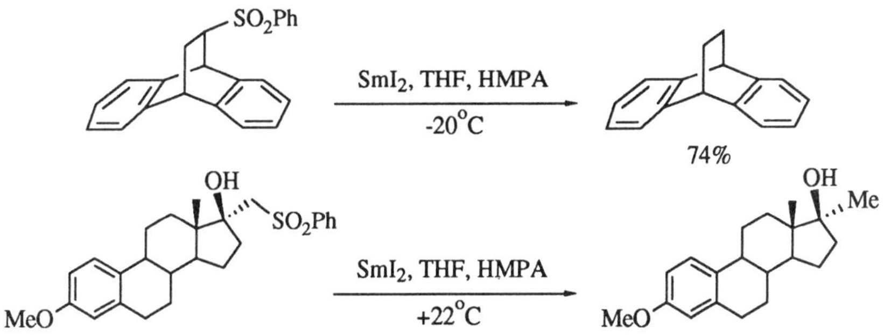

Scheme 4

The process also works for vinyl sulphones, and accommodates other functionality in the substrate. In the reaction of (9) the major product is alcohol (10) in which simple desulphonylation, and not Julia elimination, has occurred.

A variety of nickel reagents has been used to remove sulphone groups.[10] One reagent commonly used for this transformation is Raney nickel.[11] In contrast, Truce and Perry showed that sulphides can be cleaved in the presence of sulphones by use of the more selective reagent nickel boride.[12] Nickel-containing complex reducing agents (NiCRAs) are useful for the desulphonylation of saturated and unsaturated sulphones.[13] The reagents themselves are generated by combining NaH, *tert*-AmONa and Ni(OAc)$_2$ in a number of different ratios. Luh and co-workers have investigated the cleavage of C–S bonds with reagents prepared by the combination of LiAlH$_4$ with a nickel salt, e.g. Scheme 5.[14]

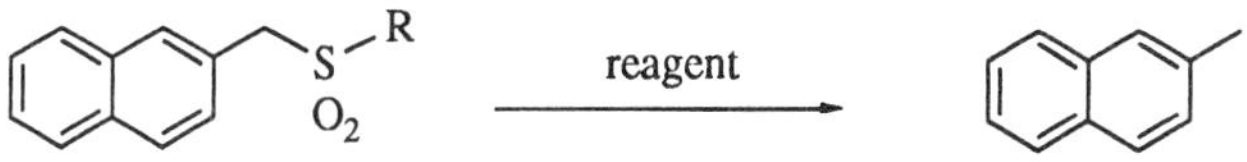

R = Me, reagent = NiBr$_2$· DME, PPh$_3$, LiAlH$_4$ (92%)
R = Et, reagent = nickelocene, LiAlH$_4$ (54%)

Scheme 5

In most, if not all, of the nickel reactions already mentioned there is evidence that the sulphone is reduced to the sulphide prior to C–S bond cleavage. Although sulphones are generally rather resistant to LiAlH$_4$ alone, they can be partially cleaved on prolonged treatment with this reagent under fairly vigorous conditions.[15] Other methods reported for desulphonylations include the use of diethyl phosphite anion under photolytic conditions,[16] and cathodic cleavage.[17]

9.1.2 Oxidative Desulphonylation

Hendrickson *et al.* first examined the conversion of triflones to ketones by way of intermediate vinyl azides, Scheme 6.[18]

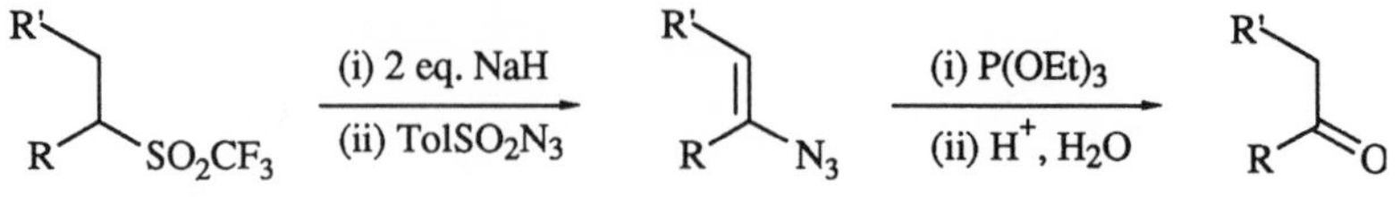

Scheme 6

The two-stage procedure can be carried out without any purification of intermediates, and gives very good overall yields. More direct methods have since been developed, involving reaction of a sulphonyl carbanion with a source of electrophilic oxygen, e.g. Scheme 7.[19]

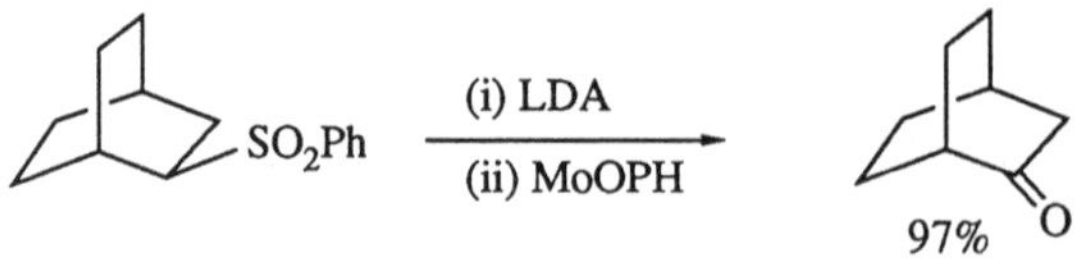

Scheme 7

The use of MoOPH in this context (a well-known oxidant for enolates) results in the direct conversion of the sulphone to a ketone. The authors also commented on their earlier attempts to use molecular oxygen for this oxidation, which resulted in a 'moderate explosion' on one occasion.[20] This method of oxidising sulphones has since been used in *trans*-decalin synthesis,[21] and in pheromone studies.[22]

When the same type of MoOPH reaction is applied to primary sulphones the corresponding aldehydes are not obtained. Instead, β-hydroxy sulphones resulting from aldol reaction of *in situ* formed aldehyde with the starting sulphone is observed, e.g. Scheme 8.[23]

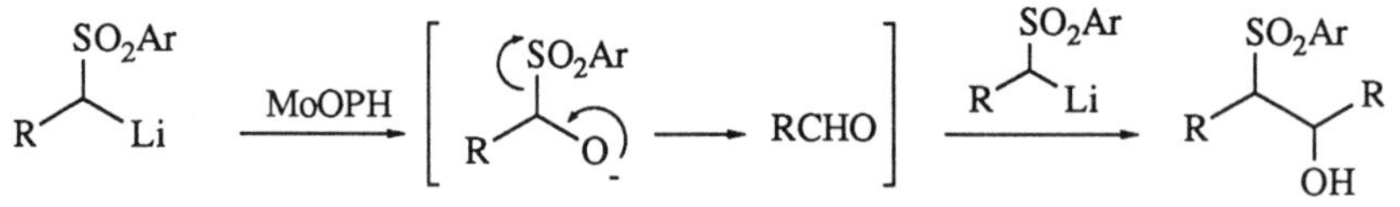

Scheme 8

Better yields of the final product are obtained by using 2 equivalents of base, and further improvement is observed on changing to LDA as base. Two additional excellent methods for oxidative removal of sulphones, which appear to be general for both primary and secondary sulphones, are shown in Scheme 9.[24,25]

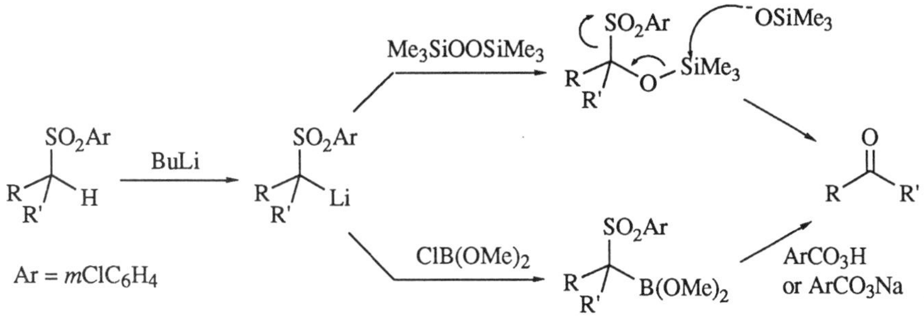

Scheme 9

The first procedure, involving the use of bis(trimethylsilyl) peroxide (BTSP) gives high yields, and is especially convenient due to the volatility of the by-products.[24] This method is also useful for preparing ^{18}O-labelled carbonyl compounds via doubly labelled BTSP. Alternatively, the reaction of a sulphonyl carbanion with chlorodimethoxyborane in THF furnishes an intermediate boronic ester which is then converted to the desired carbonyl compound using MCPBA. With unsaturated substrates unwanted side reactions can be avoided by employing the sodium salt of MCPBA in the second step.

A very interesting desulphonylation related to the last process allows oxidative desulphonylation with simultaneous C–C bond formation, e.g. Scheme 10.[26]

Scheme 10

Whilst the process is reasonably efficient using tributylborane, other reactions using mixed boranes derived from 9-BBN are less successful.

Treatment of nickel complex (**11**) with oxygen in the presence of 1M HCl results in clean conversion to dodecanal, Scheme 11.[27]

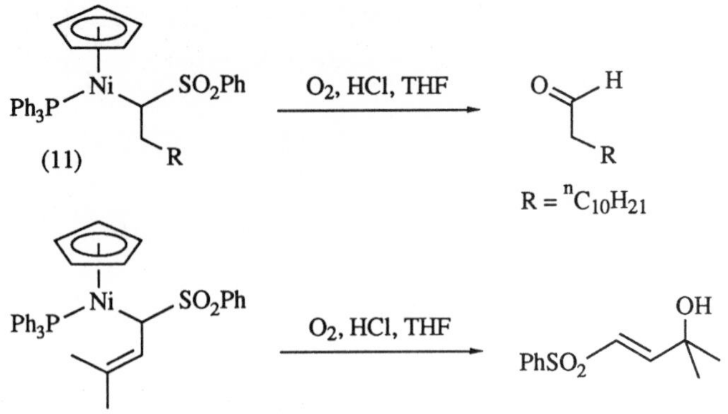

Scheme 11

The generality of this process was not determined. Indeed, complex (**12**) was shown to undergo a completely different mode of oxidation to produce alcohol (**13**).

Finally, two rather special cases of oxidative sulphone removal are highlighted in Scheme 12.[28,29]

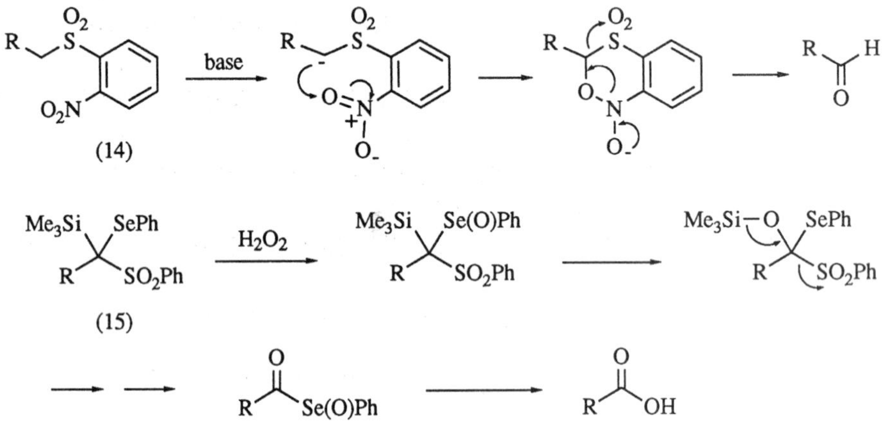

Scheme 12

In the first case the formation of an aldehyde on attempted alkylation of *ortho*-nitrophenyl sulphone (**14**) was proposed to occur by carbanion oxygenation by the neighbouring NO_2 group as shown.[28] Reaction of the densely functionalised sulphone (**15**) with hydrogen peroxide results in overall conversion to a carboxylic acid.[29] Here a sila-Pummerer process is presumably instrumental in installing a silyloxy group α to the sulphone. Decomposition in a similar fashion to that shown previously in Scheme 9, followed by oxidative removal of selenium provides the acid.

9.2 Allylic Sulphones

The desulphonylation of allylic sulphones presents special problems of regio- and stereocontrol. However, the enhanced reactivity of the allylic sulphone system, compared to a simple alkyl aryl sulphone, also provides opportunities for rather mild and versatile desulphonylation techniques. Thus allyl sulphones are amenable to desulphonylation by methods involving organometallic substitution, π-allyl complex formation, and radical chemistry, which are not effective with simple sulphones.

Despite the various special methods available for desulphonylation of allylic sulphones more mainstream methods have also been applied, e.g. Scheme 13.[30,31]

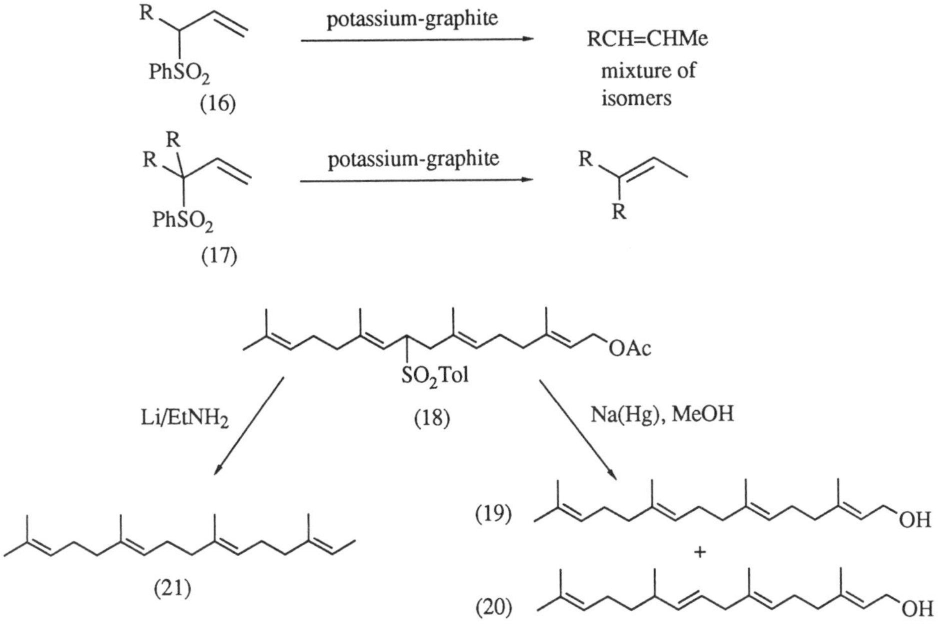

Scheme 13

In the case of sulphones (**16**) and (**17**) treatment with potassium-graphite gives rearranged alkene products corresponding to the more thermodynamically stable double-bond products.[30] Analogous desulphonylations of vinyl sulphones were also examined, with a low degree of stereocontrol being possible. The reactions of sulphone (**18**) illustrate the care which must be exercised in the choice of reducing agent.[31] Whilst sodium amalgam effects the desired transformation of (**18**) into all *trans*-geranylgeraniol (**19**) (formed in a 7:3 ratio with isomer (**20**)), reaction with lithium in ethylamine gives hydrocarbon (**21**) in 70% yield. The best conditions for obtaining (**19**) were found to involve treatment of the alcohol corresponding to (**18**) with lithium in ethylamine, with ether as cosolvent, at -78°C.

9.2.1 Sulphone Displacement by Organometallics

The reactions of a range of allylic sulphones with organometallics such as Grignard reagents and organocuprates have been examined by the groups of Julia and Masaki.[32,33] These studies revealed that regio- and stereoselective substitution of the sulphone group can often be achieved, e.g. Scheme 14.

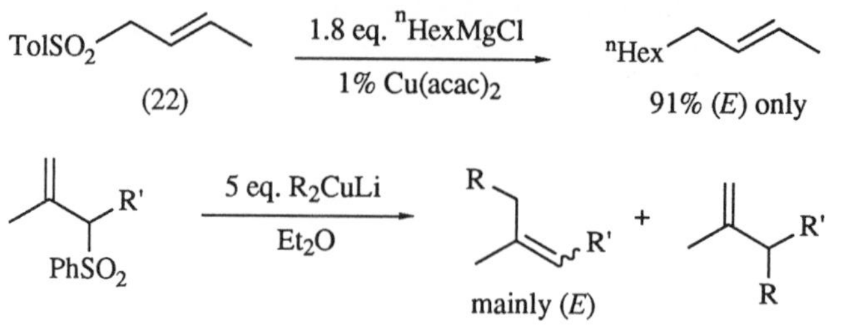

Scheme 14

In this type of reaction nucleophilic attack takes place at the less hindered end of the allylic system. Substitutions employing single geometric isomers of sulphones such as (22) occur mainly with retention of the original double-bond geometry.[32] In the report by Julia *et al.*, a critical examination of the reaction conditions required for optimal Grignard coupling was made. Best results are obtained using alkylmagnesium chlorides in THF as solvent. Except for terminal allylic sulphones the copper-catalysed Grignard method appears to be superior to the use of dialkyl cuprates.

Another report by Julia *et al.* shows that for tertiary allylic sulphones such as (23) (which cannot undergo α-deprotonation) efficient coupling with organolithium or organomagnesium reagents does not require transition metal catalysis, Scheme 15.[34]

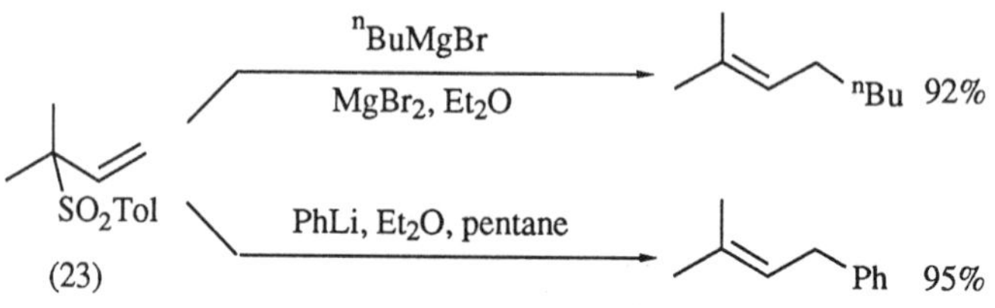

Scheme 15

This paper also describes analogous displacements using heteronucleophiles such as ArSLi, ArOMgBr and R_2NLi. A beneficial effect of mild Lewis acids such as $MgBr_2$ appears general for many such reactions.

Similar findings have been described by Trost and Ghadiri, this time employing organoaluminium compounds to effect the desired substitution, e.g. Scheme 16.[35]

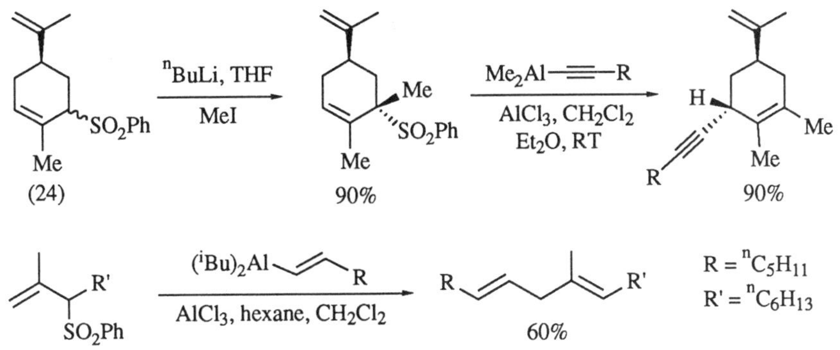

Scheme 16

This methodology is highly effective for the regio- and stereoselective introduction of alkenyl and alkynyl groups onto an allylic system, resulting in useful, 'skipped' diene and enyne products. In the two-step process starting from (24), involving sulphone alkylation and then sulphone displacement, the sulphone effectively serves as a 1,1-dipole. Related work by the same authors shows how Lewis acid activation can allow sulphone displacement in Friedel–Crafts-type alkylations, e.g. Scheme 17.[36]

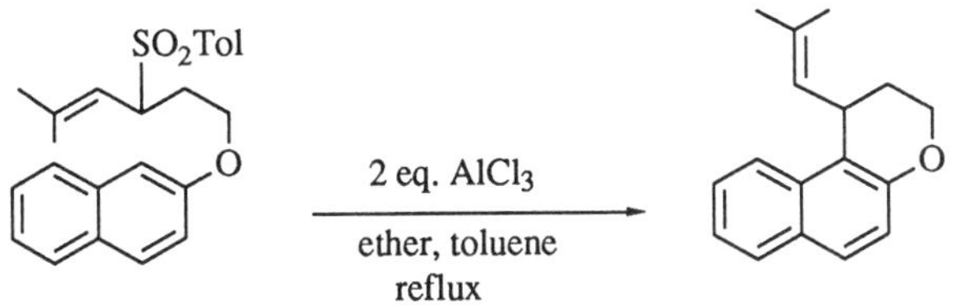

Scheme 17

Significantly, one example involved the use of a simple tertiary sulphone, rather than an allylic sulphone, as the electrophilic group.

9.2.2 π-Allyl Palladium Chemistry

In 1982 Kotake *et al.* reported a convenient new synthesis of homoallylic alcohols based on sulphone chemistry, Scheme 18.[37]

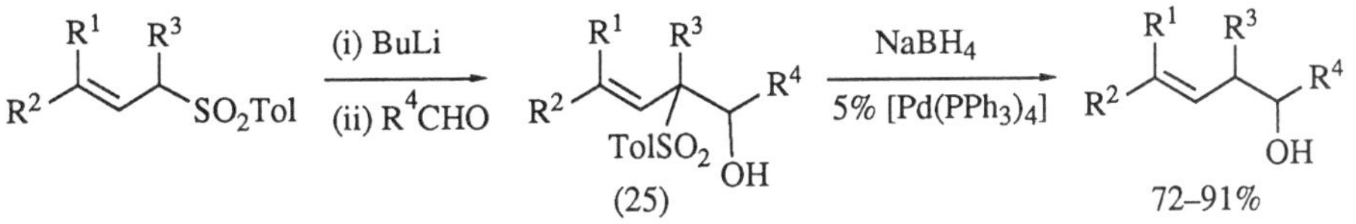

Scheme 18

Attempted desulphonylation of (**25**) with reagents such as Al(Hg), Na(Hg) or Li–$EtNH_2$ leads to double-bond migration, or retro-aldol reactions. The use of $NaBH_4$ in the presence of a catalytic quantity of Pd^0 was found to be very effective for the regioselective desulphonylation of such systems. At about the same time another report described a similar system for the reduction of allylic functional groups, including one example of an allylic sulphone.[38] This report recommends the use of more potent hydride donors, in particular $LiHBEt_3$, with palladium catalysis for such reductions. This reagent was subsequently also adopted by the Japanese group, after finding that the $Pd(PPh_3)_4$–$NaBH_4$ system is ineffectual for allylic sulphones not having hydroxyl functionality as in (**25**).[39] This later report found that a $PdCl_2$(dppp)–$LiHBEt_3$ combination is most effective for the highly regioselective desulphonylation of simple allylic sulphones, Scheme 19.

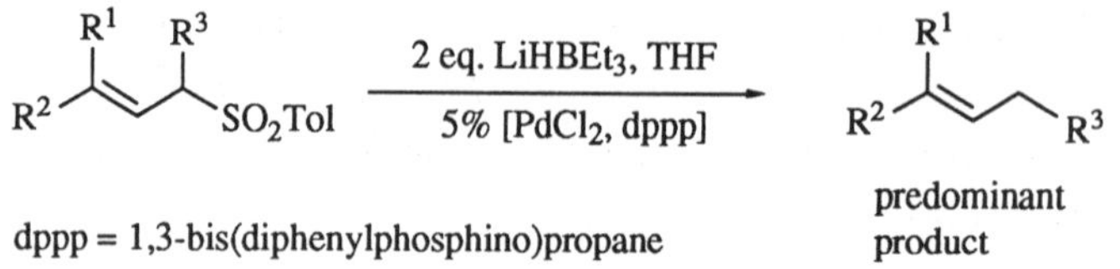
R1
R3
R2
SO2Tol
2 eq. LiHBEt3, THF
5% [PdCl2, dppp]
R1
R2
R3
predominant product
dppp = 1,3-bis(diphenylphosphino)propane

Scheme 19

The method effects reduction with very good levels of retention of double-bond position and stereochemistry. Yet another report from the same group re-examines the regiocontrol possible in desulphonylation of the aldol-type of sulphones (**25**).[40] Two methods were established for desulphonylation with complementary regiochemical results, Scheme 20.

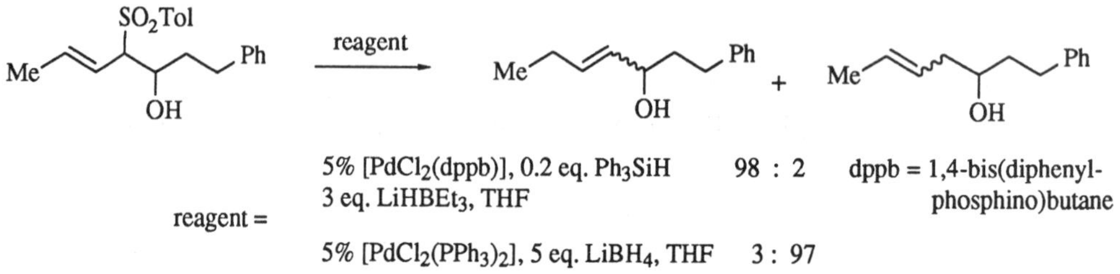
SO2Tol
Me
Ph
OH
reagent
Me
Ph
OH
+
Me
Ph
OH
reagent =
5% [PdCl2(dppb)], 0.2 eq. Ph3SiH
3 eq. LiHBEt3, THF
98 : 2
dppb = 1,4-bis(diphenyl-phosphino)butane
5% [PdCl2(PPh3)2], 5 eq. LiBH4, THF
3 : 97

Scheme 20

Yields and selectivities are generally very good, the stereoselectivities usually being better (>96% *E*) in the formation of the allylic alcohol products than their homoallylic counterparts (typically *ca.* 90% *E*).

A different type of desulphonylation can be effected from allylic sulphones of general structure (**26**) by treatment with base in the presence of [Pd(PPh$_3$)$_4$], Scheme 21.[41]

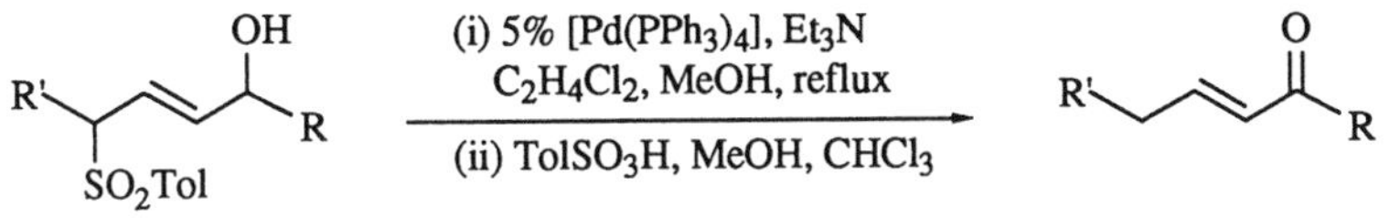

Scheme 21

The proposed reaction pathway involves elimination of sulphinic acid via an intermediate π-allyl complex. The initial product is a dienol which tautomerises to give predominantly, or exclusively, the α,β-unsaturated ketone.

The first contribution to the area of π-allyl complex formation from allyl sulphones was made by Trost's group.[42] This report showed that alkylation of soft carbon nucleophiles can be carried out by this approach, e.g. Scheme 22. As in the examples described earlier in this section nucleophilic attack usually occurs predominantly at the less hindered position.

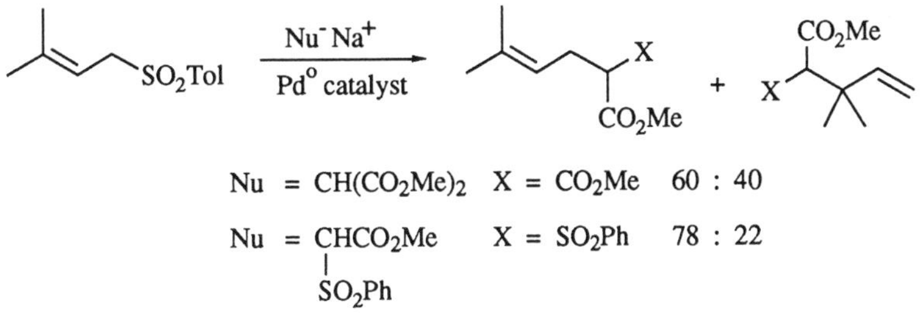

Nu = CH(CO$_2$Me)$_2$ X = CO$_2$Me   60 : 40

Nu = CHCO$_2$Me   X = SO$_2$Ph   78 : 22
        |
        SO$_2$Ph

Scheme 22

The reaction also proceeds with overall retention of configuration since both of the individual steps involved, π-allyl complex formation, and nucleophilic attack, involve inversion, Scheme 23.

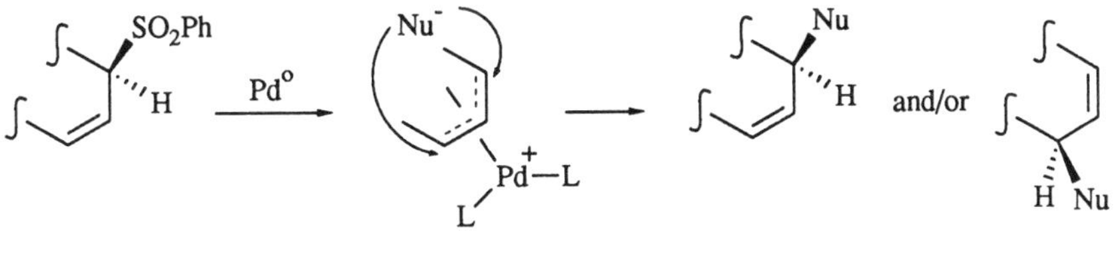

Scheme 23

Similar chemistry has also been examined using catalysis by nickel or molybdenum, e.g. Scheme 24.[43,44]

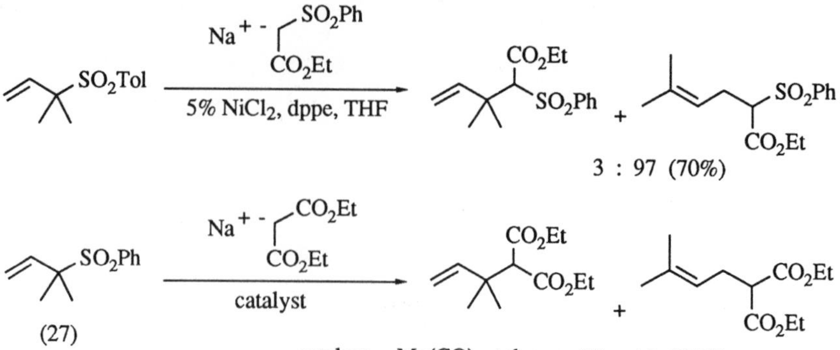

(27)

dppe = 1,2-bis(diphenyl-phosphino)ethane

catalyst =	$Mo(CO)_6$, toluene	90 : 10 (61%)
	$Pd(PPh_3)_4$, THF	68 : 32 (67%)
	$Ni(dppe)_2$, THF	39 : 61 (92%)

Scheme 24

Reactions of (**27**) with malonate were examined using several catalyst systems.[44] The molybdenum system appears very useful for directing nucleophilic attack to the more electron deficient site in the intermediate π-allyl molybdenum complex, at least with small nucleophiles such as malonate. This results in substitution at the more substituted end of the allylic unit – usually a complementary result to that obtained using palladium or nickel systems. The overall stereochemical course of the reaction is also rather more flexible than that with palladium catalysis, since epimerisation of the intermediate complexes is possible. Finally, a related report describes the use of stoichiometric $Mo(CO)_6$ for both alkylative and reductive desulphonylation of allylic sulphones.[45]

9.2.3 Pyrolytic Extrusion of SO_2

Although sulphones are thermally rather stable, under extreme conditions they can lose SO_2, and depending on the reaction conditions the released alkyl groups may recombine with the formation of a new C–C bond.[46] Much of this pyrolysis chemistry has been conducted using cyclic sulphones, and examples of this can be found in Chapter 8. However, this method is also applicable to acyclic compounds, provided that at least one of the groups attached to sulphur is allylic or benzylic. The earliest report describing the pyrolysis of a range of allylic sulphones is by La Combe and Stewart.[47] They found that sulphones of general formula (**28**) give respectable yields of sulphur-free alkenes on pyrolysis at temperatures ranging from 170°C to 390°C, Scheme 25.

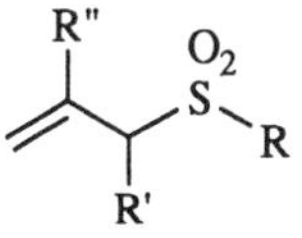

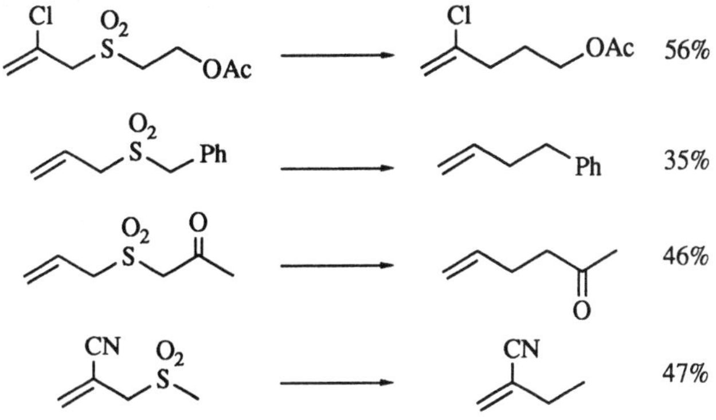

Scheme 25

As is evident from the examples shown, the reaction tolerates functionality on either of the groups attached to the sulphone. Simple dialkyl sulphones are unaffected by the reaction conditions usually employed. The two experimental procedures employed in this work use either a batch or a continuous flow technique, the temperature required for reaction depending on the structure of the sulphone. A later report describes the pyrolysis of a less extensive list of sulphones, but quotes a yield of 80% of 4,4-dimethyl-1-pentene from allyl *tert*-butyl sulphones.[48] The formation of good yields of product R–R' from sulphones RSO_2R', without substantial amounts of R–R or R'–R', points to a thermal pericyclic mechanism for these relatively low temperature pyrolyses. However, it is generally accepted that the majority of sulphone pyrolyses, especially those at higher temperatures, involve radical intermediates.[49] A few examples of the application of sulphone pyrolysis to the synthesis of terpenoids have been reported.[50]

Finally it should be noted that benzylic sulphones undergo extrusion of SO_2 in an analogous fashion to allylic sulphones, under either pyrolytic[51] or photolytic conditions.[52]

9.2.4 Free Radical Desulphonylation using Tin Reagents

Ueno *et al.* first described the use of Bu_3SnH for the regioselective desulphonylation of terminal allylic sulphones, Scheme 26.[53]

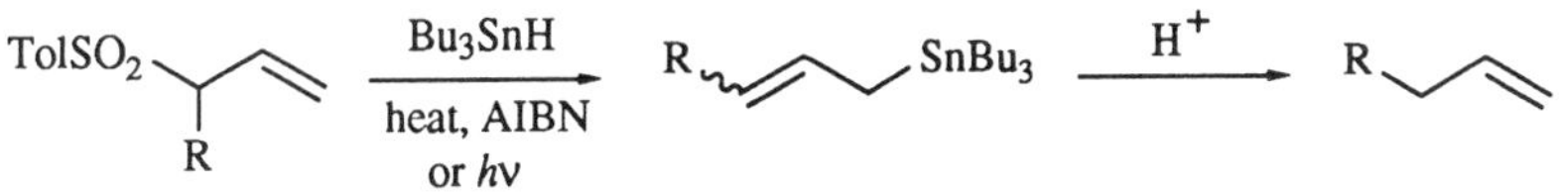

Scheme 26

Addition of tributylstannyl radical to the terminus of the double bond results in the formation of transposed allylstannane products as mixtures of stereoisomers.

Protolytic destannylation of these products with HCl or HOAc then produces the desired terminal alkenes. The method is reasonably efficient, tolerates remote unsaturation in the substrate, and if desired both steps can be carried out in one pot. The formation of allylstannane intermediates is attractive, since they can be usefully reacted with electrophiles other than a proton. An example of this strategy is found in the later report of Ueno *et al.*, in which the synthesis of the natural product lavandulol (**29**) was described, Scheme 27.[54]

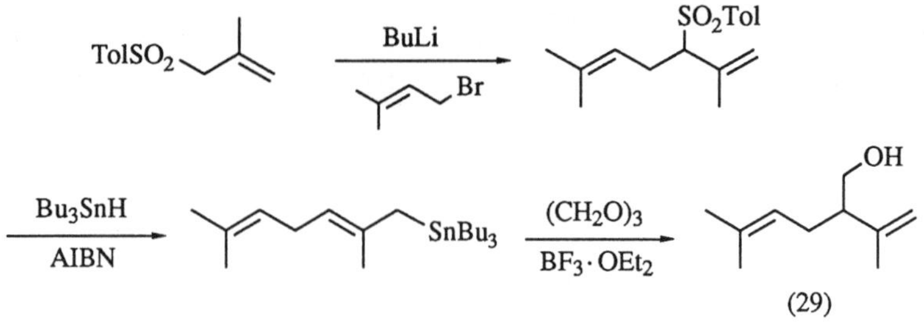

Scheme 27

In this case trioxan, activated by a Lewis acid, was found to be the best reagent for effecting hydroxymethylation of the intermediate allylstannane. Tin-mediated desulphonylation can also be used to convert certain types of β-hydroxy sulphone into 2-substituted-1,3-butadienes.[55] A slight modification of the Bu$_3$SnH chemistry has also been reported involving the use of tributyltin formate (Bu$_3$SnOCHO).[56] The tin method can also be applied to simple allylic sulphones having some additional functionality, e.g. Scheme 28.[57] In this case the 2-functionalised allylstannane products were used in a free radical C–C bond-forming coupling with a glycine derivative.

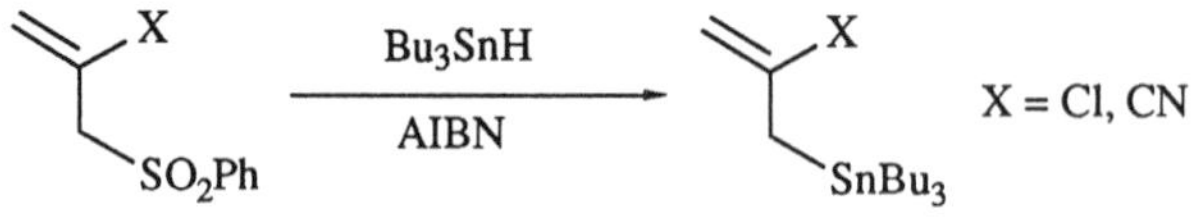

Scheme 28

9.3 Vinyl Sulphones

The ready availability of vinyl sulphones, often of fixed double-bond geometry, has stimulated a search for regio- and stereoselective methods for their desulphonylation. The available methods fall roughly into two groups: those resulting in replacement of the sulphone by hydrogen (hydrogenolysis or reductive desulphonylation), and those in which sulphone removal is accompanied by C–C bond formation (alkylative desulphonylation). Some desulphonylation chemistry involving vinyl sulphone intermediates can also be found in Chapter 7.

9.3.1 Reductive Desulphonylation

In general the reductive removal of the sulphone group from vinyl sulphones using dissolving metals or metal amalgams is not stereospecific. Pascali and Umani-Ronchi have shown that the desulphonylation of certain styryl sulphones with either aluminium amalgam or $LiAlH_4$–$CuCl_2$ results in the formation of the more stable (E)-alkene products exclusively.[58]

Julia has been a pioneer in the discovery of stereospecific methods for both the reductive and alkylative desulphonylation of vinyl sulphones. Two important general methods which fall into the former category have emerged from this work. The first involves reaction of the vinyl sulphone with ^nBuMgCl in the presence of a transition metal catalyst, Scheme 29.[59]

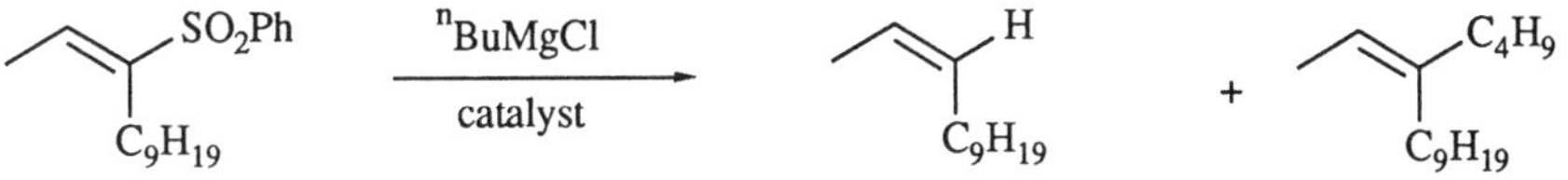

		(E) : (Z)	
		64% 2.5 : 97.5	4%
catalyst =	$Ni(acac)_2$		
	$Pd(acac)_2$, $P(^nBu)_3$	83% 1.5 : 98.5	5%

Scheme 29

Both nickel and palladium catalysts are highly effective, the yields and stereoselectivity being slightly higher when $Pd(acac)_2$ is employed in the presence of $P(^nBu)_3$ or DABCO. This chemistry was later applied to dienyl sulphones of types (**30**)-(**32**), e.g. Scheme 30.[60]

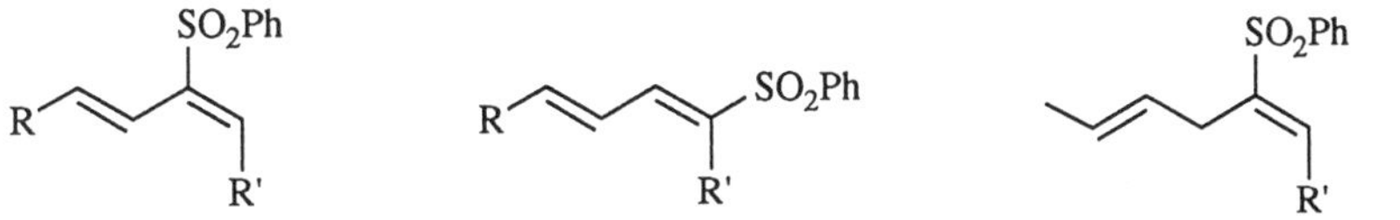

(30)   (31)   (32)

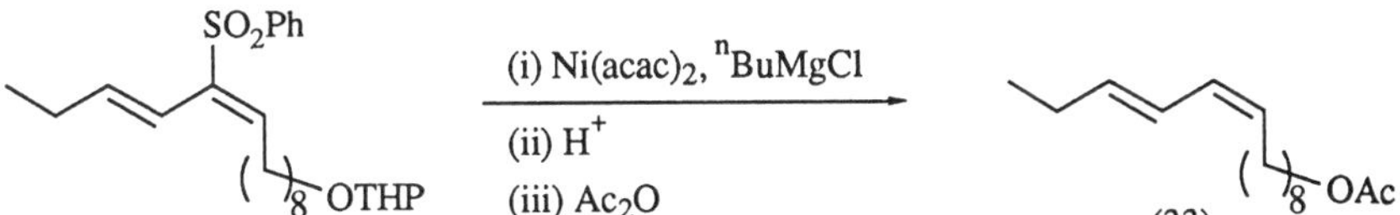

(i) $Ni(acac)_2$, ^nBuMgCl

(ii) H^+

(iii) Ac_2O

(33)

Scheme 30

Whilst transition metal-catalysed desulphonylations of sulphones of types (**31**) and (**32**) give mixed results, reactions of sulphones of structure (**30**) are notably more successful. This type of sulphone was used to synthesise various diene pheromones such as (33) in high stereochemical purity. Attempts to desulphonylate sulphones such as (30) with aluminium amalgam or buffered sodium amalgam resulted in reduction of the vinyl sulphone double bond, thus giving an allylic sulphone product.

The second general method for vinyl sulphone hydrogenolysis introduced by Julia's group involves treatment of the vinyl sulphone with sodium dithionite and NaHCO$_3$ in aqueous DMF, e.g. Scheme 31.[61]

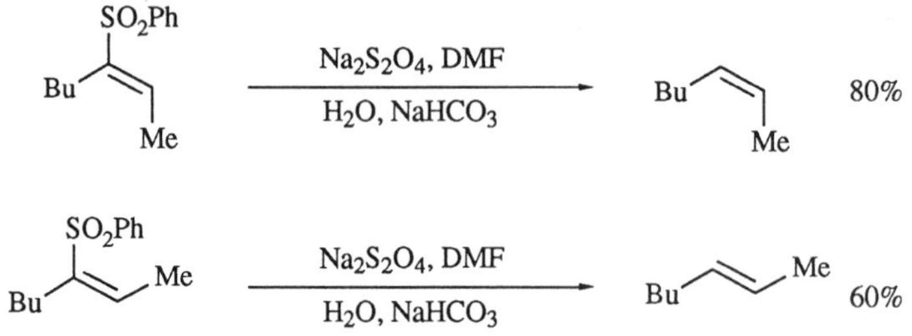

Scheme 31

The mechanism of the reaction involves the conjugate addition of HSO$_2^-$ (*syn*) followed by loss of SO$_2$ and expulsion of sulphinate in an *anti*-fashion.[62]   The method can be adapted to allow deuterolysis, simply by employing D$_2$O in the reaction medium in place of water.[63] The dithionite method was also examined with dienyl sulphones (30)-(32), with only the last class of sulphone giving satisfactory results.[60]

Huang and Zhang have described the desulphonylation of certain α-methylthio-α,β-unsaturated sulphones using sodium hydrogen telluride in ethanol at room temperature, Scheme 32.[64]

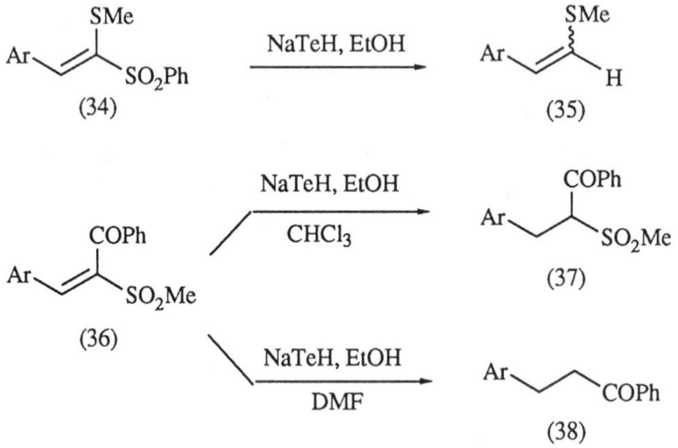

Scheme 32

Hence (34) is reacted to give (35) as an 3:1 (Z):(E) mixture of stereoisomers. The same type of reduction was also applied to unsaturated sulphones such as (36) having an additional ketone function.[65] Depending on the reaction conditions, either reduction to (37) or tandem reduction–desulphonylation to give (38) can be effected in good yield.

Eliminations of β-substituted sulphones, including β-stannyl sulphones, are discussed in detail in Chapter 7.[66] One additional type of desulphonylation of vinyl sulphones has also proved possible by treatment with Bu_3SnH under free radical conditions, Scheme 33.[67] This reaction may involve either an addition–elimination sequence, or alternatively, an electron transfer process.

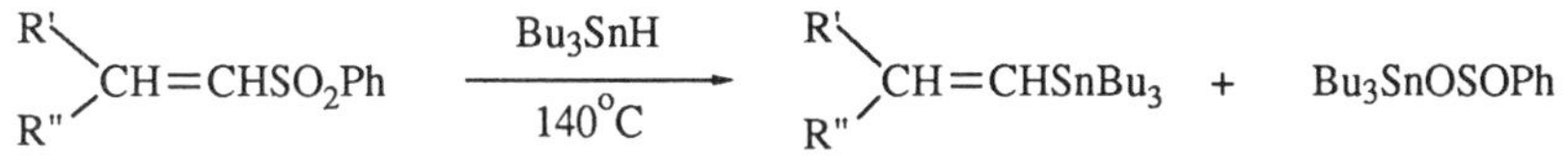

$$R' \underset{R''}{\diagup}CH=CHSO_2Ph \xrightarrow[140°C]{Bu_3SnH} R'\underset{R''}{\diagup}CH=CHSnBu_3 \quad + \quad Bu_3SnOSOPh$$

Scheme 33

9.3.2 Alkylative Desulphonylation

The reaction of methyl styryl sulphone with $^nBu_3B$ or $^iPr_3B$ yields the corresponding alkyl-substituted styrene in each case, Scheme 34.[68]

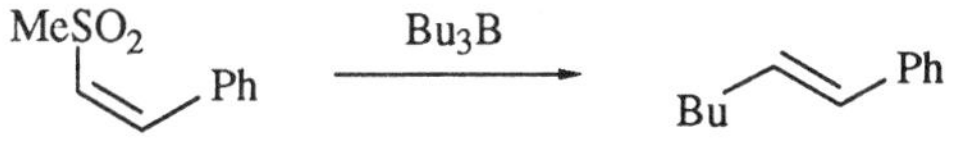

Scheme 34

The (E)-product is formed, regardless of the starting material geometry. The reaction is proposed to proceed by formation of a vinyl radical intermediate. In accord with this proposal it was found that in THF or toluene some minor products arising from combination with solvent-derived radicals are also obtained.

A rather different type of desulphonylation mediated by boron reagents has been described by Brown *et al.*[69] Treatment of a diaryl sulphone with two equivalents of lithium triethylborohydride in THF at reflux results in rapid cleavage e.g. Scheme 35.

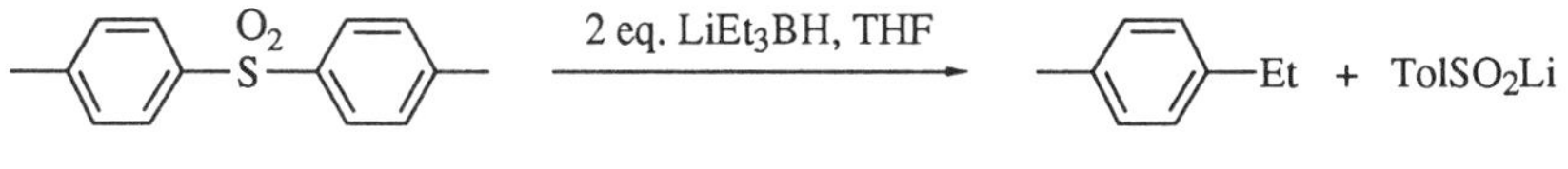

Scheme 35

Regiospecific substitution of the TolSO$_2$ group for an alkyl group from the reducing agent is observed. The reaction proceeds well with a few different trialkylborohydrides, and gives the best yields when 1-octene is added to minimise side reactions thought to be due to the diethylborane by-product. Unfortunately the method is rather less effective with alkyl aryl sulphones; methyl phenyl sulphone gives a 38% yield of ethylbenzene.

An indirect sequence for alkylative desulphonylation of methyl vinyl sulphones has been developed by Baldwin's group, allowing the synthesis of (*E*)- and (*Z*)-α-bisabolenes (**39**), Scheme 36.[70]

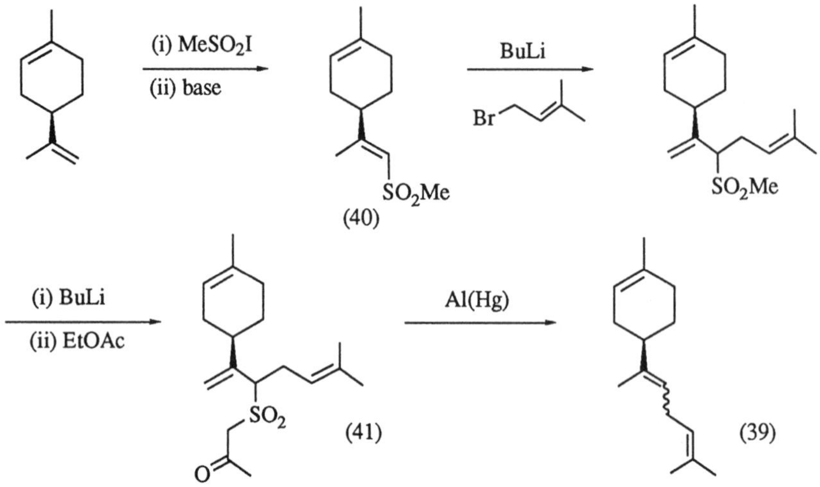

Scheme 36

The conversion of vinyl sulphone (**40**) to the mixture of stereoisomeric products (**39**) represents overall alkylative desulphonylation. The key step involves treatment of β-ketosulphone (**41**) with aluminium amalgam to give an intermediate allylic sulphinic acid which then undergoes [3,3] sigmatropic loss of SO$_2$ to give the observed products. Although the process is highly regioselective and occurs without racemisation the formation of mixtures of stereoisomers is a serious drawback. Despite this, the method is significant in effecting the desulphonylation of β,β-disubstituted vinyl sulphones, substrates which are often problematic with many of the desulphonylating systems described elsewhere in this chapter.

The most significant advances in direct stereoselective alkylative desulphonylation of vinyl sulphones have undoubtedly been made by Julia and co-workers. Thus reaction of stereochemically pure vinyl sulphones with Grignard reagents under nickel[71] or iron[72] catalysis gives substitution products, e.g. Scheme 37.

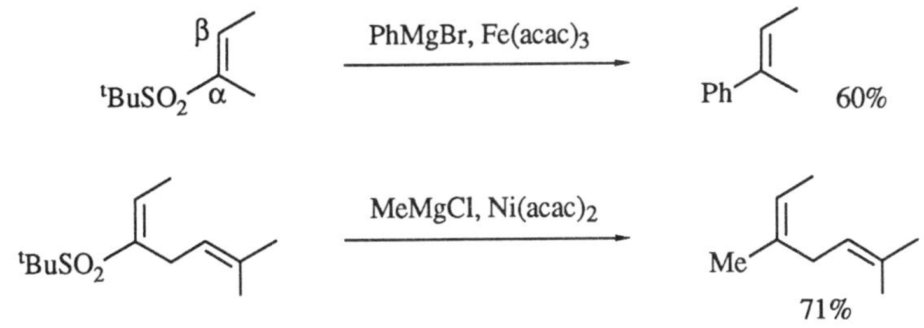
β
ᵗBuSO₂
α
PhMgBr, Fe(acac)3
Ph
60%
ᵗBuSO₂
MeMgCl, Ni(acac)2
Me
71%

Scheme 37

High yields of substitution products can be obtained from α,β-disubstituted and β,β-disubstituted vinyl sulphones using phenyl Grignard reagents with either type of catalyst. In the case of simple β-substituted sulphones Michael addition is a serious problem, however. Methyl groups are effectively introduced using the nickel catalyst, no coupled products being obtained using the MeMgX–$Fe(acac)_3$ system. Likewise, the nickel system is superior for coupling with vinyl Grignard reagents. However, the iron system is uniquely effective for coupling higher *n*-alkyl groups, including butyl, hexyl and octyl groups. In combination with $Ni(acac)_2$ such Grignard reagents effect sulphone hydrogenolysis, and not coupling, as described in Scheme 29. The hydrogenolysis is also observed using the $Fe(acac)_3$ catalyst and isopropyl Grignard reagents.

Remarkably, Smorada and Truce found that reaction of certain types of aryl alkynyl sulphones with either Grignard reagents or organolithiums results in substitution, Scheme 38.[73]

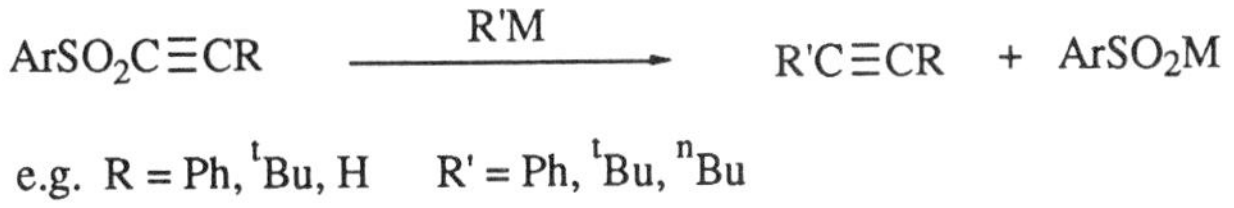
ArSO₂C≡CR
R'M
R'C≡CR + ArSO₂M
e.g. R = Ph, ᵗBu, H R' = Ph, ᵗBu, ⁿBu

Scheme 38

The reaction is limited to alkynyl sulphones not having hydrogens at the propargylic position of the R group. The nature of this reaction was subsequently examined by Eisch's group, and found to involve single electron transfer (SET).[74] The reaction was also shown to be applicable to alkylative desulphonylation of a vinyl and a diaryl sulphone, Scheme 39.

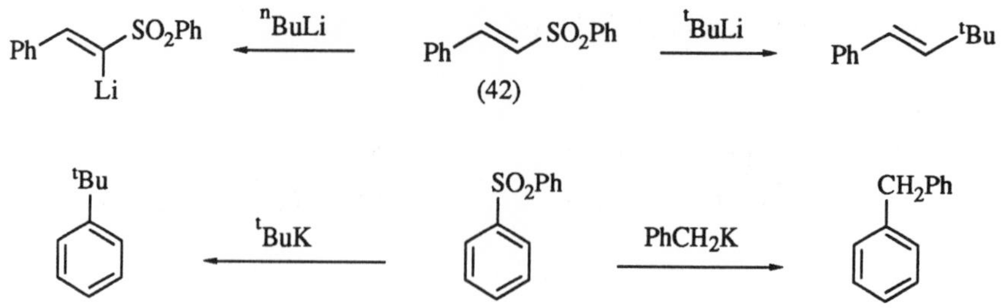

Scheme 39

Particularly interesting is the change in the reactivity of vinyl sulphone (**42**) on switching from ^{n}BuLi to ^{t}BuLi. With the former organometallic α-deprotonation of the vinyl sulphone occurs, whereas substitution is seen with the tertiary alkyl metal. Since the substitution involves SET it is not surprising that tertiary, allylic and benzylic organometallics are particularly suitable for this type of reaction.

9.4 Functionalised Sulphones

A wide range of sulphones in which additional local functionality is present have been shown to be susceptible to desulphonylation. This section describes this chemistry which encompasses methods already included in earlier sections of this chapter, as well as methods tailor-made for particular types of polyfunctional sulphone.

9.4.1 Hydrogenolysis using Metals, Metal Amalgams and Related Agents

Metal amalgams, particularly sodium amalgam, remain the most common choice for the desulphonylation of all types of sulphones. A good example is the reaction of sulphonyl aziridines such as (**43**), Scheme 40.[75]

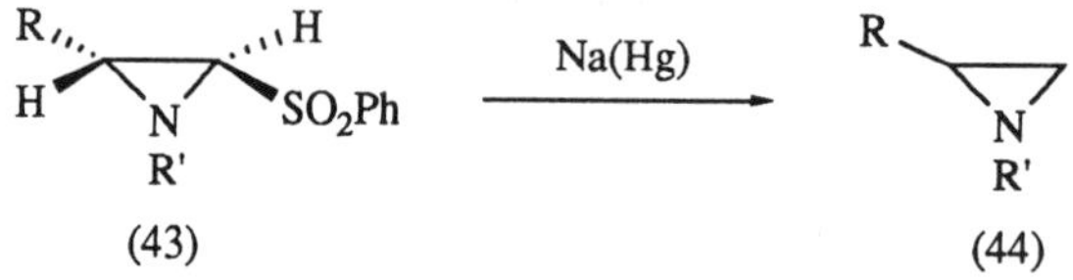

Scheme 40

Whilst sodium amalgam provides a quantitative yield of (**44**), in one case the reaction of an aziridine with LiAlH$_4$ results in competing desulphonylation and reduction to the aziridine sulphide.

Epoxides derived from allylic sulphones undergo desulphonylation with concomitant epoxide ring opening, Scheme 41.[76]

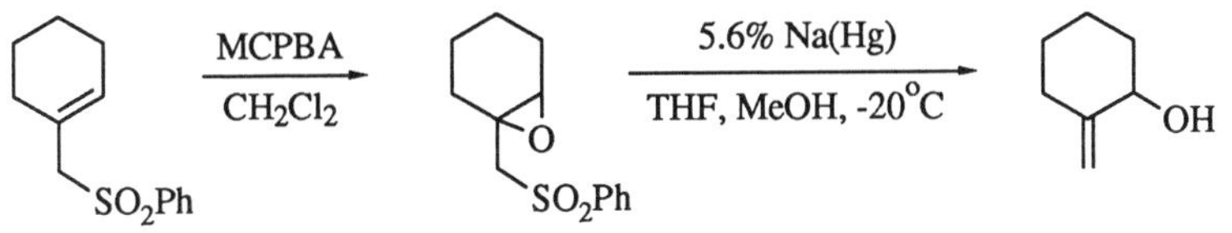

Scheme 41

The method constitutes a synthesis of α-methylene carbinols in which overall allylic transposition of the functional group has occurred.

The use of aluminium amalgam for the desulphonylation of β-ketosulphones was pioneered by Corey and Chaykovsky.[77] Recently this method was employed by Isobe's group in their total synthesis of the complex marine toxin, okadaic acid, Scheme 42.[78]

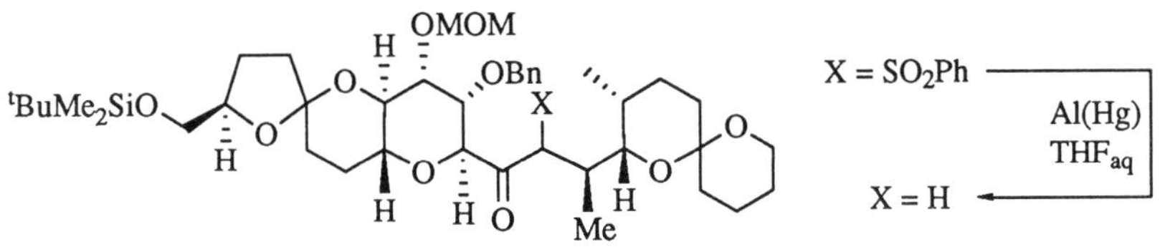

Scheme 42

Whilst effective on this crucial example, additional experiments using model ketosulphone systems indicate that cleavage of a neighbouring tetrahydropyran ring can occur on treatment with excess reagent.

The reductive desulphonylation of α-sulphonyl acetates has been studied in some detail with a range of reducing agents, Scheme 43.[79,80]

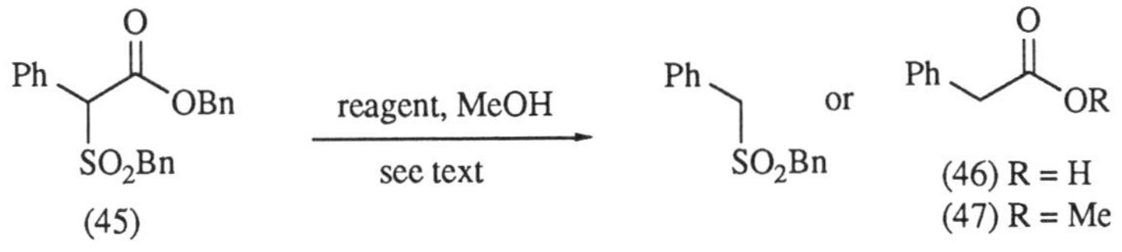

Scheme 43

Cleavage of either the sulphone group or the ester group of (45) is observed, depending on the reaction conditions. The use of buffered sodium amalgam gives dibenzyl sulphone, whilst omission of the buffer facilitates desulphonylation to give (47) by initial transesterification of (45). An efficient one-pot transesterification–desulphonylation–ester hydrolysis procedure was developed, allowing access to acids such as (46).[79] Interestingly, both treatment with Raney nickel in ethanol and simple alkaline hydrolysis of (45) also result in conversion to dibenzyl sulphone.[80] Various desulphonylations of ester sulphones using sodium and ethanol in THF were also described.

Two further examples illustrate the usefulness of both the buffered and unbuffered sodium amalgam procedures for desulphonylation of lactone sulphones, Scheme 44.[81,82]

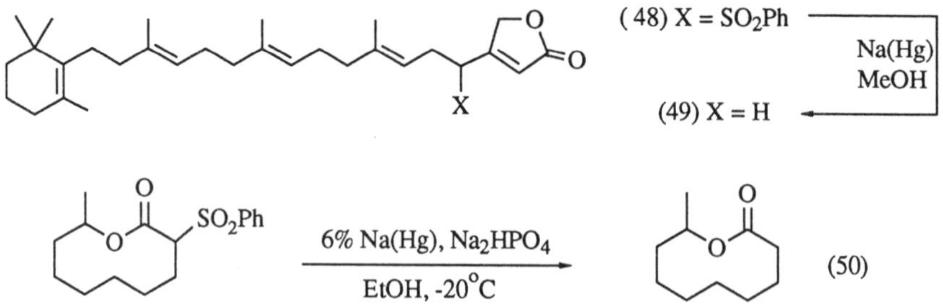

Scheme 44

The desulphonylation of the butenolide sulphone (48) enabled a total synthesis of the natural product mokupalide (49) to be achieved.[81] Whilst the example yielding (50) involves the usual phosphate buffer, another example required the use of acetic acid for optimal results.[82] Unsaturated ketosulphones have been desulphonylated with either zinc in acetic acid, lithium in liquid ammonia, or excess lithium dimethylcuprate, e.g. Scheme 45.[83]

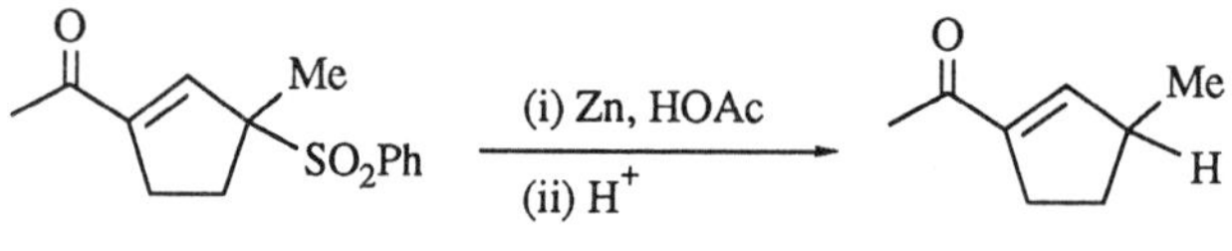

Scheme 45

The process initially gives a deconjugated enone, presumably by kinetically controlled α-protonation of an intermediate dienolate. This is then isomerised by treatment with acid to give the product shown. Zinc in acetic acid has also been used to obtain 1,4-diketones from the corresponding 2-sulphonyl-1,4-diketones.[84]

The use of the sulphone group as a regiocontrol element for the alkylation of a cyclic sulphone such as (**51**) has some drawbacks. Thus, after allylation and desulphonylation with aluminium amalgam, only a moderate yield (41%) of (**52**) was obtained, Scheme 46.[85]

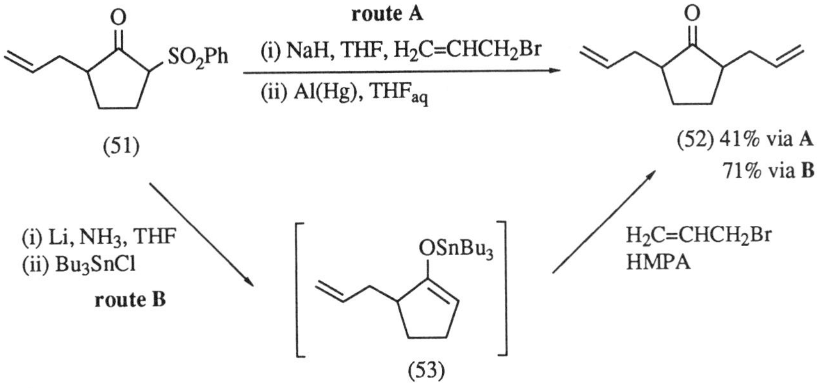

Scheme 46

A new method was developed involving regiospecific generation of the tin enolate (**53**) by treatment with lithium in ammonia and THF. Subsequent alkylation is then possible if HMPA is added, giving (**52**) in greatly improved overall yield.

A series of papers by Beau and Sinay describes some very elegant uses of the sulphone group for the stereocontrolled formation of *C*-glycosides.[86] This chemistry involves the desulphonylation of anomeric sulphones derived from 2-deoxy-D-glucose with lithium naphthalenide, e.g. Scheme 47.

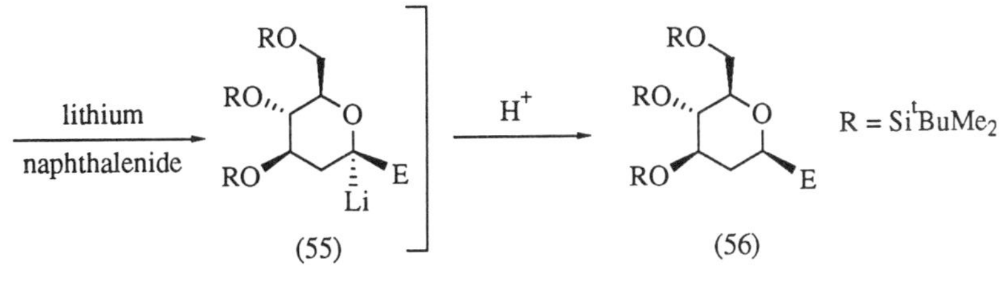

Scheme 47

Treatment of the starting sulphone (**54**) with LDA gives the corresponding carbanion which is then reacted with a range of electrophiles (D_2O, MeI, R'CHO). Addition of lithium naphthalenide then results in *in situ* desulphonylation to give products (**56**) with very high diastereoselectivity. The stereocontrol originates in the desulphonylation step, which occurs by successive single electron transfers. An intermediate anomeric σ-radical is formed in which the singly occupied orbital adopts an α-orientation due to stereoelectronic stabilisation. The anomeric substituent consequently adopts the β-orientation, and this arrangement is retained in both the subsequently formed carbanion (**55**) and derived products (**56**). A typical product is the β-'aldol' (**57**), which can be oxidised to ketone (**58**), Scheme 48.

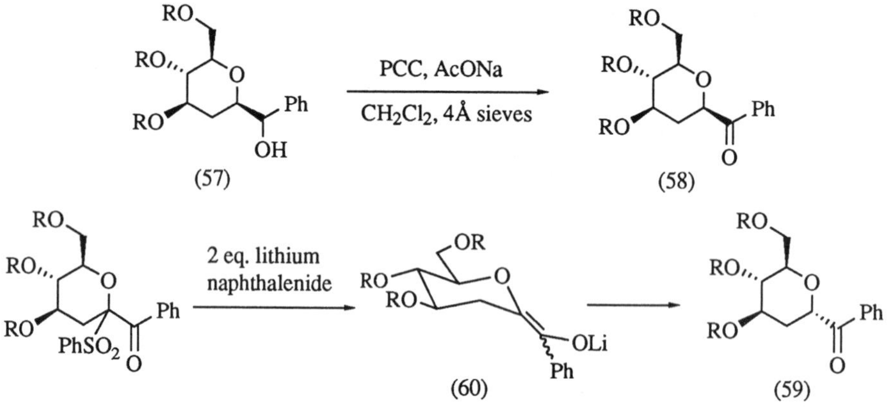

Scheme 48

A stereocomplementary synthesis of the α-keto product (**59**) is also available. In this case the stereochemical outcome is dependent on the face-selective protonation of an enolate (**60**) formed by the desulphonylation. Application of this chemistry to the C-linking of two sugar rings was also examined.

The reductive desulphonylation of a substituted (allyloxy)methyl sulphone using lithium naphthalenide has been briefly examined as an entry to [2,3] Wittig rearrangements, Scheme 49.[87]

Scheme 49

The reaction is thought to proceed via the intermediate α-oxygen-stabilised carbanion which then rearranges as expected. The sulphone was shown to undergo this reductively initiated Wittig rearrangement rather less efficiently than the corresponding sulphide.

The interesting problem of effecting stereospecific desulphonylation of an α-chiral sulphone has not been solved. A report by Grimm and Bonner indicates that the desulphonylations of certain α-sulphonyl amides using Raney nickel show complex stereochemical behaviour.[88] Results ranging from predominant inversion to predominant retention are obtained, depending upon the sulphone structure, the method of Raney nickel pretreatment, and the solvent used for the reaction. Further work in this area is desirable before any general conclusions can be drawn.

9.4.2 Alternative Hydrogenolyses and Substitutions of the Sulphone Group by Heteroatom Functionality

The success of reductive desulphonylations of sulphones with dihydropyridines depends critically on the presence of neighbouring functionality. Both α-nitro-sulphones[89,90] and β-ketosulphones[91] can be desulphonylated in this way, Scheme 50.

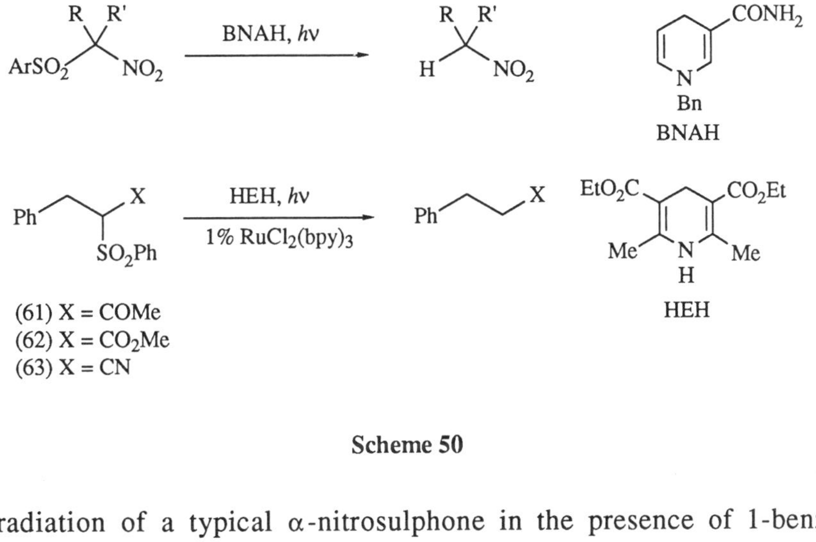

(61) X = COMe
(62) X = CO₂Me
(63) X = CN

Scheme 50

Thus irradiation of a typical α-nitrosulphone in the presence of 1-benzyl-1,4-dihydronicotinamide (BNAH) results in clean desulphonylation via radical anion intermediates.[89,90] An attempt to use buffered sodium amalgam on such substrates was unsuccessful.[90] Desulphonylation of β-ketosulphones such as (61) can be accomplished in a similar way, although in this case the more photo-stable Hantzsch ester (HEH) is preferred.[91] No reaction occurs with the related ester (62) or nitrile (63). The report by Wade's group includes two additional types of desulphonylation of α-nitrosulphones which are worthy of note, Scheme 51.[90]

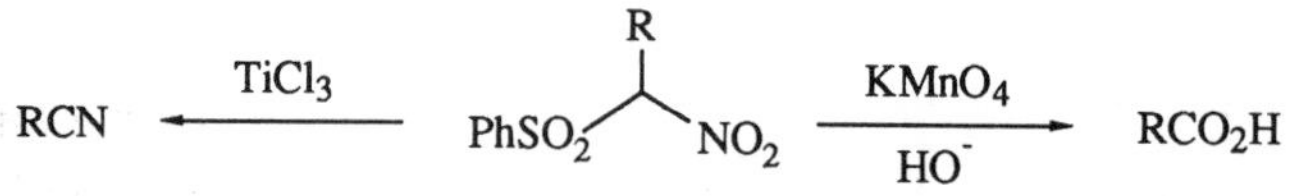

Scheme 51

Reaction with 20% aqueous $TiCl_3$ results in formation of the corresponding nitrile by initial formation of an α-sulphonyl oxime. Alternatively, oxidation with basic potassium permanganate affords the carboxylic acid.

A mild glycosidation procedure has been developed involving the displacement of glycosyl sulphones with alcohols, e.g. Scheme 52.[92]

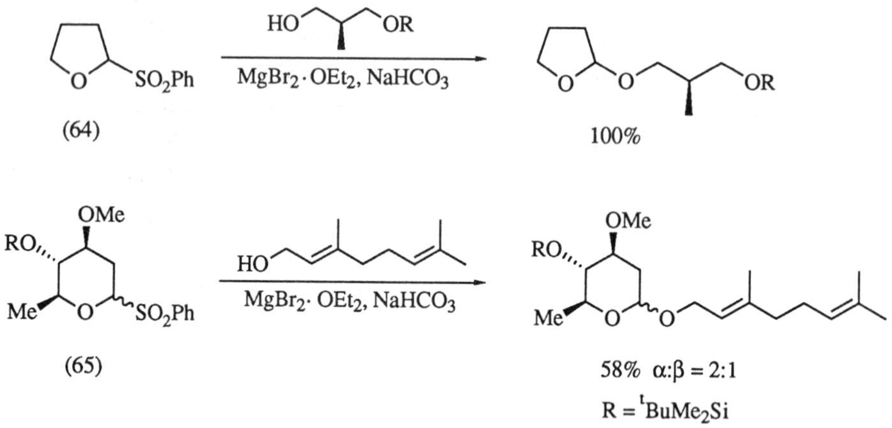

Scheme 52

The process occurs under mild conditions at room temperature for simple sulphones such as (64) and its six-membered counterpart. With more complex sulphones such as (65) the reaction is more sluggish and requires excess of reagents and more vigorous conditions (heating or ultrasonication) for best results.

Crich and Ritchie described some unsuccessful attempts to prepare orthoester analogues of this type of sulphone, i.e. (67).[93] Thus, although the halogenated derivatives (66) could be prepared, their reaction with *n*-decanol in the presence of silver(I) triflate gives, not the desired 2-arenesulphonyl-2-alkoxytetrahydropyran (67), but decyl 4-toluenesulphinate, Scheme 53.

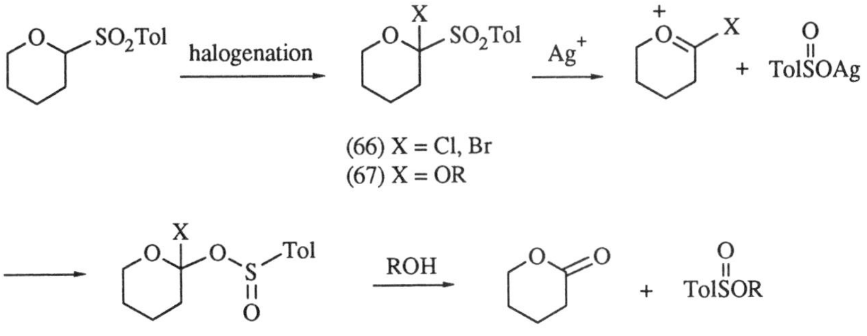

(66) X = Cl, Br
(67) X = OR

(68)

Scheme 53

This outcome was explained by invoking initial rearrangement to the sulphinate ester (**68**) before reaction with the alcohol. Internal displacement of a sulphone group from glycosyl sulphones, with concomitant formation of a 1,6-anhydro link, has also been reported, Scheme 54.[94]

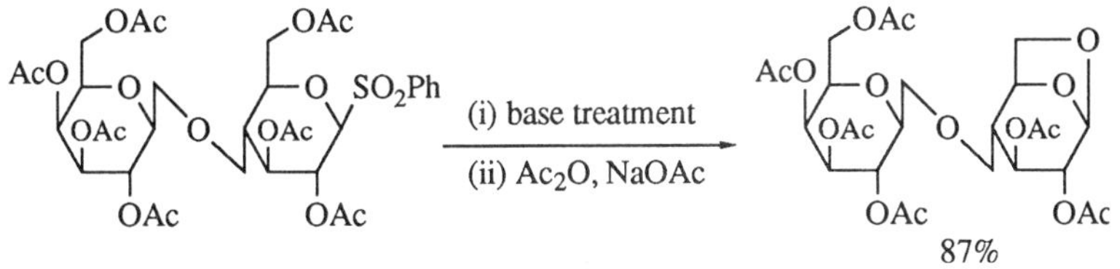

Scheme 54

On treatment with base followed by reacylation, the starting sulphone disaccharides are cleanly converted to the corresponding 1,6-anhydro disaccharide. In some cases better results can be obtained by use of the corresponding 4-chlorophenylsulphonyl group.

Certain α-(phenylsulphonylalkyl)enones undergo displacement of the sulphonyl group on treatment with nucleophiles including amines and thiolates, Scheme 55.[95]

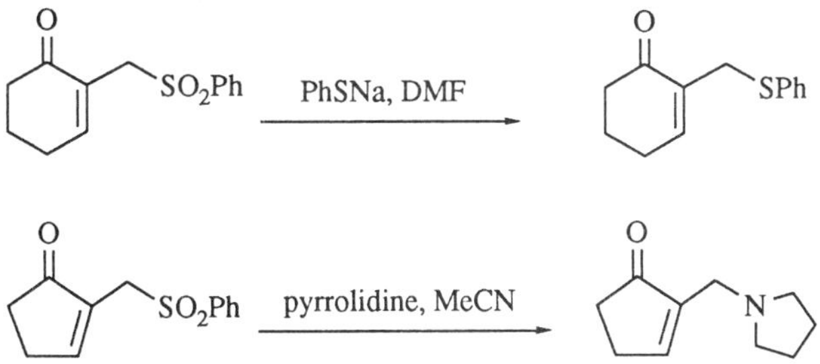

Scheme 55

These reactions probably proceed by a double S_N2' mechanism. Other reactions examined include those with ethane-1,2-dithiol sodium salt to effect annulation of a sulphur containing ring, and alkylative desulphonylations using soft carbon nucleophiles such as malonate.

Finally, a well-used desulphonylation involves the reaction of α,β-epoxy sulphones with $MgBr_2$, resulting in the formation of α-bromo aldehydes and ketones, Scheme 56.[96]

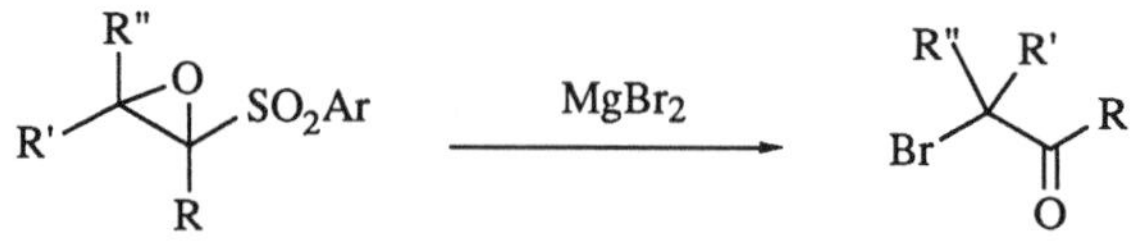

Scheme 56

The method has the advantages of being operationally very straightforward and using readily available starting materials.

9.4.3 Alkylative Desulphonylation

In addition to the substitution reactions described in the last section, glycosyl sulphones also undergo an impressive array of reactions with carbon nucleophiles, Scheme 57.[97]

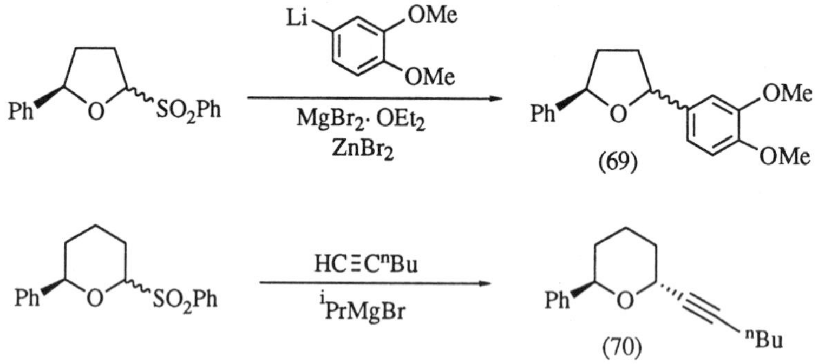

Scheme 57

The preparations of (**69**) and (**70**) illustrate some salient features of this chemistry. Firstly, a combination of Grignard reagent and zinc bromide is required for good reaction; alkyllithiums can be used only if $MgBr_2.OEt_2$ and $ZnBr_2$ are added. With five-membered ring systems the stereoselectivity is usually poor; (**69**) is formed as a 1:1 mixture of diastereoisomers. Six-membered systems give improved stereocontrol since axial attack on the conformationally locked oxonium ion intermediate is preferred. A range of aryl, heteroaryl, vinyl and alkynyl nucleophiles can be utilised.

An attractive feature of the chemistry is the ability to build in two substituents by two successive displacements, Scheme 58.

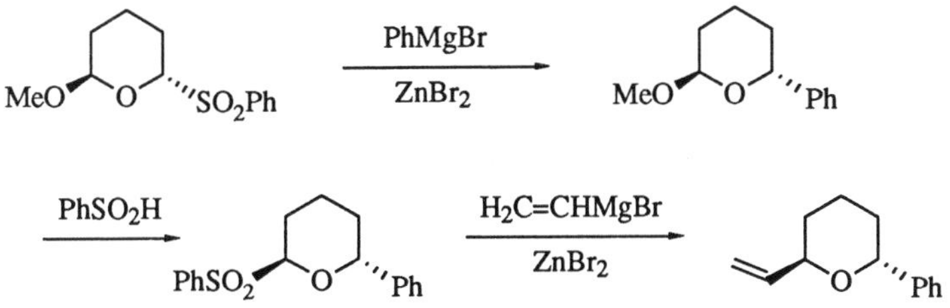

Scheme 58

Other nucleophiles, including silyl enol ethers, silyl ketene acetals, allyltrimethylsilane, trimethylsilyl cyanide and trimethylaluminium give similarly good results.

Analogous chemistry is also possible using the corresponding 2-phenylsulphonyl-piperidines and -pyrrolidines. An example is the reaction of pyrrolidine sulphone (**71**) with a silyl enol ether, thus allowing a synthesis of the simple alkaloid ruspolinone (**72**), Scheme 59.[98]

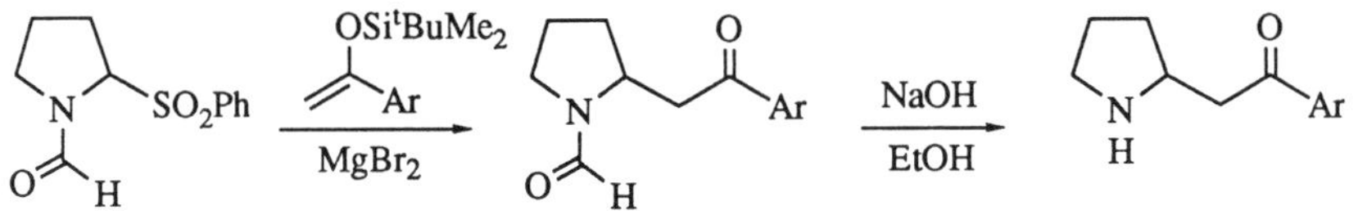

(**71**)   Ar = 3,4-dimethoxyphenyl   (**72**)

Scheme 59

Similar displacements have been carried out on other systems, notably using β-lactam sulphones, e.g. Scheme 60.[99,100]

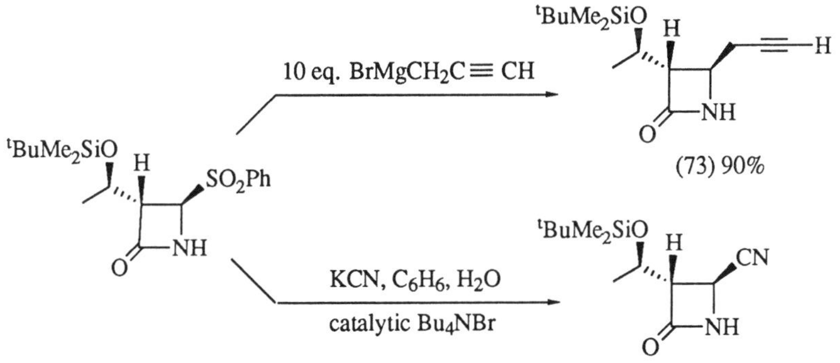

Scheme 60

Protection of the lactam nitrogen in (73) then leads to a known intermediate in a previous total synthesis of (+)-thienamycin.[99] Likewise, nitrile (74), prepared under phase-transfer conditions, serves as a carbapenem precursor.

α-Nitrosulphones undergo alkylative desulphonylation on reaction with nitronate anions or the sodium salts of some malonates and β-ketoesters, Scheme 61.[101]

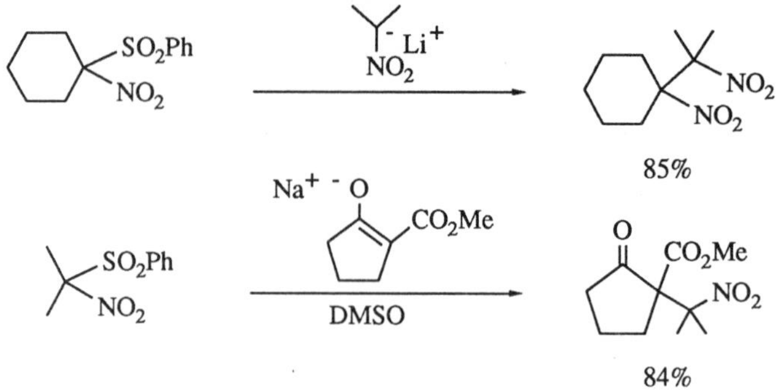

Scheme 61

The yields of substitution products are very good, the reaction proceeding by way of a chain mechanism involving both radicals and radical anions.

The appropriate positioning of a sulphonyl group on a typical Michael acceptor can allow addition–elimination reactions in which the sulphone is effectively substituted for a carbon nucleophile, Scheme 62.[102–104]

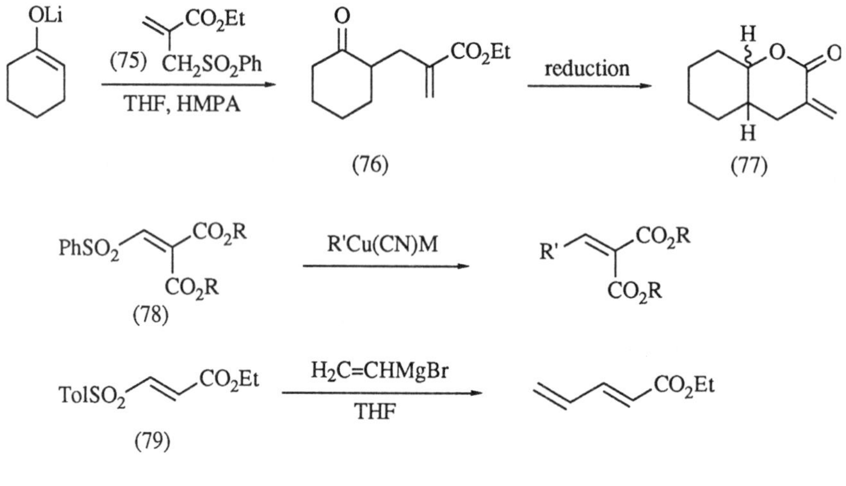

Scheme 62

The reactions of lithium enolates with sulphone (75) are similar to chemistry mentioned above in Scheme 55.[102] Here the aim was to construct ketoesters such as

(76) which could then be transformed into *exo*-methylene valerolactones such as
(77). The sulphones (78)[103] and (79)[104] both allow organometallic coupling at the
β-position of an α,β-unsaturated ester (or amide) system. Notably the use of mixed
copper–zinc organometallics in the coupling with (78) enables incorporation of
various functional groups into R'.

A related reaction has been carried out using a sulphonyl-substituted isoxazoline
(80), Scheme 63.[105]

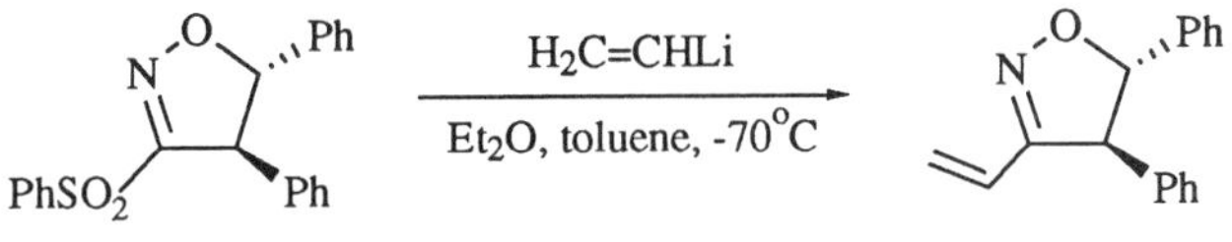

(80)

Scheme 63

In this reaction the use of the vinyllithium (prepared from tetravinyltin) gives far
superior yields to the more readily available vinylmagnesium bromide.

Finally, a rearrangement, which results in desulphonylation, has been reported to
occur on treatment of β-ketosulphones such as (81) with base, Scheme 64.[106]

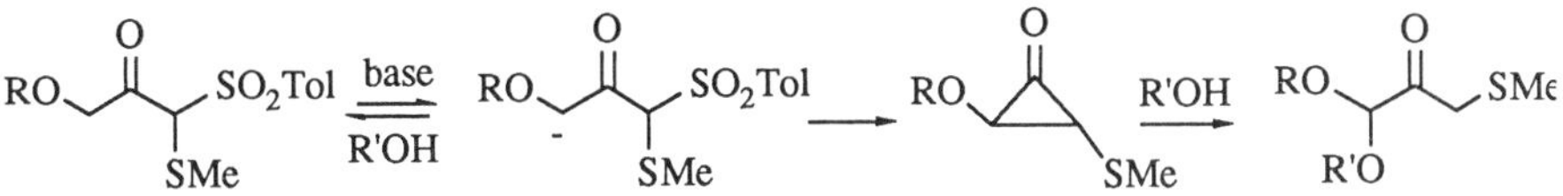

(81)   (82)

Scheme 64

Thus on reaction of (81) with alcoholic base, conversion to the ketosulphide acetals
(82) is observed. The reaction most probably proceeds as shown by an initial
cyclopropane formation somewhat reminiscent of a Favorskii reaction.

9.5 Sulphone Reduction

Most sulphones are extremely resistant to even the most powerful reducing agents.
However, under vigorous conditions reduction to the corresponding sulphides can be
effected. Bordwell and McKellin conducted the first systematic examination of the
reduction of sulphones to sulphides using $LiAlH_4$.[107] It was found that most
sulphones can be reduced in satisfactory yield by the use of excess reagent under quite
severe conditions, e.g. refluxing butyl ethyl ether (92°C). Simple acyclic and six-
membered ring sulphones require prolonged reaction times, whilst four- and five-
membered ring sulphones react much more rapidly, Scheme 65.

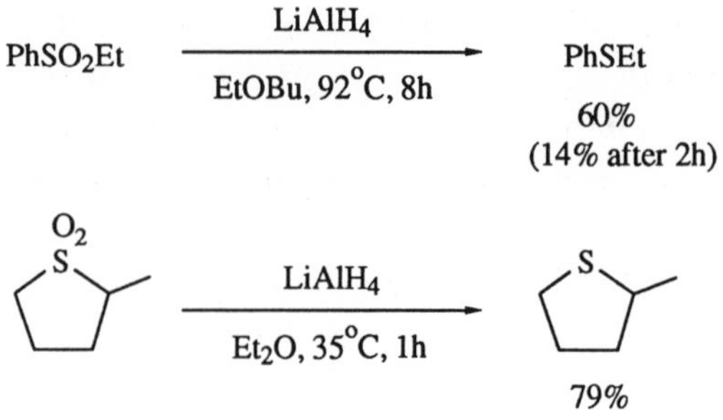

PhSO$_2$Et $\xrightarrow[\text{EtOBu, 92}^\circ\text{C, 8h}]{\text{LiAlH}_4}$ PhSEt

60%
(14% after 2h)

$\xrightarrow[\text{Et}_2\text{O, 35}^\circ\text{C, 1h}]{\text{LiAlH}_4}$

79%

Scheme 65

Subsequent studies by Gardner *et al.* revealed that this type of reduction can be carried out more efficiently (with respect to the number of hydride equivalents used) by the use of DIBAL in toluene at reflux.[108] Kemp and Buckler found that sulphone (83) gives mixtures of products when reacted with LiAlH$_4$ or DIBAL under the usual conditions, Scheme 66.[109]

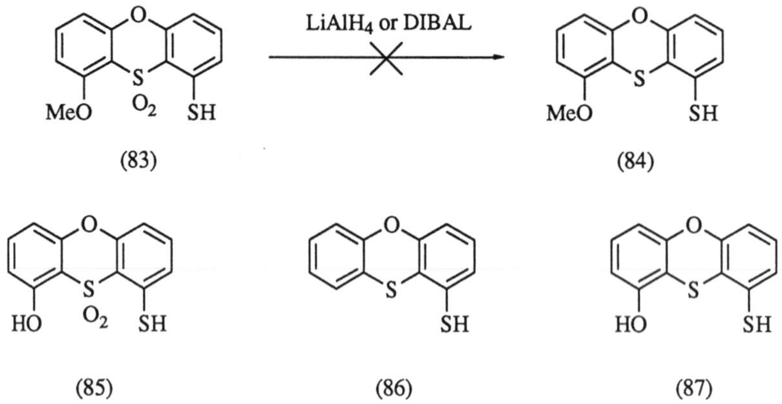

(83)

LiAlH$_4$ or DIBAL

(84)

(85)

(86)

(87)

Scheme 66

Whilst DIBAL gives traces of the desired sulphide (84), the major products are the sulphides (86) and (87). LiAlH$_4$ gives the ether cleavage product (85). Trial reactions involving the use of TiCl$_4$/LiH, Zn/HCl/HOAc, molten S$_8$ and S$_8$ in KOH were unsuccessful. However, a re-examination of the DIBAL system revealed that the use of this reagent in refluxing dioxane is uniquely effective, and provides (84) in 78% yield. The mechanism of the LiAlH$_4$ reductions is not fully understood. In the case of five-membered ring sulphones (which undergo anomalously fast reduction) it has been shown that no deprotonation α to the sulphone occurs. Thus tetrahydrothiophene 1,1-dioxide systems, deuterated at the α-position, do not undergo D–H exchange, and do not undergo epimerisation at an α-asymmetric centre.[110,111] However, other types of sulphone do undergo exchange of α-hydrogens on treatment with LiAlH$_4$, and a mechanism involving the formation of α-sulphonyl carbanions or α,α'-dianions has been proposed.[111]

The reduction of both diphenyl sulphone and dibutyl sulphone using SmI$_2$ in THF and HMPA has been described, although the yield of sulphide from the latter sulphone is only 26%.[112]

The reduction of sulphones to sulphoxides has received scant attention. Only one general procedure has been reported, which involves the hydride reduction of aryloxysulphoxonium salt intermediates, Scheme 67.[113]

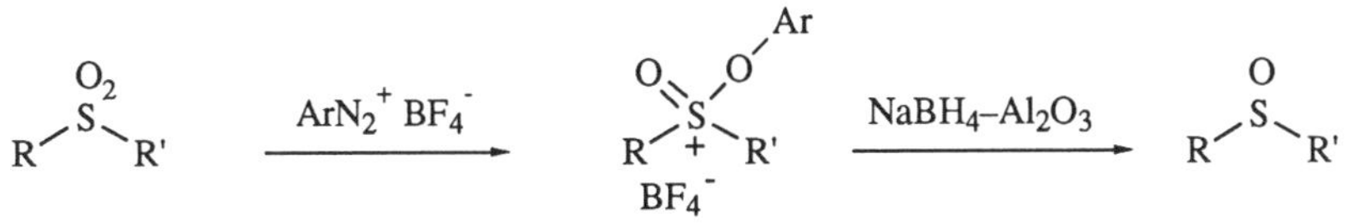

Scheme 67

The major reaction pathway involves direct hydride attack at sulphur, displacement of the aryloxy group, and loss of a proton to give the sulphoxide product. Other pathways involving α-proton removal are much less significant.

Chapter 9 References

1. W. E. Truce, D. P. Tate, and D. N. Burdge, *J. Am. Chem. Soc.,* **1960**, *82*, 2872; W. E. Truce and F. J. Frank, *J. Org. Chem.,* **1967**, *32*, 1918.
2. J. A. Marshall and D. G. Cleary, *J. Org. Chem.,* **1986**, *51*, 858.
3. L. A. Paquette, J. W. Fischer, A. R. Browne, and C. W. Doecke, *J. Am. Chem. Soc.*, **1985**, *107*, 686; see also D. K. Hutchinson and P. L. Fuchs, *J. Am. Chem. Soc.*, **1987**, *109*, 4755.
4. T. Chou and M-L. You, *Tetrahedron Lett.,* **1985**, *26*, 4495; T. Chou, S-H. Hung, M-L. Peng, and S-J. Lee, *Tetrahedron Lett.*, **1991**, *32*, 3551.
5. T. Chou and M-L. You, *J. Org. Chem.,* **1987**, *52*, 2224.
6. R. E. Dabby, J. Kenyon, and R. F. Mason, *J. Chem. Soc.,* **1952**, 4881.
7. B. M. Trost, H. C. Arndt, P. E. Strege, and T. R. Verhoeven, *Tetrahedron Lett.,* **1976**, 3477.
8. M. B. Anderson, M. G. Ranasinghe, J. T. Palmer, and P. L. Fuchs, *J. Org. Chem.,* **1988**, *53*, 3125.
9. H. Kunzer, M. Strahnke, G. Sauer, and R. Wiechert, *Tetrahedron Lett.,* **1991**, *32*, 1949; see also A. S. Kende and J. S. Mendoza, Tetrahedron Lett., **1990**, *31*, 7105.
10. G. R. Pettit and E. E. van Tamelen, *Org. React.,* **1966**, *12*, 356; H. Hauptman and W. F. Walter, *Chem. Rev.,* **1962**, *62*, 347.
11. W. A. Bonner and R. A. Grimm, in *The Chemistry of Organic Sulphur Compounds*, Ed. N. Kharasch and C. Y. Meyers, Pergamon Press, **1966**, *2*, p.35; see also L. Horner and G. Doms, *Phosphorus and Sulphur*, **1978**, *4*, 259.
12. W. E. Truce and F. M. Perry, *J. Org. Chem.,* **1965**, *30*, 1316.
13. S. Becker, Y. Fort, and P. Caubere, *J. Org. Chem.,* **1990**, *55*, 6194.
14. M-C. Chan, K-M. Cheng, K. M. Ho, C. T. Ng, T. M. Yam, B. S. L. Wang, and T-Y. Luh, *J. Org. Chem.,* **1988**, *53*, 4466; K. M. Ho, C. H. Lam, and T-Y. Luh, *J. Org. Chem.,* **1989**, *54*, 4474.
15. N. Dufort and B. Jodoin, *Can. J. Chem.,* **1978**, *56*, 1779.
16. C. Chen and L. M. Stock, *J. Org. Chem.,* **1991**, *56*, 2436.
17. B. Lamm and K. Ankner, *Acta Chem. Scand. B,* **1977**, *31*, 375; B. Lamm and A. Nilsson, *Acta Chem. Scand. B,* **1983**, *37*, 77; L. Horner and H. Neumann, *Chem. Ber.,* **1965**, *98*, 1715; O. Manousek, O. Exner, and P. Zuman, *Collect. Czech. Chem. Commun.,* **1968**, *33*, 3988; B. Lamm, *Tetrahedron Lett.,* **1972**, 1469.
18. J. B. Hendrickson, K. W. Bair, and P. M. Keehn, *J. Org. Chem.,* **1977**, *42*, 2935. For another indirect method of oxidative desulphonylation see H. Kotake, K. Inomata, H. Konoshita, Y. Sakamoto, and Y. Kaneto, *Bull. Chem. Soc. Jpn.,* **1980**, *53*, 3027.
19. R. D. Little and S. O. Myong, *Tetrahedron Lett.,* **1980**, *21*, 3339.

20. L. Horner, H. Hoffmann, G. Klahre, V. G. Toscano, and H. Ertel, *Chem. Ber.,* **1961**, *94,* 1987.
21. W. Reischl and W. H. Okamura, *J. Am. Chem. Soc.,* **1982**, *104,* 6115.
22. A. Clasen and H-D. Scharf, *Liebigs Ann. Chem.,* **1990**, 123.
23. M. Capet, T. Cuvigny, C. Herve du Penhoat, M. Julia, and G. Loomis, *Tetrahedron Lett.,* **1987**, *28,* 6273.
24. J. R. Hwu, *J. Org. Chem.,* **1983**, *48,* 4432.
25. J-B. Baudin, M. Julia, and C. Rolando, *Tetrahedron Lett.,* **1985**, *26,* 2333.
26. D. Uguen, *Bull. Soc. Chim. Fr.,* **1981**, 99.
27. M. Julia, H. Lauron, and J-N. Verpeaux, *J. Organomet. Chem.,* **1990**, *387,* 365.
28. P. S. Manchand, M. Rosenburger, G. Saucy, P. A. Wehrli, H. Wong, L. Chambers, M. P. Ferro, and W. Jackson, *Helv. Chim. Acta,* **1976**, *59,* 387.
29. M. Isobe, Y. Ichikawa, and T. Goto, *Tetrahedron Lett.,* **1981**, *22,* 4287.
30. D. Savoia, C. Trombini, and A. Umani-Ronchi, *J. Chem. Soc., Perkin Trans. 1,* **1977**, 123.
31. K. Sato, S. Inoue, A. Onishi, N. Uchida, and N. Minowa, *J. Chem. Soc., Perkin Trans. 1,* **1981**, 761; see also M. Julia and D. Uguen, *Bull. Soc. Chim. Fr.,* **1976**, 513; P. A. Grieco and Y. Masaki, *J. Org. Chem.,* **1974**, *39,* 2135
32. M. Julia, A. Righini-Tapie, and J-N. Verpeaux, *Tetrahedron,* **1983**, *39,* 3283; M. Julia, A. Righini, and J-N. Verpeaux, *Tetrahedron Lett.,* **1979**, 2393.
33. Y. Masaki, K. Sakuma, and K. Kaji, *J. Chem. Soc., Perkin Trans. 1,* **1985**, 1171; Y. Masaki, K. Sakuma, and K. Kaji, *J. Chem. Soc., Chem. Commun.,* **1980**, 434.
34. M. Julia, M. Nel, and D. Uguen, *Bull. Soc. Chim. Fr.,* **1987**, 487.
35. B. M. Trost and M. R. Ghadiri, *J. Am. Chem. Soc.,* **1986**, *108,* 1098.
36. B. M. Trost and M. R. Ghadiri, *J. Am. Chem. Soc.,* **1984**, *106,* 7260.
37. H. Kotake, T. Yamamoto, and H. Kinoshita, *Chem. Lett.,* **1982**, 1331.
38. R. O. Hutchins and K. Learn, *J. Org. Chem.,* **1982**, *47,* 4380.
39. M. Mohri, H. Kinoshita, K. Inomata, and H. Kotake, *Chem. Lett.,* **1985**, 451.
40. K. Inomata, S. Igarashi, M. Mohri, T. Yamamoto, H. Kinoshita, and H. Kotake, *Chem. Lett.,* **1987**, 707.
41. K. Inomata, Y. Murata, H. Kato, Y. Tsukahara, H. Kinoshita, and H. Kotake, *Chem. Lett.,* **1985**, 931.
42. B. M. Trost, N. R. Schmuff, and M. J. Miller, *J. Am. Chem. Soc.,* **1980**, *102,* 5979.
43. T. Cuvigny and M. Julia, *J. Organomet. Chem.,* **1986**, *317,* 383.
44. B. M. Trost and C. A. Merlic, *J. Org. Chem.,* **1990**, *55,* 1127.
45. Y. Masuyama, K. Yamada, S. Shimizu, and Y. Kurusu, *Bull. Chem. Soc. Jpn.,* **1989**, *62,* 2913.
46. F. Vogtle and L. Rossa, *Angew. Chem., Int. Ed. Engl.,* **1979**, *18,* 515; see also F. S. Guziec, Jr. and L. J. Sanfilippo, *Tetrahedron,* **1988**, *44,* 6241.
47. E. M. La Combe and B. Stewart, *J. Am. Chem. Soc.,* **1961**, *83,* 3457.

48. J. B. Hendrickson and R. Bergeron, *Tetrahedron Lett.*, **1973**, 3609.

49. For a mechanistic exploration of the photo-extrusion of SO_2 from benzylic sulphones see R. S. Givens, B. Hrinczenko, J. H-S. Liu, B. Matuszewski, and J-T. Collison, *J. Am. Chem. Soc.*, **1984**, *106*, 1779.

50. M. Julia, D. Lave, M. Mulhauser, M. Ramirez-Munoz, and D. Uguen, *Tetrahedron Lett.*, **1983**, *24*, 1783.

51. E. C. Leonard, Jr., *J. Org. Chem.*, **1962**, *27*, 1921.

52. G. E. Robinson and J. M. Vernon, *J. Chem. Soc., Perkin Trans. 1*, **1977**, 1682; see also R. F. Langer, Z. A. Marini, and J. A. Pidcock, *Can. J. Chem.*, **1978**, *56*, 903.

53. Y. Ueno, S. Aoki, and M. Okawara, *J. Am. Chem. Soc.*, **1979**, *101*, 5414.

54. Y. Ueno, S. Aoki, and M. Okawara, *J. Chem. Soc., Chem. Commun.*, **1980**, 683.

55. Y. Ueno, H. Sano, S. Aoki, and M. Okawara, *Tetrahedron Lett.*, **1981**, 22, 2675.

56. J. Nokami, T. Sudo, H. Nose, and R. Okawara, *Tetrahedron Lett.*, **1981**, 22, 2899.

57. J. E. Baldwin, R. M. Adlington, C. Lowe, I. A. O'Neil, G. L. Sanders, C. J. Schofield, and J. B. Sweeney, *J. Chem. Soc., Chem. Commun.*, **1988**, 1030.

58. V. Pascali and A. Umani-Ronchi, *J. Chem. Soc., Chem. Commun.*, **1973**, 351.

59. J-L. Fabre and M. Julia, *Tetrahedron Lett.*, **1983**, *24*, 4311.

60. T. Cuvigny, C. Herve du Penhoat, and M. Julia, *Tetrahedron*, **1987**, *43*, 859.

61. J. Bremner, M. Julia, M. Launay, and J-P. Stacino, *Tetrahedron Lett.*, **1982**, *23*, 3265.

62. M. Julia, H. Lauron, J-P. Stacino, J-N. Verpeaux, Y. Jeannin, and Y. Dromzee, *Tetrahedron*, **1986**, *42*, 2475.

63. M. Julia and J-P. Stacino, *Bull. Soc. Chem. Fr.*, **1985**, 831.

64. X. Huang and H-Z. Zhang, *Synthesis*, **1989**, 42.

65. X. Huang and H-Z. Zhang, *Synth. Commun.*, **1989**, *19*, 97; see also X. Huang and J-H. Pi, *Synth. Commun.*, **1990**, *20*, 2297.

66 M. Ochiai, T. Ukita, and E. Fujita, *J. Chem. Soc., Chem. Commun.*, **1983**, 619.

67. Y. Watanabe, Y. Ueno, T. Araki, T. Endo, and M. Okawara, *Tetrahedron Lett.*, **1986**, *27*, 215.

68. N. Miyamoto, D. Fukuoka, K. Utimoto, and H. Nozaki, *Bull. Chem. Soc. Jpn.*, **1974**, *47*, 503.

69. H. C. Brown, S-C. Kim, and S. Krishnamurthy, *Organometallics*, **1983**, *2*, 779.

70. J. E. Baldwin, R. M. Adlington, Y. Ichikawa, and C. J. Kneale, *J. Chem. Soc., Chem. Commun.*, **1988**, 702.

71. J-L. Fabre, M. Julia, and J-N. Verpeaux, *Tetrahedron Lett.*, **1982**, *23*, 2469; J-L. Fabre, M. Julia, and J-N. Verpeaux, *Bull. Soc. Chim. Fr.*, **1985**, 762.

72. J-L. Fabre, M. Julia, and J-N. Verpeaux, *Bull. Soc. Chim. Fr.*, **1985**, 772.

73. R. L. Smorada and W. E. Truce, *J. Org. Chem.*, **1979**, *44*, 3444.

74. J. J. Eisch, M. Behrooz, and J. E. Galle, *Tetrahedron Lett.*, **1984**, *25*, 4851.

75. J. M. Gaillot, Y. Gelas-Mialhe, and R. Vessiere, *Chem. Lett.*, **1983**, 1137.

76. P. J. Kocienski, *Tetrahedron Lett.*, **1979**, 441; P. J. Kocienski and J. Tideswell, *Synth. Commun.*, **1979**, *9*, 411.

77. E. J. Corey and M. Chaykovsky, *J. Am. Chem. Soc.*, **1965**, *87*, 1345; E. J. Corey and M. Chaykovsky, *J. Am. Chem. Soc.*, **1964**, *86*, 1639.

78. M. Isobe, Y. Ichikawa, D-L. Bai, H. Masaki, and T. Goto, *Tetrahedron*, **1987**, *43*, 4767; see also L. A. Paquette, H-S. Lin, and M. J. Coghlan, *Tetrahedron Lett.*, **1987**, *28*, 5017.

79. Y-C. Kuo, T. Aoyama, and T. Shiori, *Chem. Pharm. Bull.*, **1983**, *31*, 883.

80. Y-C. Kuo, T. Aoyama, and T. Shiori, *Chem. Pharm. Bull.*, **1982**, *30*, 2787. For an example of β-ketosulphone desulphonylation using Raney nickel, see S. M. F. Lai, J. J. A. Orchison, and D. A. Whiting, *Tetrahedron*, **1989**, *45*, 5895.

81. F. W. Sum and L. Weiler, *J. Am. Chem. Soc.*, **1979**, *101*, 4401.

82. B. M. Trost and T. R. Verhoeven, *J. Am. Chem. Soc.*, **1979**, *101*, 1595. For an example involving a nitrile sulphone, see J. P. Genet, A. Denis, and F. Charbonnier, *Bull. Soc. Chim. Fr.*, **1986**, 794.

83. P. T. Lansbury, R. W. Erwin, and D. A. Jeffrey, *J. Am. Chem. Soc.*, **1980**, *102*, 1602.

84. B. Koutek, L. Pavlickova, and M. Soucek, *Collect. Czech. Chem. Commun.*, **1974**, *39*, 192.

85. M. J. Kurth and M. J. O'Brien, *J. Org. Chem.*, **1985**, *50*, 3846.

86. J-M. Beau and P. Sinay, *Tetrahedron Lett.*, **1985**, *26*, 6185, 6189 and 6193.

87. B. Kruse and R. Bruckner, *Chem. Ber.*, **1989**, *122*, 2023.

88. R. A. Grimm and W. A. Bonner, *J. Org. Chem.*, **1967**, *32*, 3470; see also G. Sauer, K. Junghans, U. Eder, G. Haffer, G. Neef, R. Wiechert, G. Cleve, and G-A. Hoyer, *Liebigs Ann. Chem.*, **1982**, 431; H. Suzuki, Q. Yi, J. Inoue, K. Kusume, and T. Ogawa, *Chem. Lett.*, **1987**, 887.

89. N. Ono, R. Tamura, R. Tanikaga, and A. Kaji, *J. Chem. Soc., Chem. Commun.*, **1981**, 71; see also N. Ono, H. Miyake, A. Kamimura, I. Hamamoto, R. Tamura, and A. Kaji, *Tetrahedron*, **1985**, *41*, 4013.

90. P. A. Wade, H. R. Hinney, N. V. Amin, P. D. Vail, S. D. Morrow, S. A. Hardinger, and M. S. Saft, *J. Org. Chem.*, **1981**, *46*, 765.

91. M. Fujii, K. Nakamura, H. Mekata, S. Oka, and A. Ohno, *Bull. Chem. Soc. Jpn.*, **1988**, *61*, 495.

92. D. S. Brown, S. V. Ley, and S. Vile, *Tetrahedron Lett.*, **1988**, *29*, 4873. For some related reactions, see K. Claus, D. Grimm, and G. Prossel, *Liebigs Ann. Chem.*, **1974**, 539; J. Morton, A. Rahim, and E. R. H. Walker, *Tetrahedron Lett.*, **1982**, *23*, 4123; C. Branch and M. J. Pearson, *J. Chem. Soc., Perkin Trans. 1*, **1979**, 2268.

93. D. Crich and T. Ritchie, *Tetrahedron*, **1988**, *44*, 2319.

94. M. Funabashi and H. Nagashima, *Chem. Lett.*, **1987**, 2065.

95. R. Tamura, H. Katayama, K. Watabe, and H. Suzuki, *Tetrahedron*, **1990**, *46*, 7557.

96. F. de Reinach-Hirtzbach and T. Durst, *Tetrahedron Lett.*, **1976**, 3677.

97. D. S. Brown and S. V. Ley, *Tetrahedron Lett.*, **1988**, *29*, 4869; D. S. Brown, M. Bruno, R. J. Davenport, and S. V. Ley, *Tetrahedron*, **1989**, *45*, 4293; see also M. J. Davies, C. J. Moody, and R. J. Taylor, *Synlett*, **1990**, 93.

98. D. S. Brown, T. Hansson, and S. V. Ley, *Synlett,***1990**, 48; D. S. Brown, P. Charreau, and S. V. Ley, *Synlett,***1990**, 749.

99. M. Shibasaki, A. Nishida, and S. Ikegami, *J. Chem. Soc., Chem. Commun.*, **1982**, 1324; see also A. Nishida, M. Shibasaki, and S. Ikegami, *Tetrahedron Lett.*, **1984**, *25*, 765; H. Maruyama and T. Hiraoka, *J. Org. Chem.*, **1986**, *51*, 399.

100. K. Hirai, Y. Iwano, and K. Fujimoto, *Tetrahedron Lett.*, **1982**, *23*, 4025.

101. N. Kornblum, S. D. Boyd, and N. Ono, *J. Am. Chem. Soc.*, **1974**, *96*, 2580.

102. E. Ghera, T. Yechezkel, and A. Hassner, *J. Org. Chem.*, **1990**, *55*, 5977.

103. C. E. Tucker, S. A. Rao, and P. Knochel, *J. Org. Chem.*, **1990**, *55*, 5446.

104. C. Najera and M. Yus, *Tetrahedron Lett.*, **1989**, *30*, 173.

105. P. A. Wade, J. F. Bereznak, B. A. Palfey, P. J. Carroll, W. P. Dailey, and S. Sivasubramanian, *J. Org. Chem.*, **1990**, *55*, 3045.

106. K. Ogura, T. Uchida, T. Tsuruda, and K. Takahashi, *Tetrahedron Lett.*, **1987**, *28*, 5703.

107. F. G. Bordwell and W. H. McKellin, *J. Am. Chem. Soc.*, **1951**, *73*, 2251.

108. J. N. Gardner, S. Kaiser, A. Krubiner, and H. Lucas, *Can. J. Chem.*, **1973**, *51*, 1419.

109. D. S. Kemp and D. R. Buckler, *J. Org. Chem.*, **1989**, *54*, 3647.

110. T. A. Whitney and D. J. Cram, *J. Org. Chem.*, **1970**, *35*, 3964.

111. W. P. Weber, P. Stromquist, and T. I. Ito, *Tetrahedron Lett.*, **1974**, 2595.

112. Y. Handa, J. Inanaga, and M. Yamaguchi, *J. Chem. Soc., Chem. Commun.*, **1989**, 298.

113. I. W. J. Still and F. J. Ablenas, *J. Org. Chem.*, **1983**, *48*, 1617; I. W. J. Still and S. Szilagyi, *Synth. Commun.*, **1979**, *9*, 923.

INDEX